CAMBRIDGE LIBRARY COLLECTION

Books of enduring scholarly value

Life Sciences

Until the nineteenth century, the various subjects now known as the life sciences were regarded either as arcane studies which had little impact on ordinary daily life, or as a genteel hobby for the leisured classes. The increasing academic rigour and systematisation brought to the study of botany, zoology and other disciplines, and their adoption in university curricula, are reflected in the books reissued in this series.

British Fossil Brachiopoda

British palaeontologist Thomas Davidson (1817–85) was born in Edinburgh and began his studies at the city's university. Encouraged by German palaeontologist Leopold von Buch, he began to study brachiopod fossils at the age of twenty, and he quickly became the undisputed authority. He was elected fellow of the Geological Society of London in 1852, receiving the Wollaston medal in 1865. He became a Fellow of the Royal Society in 1857. Published between 1850 and 1886, this six-volume work became the definitive reference text on the subject. It includes more than two hundred hand-drawn plates and a comprehensive bibliography. This volume, the first of six, includes an essay on the terebratulids by Richard Owen, an analysis of brachiopod shell structure by W. B. Carpenter and a guide to classification by Davidson himself. The rest of the volume describes Cretaceous, Tertiary, Oolitic and Liasic brachiopod species.

Cambridge University Press has long been a pioneer in the reissuing of out-of-print titles from its own backlist, producing digital reprints of books that are still sought after by scholars and students but could not be reprinted economically using traditional technology. The Cambridge Library Collection extends this activity to a wider range of books which are still of importance to researchers and professionals, either for the source material they contain, or as landmarks in the history of their academic discipline.

Drawing from the world-renowned collections in the Cambridge University Library, and guided by the advice of experts in each subject area, Cambridge University Press is using state-of-the-art scanning machines in its own Printing House to capture the content of each book selected for inclusion. The files are processed to give a consistently clear, crisp image, and the books finished to the high quality standard for which the Press is recognised around the world. The latest print-on-demand technology ensures that the books will remain available indefinitely, and that orders for single or multiple copies can quickly be supplied.

The Cambridge Library Collection will bring back to life books of enduring scholarly value (including out-of-copyright works originally issued by other publishers) across a wide range of disciplines in the humanities and social sciences and in science and technology.

British Fossil Brachiopoda

Volume 1: Tertiary, Cretaceous, Oolitic, and Liasic Species

Thomas Davidson
With Introduction By
Richard Owen &
W. B. Carpenter

CAMBRIDGE
UNIVERSITY PRESS

CAMBRIDGE UNIVERSITY PRESS

Cambridge, New York, Melbourne, Madrid, Cape Town,
Singapore, São Paolo, Delhi, Tokyo, Mexico City

Published in the United States of America by Cambridge University Press, New York

www.cambridge.org
Information on this title: www.cambridge.org/9781108038171

© in this compilation Cambridge University Press 2011

This edition first published 1851-55
This digitally printed version 2011

ISBN 978-1-108-03817-1 Paperback

BRITISH
FOSSIL BRACHIOPODA.

BY

THOMAS DAVIDSON, F.G.S.,

MEMBRE ÉTRANG. DE L'INSTITUT DES PROVINCES; OF THE GEOLOGICAL SOCIETY OF FRANCE; LINNÆAN OF NORMANDY; ROYAL SOCIETY OF SCIENCES OF LIÉGE; ZOOLOGICAL SOCIETY OF VIENNA, ETC.

VOL. I.

TERTIARY, CRETACEOUS, OOLITIC, AND LIASIC SPECIES.

WITH

A GENERAL INTRODUCTION:

I. ON THE ANATOMY OF THE TEREBRATULA.
BY PROFESSOR OWEN, M.D., LL.D., F.R.S., L.S., G.S., &c.

II. ON THE INTIMATE STRUCTURE OF THE SHELLS OF BRACHIOPODA.
BY PROFESSOR W. B. CARPENTER, M.D., F.R.S., G.S., &c.

III. ON THE CLASSIFICATION OF THE BRACHIOPODA.
BY THOMAS DAVIDSON, F.G.S., &c.

LONDON:
PRINTED FOR THE PALÆONTOGRAPHICAL SOCIETY.
1851—1855.

PREFACE.

Of all the species composing the SUB-KINGDOM MOLLUSCA, none are, perhaps, more varied, more elegant in their shapes, or more abundantly distributed, than those to which the term BRACHIOPODA or PALLIOBRANCHIATA has been applied: they are found in the oldest deposits at present known to contain vestiges of animal life, and have continued to exist, some in similar, but many under different shapes to the existing forms, through the long successive periods which lead us to the present time. Their value to the geologist is consequently very great; and, as they so commonly fall under his hammer, where other classes are often but sparingly represented, they must, therefore, be looked upon as excellent data for the age of the deposit; for, although some few individual forms pass from one stage to the other, the generality are limited to defined horizons. The study of so important an ORDER has, within the last fifteen or twenty years, particularly interested the Palæontologist who regarded something more than *external shape*. His efforts, in this direction, have tended to acquire an intimate acquaintance with the inhabitants of so remarkable a shell; and we have become more and more convinced that to arrive at so desirable a knowledge, it was absolutely necessary to call upon ZOOLOGY to point out, from a minute study of the animal, of the few species and genera still existent, the use and origin of those remarkable calcified processes and varied impressions visible in the numerous forms now extinct.

Having devoted many years to the careful investigation of the subject under consideration, and feeling desirous to contribute the little in my power to the advancement of the views so laudably put forth by the Palæontographical Society, in 1849 I proposed to the Council to do my best to prepare for them a general work on all the British species of Brachiopoda hitherto discovered,[1] arranged in seven distinct parts, which

[1] From being almost a stranger in the country at the period above mentioned, on account of a residence of many years in different parts of the continent, I did not sufficiently estimate all the difficulties I should have had to contend with in preparing so extensive and difficult a work *entirely on British species*. I had not only to become acquainted with the local geologists and their collections, but also with a

1

would correspond to the great geological divisions or systems now in general use, viz. :—

1. TERTIARY.   4. PERMIAN.

2. CRETACEOUS.   5. CARBONIFEROUS.

3. OOLITIC.   6. DEVONIAN.

7. SILURIAN.

So that each Part would form a complete Monograph zoologically arranged as far as the state of our present knowledge will admit.[1]

It has also been considered desirable to preface the main subject of this work by some preliminary observations on the class in general; in order to enable the reader better to understand why certain sections and names have been rejected, or admitted in preference to others; and it has afforded me no small pleasure to find that Professor Owen and Dr. Carpenter were likewise disposed to annex their recent and most important anatomical and structural investigations to the general Introduction I was then preparing. And, in accepting so liberal an offer from those eminent and justly celebrated zoologists, I felt convinced that I was attending not only to the best interests of Science, but also to those of the Society,—interests which it is the main object and desire of all the members to advance.

THOS. DAVIDSON.

LONDON; *January 4*, 1853.

---

great number of the localities themselves: and, from having been at the commencement more prepared with the Oolitic species, I deemed it advisable to begin with them, especially, as by so doing, no subsequent inconvenience would be incurred by the members of the Society. I must also avail myself of this opportunity to express my warmest thanks to all those local geologists who have so kindly and liberally assisted in the prosecution of my investigations; many of these gentlemen I have already had the pleasure of naming in portions of the work already published, and, in addition, I have now the satisfaction of mentioning Professor Owen, Dr. Carpenter, Drs. Gray, Volborth, and Perier; Professors Beyrich and Kutorga; Col. Helmersen; Messrs. Salter, Cuming, Hanley, de Lorière, de Hagenow, and Michelotti; and I am most particularly indebted to Professor King, Mr. Woodward, and M. Suess, of Vienna, for much valuable information, and for the kind interest they have shown in the progress of the work.

[1] It has been estimated that the work now in progress will require from eighty to one hundred plates, and it was found desirable to divide the subject into two volumes, the first will be devoted to the *Tertiary, Cretaceous,* and *Oolitic* species; the second, to the *Palæozoic* forms.

INTRODUCTION.

CHAPTER I.

ON THE ANATOMY OF THE TEREBRATULA.

By PROFESSOR OWEN, F.R.S., F.G.S., &c.

WITH the view of extending the knowledge of the organisation of the Brachiopoda, the unfolding of which Pallas and Cuvier had so well begun—the one in the description of his so-called *Anomia*,[1] the other by his admirable anatomy of the *Lingula anatina*,[2]—I communicated, in 1833, an account of my dissections of some species of *Terebratula* and *Orbicula* to the Zoological Society of London,[3] and have since availed myself of every opportunity of completing the anatomy of this very beautiful and singular class of Mollusks.

Cuvier had shown, in *Lingula*, a condition of the respiratory organ, which might be paralleled with one of the transitory states of that organ in the Lamellibranchs, that, viz. in which the rudimental gills appear as processes from the inner surface of the pallial lobes, and in which the distinction, whether morphological or physiological, of the gills and mantle is not fully established: the modifications of the breathing organ in both *Terebratula* and *Orbicula* exhibited a more interesting condition, comparable to a still earlier stage of the respiratory system in the embryo Lamellibranch, that, viz. in which the vessels of the pallial lobes have not begun to bud out in parallel rows of vascular loops,—the first stage in the formation of gills, and the one at which it is arrested in the *Lingula*. Notwithstanding, therefore, the manifestation of many beautiful modifications of structure, which seemed to render the organisation of the Brachiopod more complex than that of the ordinary Bivalve, I was led, in 1833, to regard the latter as standing higher in the Acephalous series, and to place the *Brachiopoda* between the *Lamellibranchiata* and the *Tunicata*: subsequent investigations, especially those recorded in the present Memoir, have confirmed me in that view. I am the more desirous to repeat these convictions in reference to the

[1] Miscellanea Zoologica, 4to, 1775, p. 182, (*Anomiarum Biga*.)

[2] Annales du Muséum d'Hist. Nat., 4to, tom. i, p. 69 (1802). Mémoires pour servir à l'Histoire et l'Anatomie des Mollusques, 4to, 1817.

[3] Transactions of the Zoological Society of London, 4to, vol. i, p. 145 (1835).

position and affinities of the acephalous Mollusks, because my statement as to the value of the group *Brachiopoda*, in comparison with the group *Lamellibranchiata*,[1] has been misunderstood by some esteemed contemporaries on the continent. I believe the *Brachiopoda* and *Lamellibranchiata* to have equal claims to be considered as distinct groups, whether called classes, sub-classes, or orders of *Acephala ;* but, I regard the group *Brachiopoda* as exemplifying a less advanced grade of organisation than the *Lamellibranchiata*, in which the lowest forms, as, *e. g. Anomia* and *Ostrea*, show the nearest affinities to the *Brachiopoda*. Those monomyary or unimuscular bivalves become fixed, like the *Orbiculæ* and *Terebratulæ*, whilst the higher Lamellibranchs exhibit, with a progressive development and increased complexity of the muscular system, powers of locomotion more or less active and varied ; some, *e. g.* the Solens, burrowing ; others, *e. g.* the Cardiums, progressing by short leaps, whence the name *subsilentia*, given to the group by Poli ; others, again, *e. g.* the Pectens, are said to swim by violent flapping movements of their valves, whence these light and richly-painted bivalves have been called " Sea-butterflies."

The Brachiopoda, being deprived of the power of locomotion, have the development of their respiratory system arrested at a corresponding low grade ; the *Lingulæ*, which have the largest and most flexible peduncle, being the only forms that show distinct rudiments of gills. In the *Terebratulæ*, the mantle-lobes, besides their ordinary office, perform by their rich vascularity the breathing function. The admirable microscopical researches of Dr. Carpenter and Prof. Quekett have demonstrated that close and intimate adhesion of the mantle to the shell, which was noticed in my earlier dissections of the *Terebratula*,[2] to be due to the penetration of the pores of the shell by minute tubular membranous processes, which they believe to be glandular cæca.[3] These processes may perform an excretory function, and be associated in that action with the depurative respiratory office of the mantle, the probable condition of their development being the low grade of the proper branchial organisation, or they may take some share in the formation of the shell itself, necessitated by the modification of the mantle to subserve respiration.

[1] " M. Owen a été le premier à faire remarquer que les Brachiopodes ne doivent pas être envisages comme une classe à part, mais qu'ils peuvent être convenablement rangés sur la même ligne que les Monomyaires et les Dimyaires." (Agassiz, in 'Poissons Fossiles' of the "Old Red," 4to, 1845, p. 15.) Compare, however, 'Zool. Trans.,' vol. i, 1835, p. 159 :—

" In all the essential points, the *Brachiopoda* closely correspond with the *Acephalous Mollusca*, and I consider them as being intermediate to the *Lamellibranchiate* and *Tunicate* orders ; not, however, possessing, so far as they are at present known, distinctive cnaracters of sufficient importance to justify their being regarded as a distinct class of Mollusks, but forming a separate group of equal value with the *Lamellibranchiata*."

[2] Trans. Zool. Soc., vol. i, p. 147.

[3] Carpenter, Report on the Microscopic Structure of Shells, Part II, Trans. Brit. Association, 1847, p. 93. " In these tubes, as will hereafter be shown, certain cæcal appendages of animal membrane are situated."— Quekett, 'Histological Catalogue,' vol. i, 1850, p. 270. In the same work it is shown that " each perforation has a series of radiating lines or tubes on its outer margin," (p. 270). The corresponding parts of the membranous tubes would resemble a terminal brush of vibratile cilia.

To the question proposed to me by their discoverer, " Can they have any analogy to the hepatic papillæ of the Nudibranchiates?" I could only reply that, the pallial lobes are so distinct from the visceral mass, as compared with the skin of the Nudibranchiates, that they cannot have any special functional relation to the liver. And, in regard to the generative system, the membranous tubes penetrating the shell-pores in the *Terebratula flavescens* are so much more minute than the ova discernible in the ramified ovaria, and their presence is so equally manifested over the non-ovarian and ovarian parts of the mantle—the same remark being equally applicable to the pallial lobes and ramified testes in the male—that one cannot connect them, as subordinately related in function, with the generative organs.

The difficulty of satisfactorily assigning the final cause or purpose of the microscopic pallial processes is increased, as in many analogous cases, by the non-development of the organs in question in certain *Brachiopoda*, e. g., the recent *Atrypa* and the fossil *Rhynconellidæ*, which latter, from the general analogy of the structure of their shell, might be supposed to have the respiratory organs at the same low degree of development as the *Terebratulidæ*, and to have the same need for the minute calcifying or excretory ciliated shell-tubes.

Owing to this organic connection between the pallial lobes and the shell, which is particularly close towards the periphery of the lobes, I have been compelled, in some of my dissections, to sacrifice the shell, and remove it piecemeal, in order to preserve and show the characters of the external lobes of the mantle entire. The difficulty of removing the imperforate valve is increased by the median crest, *d*, fig. 1, p. 9, and the attachments of the loop, *ib.*, *e e*, both of which require to be broken close off from the imperforate valve, and left inclosed in the folds of the mantle that invest them, in order to obviate laceration of its outer surface.

The dorsal lobe, D *p*, of the mantle, thus exposed in the *Terebratula flavescens*, shows the single pair of wide pallial sinuses, 5, containing the correspondingly ramified testes or ovarium, according to the sex; the sinuses and their contents extending from the outer side of the hepatic portion of the visceral mass, forwards, along the outside of the great anterior muscular impressions, to near the anterior border of the mantle: the sinus is equidistant from the median line and the lateral border of the mantle, in each half of the lobe: the marginal ramifications of the sinus, the large sheathed cilia, and the smaller marginal cilia, correspond in structure with the same parts in *Ter. chilensis*.[1] Immediately behind a line equally bisecting transversely the mantle lobe in question, are the expanded extremities of attachment of the two divisions, *o' p'*, fig. 1, p. 9, of the *adductor longus* muscle, (" anterior pair of muscles arising from the imperforate valve," tom. cit., 1833, p. 161, pl. xxii, fig. 6, *Ter. chilensis*.)

In the space behind these muscles, and between them and the *retractores superiores*, the two lateral masses of hepatic cæca are clearly visible beneath the thin transparent pallial covering. Behind the *retractores superiores*, (" posterior pairs of muscles arising

[1] Trans. Zool. Soc., pp. 147, 148, pl. xxii, fig. 11.

from the imperforate valve," tom. cit., 1833, p. 161, pl. xxii, fig. 6, *Ter. chilensis,*) near the middle line, and close to the hinder margin of the dorso-pallial lobe, are the short, approximate tendinous insertions of the *adductor brevis, q',* and the line of attachment of the almost tendinous *cardinales* muscles, *u'.*

The exterior surface of the ventral lobe, when exposed with the same care, shows the two pairs of pallial sinuses and ramified generative lobes. The muscular extremities perforating this lobe are more aggregated than in the opposite lobe : the anterior, forming a pair, of a pyriform shape, with the great end forwards, belong to the *adductor brevis, q',* fig. 2, p. 9, (" anterior pair of muscles arising from the perforate valve," tom. cit., 1833, p. 161, pl. xxii, fig. 5, *Ter. chilensis*): external and posterior to these is a second pair of muscular attachments belonging to the '*retractor inferior,*' *ib. s',* (" posterior pair of muscles of the perforate valve," tom. cit., 1833, p. 162, pl. xxii, fig. 5 ;) in the median interspace of the foregoing muscles is the common attached extremity of the *adductores longi, ib. o'* : of a pyriform figure, with the great end backwards, and emarginate or notched at the middle from the lodgment of the anal end of the intestine, *ib. ψ.* Immediately behind this part are the ends of the small pair of *cardinales* muscles, *ib. u',* and behind these the transversely extended glistening surface of attachment of the musculo-aponeurotic fibres of the sheath of the peduncle (*capsularis r'*).

As in the *Terebratulæ* described in my former Memoir, the pallial lobes, with the exception of the small part at the hinge of the shell, are free in the rest of their circumference, and inclose a wide cavity to which the sea-water is admitted, and in which float freely the long filaments of the fringed arms. The long adductors and expanded origins of the short adductors appear also to cross the pallial cavity, but they are inclosed by a delicate transparent duplicature of the pallial membrane : a similar duplicature invests the beginning of the œsophagus, and is bent around the terminal part of the intestine, being reflected upwards a little above the vent, which opens, like the mouth, into the pallial interspace, immediately behind the ventral insertion of the long adductors. The visceral mass, consisting of the alimentary canal, pl. i, fig. 1, τ, φ, χ, liver, ω, hearts, 1, and looped origins of the ventral ovaria or testes, pl. iii, fig. 1, 12, occupies the small space between the long adductors, *o, p,* the hinge, and the peduncle.

MUSCULAR SYSTEM.

I have but little to add to the illustrations of this system derived from the dissections of *Ter. chilensis, Ter. Sowerbii,* and *Ter. psittacea,* in my former Memoir; but figures are subjoined to render its anatomy better understood. Taking the ventral or perforated valve, to which the peduncle is directly attached, as the more fixed point, the *adductores longi* (Pl. i, fig. 2, *o, p,*) arise from a single pyriform area at the middle of that valve, v, a little behind a line transversely bisecting it ; the fibres soon become tendinous, converge, and group themselves into two lateral muscles forming a pair ; each of these muscles as it approaches the opposite valve expands and subdivides into an anterior and posterior portion; glides between the stomach and the crus of the calcareous loop of its own side, and is attached by the double expanded insertion into the dorsal or perforated valve, D: for the anterior division, I propose the name of 'adductor longus anticus,' *o*, for the posterior one, that of 'adductor longus posticus,' *p*; but both of the so-divided adductors, by reason of their ventral confluent attachment, may be regarded as constituting one quadricipital muscle. The action of this complex muscle is directly to close or adduct the valves, in which action it will slightly compress the hepatic lobes and stomach.

The *adductor brevis, ib. q*, forms a symmetrical pair, having their expanded disc of attachment to the ventral valve extended somewhat in advance of the confluent insertion of the preceding muscles: the fibres of each pass obliquely backwards, and converge to a small round shining tendon, the tendons passing on each side the intestine to be inserted close together into the cardinal process of the dorsal valve, D. Their action will be to adduct the valves; but with a more oblique movement bearing upon the ventral valve, v, which they, as it were, help to suspend from the hinge, as from a fixed point.

A third pair of carneo-aponeurotic muscles, *u*, which pass from valve to valve, are attached by both extremities nearer their line of junction, serve to strengthen the hinge, and, by compressing the sides of the base of the peduncle, may aid in protruding that part after it has been forcibly retracted: the muscles in question, which I propose to call '*musculi cardinales*,' or hinge muscles, are attached by their smaller and most tendinous extremity to the linear ridge between the hinge-teeth cavities, fig. 1, p. 9, *b*, and hinge-plate, *ib. c*, in the imperforate valve; they arise by their larger and more fleshy ends from the imperforate valve, close together, behind the common attachment of the *adductores longi*, the rectum alone intervening; some of their fibres appear to be lost upon the sides of the sheath of the peduncle.

The proper muscles of the peduncle consist of two pairs, for its retraction and attachment to the valves; and of some circular or transverse fibres of the sheath, which, though for the most part of an aponeurotic character, appear to be arranged so as to act as compressors and elongators, or protrusors, of the peduncle.

The name *retractor inferior*, pl. i, fig. 2, *s*, is given to a pair of muscles which arise from the ventral valve by a thick carneous end, exterior to the 'adductores longi' and 'brevis:' the fibres pass obliquely backwards, and rapidly diminish to a tendon which penetrates the upper and lateral part of the sheath of the peduncle, and the terminal fibres of which appear to constitute part of the peduncle itself. This pair of muscles serves to suspend the Terebratula by means of the perforated valve to the peduncle, and forms the most direct retractor of that part, and consequently the chief agent in such limited movements, as the fettered state of the shell will allow.

The name '*retractor superior*,' *ib. t*, is applied to a pair of muscles which have a broad subtriangular carneous origin from the hinge-plate, and a strong aponeurosis extending therefrom to the crus of the calcareous loop, *ib.* fig. 3, *t, t*: the fibres curve over the sides of the swollen part of the capsule of the peduncle, penetrate the capsule, interlace with the inserted fibres of the inferior retractor, and terminate for the most part in the peduncle.

Some not very clearly defined, partly carneous, chiefly tendinous, fibres, which interlace, running mostly in the transverse direction upon and in the capsule of the peduncle, and make up, in fact, a chief part of its substance, have a transversely oblong surface of attachment or strong adhesion to the lower part of the bent conical prolongation of the ventral valve lodging the peduncle : this surface appears in the exterior of the soft parts of the Terebratula, behind the origins of the *cardinales* muscles, fig. 2, p. 9, *r*. I have proposed the name of '*capsularis*,' for the sum of the carneo-tendinous fibres which have the attachment in question. In conjunction with the completely encircling fibres of the peduncular capsule, they must compress and elongate the peduncle.

In addition to the carneo-tendinous fibres arranged in the more or less definite masses above-described, there must be enumerated in the muscular system the double spiral fibres which form the muscular wall of the canal, pl. i, fig. 1, and pl. iii, fig. 2, *z*, traversing the stems of the fringed arms; and the muscular fibres of the pallial lobes, which latter are extremely feebly developed, and recognisable only near the periphery. Thus, to recapitulate the designations of the several muscles in the *Terebratula*, as demonstrated by dissections of the *Ter. chilensis*, *Ter. psittacea*, and *Ter flavescens*, there may be enumerated the—

> *Adductor longus anticus,*
> *Adductor longus posticus,*
> *Adductor brevis,*
> *Cardinalis,*
> *Retractor superior,*
> *Retractor inferior,*
> *Capsularis,*
> *Brachial muscles,*
> *Pallial muscles.*

The first seven muscles leave more or less recognisable impressions on the interior of the valves : the marginal muscles of the mantle are too feebly developed to mark the shell, as it is impressed in the Lamellibranchiate bivalves.[1]

In the subjoined cuts the muscular impressions and some other parts of the valves are indicated by the following letters :—Fig. 1 (Dorsal Valve) : *a*, cardinal or hinge-process; *b*, depression for hinge-teeth; *c*, hinge-plate; *e*, crus or origin of calcareous loop; *e'*, crural process; *f*, produced plate—*g*, reflected plate—of calcareous loop; *o'*, impression of *adductor longus anticus*; *p'*, impression of *adductor longus posticus*. Fig. 2 (Ventral

---

[1] Most of the above details of the muscular system were communicated by me to the Meeting of the Italian naturalists at Naples in 1846; and an abstract was left for publication in the 'Annals' of that Meeting.

Valve) : *i*, perforation for passage of peduncle ; *k*, accessory shelly piece, called ' thecidium ;' *l*, hinge-tooth ; *q'*, impression of *adductor brevis*; *s'*, impression of *adductor inferior*; *o'*, impression of confluent attachments of the *adductores longi*; *u'*, impressions of the *cardinales* or hinge-muscles ; *r'*, impression of the *capsularis*; ψ, the position of the anus.

FIG. 1.

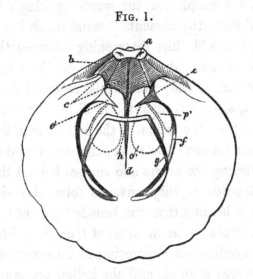

Inner surface, with calcareous loop, and muscular impressions
of the Dorsal Valve, *Terebratula flavescens*, (magn.)

FIG. 2.

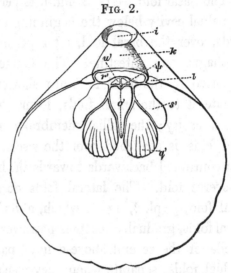

Inner surface and muscular impressions of the Ventral
Valve, *Terebratula flavescens*, (magn.)

With respect to the fringed arms, I have very little to add to the descriptions in my earlier Memoir. The modifications of the calcareous support—a kind of internal skeleton, fig. 1, p. 9, and pl. 1, fig. 2,—have been fully and accurately detailed by my able colleague, Mr. Davidson, in his communications to the ' Annals and Magazine of Natural History,' and in the Monograph to which the present notes are prefixed. I shall only, therefore, append some remarks on the arrangement of the soft parts supported by the calcified loop, as they exist in the *Terebratula flavescens*. In this, as in other *Terebratulæ*, the fringed portions of the arms are not immediately supported by the loop, this serves only for the attachment of the thin but firm aponeurotic membrane which forms the true basis of support of the beautiful and peculiar organs in question.

The crura, *e*, fig. 1, of the shelly-loop, are attached to the hinge-plates of the dorsal valve : from each crus is continued a produced and a reflected plate, which latter are united by a transverse portion completing the loop : a short descending process, *e'*, is sent off near the basal attachment of each crus. The produced plate, *f*, pl. 1, fig. 2, after advancing forwards to about three fourths of the antero-posterior diameter of the shell, bends towards the ventral valve to form the reflected plate, *g*, and both portions are also bent with their convexities outwards. The brachial aponeurosis, *v*, pl. 3, figs. ii and iii, which incloses the loop and its processes, is continued transversely from one crus to the other, near the hinge, forming the ' basal fold,' *w*, pl. 2, fig. i ; passes from the produced to the reflected plates, binding them closely together, and forming the ' lateral folds,' *a, a*, pl. 3, fig. iii ;

2

is next continued forwards from the basal fold to the transverse end of the loop, and is thence continued forwards, independently of the loop, between the bases of the spiral portions of the arms, uniting the right to the left throughout the extent of the spire, forming the 'spiral fold,' *y*, pl. 2, fig. i, of the brachial aponeurosis.

The basal fold, *w*, pl. 2, fig. i, is perforated by the œsophagus, the mouth opening into the pallial cavity below the beginning of the spiral fold : the alimentary canal bends backwards, over the basal fold, after perforating it ; the fold thus intervening between the œsophagus and intestine. The filaments given off from the beginning of the arms, supported by the basal fold, are shorter than the rest. The aponeurosis, reflected from the lateral to the spiral folds, forms the fore-part, *v, v*, pl. 3, figs. ii and iii, of the small visceral cavity ; the pallial membrane is continued from the union of the lateral with the basal folds, is stretched over the visceral interspace between the advancing crura, and is then continued backwards towards the hinge, protecting the hearts and sinuses behind the transverse fold. The lateral folds decrease in breadth as they advance forwards : the spiral fold, *y*, pl. 2, fig. 1, which, at its beginning, is broader than the broadest part of the lateral folds, gradually becomes narrower as it approaches the termination of the spire. The lamellæ of the several above-defined parts of the aponeurotic supporting and connecting brachial folds, separate when they reach the calcareous plate, *a*, and the hollow muscular stems, *z, z*, pl. 3, fig. ii, of the fringed arms, closely surround and adhere to these stems, being continued thence upon the border formed by the roots of the fringe-filaments, where the aponeurotic character is exchanged for that of a delicate membrane, which is finally lost upon the filaments themselves.

The fibres of the muscular walls of the brachial canal, pl. 3, fig. ii, *z*, are beautifully arranged in a decussating double spiral, evidently adapted for compressing the contained fluid, and thereby reacting upon the arm of which the muscular canal forms the base. In the *Terebratulæ*, like *Ter. psittacea*, with free multispiral brachia, the fluid of the canal being acted upon by the spirally-disposed muscles composing its parietes, is forcibly injected towards the extremity of the arm, which is thus unfolded and protruded outwards. In the species resembling *Ter. flavescens*, the spiral portion may also be, in like manner, so far unfolded as to react upon the closed valves of the shell.

What power the animal may possess of further unfolding and protruding the free extremity of the united spiral portions of the arms, can only be determined after careful observation of the living *Terebratulæ* in their native localities. The structure of the parts in question led me to note, in 1833, the important difference between *Lingula* and those species of *Terebratula* which resemble *Ter. chilensis* in the structure of the arms, " since, from their attachments, they are fixed and cannot be unfolded outwards as in *Lingula*."[1] Mr. Davidson has expressed a similar opinion in his comprehensive and able " Classification of recent Brachiopoda,"[2] in reference to the spiral portion of the fringed arms ;

[1] Zool. Trans., vol. i, p. 149.
[2] Annals and Mag. of Nat. Hist., t. ix, 2d ser., 1852, p. 363.

observing with respect to the *Ter. australis*, *Ter. vitrea*, and other species of his genus *Terebratula* proper, that, "from its texture and relations it never could be moved or unrolled at the will of the animal;" and he adds, "the sole use of the folding of the arms is to give increased surface for the disposition of the cilia." By the latter term, the author means what I have here and elsewhere called the 'filaments,' forming the fringe. But comparison of the parts in action of a living Mollusk, with the same parts in a dead specimen stiffened in alcohol, has in most cases tended greatly to expand the observer's notions of the powers and uses of such parts derived from dissections; and it would be rash to conclude that so well-organised and beautiful a mechanism for acting on the fringed arms in the direction of their length, as is shown in figures ii, pl. 2, and ii, pl. 3, should exist there without a purpose, as would be the case were the folding of the arms solely for the purpose of augmenting the surface for the attachment of the filaments. In my earlier Memoir, while appreciating and admitting the great difference in the arrangement of the arms for the facilitating their movements of extension, in different species of *Terebratulæ*, since raised to the rank of genera, other uses of the foldings of the loops had also suggested themselves. "The muscular stem, by means of its attachment to the calcareous loop has the power of acting upon that part to the extent its elasticity admits of, which is sufficient to produce such a degree of convexity in the reflected part of the loop, as to cause it to press upon the perforated valve, and separate it slightly from the opposite one."[1] Observations on living Mollusks are, however, essential to the formation of adequate and exact ideas of the uses of parts of the several muscular parts of their organisation.

The fringe-filaments, γ, δ, pl. 3, figs. ii and iii, of the produced and reflected portions of the brachia, are in a single series; they are compressed, very narrow, close set, with their flat sides towards each other, very gradually tapering to the extremity which is slightly bent, the rest of the filament being usually straight: those, ε, of the spirally-disposed portion are split at the end, and the split is deeper as the filaments are situated nearer the end of the spire, where they appear thereby to be arranged in a double row. In most specimens, the filaments of the spire incline towards each other, and meet at their extremities, inclosing a triangular space, as shown in pl. 3, fig. ii, in which the end of the spire has been cut off.

NERVOUS SYSTEM.

The nervous system of the Terebratula consists of three principal parts or systems, the "pallial," "brachial," and "visceral." The roots or origins of these three systems centre in the œsophageal ring, ξ, pl. 2, fig. i.

This annular centre is situated in and defended by the basal fold of the brachial aponeurosis, *ib. w*, surrounding the aperture by which the œsophagus penetrates the visceral chamber. A few very delicate filaments pass off from the part of the ring which is turned

[1] Tom. cit., p. 150.

towards the aperture of the shell, and which part, from the downward bend of the mouth, is made anterior, instead of being, as in the normal position of the mouth, dorsal. These filaments are lost in the beginning of the spiral connecting fold, *w, y*, the brachial spire thus receiving some of their nerves from the same part of the œsophageal ring, as the antennæ do in insects, and the cephalic tentacula in the higher Mollusks. The two chief nerve-trunks from the ring, ξ, figs. i and ii, come off from its lower and lateral angles, a very slight swelling, hardly to be called a ganglion, occurring at their origin. Each of these trunks quickly divides, one division, *ι*, going to the mantle, the other, *π*, penetrating the base of the fringed arm of its own side. The pallial trunk is the largest; it soon divides to supply the upper and lower mantle-lobes of its own side: the course of that distributed upon the dorsal lobe is shown in pl. 2, fig. ii.

The dorso-pallial nerve, *μ*, extends a short way upon it, and then expands into what appears to be an oblong narrow ganglion; but which is a loop formed by the slight divarication and reunion of the fibres of the trunk. From this loop, *ν*, most of the pallial filaments diverge. In their course towards the margin of the lobe they cross obliquely the great pallial sinus, and give off branches, ξ, most of which correspond with the branches of the sinus, these branches subdividing, and their ramifications appearing to unite in a common circumpallial nerve, *o*, which runs along the inserted bases of the marginal cilia. The brachial nerves, *π*, fig. 1, may be traced some way along the muscular canal of the fringed arms.

Two delicate filaments which traverse part of the visceral chamber come off from the lower part of the œsophageal circle near the origins of the great pallio-brachial trunks; they probably supply the muscles which traverse the visceral chamber, as well as the hearts and alimentary canal; but the indications of this part of the system were too faint and uncertain, in *Terebratula*, for safe depiction.

In *Lingula* they are more distinct, and are shown in fig. iii. The nerves, marked *ρ*, come off from the subœsophageal ganglion, diverge as they pass backwards along the visceral chamber, then converge to their insertion in the anterior muscles. The nerves, *σ*, come off directly from the subœsophageal ganglions, and run more parallel as they pass along the ventral aspect of the anterior muscles to go to the posterior muscles. More delicate filaments pass to the alimentary canal and the hearts. *Lingula* has also the pallial and brachial systems of nerves as well developed as in *Terebratula*. I have not been able to detect any traces of special sense-organs in the *Brachiopoda*.

DIGESTIVE SYSTEM.

The mouth, pl. 1, figs. 1 and 4, *τ*, of the Terebratula, opening downwards, as before remarked, at the base or beginning of the passage formed by the spiral brachial fold and the converging fringes of the brachial spire, has a tumid and sub-bilobed upper or anterior lip, and a thinner and broader lower or posterior lip, which is attached to the basal portion of the fringed arms uniting the central portions. The swollen margins of the mouth are formed by both muscular fibres and secerning cells: there are no rudiments of a maxillary or dental apparatus. The organic molecules subserving the nutrition of the *Brachiopoda* are brought by ciliary action within reach of the mouth, are there seized and swallowed. The œsophagus, v, is short, of uniform diameter, has a delicate membranous outer tunic, a muscular coat, and a thicker epithelial lining than that of the rest of the alimentary canal. It inclines slightly forward as it ascends, between the anterior portions of the liver, to terminate in the stomach: it is slightly constricted at its termination, *ib.* fig. 4: there is no valvular structure at the cardiac aperture. The stomach, ϕ, is a simple oblong cavity, swelling out slightly at the cardiac end, where it receives the biliary secretion, continuing thence for some way of the same diameter, which is rather more than half its length; and gradually contracting to the pylorus. Its tunics consist of an outer membranous, a muscular, and a smooth inner mucous, coat, the epithelium of which is more delicate than in the œsophagus. The whole cavity is bent down at an acute angle with the œsophagus, and the cardiac half is buried in the large granular liver, *ib.* fig. 1, ω. There is no valvular structure at the pylorus: but in some specimens it presented a slight circular constriction. The intestine, χ, is short, straight, and is continued downwards and a little backwards, in a line with the pyloric part of the stomach to the interspace between the attachments of the adductores longi and cardinales to the ventral valve, where the minute vent, ψ, opens into the pallial cavity, *ib.* fig. 1. It does not perforate the capsule of the pedicle. The fæces are carried out by the pallial and brachial currents. The intestine is enveloped to within a very short distance of the vent by an extremely delicate venous sinus, pl. 3, fig. 1, 8, the outer wall of which is connected with the plicated auricles, *ib.* 1, 1, situated one on each side of the gut, a little above its middle. The muscular tunic of the intestine presents the same uniform thickness as that of the stomach. The muco-epithelial lining membranes are disposed in very delicate transverse plaits, pl. 1, fig. 4, χ.

The liver, ω, is about three times the bulk of the stomach, and forms the most conspicuous of the chylopoietic viscera, when the abdominal cavity is exposed. It consists of very numerous ramified follicles, the terminal ones of which are of equal size, and their round closed ends give the apparently granular exterior surface to the gland: the intimate structure of the hepatic follicles in *Ter. flavescens*, agrees with that described in my earlier Memoir, in *Ter. psittacea* and *Ter. chilensis*. There is no natural division into lobes; a slight pressure suffices to displace groups of follicles, which then assume the lobular character. The ducts form the common stems of the manifold ramifications, and they are usually two in number, communicating, each by a distinct aperture, with the cardiac end of the stomach. The liver, as in other Mollusks, is supplied by ramifications of a hepatic artery, the blood thence passing into veins which speedily expand into wide shapeless sinuses, communicating, like the intestinal sinus, with the plicated auricles.

VASCULAR SYSTEM.

The vascular system of the *Brachiopoda* is peculiarly well adapted to demonstrate that remarkable condition of the venous system in the Molluscous class which, when first and somewhat imperfectly observed, gave rise to the idea of the blood being extravasated into the lacunæ of the viscera, and the interstices of other soft parts and tissues. It was on that account that I was induced to communicate a note, with a few illustrations of the real state of the venous system, as exemplified in the *Terebratula* and *Lingula*, to my esteemed and distinguished friend and colleague in the French Academy of Sciences, Dr. Milne Edwards, whose descriptions of an analogous condition of the venous system discovered by him in other Mollusks might have been interpreted, through the ambiguity of certain terms and expressions, as giving countenance to the theory of extravasation, which had been at that time propounded under the term 'phlebenterism.' In the Number of the 'Annales des Sciences Naturelles' for March 1845, *e. g.*, M. Edwards, in his 'Report on the Results of a Voyage to Sicily,' writes "Chez les Mollusques, même les plus parfaits, le système des vaisseaux à l'aide desquels le sang circule dans l'économie est plus ou moins incomplet; de sorte que, dans certains points du cercle circulatoire, ce liquide s'épanche dans les grandes cavités du corps ou dans les lacunes dont la substance des tissues est creusée." —p. 139. Mere words rarely turn men from wrong opinions, at least in a science of observation; but a step, I thought, would be gained in the right direction, if the things, with what might be their more correct interpretation and designations, were placed before the eyes of the accomplished author above cited. The desired result was not, however, attained; for, in a subsequent number of the 'Annales des Sciences,' May, 1845, p. 292, in which my Paper 'On the Circulation of the Brachiopoda' appeared, the able Editor reproduces his former statement in reference to the circulation in the Mollusks; and, by way of contrast with what he continued to regard as the less correct view, he proceeds to remark: "Effectivement, dans les ouvrages les plus récents sur ces matières, on dit que cet appareil est *un systéme de vaisseaux clos dans lequel le sang de tout le corps est enfermé:*" reference being made to 'Owen, Lectures on Comparative Anatomy and Physiology of the Invertebrate Animals,' p. 13, London, 1843.

I believe, however, that it is now admitted by most, if not all Malacologists, that a system of arteries and veins is not the less 'a closed and continuous system of circulation,' because the veins exchange their cylindrical for less regular forms, with a concomitant evanescence of all the coats save the innermost tunica propria.

Such is the condition of the major part of the venous system in the *Terebratulæ*. The hearts are two in number, and distinct; they consist, as in other Brachiopoda, each of an auricle and a ventricle, and are situated, in the *Ter. flavescens*, at the back part of the visceral cavity, on the dorsal aspect of the intestine, one on each side of its upper or anterior half.

If the dorsal valve and corresponding lobe of the mantle be removed, as in pl. 3, fig. iii, and the 'musculi retractores superiores,' *t*, be gently divaricated, or the mesial fasciculi carefully removed, as in fig. ii, the delicate membrane of the venous sinuses, *ib.* fig. ii, 7, and fig. i, 6, communicating with and closing the basal apertures of the auricles, 1, is immediately exposed; and is so transparent as to permit the plicated structure of those cavities to be clearly seen.

If the viscera be exposed by a side view, as in pl. 1, fig. 1, the heart of the side exposed will be seen behind the beginning of the intestine, as at 1, fig. 1.

The ventricle, 2, pl. 3, fig. 1, in each heart, is a smooth oval feebly-muscular cavity, which transmits the blood by two arteries,—the largest to the two halves of the mantle-lobes nearest the ventricle, the smallest to the viscera and muscles. In the mantle the arteries terminate at the periphery in the circum-pallial vein or sinus, 4, pl. 2, fig. ii, from which the branches, *ib.* 5, 5, of the large pallial sinuses commence. These sinuses, although the cylindrical is exchanged for a flattened depressed figure, retain enough of the more normal character of the veins to be recognised distinctly as such. There are two of these branched sinuses, 5, in the dorsal lobe, or that which lines the imperforate valve, and four, 6, in the ventral lobe, or that which lines the perforated valve. Their longitudinal course and relative positions so closely correspond with those which I have described and figured in *Terebratula chilensis*,[1] as to preclude the necessity of repeating either description or figures in this place. The two sinuses in each half of the ventral lobe unite into one, and this trunk unites with the base of the single sinus of the corresponding half of the opposite mantle lobe; the common sinus, so formed, at the back part of the visceral chamber receiving the venous blood from other sinuses there that fill, line, and seem to form, the visceral or peritoneal cavity. The common sinus so constituted, on each side of the intestine, terminates by being continued into the margin of what seems to be the widely-open auricle, as seen through the transparent parietes of the sinus, 1, pl. 3, fig. 1. The auricular cavity would, however, be more correctly described as a closed one, consisting at the half next the ventricle of a beautifully plicated muscular coat, in addition to the membranous one, but at the other half, next the venous sinuses, of venous membrane only : the latter might be termed the auricular sinus, the former the auricle proper. The proper auricle, then, *ib.* 1, 1, presents the form of an oblong depressed cone, attached by its apex to the ventricle, the apex being penetrated by the auriculo-ventricular aperture, and adherent by its base to the auricular sinus, which might be said to conduct into the common peritoneal cavity, since the delicate tunics of the visceral sinuses appear to take the place of a peritoneum. The muscular walls of the auricles are thin and delicate, but exhibit two layers of fibres, the exterior ones being disposed transversely or circularly, the interior ones longitudinally, the latter being the most delicate and radiating from round the perforated apex to the plicated circumference of the auricle. These muscular walls are beautifully disposed in

---

[1] Tom. cit., pp. 147, 148, pl. xxii, figs. 5 and 6.

small plaits or folds, also radiating from the apex: and the longitudinal or radiating folds are puckered by transverse folds, 1, pl. 3, fig. 1. The figures ii and iii, pl. 3, which are magnified to the same degree, exemplify two of the different states in which the plicated auricles, 1, are found in different individuals, and so far exemplify the extent of dilatation and contraction of which these complex cavities are susceptible; but they have presented more extreme differences of size in other specimens, and they must possess considerable powers of altering their capacity. It is, therefore, probable, that, when the circulating fluid is accumulated in unusual quantity in the pallial and visceral sinuses, the longitudinal auricular fibres evert and expand the margins of the basal aperture to which the delicate tunic of the sinuses is attached, as shown in fig 1, 1, 5, 6, that the fluid is drawn by a kind of vermicular movement or peristaltic suction into the auricles, 1, 1, and is thence propelled by successive contraction of the circular fibres into the ventricles, 2, 2. From the ventricles the blood is driven through the ramifications of the pallial and visceral arteries again into the more or less irregular and capacious sinuses, and so returns slowly back to the heart. In this circulation there is no actual extravasation: I find, after the most patient scrutiny, no evidence of an escape of the blood into mere lacunæ or interspaces excavated in the tissues of other and surrounding systems; but the result of such scrutiny has been, invariably, the detection of the continuation and expansion of the proper tunic of the veins into such seeming lacunæ or interspaces. And although, in the wide clefts between the viscera and muscles in the abdominal chamber, the tunic of the venous sinuses is disposed like a peritoneum, seems to perform, also, the function of a peritoneum, and the contained fluid, that of peritoneal serum, in addition to their own more proper and important offices, —and although an anatomist might be permitted, by such similarity of function, to call the cavities of the sinuses 'intervisceral lacunæ,' and the walls of the sinuses 'peritoneum,' —yet, if he were guided in his nomenclature by considerations of homology instead of analogy, he would more correctly term them 'abdominal venous sinuses' and 'venous tunic' respectively. In either case, as a matter of fact, there is no real or essential departure from a circulation in a closed system of arteries and veins; but only a morphological departure from the typical character of the organs of circulation,—an extreme one, it is true, but little likely to lead astray the zootomist who had been prepared for such formal modifications of the vascular system by the discoveries of Hunter, as they are manifested in his preparations, and by the descriptions and figures of such preparations, of the venous system in the classes of Insects and Crustaceans.

As these clearly enunciated discoveries have been overlooked by the continental Physiologists, who have been stimulated by recent discussions on 'phlebenterism' to take a retrospect of the history of the progress of anatomical research in the field of the vascular system of the Invertebrata,[1] it may not be without its use, and it is but a mere act of justice to our own great anatomist, to call attention here to his interpretation of

---

[1] Rapport sur Phlébentérisme, par M. le Dr. Robin, 8vo, 1851.

the facts which have recently misled some otherwise valuable labourers in that field, into phlebenteric notions.

"In the winged insects, which have but one heart, as also but one circulation, there is this heart answering both purposes, (viz., the corporeal and pulmonary circulations)." (Hunter on the 'Blood and Inflammation,' 4to, 1796, p. 134.) "When the veins entering into the heart are small, in comparison with the quantity of blood which is wanted in the ventricles, there we have an auricle; but where the veins near the heart are large, there is no auricle, as in the lobster, and generally in insects. In the snail, where the veins in common are large, yet as they are small where they enter the heart, there is an auricle." (Op. cit., p. 138.) But what, it may be asked, did Hunter mean by "large veins" in insects, the lobster, and the snail. He has left us the explanation of his meaning in his observations 'On the Circulation in Insects,' where he writes:—

" OF THE VEINS."

"The veins of the insect would appear to be simply the cellular membrane; but they are regularly formed canals, although not so distinctly cylindrical canals, as in the quadruped, &c., nor branching with that regularity. They would appear to be, or to fill up, the interstices of the flakes of fat, air-cells, muscles, &c., and, therefore, might be called, in some measure, the cellular membrane of the parts." ('Hunterian MS. Catalogue,' printed in the 'Physiological Catalogue,' tom. ii, p. 31, (1834.)

Baron Cuvier, as is well known, entertained with regard to the vascular system of Insects, ideas closely akin to those which some of his pupils have more recently expressed by 'phlebenterism,' after modifying M. Quatrefage's original meaning of the term; for Cuvier supposed that the whole of the blood of Insects stagnated in the lacunæ or cellular interspaces of the several organs: he was consequently led to deny that insects possessed a true circulation, or that the dorsal tube—"heart, extending through the whole length of the animal," (Hunter, op. cit., 1793, p. 137,) acted as a heart. The more truthful views of Hunter, based on the analogy of the already commencing irregularity and extent of the venous sinuses in the lobster and snail, have been amply confirmed by the researches of Professor Carus, on the 'Circulation of the Blood in the larvæ of Ephemerides and Libellulæ.'[1]

With regard to the Crustacea, Hunter, who left preparations, and a beautiful series of drawings, illustrative of the circulating system in the Lobster (*Astacus marinus*), thus describes the latter:—

"OF THE VEINS IN THE LOBSTER.

" The veins in this class of animals, as in the winged Insect, &c., are principally in the form of large irregular cells, as if the cellular or investing membrane of the animal contained the venal blood; and, when injected, we find the injection principally in large

---

[1] Blutkreislaufes in den Larven netzflüglicher Insekten, 4to, 1827.

masses." He then, referring to his figures, describes the different sinuses, as "*a*, a large mass of vein, lying on the stomach : *b*, another mass similar to the above, lying principally on the heart, which might almost be considered an auricle, as from it are openings into the ventricle." (Hunterian MSS., in 'Physiological Catalogue of the Museum of the Royal College of Surgeons,' vol. ii, 1834, p. 138.)

The conditions of the vascular system in insects, crustaceans, and the snail, enunciated in brief but clear general terms in the work 'On the Blood,' 4to, 1793, and exemplified by preparations, drawings, and manuscript descriptions left by Hunter in his museum, at his demise in the same year, appear to have passed unappreciated abroad, until the successive discoveries of analogous structures in other invertebrata, or the rediscovery of the same structures in the same species which Hunter had dissected, had been made.

Baron Cuvier appears to have been the first to recall the attention of comparative anatomists to this diffused and expanded condition of the venous system in his dissection of the *Aplysia ;* but he mistook the expanded sinuses which fill the abdominal cavity, like those in the Terebratulæ, for that cavity itself. Describing the 'venæ cavæ,' which perform the office of the branchial arteries, he writes :—"Leurs parois se trouvent formées de rubans musculaires transverses et obliques, qui se croisent en toutes sortes de sens, mais qui laissent entre eux des ouvertures sensible à l'œil, et encore plus à toutes les espèces d'injections, et qui établissent une communication libre entre ces vaisseaux et la cavité de l'abdomen ; de manière que les fluides contenues dans celui-ci pénètrent aisément dans ceux-la, et réciproquement."—"Il résulte toujours que les fluides épanchés dans la cavité abdominale peuvent se mêler directement dans la masse du sang et être portés aux branchées et que les veines font l'office des vaisseaux absorbants. Cette vaste communication est sans doute un première acheminement à celle bien plus vaste encore que la nature a établie dans les insectes, où il n'y a pas mêmes de vaisseaux particuliers pour le fluide nouricier,"[1] (pp. 299, 300.)

Jurine, in 1806, observing living specimens of a minute crustacean (*Argulus foliaceus*) which were asphyxiated by a few drops of alcohol added to the water they were in, traced the course of the circulation, which he describes, remarking :—J'ai evité d'employer le mot *vaisseau* pour désigner les conduits dans lesquels le sang circule, et que j'ai remplacé ce mot, tantôt par celui de *colonne*, tantôt par celui de *rameau*. Les raisons qui m'ont engagé à le faire reposent sur la manière dont s'opère cette circulation. En effet, le sang chassé dans la partie antérieure du teste paroît s'y répandre et s'y disséminer de manière à faire croire que les globules sanguins sont dispersés dans le parenchyme de ces parties, plutôt que d'être contenus dans des vaisseaux particulières. Je ferai cependant observer qu'il existe dans ce liquide quatre espèces de courans qui ferment les quatre rameaux dont j'ai parlé plus haut, et que, dans les ailes comme dans le queue, la circulation ne se fait pas d'une manière aussi diffuse que dans la partie antérieure du

---

[1] Mémoire sur le Genre *Aplysia*, &c., in 'Annales du Muséum,' tom. ii, 1803, p. 287.

teste, le liquide globuleux paroissant y être renfermé dans une espèce de large canal pratiqué dans le parenchyme de ces parties." (p. 439.)[1]

It was, probably, owing to the misinterpretation of the remarkable facts which Cuvier had observed, that Meckel, notwithstanding the analogous structures, in the meanwhile pointed out by Jurine in the *Argulus*, and by Gaspard[2] in the *Helix pomatia*, was led to deny the existence of the apertures of communication observed by Cuvier in the large muscular venæ cavæ of the Aplysia. In 1832, however, I detected a structure in the venous system in the *Nautilus Pompilius*, closely analogous to that which Cuvier had pointed out in the Aplysia: but, having traced the continuity of the proper lining tunic of the great muscular vena cava, through the apertures in that coat, with a similar membrane lining the abdominal cavity, I was led to describe the tunic of the abdominal sinus as the peritoneum. "There are several small intervals left between the muscular fibres and corresponding round apertures in the membrane of the vein and in the peritoneum, so that the latter membrane is continuous with the lining membrane of the vein."[3] And, after referring to the analogous structure in the Aplysia, I add that this correspondence leads to the "suspicion that it may be more generally found on a further and more diligent investigation of the venous system in this remarkable class of animals."

Ten years later, M. Pouchet,[4] Professor of Zoology at Rouen, demonstrated in *Limax*, a structure answerable to that which M. Gaspard had described in *Helix*: viz., that the blood passed from the arterial capillaries into the visceral cavity, whence it was received by particular orifices into the veins that carried it directly to the pulmonary chamber, ramifying there like a 'vena portæ' before returning to the heart.

In 1834, the same year in which I edited the figures and descriptions by Hunter of the Circulating System in the Lobster, M. Milne Edwards recorded his examination of the same system in the same species, in the 'Histoire Naturelle des Crustacés,' vol. i, p. 101. He there describes the expanded venous sinuses, as "plutôt des lacunes situeés entre les divers organes que des canaux à parois bien formées." (Op. cit., p. 102.) The term 'lacunes' is also adopted for that of 'venous sinuses' by the Editor of the posthumous edition of the 'Leçons d'Anatomie Comparé of Cuvier,' t. vi, 1839, pp. 504, 505.

In 1843, M. Quatrefages[5] believed that he had discovered a Mollusk, his *Eolidina paradoxum*, in which the organs of circulation were reduced to a univentricular heart and a system of arteries. "Le système veineux," he writes, "manque entièrement. Il est en quelque sorte remplacé par des lacunes du tissu aréolaire." In the same nudibranchiate

[1] Sur l'*Argule foliacé*, in 'Annales du Muséum,' tom. vii, p. 431, 1806.

[2] Recherches sur la Physiologie de l'Escargot des Vignes (*Helix pomatia*), 'Journal de Physiol. de Magendie,' 1822, t. ii, p. 295.

[3] Owen, on the Pearly Nautilus, &c., 4to, 1832, p. 28.

[4] Recherches sur l'Anat. et la Physiol. des Mollusques, 4to; Rouen, 1841.

[5] Annales des Sciences Nat., 1843, t. xix, Comptes Rendus, 1843, p. 1124.

the ramifications of the alimentary canal are also described as penetrating the branchiæ,
"where the chyle was directly submitted to the atmospheric influence exercised by the
surrounding water." To this supposed condition of the digestive, circulating, and respiratory systems, he gave the name, 'Phlebenterism,' and proposed thereon some corresponding changes in the classification of the mollusca.

In the 'Report' on this and other memoirs of M. Quatrefages on the Nudibranchiate Mollusks, by M. Milne Edwards, that distinguished professor adopts the
mode of interpreting the modification of the venous system, and applies it to the
Crustacea. "Il existe" (dans le genre Eolidine) un cœur et des artéres bien constitués,
mais pas des veines proprement dites, et le sang ne revient des divers parties du corps
que par un système des lacunes irregulières, disposition tout-à-fait analogue à celle dont
les Crustacées nous avaient deja fourni un example. Enfin dans d'autres espèces, que
M. Quatrefages a découverte sur les côtes de la Bretagne, le cœur et les artères
disparaissent à leur tour; de sorte que la circulation devient des plus incomplètes et
resemblent à celle qu'on apperçoit chez les Bryoozoaires." Subsequent researches by
Messrs. Embledon and Hancock demonstrated, however, the existence in the genera
Chalide and *Actéonie*, of both heart and arteries; and the same careful anatomists confirm
the observations of M. Souleyet,[1] that the veins were not wanting in the nudibranchs, but
had only undergone that modification of form to which the term sinuses is more
properly given. Not any of these authors have, however, published exact and recognisable representations of the relations of the attenuated tunica propria of the veins, to the
interspaces or laminæ which that tunic lines in forming the large and irregular sinuses in
which the blood is diffused. With a view of supplying this desideratum, further illustrations of the circulating system of the Brachiopoda are added here to those which were
published in my earlier memoirs.[2] Figure 1, pl. 3, more especially will show that the venous
system of the *Terebratula*, like that of other mollusks, is continuous, *i. e.* forms a part by
continuity of tissue, with the rest of the circulating system. Thus all ulterior researches
rightly interpreted, since the time of Hunter, have served to confirm and establish the
accuracy of his appreciation and explanation of the facts he discovered in insects and
crustaceans, and to extend our knowledge of the same modification of the venous
system as it exists in the general series of mollusks.

The absorbent system, as Cuvier truly states,[3] is wanting in mollusks. In the
comparatively low organised form under consideration (*Terebratula*), the chyle or product
of digestion must transude through the intestinal tunics into the intestinal sinus, pl. 3,
fig. 1, 8 : all the fluid answering to lymph, from other parts of the organization will be

---

[1] Observations sur les Mollusques Gasteropodes désignés sous le nom de 'Phlebentères' par M. de
Quatrefages; Comptes Rendus des Sciences, &c., 1844, t. xix, p. 355.

[2] Trans. Zool. Soc., 1833, pl. xxii, figs. 7, 8, 11; pl. xxiii, figs. 11, 16.—Annales des Sciences Nat.,
pl. iv, figs. 8, 9, and 10.

[3] Tom. cit., p. 300.

received into the intervisceral, 9, intermuscular and pallial, 5, 6, sinuses, which will perform the double but closely allied functions of veins and absorbents.

The streams of sea-water excited and maintained by the complex ciliated structure of the mantle, will effect the requisite respiratory changes as they course over the delicately coated, branched, sinuses of the pallial lobes; these, therefore, form the chief seat of the breathing function; but, wherever similar currents come in contact with the vascular system, to that extent the respiratory operations will be diffused.

GENERATIVE SYSTEM.

The generative organs have presented the same form in all the individuals of the different species of *Terebratulæ* examined by me. This form corresponds with that of the pallial sinuses in which they seem to be situated. The figures 15 and 16 *r, r,* pl. 22, 'Zool. Trans. vol. I,' from the *Terebratula Sowerbii,* represent so closely the general form and disposition of the same organs in *Terebratula flavescens,* that it seems needless to repeat the figures of them in the latter species.

I have observed, however, a manifest difference of texture and colour in the generative organs of different individuals of *Ter. flavescens* collected at the same period in the neighbourhood of Port Jackson, by my friend Mr. George Bennett, F.L.S., and transmitted in the same preserving liquid. In some specimens the organs are better defined, more compact, and of a paler colour: in others the organs are broader or more diffused, and of a deeper yellow colour.

On a microscopic examination of the first kind, they are found to consist of an aggregate of minute cells, closely resembling those in the half mature testes of the common oyster. I conclude, therefore, that they are the male generative organ in the *Terebratulæ,* and that the individuals of this genus are diœcious, not, as I formerly supposed, simple hermaphrodites. The true ova are very plainly manifested in the broader and deeper-coloured dendritic organs. They are developed, like the sperm-cells, between the outer layer of the mantle and the delicate tunic of the venous sinus, and protrude into that cavity, pushing its lining membrane inwards. The generative organs, in the male, as in the female, developed in the ventral lobe of the mantle, commence—if we may term the part next the hinge of the shell their beginning—by a loop, pl. 3, fig. 1, 12, on each side of that lobe, situated at the point of bifurcation of the two sinuses; these loops are shown in fig. 1, pl. 3, the inner layer of the mantle with the adherent tunic of the sinus being reflected to show the looped portions of the ovaria, 12, 12. In the dorsal lobe the generative organ commences at about the same distance from the hinge, by a simple obtuse extremity; and follows, as it advances, the ramifications of the sinus in which it is lodged. There are thus four distinct ovaria or testes; two in each mantle lobe; those of the ventral lobe being doubled, or bent upon themselves, near the cardinal attached border of the lobe. It may be presumed that the embryos developed under the influence of their suspension in the aerated and nutrient fluids of the pallial sinuses, escape at a certain stage of that development, by dehiscence, at or near the free ciliated border of the pallial lobes. The course and phenomena of this development would form a most interesting and acceptable subject to a competent microscopical observer favorably situated for pursuing it throughout the breeding season of the *Terebratulæ.*

I subjoin a few stages observed in ova taken from the ramified ovarium of a *Lingula,* preserved in spirits.

Pl. 1, fig. 7, *a,* is an impregnated ovum, in which the germinal vesicle or vesicles have disappeared, and the germ-mass has been formed, occupying the entire ovum, which has assumed an oblong form, a peripheral stratum of the derivative germ-cells was more compact and of a somewhat lighter colour than the central mass.

b, shows the formation of a smooth membrane, probably covered by ciliated epithelium, around the germ-mass.

c, is a transverse section of such an ovum showing its triedral figure.

d, is an ovum further advanced, with the rudiment of a peduncle.

e, an embryo with the peduncle more produced. I could distinguish no organs in these embryos of the *Lingula :* there was no trace of shell.

The dissections from which Mr. Scharf's drawings, engraved in the subjoined plates, were taken, were prepared by my former anatomical assistant, Mr. Henry Goadby, to whose skill in the manipulation of minute objects I am more especially indebted for the demonstration of the nervous system, as shown in figs. 1 and 2, pl. 2.

CHAPTER II.

ON THE INTIMATE STRUCTURE OF THE SHELLS OF BRACHIOPODA.

BY PROFESSOR CARPENTER, M.D., F.R.S., &c. &c.

THE shells of the Brachiopoda, whether recent or fossil, present most interesting subjects for microscopical investigation; their structure being, in almost every instance, quite distinct from that of the shells of the Lamellibranchiata, or Gasteropoda, and being so characteristic, that even minute fragments may be referred with certainty to this group, unless their texture has been altered by metamorphic action. Such action appears to have so constantly taken place in the shells preserved in the Palæozoic rocks, that, with regard to some of those genera which are limited to that series, such as *Pentamerus, Calceola,* or *Productus,* it is impossible to say with positive certainty how far their structure is conformable to the general type of the group, since this is but very obscurely shadowed forth in them. But, since we find *Terebratulæ, Rhynconellæ,* and *Spirifers* of those formations presenting the very same deficiency in characteristic structure with that which is encountered among the genera previously named, we seem fully entitled to affirm, that the absence of such structure in the *fossils* of the strictly-Palæozoic genera is not to be taken as any indication that the *living* shells departed from the general type; whilst, on the other hand, the perfect conformity to that type, of such structure as *can* be distinguished in them, is a positive indication that the shells of these genera were constructed upon the same plan with those of the existing Brachiopoda, as their other affinities would render probable. The following statements, then, which are chiefly based on the examination of *recent* Brachiopoda, may be considered to apply equally to the *extinct,* until any proof shall have been given to the contrary.

There is not, in the shells of Brachiopoda, that distinction between the *outer* and *inner* layers, either in structure or mode of growth, which prevails among the ordinary bivalves; and it seems obvious, both from the nature of the shell substance, and from the mode in which it is extended, that the whole thickness of the Brachiopod shell corresponds with the outer layer only in the Lamellibranchiata, (Pl. IV, fig. 1.) For, as will be presently shown, it must be regarded in the light of a *calcified epidermis,* like the prismatic external layer of Pinna or Avicula;[1] and it is augmented by an addition *to its*

[1] See the author's Memoir on the Microscopic Structure of Shells, in the Report of the British Association for 1844, p. 6.

margin (as seen at *a a, b b, c c*), as is the case with the outer layer of Lamellibranchiate shells generally, whilst *their* inner layer, formed by the calcification of the basement-membrane of the mantles, is augmented by the interposition of new lamina between the animal and *the whole previous external surface* of the shell, such new lamina being extended beyond the preceding so long as the shell continues to grow, and bearing on its projecting part the marginal addition which has been made to the external layer.[1]   I have occasionally met with a second layer, in recent *Terebratulæ*, within the earlier-formed portion of the shell, but confined to only a part of the surface, instead of extending beyond it.   Such a layer (represented at Pl. IV, fig. 1, *d d,*) is of precisely the same texture with the remainders of the shell, and appears to be simply destined to afford a mechanical support which its older portion may require, when this has been greatly extended by new growths.   Several such layers present themselves in the shell of *Crania* (Pl. V, fig. 8), whose upper valve is formed upon the plan of that of Patella, and is increased, both in thickness and diameter, by new formations successively added to the internal surface, and extending beyond the margins of the old.   A like succession of layers, moreover, is easily to be distinguished in the massive shells of many fossil Gasteropoda, in consequence of their ready separation from one another; but in these, as in the preceding cases, they arc all repetitions of the first or outermost layer, and do not present any of that difference in intimate texture which almost invariably distinguishes the inner and outer layers in Lamellibranchiate shells. Moreover, in all those Brachiopodous shells which are perforated by the vascular (?) canals presently to be described, the layers are alike perforated by them, although the passages are usually much smaller in the internal or last-formed layers than they are in the outer and older, as is well seen in the portion of the shell of Terebratula delineated in Pl. IV, fig. 3.   And when the passages themselves are large, as happens in the internal layers of the shell of Crania (Pl. V, fig. 8), their openings upon the inner surface are frequently so small as to escape observation, when it is examined with a hand-magnifier only.[2]

The "lines of growth" which are so obvious on the external surfaces of the shells of many Brachiopoda do not by any means constantly correspond with interruptions in continuity of structure as displayed in vertical sections of the shell; thus in Pl. I, fig. 1, there is seen at *c* such an inequality of surface as appears to indicate a marginal addition, yet the real extension, as marked by the line *c c*, appears to have taken place at some little

---

[1] Op. cit., 1847.

[2] It is apparently from observations of this kind, that Mr. Daniel Sharpe has stated, in his account of the Genus *Trematis* (Quarterly Journal of the Geological Society, vol. iv, p. 67), that "an inner layer of unpunctuated shell lining an outer punctuated layer, is a common occurrence among the Brachiopoda; it may be especially observed in the flat species of Orthis, and in many Leptænæ, covering all the central parts of the shell, and leaving the punctuations open only round the margin.   Among the Productæ and some species of Chonetes, the punctuations are closed up everywhere, except at the edge, by a gradual deposit of shelly matter in the interior.   In *Crania, Thecidea,* and some recent *Terebratulæ,* the punctuations can only be seen in the interior round the edge of the shell; but in the majority of the recent species of Terebratula, they are equally visible over the whole shell."   This statement will be perfectly correct, if

distance from this.   From the indistinctness of the *breaks* between the older and the newer portions of the shell, both here and elsewhere, I am inclined to think that the formative action must be more constant and gradual among the Brachiopoda, than it seems to be among the Mollusca generally.

Excluding the *Discinidæ* and *Lingulidæ*, whose shells are almost entirely composed of a horny animal substance, the shells of Brachiopoda may be said to be almost exclusively calcareous in their composition; the proportion of animal matter which they contain, being far less than that which exists in the shells of most Lamellibranchiate bivalves.   When a portion of the shell of a recent Terebratula which has been preserved in spirit, is submitted to the action of dilute acid, the greater part of it is speedily dissolved with effervescence; and in the membranous residue I have not been able to distinguish with certainty and uniformity more than two layers, both of great tenuity, but one of them considerably thicker than the other.   This thicker layer, which is usually of a yellowish hue, obviously covers the exterior in the manner of the *periostracum* of most Lamellibranchiate shells, although more closely incorporated with its substance; whilst the thinner layer, which is quite pellucid and colourless, appears to have lined its interior, this also being so closely incorporated with the shelly texture, as not to be separable from it without the complete destruction of the latter.   I have not been able to discern any distinct *structure* in either of these membranes, although they exhibit faint markings, which I am disposed to regard as nothing else than the impressions left upon them by the peculiar shelly texture with which they are in contact.   Intermediate layers are occasionally to be met with, sometimes resembling the inner membrane, and sometimes the outer, in their characters.   A repetition of the inner membranous layer may of course be expected wherever a new layer of shell is formed within the old; and I have found that such a layer is sometimes furnished, in Crania, though not in Terebratula, with its own periostracum.

It is, however, in the texture of the substance which intervenes between these layers of membrane, that the most characteristic peculiarity of the shells of the Brachiopoda exists. In all the recent *Terebratulidæ* and *Rhynconellidæ*, and in all the fossil specimens of those groups, as well as of *Spiriferidæ*, *Strophomenidæ*, and *Productidæ*, in which there is no indication of metamorphic action, the shell is found to consist of flattened prisms, of considerable length, arranged parallel to each other with great regularity, and at a very

we substitute for the term " a layer of unpunctuated shell," " a layer of shell having *minute punctuations*." When the uses of these passages (which are very different from what Mr. Sharpe supposed them to be) come under discussion (p. 29), it will be seen to be very improbable that they should in any instance be closed-up from the *inside*.   No such statement can be justified, save by a careful microscopical examination of a *transparent* section of the substance to which it relates, especially in *fossil* specimens, in which, from the channels in the shell being filled with the matrix, their orifices (the punctuations) are often almost indistinguishable, even when of considerable size (see p. 31).   And all my own observations— ome of them made upon specimens kindly supplied to me by Mr. Sharpe—lead me to believe that the passages always retain their internal openings (as seen in Pl. V, fig. 8), although these may be of very minute size.

acute angle—usually only about 10° or 12°—with the surfaces of the shell. These prisms range from about 1-2000th to 1-600th of an inch in breadth, the average being probably about 1-1000th of an inch, and are not above one tenth of the last-named amount in thickness. With regard to their length, I am not able to speak positively, as I have never succeeded in tracing any single prism continuously from one of its extremities to the other; but I have frequently met with prisms, broken at one end, whose length measured 1-50th of an inch. It sometimes happens that a shell of a recent *Terebratula*, which has been kept for some time in spirit, can be readily resolved into such prisms, even by rubbing portions of it between the fingers (Pl. I, fig. 5); but more commonly they are best made apparent, by cleaving the shell in the plane in which they lie. Most Brachiopodous shells readily admit of this kind of cleavage; but it is more easily practised with the *Rhynconellidæ*, *Spiriferidæ*, and *Strophomenidæ*, than it is with the Terebratulidæ. When their natural laminæ thus obtained are examined microscopically, the flattened prisms are seen to lie side by side with great regularity (Pl. IV, fig. 6; Pl. V, figs. 4, 6), usually overlapping one another at their edges; they are sometimes gently curved *laterally*,[1] but they never seem to depart widely from the straight line in that direction. The laminæ which they thus form, however, are often far from being flat; and this, which is particularly the case in those parts of the shell whose curvature is great, is due to the inflexion of the prisms in the direction of the *thickness* of the shell; so that, when seen in a vertical section, they often present considerable curvatures, especially towards their extremities, when these abut upon the perforations for the vascular canals (Pl. IV, fig. 4; Pl. V, fig. 14). The *internal surface* of the shell derives a most peculiar appearance from the 'cropping-out' of these prisms at the acute angle just mentioned; for it presents a very regular imbricated aspect (Pl. IV, figs. 7, 10, 11, 14; Pl. V, figs. 4, 5), each of the imbrications being the rounded extremity of one of the flattened prisms. The *external surface*, owing probably to the incorporation of the periostracum with the terminations of the prisms, does not distinctly show any such arrangement; but, when a little abraded, as happens in most fossil specimens, it very commonly exhibits the ends of the prisms cropping-out upon the surface at an extremely slight inclination, but without any regularly-formed terminations. When a *section* of the shell is made more or less parallel to its surface, this may traverse the prisms for a time in their own plane, or nearly so, and they will then present almost the same appearance as they do in a natural lamina; but it may cross the prisms more or less obliquely (Pl. V, fig. 3), and the appearances presented will vary according to the degree of that obliquity. Not unfrequently, owing to the curvature which has been already noticed as presented by the prisms, a section will traverse them in one part almost in the plane of their length, and will cross them very obliquely at another. A great variety of appearances may thus be presented by the shells of

---

[1] That is, when considered in relation to the shell of which they form part, their general line of direction being from the umbo towards the margin.

Brachiopoda, which are at once understood when the form and arrangement of their component prisms have been duly studied.[1]

The question now arises,—what is the *histological* character of these prisms, and in what way are they formed? Considering that they bear a certain degree of resemblance, both in form and arrangement, to the component prisms of the shells of Pinna, &c. (their chief differences being that they are flattened, instead of being hexagonal or nearly so, and that they lie at a small inclination to the surfaces of the shell, instead of passing vertically from one surface to the other), it might be not unfairly surmised that they are formed, like these, as *elongated cells.* It seems at first sight an objection to this view, and was long felt as such by myself, that no trace of cellular membrane is discoverable after the action of acid upon the prismatic substance; but this is true of a great many other shells also, in which, as I think I have shown to be almost certain, if I have not absolutely demonstrated,[2] the shelly texture was originally of a truly cellular nature. Now, it is a very curious circumstance, that in the shell of *Crania,* the greater part of which presents none of the characteristic Brachiopodous structure, there should be one portion,—that to which the muscles (or rather their tendons) are attached,—which is distinctly prismatic, the prisms having very much the appearance of those of *Terebratulæ,* but being arranged vertically to the surface of the shell as in *Pinna* (Pl. V, fig. 8, *a, b*): and that in this structure, a series of elongated cells may be distinctly traced, standing parallel to each other, and presenting their prismatic extremities on the inner surface of the shell (Pl. V, fig. 8); these retaining their structure and arrangement in the decalcified shell (Pl. V, fig. 13). And thus the presumption becomes greatly strengthened, that the component prisms of Terebratulæ, Rhynconellæ, &c. were originally formed within a cell-membrane, and have taken the shape of the cell-cavity within which they were moulded. Further, the component prisms of the shell of *Rhynconella octoplicata* exhibit a remarkable double-oblique striation (Pl. V, fig. 6), which somewhat resembles the transverse striation seen on the prisms of Pinna, and which suggests the idea that each of its long prisms is formed by the coalescence of a series of flattened cells linearly arranged, as I have shown to be the case with those of Pinna.[3]

In a large proportion of the species of *Brachiopoda,* the shell is traversed by a system of canals, which pass from one surface to the other, for the most part in a nearly vertical direction, and at tolerably regular intervals. The diameters of these canals differ greatly in the several species which exhibit them, as do also their distances from each other, and, consequently, the number which present themselves within a given area; as will be seen by comparing together figs. 6, 7, 8, 11, 12, 14 of Plate IV, and figs. 1, 2, 3 of Plate V,—

---

[1] The peculiar structure of the shells of Brachiopoda is liable to be much obscured by mounting the sections in Canada balsam, which goes far to obliterate the lines of division between the prisms. It is on this account alone, that the sections delineated in figs. 1 and 2 of Pl. V do not exhibit the same appearance as fig. 3.

[2] See my Memoir on Shell-Structure, in the 'Report of the British Association for 1847,' p. 98.

[3] Op. cit., 1844, p. 8, and 1847, p. 96.

all of which are drawn under the same magnifying power, namely 100 diameters. Among the recent Terebratulæ, there is none which have canals of larger size or more closely set than *Waldheimia australis* (Pl. IV, figs. 6, 7, 8), their average diameter being about $\frac{1}{800}$th of an inch, and the distance of their centres about $\frac{1}{400}$th. It is in *Terebratulina caput-serpentis* (Pl. IV, figs. 11, 12), on the other hand, that the canals are of smallest dimensions, their largest diameter being about $\frac{1}{1500}$th of an inch, whilst their average distance from each other is about the same as in the preceding case,—their regular arrangement, however, being so modified, that the external orifices are principally seen upon the elevated parts of the plications (fig. 12), whilst they open internally in similar rows (fig. 11). Among the fossil species of Terebratulæ, *Ter. bullata* (Pl. V, fig. 1) is among the most remarkable for the size of its canals, their diameter being about $\frac{1}{400}$th of an inch; whilst in *Ter. lima* (Pl. V, fig. 2), whose shell-canals are smaller than those of any Terebratula, recent or fossil, that I have examined, their diameter is scarcely $\frac{1}{2000}$th of an inch. The relative distances of the centres of the canals, however, are nearly the same in these two species, averaging about $\frac{1}{250}$th of an inch, so that about the same number may be counted in any given area; thus showing that there is no fixed relation in the development of the canal-system, between their *size* and their *number*. The largest canals which I have met with in any Brachiopod shell, are those of *Spiriferina rostrata* (Pl. V, fig. 3), these being about $\frac{1}{300}$th of an inch in diameter. Their number in a given area, however, is much less than it is in the Terebratulidæ generally, the distances of their centres averaging about $\frac{1}{150}$th of an inch; so that the proportion which their united areæ bear to that of the shell is not very different from that which obtains in *Terebratula australis*, or *Ter. bullata*,—the diameter of each canal being, in these species, from about one-half to five-eighths of the distance between the centres of the contiguous canals.

In all these statements, however, the *largest* diameter of the canals has been alone considered; and this, in the *Terebratulidæ*, is usually near the *exterior* of the shell, the canals expanding as they approach this, so as usually to present somewhat of a trumpet-form (Pl. IV, figs. 3, 4). Hence, the diameter of the internal orifices of the canals is frequently not more than one-half, sometimes not more than one-third, of that of the exterior; and it is still further reduced, when a new layer of shell has been formed within the preceding (as already noticed, p. 24). Of this, a good illustration is shown in Pl. IV, fig. 3, where *d d* marks the plane of junction between the new layer and the old (its position in the shell being seen at *d d*, fig. 1), below which the canals are seen to be con- tracted to about one-half the diameter of their ordinary internal terminations. Sometimes the canals are seen to bifurcate in their passage from the interior towards the exterior of the shell (Pl. IV, fig. 4); this bifurcation, however, is not common in Terebratulidæ. The external termination of each canal is covered with a sort of discoidal operculum (Pl. IV, fig. 8), which is much thicker and more opaque than the periostracal membrane; to this it usually remains adherent when the calcareous shell has been dissolved away by dilute acid, but it is occasionally found to be detached from it, so that the adhesion can scarcely be complete. In many Terebratulæ, these disks are seen to be surrounded by delicate radiating lines, as was first pointed out by Mr. Quekett;[1] the nature of these is somewhat obscure, but from the appearances presented by certain preparations made from young shells by Mr. Quekett, I am disposed (with him) to consider them as *cilia*, whose office it is to produce currents of water over the extremities of the cæca, for the purpose of aerating their contents.

The case is very different, however, with the *Craniadæ*; for here the canals, instead of enlarging as they approach the exterior, subdivide in an arborescent manner (Plate V, fig. 8); and thus, instead of presenting on the external surface a set of definite discoidal terminations, their ultimate ramifications alone are seen (Pl. V, fig. 12). The largest diameter of the canals is here at their *internal* orifices, save where these are contracted by the interposition of a new internal layer (as already noticed, p. 24), between the preceding and the surface of the mantle, in the manner shown in fig. 8.

When it was first discovered that the "punctuations" which had been previously noticed upon the surface of many Brachiopodous shells, are really the orifices of canals penetrating the whole thickness of the shell,[2] it was natural to inquire what is their probable function in the economy of the animal; and at the time when it was not known that they are closed externally by a discoidal operculum, the opinion seemed not an improbable one, that they are in some way connected with the function of respiration, serving to introduce water into contact with the surface of the mantle which lines the shell. Subsequent research,[3] however, showed that they are occupied in the living animal by cæcal tubuli, which extend into the shell from the membranous layer that is incorporated with its inner surface. These are best seen by exposing a portion of a shell that has been preserved in spirits, with the animal, to the action of dilute acid; for attached to the membranous films that form the residuum, are then found a multitude of cæca, more or less crowded with yellowish-brown cells (Plate IV, fig. 9), and corresponding in size and arrangement with the passages in the shell. If the process of decalcification be stopped at a fitting time, a lamina of shell may be preserved, with the tubular cæca corresponding to its whole previous thickness projecting from its surface (Pl. IV, fig. 13); thus conclusively proving that their true place is actually in the midst of the shell-substance. Similar cæca, less crowded with cells, may be readily detected in the membranous residuum of a recent *Crania* (Plate V, fig 11).

It is of considerable importance to the due estimation of the zoological value of the character afforded by the presence or absence of the perforations, that the physiological nature of these cæca should be clearly determined. I regret that I have not yet been able

[1] Histological Catalogue of the Museum of the Royal College of Surgeons, vol. i, p. 270.

[2] This discovery was first announced in the author's Paper on the Microscopic Structure of Shell, read at the Royal Society, Jan. 17, 1843, of which an abstract was published in the 'Ann. of Nat. Hist.,' Dec., 1843.

[3] See the author's Memoir in the 'Reports of the British Association for 1847,' p. 93.

to succeed in obtaining a clear elucidation of this point; still, I cannot but think that a near approximation may be made to it. I have already mentioned, that with the internal surface of the shell is incorporated a membrane, of which the cæcal tubuli are prolongations; and I have found the free surface of this membrane to be almost covered with cells, resembling both in size and appearance those contained in the cæca. I have found similar cells upon that surface of the mantle which is applied to the interior of the shell, and which (as long since remarked by Professor Owen) possesses a degree of adhesion to it, that does not exist among the Lamellibranchiata. Hence I am much inclined to believe that, as was first suggested to me by Mr. T. H. Huxley, the space between the mantle and the proper lining membrane of the shell is, in reality, a great vascular sinus; and that the membranous cæca opening-off from that space are really to be considered as vascular processes, analogous to those which present themselves in the "test" of many Ascidians.[1] The cells which are contained in these cæca, and which are found so abundantly on the contiguous surfaces of the mantle and the interior shell-membrane, must then be considered as blood-corpuscles.

It can scarcely be supposed that these vascular prolongations can have any relation to the nutrition of the shell; since we find them absent in a considerable number of the shells of Brachiopoda, which in every other respect are formed upon the same type with those wherein these cæca abound. We can scarcely do otherwise, I think, than regard their presence or absence as bearing a relation to the general economy of the animal; and they must consequently be admitted as furnishing a character of considerable importance in a systematic point of view. We shall find, accordingly, that (so far as yet known) they exist in all the true *Terebratulidæ*, both recent and fossil; and that (with the same reservation) they are absolutely and entirely wanting in all the true *Rhynconellidæ*, both recent and fossil; so that by this character alone, any member of the former family might be safely diagnosed from any member of the latter.

As frequently happens, however, in other departments of Zoology, a character which

---

[1] The peculiar appearance of these cæca and their contained cells, led me at one time to surmise that they might possibly be the extremities of the ramified ovaria, which extend themselves through the mantle; or that, like the glandular clusters in the dorsal papillæ of the Eolidæ, they might constitute a subdivided liver. Neither of these views, however is consistent with the entire disconnection of the cæca with the proper mantle, and the animal it encloses; and, moreover, as Professor Owen has remarked to me, "these palleal processes are so much more minute than the ova discernible in the ramified ovaria, and their presence is so equally manifested on the ovarian and non-ovarian parts of the mantle, that no ground is to be got for associating them in speciality of function with the generative system; whilst the palleal lobes are so thoroughly distinct (in comparison with the skin of Nudibranchs) from the visceral mass, that they cannot have any special relation to the liver." I mention these surmises and their refutation, merely to assist in determining, *par voie d'exclusion*, what is the real character of these palleal processes; it being often easier in physiology to determine what a thing or function *is not*, than what it *is*. Professor Owen permits me to state, that he considers the view given in the text to be not discordant with the results of his researches on the vascular system of the Terebratulidæ.

is of typical value as affording an absolute distinction between certain families, possesses a greatly inferior value as regards others; thus we shall find that both *Spiriferidæ* and *Strophomenidæ* contain perforated and non-perforated shells; and it will be for further investigation to determine, whether the character has here a *generic* value (serving, for example, to distinguish Spiriferina from Spirifer, Orthesina from Orthis), or whether it must here rank as a mere specific difference.   The details which I shall presently furnish, may help towards the determination of this point; but it will require a far more complete and extended investigation than I have had either time or opportunity for making, to determine the real import of this character in every group in which it seems to be otherwise than constant.

Such an investigation, however, to be of any value, must be made with great care; *superficial* observations being here worse than useless, because they are liable to give the most erroneous results.   The perforations in the shells of fossil Brachiopoda are very liable to be filled-up by the matrix; and such a smoothness and apparent continuity may thus be given to their surfaces, that no "punctuations" may be perceptible by the microscopist, still less by the observer who is armed only with a hand-magnifier.   On the other hand, the surface of some shells is marked in such a manner, as to indicate the presence of the perforations with almost absolute certainty to any one who has not learned caution from previous experience; yet the infallible test of a transparent section, examined under a sufficient magnifying power, makes it evident that there is not the slightest vestige of real perforations, and that the supposed "punctuations" are nothing else than surface-markings, analogous to the "sculpture" of other shells.   Of the importance of such precautions, a very remarkable example is afforded by the existence of perforations in *Strigocephalus*, which has been uniformly regarded as a non-punctuated shell, and placed among the Rhynconellidæ, whilst its real position (as the presence of perforations would indicate) has been found by Mr. Davidson to be among the Terebratulidæ. Conversely, the perforations are really wanting in *Porambonites reticulata*, notwithstanding that there is a most decided and regular "pitting" of the surface of the shell, extremely resembling the large punctuations of some of the Terebratulidæ and Spiriferidæ; the place assigned to this genus on other characters being among the Rhynconellidæ, the existence of perforations in its shell would have been anomalous.[1]

[1] I lay the more stress on the importance of this method of examination, because an attempt has been made by Professor King (Monograph of the Permian Fossils of England) to put aside my former observations, made with *adequate microscopic power* upon *transparent sections* of the shells, as altogether valueless, on the strength of an examination of their *surfaces* with a *Stanhope-lens*.   "Dr. Carpenter," he says (p. 110, note), "places Hypothyrises (Rhynconellæ) in his non-perforated division of the Brachiopods; but punctures, although much more minute than those in the Terebratulidæ, occur in every species that has passed under my notice.   Punctures also occur in *Productidæ* and *Spiriferidæ*; in short, *I doubt their absence in any Brachiopod whatever.*"—Now I have Professor King's own authority for stating, that he has relied *entirely* on superficial observations made with a hand-magnifier, than which, as I have shown above, nothing can be more fallacious; and further, that he has never himself examined the

The foregoing statements will be found to apply, more or less fully, to all the families into which the Brachiopoda are subdivided by Mr. Davidson, except the *Discinidæ* and *Lingulidæ*, in whose shells a different type of structure prevails, which will be described in the special account of those families.

Family—TEREBRATULIDÆ.

The shells of all the genera included by Mr. Davidson in this family, exhibit the structure previously described, in its most characteristic form; and all of them are perforated by the system of passages which give place to the cæcal prolongations of the lining membrane. It is interesting to remark, that these passages do not exist in the calcareous 'loop,' whose structure is in other respects precisely the same as that of the shell. (Pl. IV, fig. 2, *a a*, and fig. 10.) The following is the list of species, recent and fossil, which I have examined; the former including all those contained in the unrivalled collection of Mr. Cuming, to which that gentleman has kindly allowed me free access for that purpose. Of the recent species marked *, I have been enabled to make transparent sections, chiefly by the liberality of Mr. Cuming in furnishing me with specimens for that purpose; the remainder have been determined by the examination of the surface alone; but the source of fallacy which has been already shown to render this method insufficient as regards the *fossil* Brachiopoda, does not exist with respect to the *recent*, since it is impossible for any practised observer to confound, in the latter, *superficial depressions* with *actual perforations*. Of *all* the *fossil* species enumerated under this and other heads, I have made *transparent sections;* deeming it unsafe to speak in regard to them from superficial examination alone.

TEREBRATULA.—Recent, *T. dilatata, globosa, uva, vitrea;*—Fossil, *T. Bentleyi, biplicata, bullata,*[1] *carnea, coarctata, fimbria, globata, grandis, intermedia, maxillata, perovalis, punctata, sacculus, Salteri?* (see Retzia, p. 34), *sphæroidalis, squamosa.*

WALDHEIMIA.—Recent, *W. australis* *,[2] *californiana, cranium, flavescens, Grayii, lenticularis, picta;*—Fossil, *cornuta, digona, impressa, numismalis, oblonga, obovata, ornithocephala, resupinata.*

recent *Rhynconella psittacea,* of whose structure I had given illustrations, in the Memoir with which Professor King must have been acquainted (Reports of British Association for 1844), sufficiently ample, it might have been thought, to make him pause before committing himself to such a sweeping assertion. To myself, personally, it is a matter of entire indifference, whether Professor King does, or does not, admit the correctness of my observations; but I would submit that the interests of science are not very likely to be promoted, by this easy setting-aside of observations, made with every advantage of first-rate instruments and careful preparation of specimens, in favour of glances with a hand-magnifier at shells whose surfaces are peculiarly liable to present deceptive appearances, the examination being confined to their exterior, and no adequate means being taken to examine their intimate structure and arrangement. When Professor King shall have demonstrated the existence of a system of shell-canals in *Rhynconella psittacea,* it may be reasonably permitted him, " to doubt the absence of perforations in any Brachiopod whatever."

[1] Pl. V, fig. 1.   [2] Pl. IV, figs. 1—10.

Many of these might probably be distinguished from each other by the size and arrangement of the shell-canals; but until it shall have been determined, by the examination of numerous specimens, what are the limits of variation in this respect, it would be unsafe to employ this character for the diagnosis of species, except where a very strongly-marked difference presents itself.   One of the most remarkable departures from the usual regularity of their arrangement, is presented by *Ter. coarctata*, the surface of whose shell is covered with square or oblong elevations, arranged in rows, and occasionally running together into ridges; for the canals usually pass towards the highest parts of these elevations, so that their orifices are there crowded together, whilst in the intermediate furrows there are scarcely any to be seen.

TEREBRATULINA,—Recent, *abyssicola, angusta, cancellata, candida, caput-serpentis* *,[1] *Cumingii, japonica, septentrionalis.*

TEREBRATELLA,—Recent, *Bouchardii, chilensis, coreanica, Cumingii, dorsata, Evansii, flexuosa, inconspicua, labradorensis, rubella, rubicunda* *, *sanguinea, spitzbergensis, zelandica;*—Fossil, *hæmispherica?*

MAGAS,—Fossil, *pumilus.*

BOUCHARDIA,—Recent, *tulipa* *, Blainv. $=$ (*rosea,* Humph.)

KRAUSSIA,—Recent, *cognata, Deshaysii, Lamarckiana, pisum, rubra* *.

MEGERLIA,—Recent, *pulchella* *, *truncata* *;[2] Fossil, *lima.*[3]

The greater part of the internal surface of the shell *Megerlia truncata* is studded with little papillary eminences, on which the shell-canals do not open. (Plate IV, fig. 14.[2])

MORRISIA,—Recent, *anomioides* $=$ (*depressa,* Forbes.)

ARGIOPE, — Recent, *cistellula, cuneata, decollata, lunifera, neapolitana;*—Fossil, *decemcostata.*

STRIGOCEPHALUS,—Fossil, *Burtini.*   (See p. 31.)

Genus—THECIDEA.

In this genus—which, if associated with the *Terebratulidæ* at all, must be considered as a very aberrant form of that family—the structure of the shell is much less characteristic than it is in the *Brachiopoda* generally, and much less distinguishable from that of ordinary bivalves.   From appearances which I have occasionally detected, I am inclined to believe that the shell is originally formed on the same plan with that of other *Brachiopoda*, but that the component prisms subsequently coalesce, so as to form a more compact texture.   The shell-canals are very obvious in the recent *Thecidea mediterranea* *, opening

---

[1] Pl. IV, figs. 11—13.

[2] By a mistake on the part of the artist, these eminences are not made to exhibit the imbricated arrangement of the extremities of the prisms, which they present in common with the remainder of the internal surface of the shell.

[3] Pl. V, fig. 2.

upon its external surface in rows along the ridges (Plate V, fig. 10). They are also very evident in the fossil *Th. leptænoides;* but I am not by any means sure of their presence in *Th.mayalis, Moorei, sinuata,* and *triangularis.* In the specimens of these which I have examined, however, the shell was so far from being well preserved, that I do not feel inclined to speak confidently upon the point.

FAMILY—SPIRIFERIDÆ.

In all the genera of this family, the structure of the shell is completely conformable to the general description already given ; but the system of shell-canals no longer exhibits the same constancy, being, in fact, more generally *absent* than present.

SPIRIFER.—In this genus, as usually constituted, there is a great diversity in regard to the presence or absence of the perforations ; but Mr. Davidson has been led, by considerations altogether independent of them, to divide it in such a manner that the perforated and non-perforated species fall into distinct sections. Thus, he makes his genus *Spirifer*-proper to consist of those species which, like *Sp. striatus,* are unpunctuated, and without a mesial septum in the larger or ventral valve. Under this head rank *Sp. comprimatus, concentricus, speciosus,* and *Verneuilli,* which I can state with confidence to be non-perforated (notwithstanding Professor King's assertion to the contrary), not the least trace of passages being discernible in sections which display the texture of the shell in the greatest perfection. Under the sub-genus *Cyrthia,* he places these unpunctuated species, which, like *C. trapezoidalis,* have a perforation for the passage of a pedicle ; besides this species, I have examined *C. cuspidata,* and am fully satisfied that in neither of these do any perforations exist. Under the sub-genus *Spiriferina,* on the other hand, he ranks all the species which, like *Sp. rostrata,* are punctuated, and which have a largely-developed mesial septum in the ventral valve. I have determined the existence of perforations in *Sp. Demarlii, heteroclyta, rostrata,*[1] and *Walcotii.*

ATHYRIS.—In this genus, also, it is proposed by Mr. Davidson to separate the perforated and non-perforated species, which fall into different sub-generic sections respectively, according to other features in their structure. The typical species of the genus, such as *A. herculea* and *A. tumida,* I have ascertained to be non-perforated. The same is the case with *concentrica,* which Mr. Davidson now ranks under the sectional type *Spirigera,* with other unpunctuated species. On the other hand, the punctuated species, such as *esquerra* and *melonica,* in which I have found the perforations to exist, are placed by him under the sectional type *Retzia;* and he thinks it probable that *Ter. Salteri* will have to be transferred to this sub-genus.

UNCITES.—Of this genus, I have examined only *U. gryphus,* which is not perforated.

ATRYPA.—In none of the species of this genus which I have examined, namely *A. affinis, marginalis,* and *reticularis,* is there the least trace of perforations.

[1] Pl. V, fig. 3.

Family—RHYNCONELLIDÆ.

In the shells of this family, I believe the *absence* of the perforations to be a character as constant, as their presence is in the *Terebratulidæ*. No one who examines the shells of the recent *Rhynconella psittacea* and *Rh. nigricans*, even in the most superficial manner, can have any hesitation in recognizing the entire absence of the superficial "punctuations" which mark the orifices of the shell-canals in the recent species of *Terebratulidæ*;[1] and the most careful microscopic examination of these sections of the shell, taken from any part, and in any direction, does but confirm this conclusion. Thus, in Plate V, fig. 4, is shown a considerable area of the interior of the shell (*a, a*), which displays the usual imbricated arrangement of the extremities of its component prisms, shown on a larger scale in fig. 5; whilst these prisms are exhibited in their longitudinal aspect by a fracture of the shell at *b, b*. In neither part can the least trace of perforations be seen.[2] In all other respects, the intimate structure of the shell corresponds precisely with that of *Terebratulidæ*; but it may be mentioned that the prismatic laminæ are less adherent to each other than in the perforated shells, so that they are readily split asunder, this being the case with the fossil no less than with the recent species.

RHYNCONELLA.—Recent, *psittacea* *,[3] *nigricans* *;—Fossil, *acuta, concinna, decorata, depressa, inconstans, lata, nucella, obsoleta, octoplicata,*[4] *plicatella, pygmea, rostrata, spinosa, subplicata, tetraedra.*—The only peculiarity which I have met with among these, is a peculiar oblique striation of the prisms of *Rh. octoplicata*, already described (p. 27).

CAMEROPHORIA *Schlothemii*, not perforated.

PENTAMERUS *Knightii*, not perforated.

PORAMBONITES.—It is not a little remarkable that in this genus the shell should be most unequivocally *non-perforated*, notwithstanding that the punctuation of the surface, in one species especially, has led to the supposition that large "pores" exist, whence, I presume, the generic name has been derived. The most careful microscopic examination of transparent sections, fails to bring into view any perforations either in *P. æquirostris* or in *P. reticulata*.

---

[1] In Mr. G. B. Sowerby's 'Monograph of the Genus *Terebratula*,' in which these two species are described under that generic designation, they are spoken of as "the only species that are not punctuated."

[2] I cannot refrain from here again expressing my astonishment, that any systematist should venture to affirm the universal existence of perforations in the shells of Brachiopoda, without having examined one of the most common of the recent types of the group, in which the absence of such perforations had been specified as a distinctive character by competent observers.

[3] Pl. V, figs. 4, 5.   [4] Pl. V, fig. 6.

Family.—STROPHOMENIDÆ.

In this family, as in *Spiriferidæ*, there is a want of constancy as regards the shell-canals; and further investigation is much needed to determine how far the character which is furnished by their presence or absence corresponds with those other characters on which generic distinctions may be most securely based.—The results of the examinations which I have hitherto made are as follows:—

ORTHIS.—The prevalent character in this genus is that of perforation, which presents itself in the following species: *biloba, canalis, elegantula, filiaria, hybrida, Michelini, resupinata, striatula, testudinaria.* Of nearly all these it is particularly noticeable that the openings of the perforations on the external surface of the shell are arranged in regular rows on the ridges of the plications. The only species in which I have not found the perforations to exist, are *O. biforata* and *calligramma.*

ORTHESINA.—So far as I at present know, the shells of this genus are *not* perforated. The species which I have examined are *O. ascendens, elegans* [?], and *hemipronites.*

STROPHOMENA.—Here, on the other hand, the perforation of the shells seems to be the rule; but the perforations are unusually remote from each other (Pl. V, fig. 15 *a*), and the laminæ of the shell are bent down round the margin of each, so as to give it a somewhat infundibular character in vertical or oblique sections (fig. 14 *b b*, fig. 15 *b b*). The shells of this genus which have fallen under my examination, have for the most part undergone so complete a metamorphosis, that their original texture is almost entirely obscured; and even the perforations, when filled-up by a matrix of nearly the same appearance with the metamorphic shell, may pass undetected, save when carefully looked for: I have found their existence to be sometimes more certainly revealed by *vertical* than it is by *horizontal* sections. The species which I have examined are *alternata, depressa, funiculata,* and *planoconvexa;* in all of which the perforations are clearly distinguishable.

LEPTÆNA.—In one of the most characteristic species of this genus, *L. transversalis,* the shell is obviously traversed by perforations, resembling those of *Strophomena depressa* in their size, distribution, and arrangement. I have found similar perforations in *L. semiovalis.* On the other hand, I have examined specimens of *L. Davidsoni* and *L. oblonga* without being able to recognise them; in these specimens, however, the shell had undergone such a complete metamorphic action, that it would be very unsafe to affirm the non-existence of perforations in these two species.

Family.—PRODUCTIDÆ.

In all the genera of this family large perforations exist, resembling those of *Strophomena depressa* in their general aspect, and in the infundibular arrangement of the laminæ of the shell around them.   Where the shell is furnished with spines, as is especially the case with *Productus horridus*, the perforations are continued into them; and such passages are of more than the average dimensions.   The shells of this group have for the most part undergone such a metamorphic action, that I cannot speak confidently with regard to the conformity of their texture to that of the Brachiopoda generally; but for the reasons already given (p. 24), I have no reason to believe it to have been different.   I am disposed to think, however, that in some of the massive shells of this group, which appear to have been formed by the addition of successive laminæ to the interior, the internal laminæ were sometimes destitute of perforations—at least at a distance from the margin—as stated by Mr. D. Sharpe.   Still, unless this point can be satisfactorily made out in shells whose original texture have been well preserved (if such should fortunately present themselves), it would be unsafe to make a positive affirmation regarding it.

PRODUCTUS.—Of this genus I have examined the following species:—*Cora, Flemingii, horridus*, and *striatus*, all of which have well-marked perforations.

STROPHALOSIA.—The two species of this genus which I have examined, viz., *Sedgwickii* and *sub-aculeata*, are both perforated like Productus.

CHONETES.—Of this genus, also, I have examined two species, *armata* and *lata*, both of which are perforated.

Genus—CALCEOLA.

Though I have examined several specimens of *Calceola Sandalina*, I have not met with one in which the texture was sufficiently well preserved to enable me to speak positively as to its character.   I have no reason, however, to regard it as having departed from the ordinary type; and I have not been able to detect any distinct traces of perforations.

Genus—CRANIA.

The structure of the shell in this genus, as determined by the examination of the recent *Crania norvegica*, is widely different from that of Brachiopoda generally.   Instead of a series of flattened prisms arranged with great uniformity, we only meet with a substance which does not present any regularity or distinctness in the arrangement of its components, but which is not at all unlike that of which many Lamellibranchiate shells are composed, and may probably, like it, be regarded as having been originally formed by the coalescence of cells, which were destitute of any constancy in size, shape, or general arrangement.   But whilst departing from the general Brachiopodous type in this respect, the shell of *Crania* is quite conformable to it, in being penetrated by canals which are prolonged from the lining membrane of the shell, and which pass towards its external surface. These differ from those of Terebratulæ, however, in not arriving at that surface, and in breaking-up into minute subdivisions as they approach it (Plate V, figs. 8, 12). They usually open near the internal margin of the valves, by orifices so large as to be apparent to the naked eye (fig. 7); but nearer the central part of the valves, their orifices are frequently so minute as not to be readily discernible. This is in consequence of the formation of an additional lamina within the old one, and of the contraction of the canals in their passage through it, as is seen in the lower part of the vertical section shown in Plate V, fig. 8. The texture of the shell, at the part where the muscles are attached to it, is distinctly prismatic, the prisms being arranged vertically, so that their polygonal extremities alone present themselves at this part of the internal surface, (Plate V, fig. 9), the openings *a, a, a,* of the shell-canals being seen even here. It is curious to observe that this prismatic substance presents itself in the external and older layers of the shell, in nearly the same situation as in the most internal or newest, (as shown at *a, b,* Plate V, fig. 8); although several intervening laminæ may have been formed, some of them separated from the rest by a renewal of the periostracum, and the shell extended by them as well as thickened. When a portion of the shell of Crania has been decalcified by dilute acid, a membranous residuum is obtained, having the cæcal prolongations in connection with it (Plate V, fig. 11); these closely resemble the cæca of Terebratulæ, but are less filled-up with cells. The prismatic substance to which the muscles are attached, yields a definite animal basis of a cellular character (Plate V, fig. 13), closely resembling that of the *Margaritaceous* family, but more delicate and of smaller dimensions, the average diameters of the prisms being not above 1-1800th of an inch.

The fossil specimens of Crania which I have examined, namely, *C. antiqua* from the chalk, and *C. antiquior* from the Great Oolite, present the same characters, so far as the state of preservation of the shell allows them to be discerned. It may be remarked that from the filling-up of the perforations with a matrix harder than the shell, and from the abrasion of the latter, the large orifices of the perforations upon the inner margin of the shell are frequently marked by papillary elevations, instead of by depressions. This appearance is not unfrequently presented on the surface of fossil Terebratulæ.

Genus SIPHONOTRETA.

This genus is remarkable for the penetration of its shell (of the intimate texture of which I cannot speak, on account of the metamorphic condition of the species which I have examined) by canals which have the same general arrangement with those of Productidæ, and which, as in that family, pass up into the spinous or verrucose out-growths from the shell.[1] In a vertical section which I have made of the very thin shell of

---

[1] See Dr. S. Kutorga's Memoir 'Ueber die Siphonotreteæ,' &c.; St. Petersburg, 1848.

verrucosa, these passages appear to be prolonged *inwards*, by infundibular extensions of the shell, which project for some distance into its cavity, as shown in the annexed woodcut.

FAMILY.—DISCINIDÆ AND LINGULIDÆ

The structural character of the shell, in these two families, is altogether peculiar. For in the first place it is composed of a great number of very thin laminæ, arranged parallel to the surface; to which plan of structure we have seen that an approach is presented in Crania.   Next, the substance of these shells (in the recent species at least) is horny, instead of being calcareous, thus resembling the periostracum of other shells; and further, it is penetrated by minute tubuli of extreme minuteness, resembling those of dentine, instead of by large perforations. These tubuli run parallel to each other (Plate V, fig. 16, *a a*), usually in an oblique direction, though near the margin of the shell they may often be seen to be parallel to the surface.   In *Discina* they are commonly arranged in fasciculi, so that their transverse section presents a series of clustered dots, as seen at *b b*, sometimes, however, presenting the arrangement seen at *c;* both of these plans suggest the idea of a cellular structure as the original basis, but of this I have not been able to obtain any satisfactory evidence.

Attention has been drawn by Mr. Daniel Sharpe to an apparent departure of a very remarkable kind from the usual plan of structure, in the genus *Trematis*,[1] the shell of which he describes as composed of two layers, the external one being regularly punctuated, like that of the *Terebratulidæ*, whilst the internal is destitute of all trace of punctuations.   This he endeavours to show to be only an extension of a plan of structure common among other *Brachiopoda*, namely, that the internal layer is frequently imperforate towards the centre of the valve.   On this general statement I have already had occasion to remark (p. 24, note); and have now to show that it is altogether inapplicable to the case before us.   For I have ascertained, by careful examination of the shell of *Trematis*, that it is formed on the general plan of the shells of the *Discinidæ;* and that it has not in reality the slightest relation to the Terebratulidan type.   The external markings are of the most superficial character possible, and no more indicate *perforations*, than do those of *Porambonites*.   They are confined to the outermost of the *numerous* thin laminæ of which the shell is composed; and although I cannot detect in these the minute tubular structure which is characteristic of the group, yet its absence is fully accounted for by the metamorphic condition of the shell, a like obliteration having occurred in fossil *Discinæ* and *Lingulæ*.

In bringing to a conclusion the present account of the structure of the shell in this most interesting group of Mollusca, I cannot but express my regret that it has not lain in my power to render it more complete. Those who seek for assistance in classification from the characters which I have furnished, will, I fear, complain with much reason of the small proportion of species which I have submitted to examination, especially in some of those generic types (as *Spirifer, Athyris,* and *Orthis*) whose variability in this respect makes such an examination especially important. I may be permitted to remark, however, that the labour of preparing microscopic sections (the only mode, as I have shown, which can give satisfactory results in regard to fossil *Brachiopods*) is by no means trifling, and that the results here given are based on the examination of no fewer than *three hundred* such preparations. No doubt, I venture to think, can now remain as to the importance, as a natural-history character, of the perforation or non-perforation of the shell, for the reception of the vascular processes of the animal,—a feature in its structure which I may claim to have been the first to demonstrate; and I must leave the more extensive determination of it, as regards individual species, to be carried out by those who possess more time and opportunity for such a work than I can command.

CHAPTER III.

ON THE CLASSIFICATION OF THE BRACHIOPODA.

By THOMAS DAVIDSON, F.G.S., &c.

Sub-Kingdom—MOLLUSCA, *Cuvier.*

CLASS — BRACHIOPODA, *Cuvier.*[1]   *Dumeril.*[2]

ORDER—PALLIOBRANCHIATA, *Blainville.*[3]

THE account of the gradual progress of researches in this class having been already detailed by Baron Von Buch,[4] Messrs. d'Orbigny,[5] King,[6] de Verneuil,[7] Gray,[8] M'Coy,[9] de Koninck, and others, I will pass lightly over the literature of the subject, and simply recapitulate, in a condensed shape, its most prominent features.[10]

All naturalists seem disposed to admit that Fabius Columna was the first author who called attention to some of the shells we are about to describe, and which he distinguished by the denomination of *Concha Anomya,*[11] and in his figures it is not difficult to recognize several well-known species belonging to the genera *Terebratula* and *Rhynchonella.* The *Concha Anomia* was admitted by Lister and other authors, but in 1696 Llhwyd having proposed the name of *Terebratula*[12] for the same shells, this last denomination was preferred by the generality of subsequent writers to that of ANOMIA, again adopted by

---

[1] Ann. du Mus., vol. i, p. 44; and in Roissy Moll., vol. vi, p. 460, 1805.—*Brachiopoda,* from the two variously curved and cirrated arms or labial appendages.

[2] Trait. Element., 1806.—Lamarck; Phil. Zool., 1809.

[3] Dic. des Sc. Nat., t. xxxii, p. 298, 1824.—*Palliobranchiata,* from the respiratory system being combined with the mantle, on which the vascular ramifications are distributed.

[4] Über Terebateln, 1834 ; and Mém. de la Soc. Géol. de France, vol. iii, 1st ser., p. 106, 1838.

[5] Considerations Zool. et Géol. sur les Brach. Pal. Franç. Ter. Crétacés, vol. iv, p. 281, 1847.

[6] Ann. and Mag. of Nat. Hist., vol. xviii, July, 1846; and a Monograph of English Permian Fossils, p. 67, 1849.

[7] Geol. of Russia, vol. ii, 1845.

[8] Ann. and Mag. of Nat. Hist., p. 435, 1848; and Cat. of the Brach. in the Brit. Mus., 1853.

[9] Synopsis of the Carb. Limest. Fossils·of Ireland, p. 102, 1844; and Brit. Pal. Foss. in the Cambridge Museum, 1852.

[10] The Council of the Palæontographical Society having expressed a wish that we should condense our subject as far as possible, it has been found necessary to leave out many interesting details which would have extended this introduction far beyond the limits assigned. (23d July, 1852.)

[11] De purpura, 1616.   [12] Lithophylacii Britannici Ichnographia.

Linnæus in 1758. The term *Anomia* was, however, afterwards applied to a small group of Lamellibranchiate shells, distinct from that intended by the original denominator.

In 1712, Morton had perceived the necessity of separating the plaited Terebratulæ from the smooth ones, but did not propose any distinct generic appellation;[1] and for several years after Columna's first descriptions, but little advance was made towards classification, the few known species having been allowed to remain confounded among the other shell-fish.

Cuvier, with his acute powers of perception, after having examined the animal of *Lingula*,[2] as well as the figures of the so-called *Patella Anomala* of Müller,[3] *Criopus* of Pallas,[4] and excellent illustrations of *Ter. caput-serpentis* by Gründler,[5] proposed to create for those animals a distinct class among the Mollusca, to which Dumeril applied the name of BRACHIOPODA; so that the History of the ORDER may with truth be said almost to date from the commencement of the present century; and when we look back to all that has been achieved since that short period, we may fairly hope that by the expiration of the present century, the history of the class may have nearly attained its fullest development. At the time of Cuvier's first researches, but very few species were discovered, still these had been subdivided into several excellent genera, because the differences existing between the *Terebratulæ, Lingulæ,* and *Craniæ* (*Orbiculæ*), had not escaped the scrutiny of early investigators.

The vast impulse given to geological researches since the year 1800, rapidly filled the collectors' cabinets with a rich harvest of new and undescribed forms, many of which, by presenting marked differences from those already known, tempted Palæontologists to augment the number of genera, from its having been found necessary *to separate that which is fundamentally different, as well as to unite that which is really similar;*[6] nor is it surprising to find in such attempts, at times conducted by inexperienced hands, that numerous errors should have taken root, which time and sub-

---

[1] Nat. Hist. of Northamptonshire.

[2] Mémoire sur l'Anatomie de la Lingule, 'Mém. du Mus.,' vol. i, 1802.   [3] Zoologia Danica, 1781.

[4] Test. Sicil., 1792.   [5] Naturforscher, vol. i, 2d part, p. 86, tab. iii, figs. 1—6, 1774.

[6] This expression is forcibly made use of by Professor M. Edwards and Jules Haime, in their excellent work on 'British Fossil Corals.' The confusion that has existed, and still exists, is chiefly caused by the value of the terms *Family, Genus,* and *Species,* being by many authors viewed in a different sense; thus the *Genera* of some are the equivalents of the *Families* of others, so that a continual controversy naturally arises. The excellent observations on this point, published by Mr. Strickland, in p. 217 of his report, before the British Association for 1844, should be duly weighed and considered: he observes that, "all groups of the same rank are supposed in theory to possess characters of the same value or amount of importance, and the object of the naturalist should be, to bring them as nearly as possible to this state of equality. It must, indeed, be admitted, that no certain test seems to have been discovered for weighing the value of Zoological characters. The importance of the same characters manifestly varies in different departments of nature, and must therefore be estimated by moral rather than demonstrative evidence. The real test of the value of a structural character ought to be its influence on the economy of the living animal; but here we too often have to lament our ignorance, or our false inductions, and in many cases we are wholly unable to

sequent researches alone will remove; and because such have occurred in the infancy of the science, ought we to cancel all that fifty years of arduous researches has brought to light, and retrace our steps, as some would wish, to the period of Linnæus? *No one has, as yet, nor will any one for some years to come, be able to produce a complete and satisfactory classification of the numberless species and varieties composing the order;* because, to hope to arrive at such a condition, it is absolutely necessary to be thoroughly acquainted with the interior arrangement and other characters of *all the species* one has to classify, and I am sorry to say we are still *very far* from having attained so desirable a state; all we can do, therefore, in the mean time, is to assemble conscientious observations, and gradually improve the general views in circulation at the time being.

Soon after the publication of Cuvier's discoveries, several new and excellent genera were from time to time introduced by Sowerby, Lamarck, Defrance, Fischer de Waldheim, Dalman, and others, which, with their dates and synonyms, will be found recorded under the heads of the different genera.

In 1818, Lamarck regarded the Brachiopods as forming the greater part of his 3d section of the Conchifères Monomyaires:—

1. LES RUDISTES.   1, *Spherulite;* 2, *Radiolite;* 3, *Calceole;* 4, *Birostrite;* 5, *Discine;* 6, *Cranie.*

2. The BRACHIOPODA.   1, *Orbicule;* 2, *Terebratule;* 3, *Lingule.*[1]

In 1824, M. de Blainville placed at the head of his ACEPHALOPHORES an order, which he named *Palliobranchiata*, as a substitute of *Brachiopoda*.[2]   He divides this order as follows:—

1. SYMMETRICAL SHELLS, *Lingula, Terebratula, Thecidea, Strophomena, Plagiostoma, Dianchora,* and *Podopsis.*

2. UNSYMMETRICAL SHELLS, *Orbicula* and *Crania.*   The genus *Calceola* he places in

detect the relations between structure and function.   More definite principles of classification may hereafter be discovered, and meantime all that we can do is to arrange our systems according to sound reason, and without theoretical prepossession.   By care and judgment much may be done to give greater regularity and exactness to our methods of classification, either by introducing new groups where the importance of certain characters requires it, or by rejecting such as have been proposed by others on insufficient grounds.   At the present day, many authors are in the habit of founding what they term *new Genera* upon the most trifling characters, and thus drowning knowledge beneath a deluge of names. . . .   In the sub-dividing of larger groups into genera, even in the strictest conformity with the natural method, there is evidently no other rule but *convenience* to determine how far this process shall be carried. . . .   Nature affords us no other test of the just limits of a genus (or indeed of any other group) than the estimate of its value which a competent and judicious naturalist may form.   The boundaries of genera will therefore always be liable to fluctuate. . . .   The only remedy for this excessive multiplication of genera is for subsequent authors, who think such genera too trivial, not to adopt them, but to adopt the old genus in which they were formerly included."

[1] Hist. des Animaux sans Vertèbres.

[2] Dic. des Sciences Nat., t. xxxii, p. 298 ; and Manuel Malac., &c.

his second order, RUDISTES; but, as justly observed by the author of the 'Paléonto-logie Francaise,' De Blainville's classification varies but little from that by Lamarck; and one is at a loss to imagine what relation the celebrated malacologist could have found between *Plagiostoma*, *Dianchora*, and *Terebratula*. To the present day some authors are disposed to consider the *Rudistes* as a portion of the Brachiopoda.[1] While others, and by far the greater number, repudiate such a conclusion.[2]

Eight or nine years after the publication of M. de Blainville's observations, Baron Leopold von Buch issued his very remarkable work on the 'Classification of the Terebratulæ;'[3] which materially contributed to the subsequent advancement of the science. The Baron, founding his arrangements on the position and shape of the foramen, establishes two prin-cipal divisions. In the first he places those forms attached by a pedicle issuing between the valves (*Lingula*), those with one valve perforated (*Terebratula Delthyris*), and those which he considers to be imperforated (*Calceola* and *Leptæna*). In the second division he admits those species presenting a perforation through the middle of the lower valve (*Discina*), or fixed by the substance of the same (*Crania*).

About the same period, or shortly after, M. Deshayes proposed to divide the Brachiopoda into two groups. In the first he included the *articulated*, in the second the *unarticulated* species;[4] but while geologists and conchologists were thus busily engaged framing classifications on mere external and often deceptive appearances, our celebrated anatomist, Professor Owen, was at work on the animal of some of the recent species,[5] which tended to consolidate and improve the foundation of the science by bringing to light important zoological evidence.

From 1834 to 1844[6] no very important change took place in the nomenclature, but many new forms and interiors had been discovered which increased our knowledge of the genera then existing, and pointed out the necessity for the creation of several others which were introduced with more or less success by Prof. M'Coy, M. de Verneuil, Prof. King, M.

[1] Among these we may mention Goldfuss, D'Orbigny, Gray, &c.

[2] M. Deshayes strongly opposes the annexing of the Rudistes to the Brachiopoda.

[3] Since the Baron's researches much additional information has been obtained regarding the mode of existence in the different species, and it seems now generally admitted, that no satisfactory arrangement can be framed on the single evidence of the above-mentioned character. All Brachiopoda have lived either free or fixed to submarine bottoms, by the means of a muscular pedicle, or by the substance of their lower valve. Some species were attached during all their existence, others only when young, dispensing with their attachment at a more advanced stage of life; and in many examples of the same species the aperture is entirely cicatrized in the adult.

[4] Nouv. ed. des Anim. sans Vert. de Lamarck, vol. vii, 1836.

[5] Trans. of the Zool. Soc., vol. i, 2d part, p. 141, 1835; and Ann. des Sciences Nat. Zool., vol. iii, p. 315, 1845.

[6] In M. Deslongchamp's 'Tableau Synoptique d'un Arrangement Systématique des Brachiopodes Fossiles du Calvados,' 1837: the perforation and deltidium are the chief characters made use of in this classification. In 1841, Professor Phillips proposed the following arrangement of the Brachiopoda,

d'Orbiguy, Mr. Sharpe, M. Bouchard, and others. It was also in 1844 that Dr. Carpenter's first important investigations on the shell structure appeared;[1] and in 1846 Professor King proposed to arrange the PALLIOBRANCHIATA into ten families and twenty-two genera, viz.:[2]—

Fam. OBOLIDÆ Genus.   *Obolus.*
  „   LINGULIDÆ. G. *Lingula.*
  „   ORBICULIDÆ. G. *Orbicula.*
  „   CALCEOLIDÆ. G. *Calceola.*
  „   STROPHOMENIDÆ. G. *Strophomena, Orthis, Leptæna, Chonetes*
  „   PRODUCTIDÆ. G. *Productus, Strophalosia.*
  „   TEREBRATULIDÆ. G. *Terebratula, Hypothyris, Pentamerus, Camarophoria, Uncites.*
  „   SPIRIFERIDÆ. G. *Spirifer, Atrypa, Martinia, Strigocephalus.*
  „   THECIDEIDÆ. G. *Thecidea.*

in his 'Figures and Desc. of the Pal. Fossils,' p. 54, but the author did not completely apply his own views :

"BRACHIOPODA—

Valves free ; attachment by exserted muscle.
   Valves equal . . . . *Lingula.*
Valves unequal ; larger valve imperforate (Athyridæ).
   No cardinal area . . . . *Producta.*
   A cardinal area . . . . *Calceola.*
Large valve perforated in or under the beak.
   Perforation reaching to the hinge line (*Delthyridæ*)
   Cardinal area more or less common to both valves . . . *Orthis.*
Cardinal area confined to the larger valve.
   Internal plates of the larger valve separate . . . . *Spirifera.*
Internal plates of the larger valve united on the mesial line of the shell.
   Plates narrow . . . . *Strigocephalus.*[a]
   Plates very broad . . . . *Pentamerus.*
   Cardinal area obsolete, beak incurved over a minute perforation, which is
      often obtect, or merely serves to receive the beak of the smaller valve . *Cleiothyris.*[b]
   Perforation not reaching to the hinge line (*Cyclothyridæ*[c])
   Beak truncate, perforate . . . . *Epithyris.*
   Beak acute, the perforation below it . . . . . . *Hypothyris.*

[a] The difference between *Strigocephalus* and *Pentamerus* appears to me not very important!

[b] The term Atrypa (α, privative, and τρυπα, foramen,) is objectionable ; Cleiothyris (κλειω, claudo, θυρα, janua,) would be preferable, and with the terms *Epithyris* and *Hypothyris* might console us for the loss of *Terebratula*, which in Von Buch's view includes the three groups.

[c] In the sense of encircled."

---

[1] Report of the British Association for 1844 and 1847.

[2] Annals and Mag. of Nat. Hist., vol. xviii, p. 83, 1846, at which period the learned Professor likewise published his views on the muscular system in *Terebratula*.

And in the same year, Mr. Morris published his interesting paper *on the subdivision of the Terebratulæ.*[1]

In 1847, M. d'Orbigny presented his views on the classification of the Brachiopoda,[2] chiefly founded on the presence or supposed absence of labial appendages in some of the species, and therefore divides the class into two orders, I. BRACHIDÆ; II. CIRRHIDÆ; but from perhaps not having been able to study the animal of *Argiope* and *Thecidea,* or mistaking the brachial disk of the two last-named genera for the mantle, unites them most improperly to the Rudistes which constitute his *second order;* both *Argiope* and *Thecidea* are essentially related to the *Terebratulidæ,* or to his first division of True Brachiopoda. M. d'Orbigny's arrangement is as follows:—

I. ORDER—BRACHIOPODES BRACHIDES (*Brachidæ*).

1. Fam. LINGULIDÆ. Genus 1, *Lingula;* 2, *Obolus.*
2.   „   CALCEOLIDÆ. G. 1, *Calceola.*
3.   „   PRODUCTIDÆ. G. 1, *Productus;* 2, *Chonetes;* 3, *Leptæna.*
4.   „   ORTHISIDÆ. G. 1, *Strophomena;* 2, *Orthisina;* 3, *Orthis.*
5.   „   RHYNCHONELLIDÆ. G. 1, *Hemithyris;* 2, *Rhynchonella;* 3, *Strigocephalus;* 4, *Porambonites.*
6.   „   UNCITIDÆ. G. 1, *Uncites;* 2, *Atrypa;* 3, *Pentamerus.*
7.   „   SPIRIFERIDÆ. G. 1, *Cyrtia,* 2, *Spirifer;* 3, *Spiriferina;* 4, *Spirigerina;* 5, *Spirigera.*
8.   „   MAGASIDÆ. G. 1, *Magas;* 2, *Terebratulina.*
9.   „   TEREBRATULIDÆ. G. 1, *Terebatula;* 2, *Terebratella;* 3, *Terebrirostra;* 4, *Fissurirostra.*
10.   „   ORBICULIDÆ. G. 1, *Siphonotreta;* 2, *Orbicella;* 3, *Orbiculoidea;* 4, *Orbicula.*
11.   „   CRANIDÆ. 1, *Crania.*

II. ORDER—BRACHIOPODES (*Cirrhidæ*).

1. Fam. THECIDÆ. G. 1, *Megathyris;* 2, *Thecidea.*
2.   „   CAPRINIDÆ. G. 1, *Hippurites;* 2, *Caprina;* 3, *Caprinula;* 4, *Caprinella.*
3.   „   RADIOLIDÆ. G. 1, *Radiolites;* 2, *Biradiolites;* 3, *Caprotina.*

In this plan, besides the very objectionable and wide separation of *Argiope* (= *Megathyris*) and *Thecidium* from the *Terebratulidæ,* the position of *Stringocephalus* is certainly out of place in the Family *Rhynchonellidæ,* where the author might have admitted *Pentamerus,* and by placing *Uncites* and *Atrypa* (Dalman) in the Family *Spiriferidæ* would have dispensed with his 6th division. *Magas* and *Terebratulina* should likewise have been added to the *Terebratulidæ.*

---

[1] Quarterly Journal of the Geol. Soc., vol. ii, p. 382, 1846.

[2] Comptes rendus de l'Acad. des Sciences, 1847.—Ann. des Sc. Nat., 3d series, Zool., vol. viii, p. 141, pl. vii, 1848.—Paléont. Franç. Ter. Crétacés, vol. iv, p. 281, 1848.—Cours Elémentaire de Pal., vol. ii, p. 79, 1852; and Prodrome.

In the same year M. Barrande divided the Brachiopoda of Bohemia into *Terebratula, Stringocephalus, Pentamerus, Thecidea, Spirifer, Orthis, Leptæna, Chonetes, Productus, Crania, Orbicula,* and *Lingula.*[1]

In 1848, Dr. Gray proposed his new arrangement,[2] dividing the class into five orders, which may be analysed as follows:—

BRACHIOPODA. 1. Sub-Class—ANCYLOPODA.

I. Order—Ancylobrachia. 1, Fam. *Terebratulidæ.*
II. Order—Cryptobrachia. 1, Fam. *Thecideidæ* (Argiope *and* Thecidea).

2. Sub-Class—HELICTOPODA.

III. Order—Sclerobrachia. 1, Fam. *Spiriferidæ;* 2, Fam. *Rhynchonellidæ, Rhynchonella, Camarophoria, Uncites? Trigonosemus, Rhynchora, Pygope, Delthyridea, Pentamerus.*
IV. Order—Sarcicobrachia. 1, Fam. *Productidæ, Productus, Strophalosia, Chonetes, Leptæna, Orthis, Strophomena,* and *Calceola.*
  2, Fam. *Craniadæ.*
  3, „ *Discinidæ.*
  4, „ *Lingulidæ.*
V. Order—Rudistes.

Dr. Gray's scheme appears objectionable on account of the perhaps uncalled for subdivision of the class into five orders; thus, for example, the second order, *Cryptobrachia,* seems completely unnecessary. The position of several of the families and genera is also incorrect; as instances of this we find in Order III, among the *Rhynchonellidæ, Uncites,* whose place is in the *Spiriferidæ; Trigonosemus, Pygope, Rhynchora,* and *Delthyridæa,* should have been classed in the first order with the Family *Terebratulidæ,*[3] where *Argiope* ought likewise to have been admitted. *Calceola,* quite an anomalous shell among the Brachiopoda, is placed in the same family as *Productus,* although possessing none of the essential characters of the last-named genus.

In the 'Index Palæontologicus,' 1849, M. Bronn divides the class into 29 genera, several of which are, however, only synonyms.

We now arrive at Professor King's second scheme, where the learned Palæontologist adopts three of Dr. Gray's orders, and divides the class into 16 families and 49 genera, as follows:—

[1] Über die Brachiopoden aus den Naturwissenschaftlichen Abhandlungen, 1847.
[2] Ann. and Mag. of Nat. Hist., vol. ii, p. 435, 1848.
[3] These corrections have been introduced in the Catalogue of the Brachiopoda in the Brit. Mus., 1853.

Order—SARCICOBRACHIA.

1. Fam. LINGULIDÆ. Genus *Lingula.*
2.   ,,   OBOLIDÆ. G. *Obolus.*
3.   ,,   CRANIADÆ. G. *Crania, Siphonotreta, Criopus.*
4.   ,,   DISCINIDÆ. G. *Discina, Orbiculoidea, Trematis.*
5.   ,,   CALCEOLIDÆ. G. *Calceola.*
6.   ,,   DAVIDSONIDÆ. G. *Davidsonia.*
7.   ,,   PRODUCTIDÆ. G. *Productus, Strophalosia.*
8.   ,,   STROPHOMENIDÆ. G. *Strophomena, Leptæna, Chonetes, Orthis, Streptorhynchus, Orthisina, Dicælosia, Platystrophia.*

Order—SCLEROBRACHIA.

9. Fam. HYPOTHYRIDÆ. G. *Isorhynchus, Hypothyris, Camarophoria, Uncites, Pentamerus.*
10.   ,,   SPIRIFERIDÆ. G. *Atrypa, Athyris, Cleiothyris, Retzia, Delthyris, Trigonotreta, Spirifer, Martinia.*

Order—ANCYLOBRACHIA.

11. Fam. STRIGOCEPHALIDÆ. G. *Stringocephalus.*
12.   ,,   TEREBRATULIDÆ. G. *Epithyris, Terebratella, Terebratula, Pygope, Eudesia, Megerlia, Waldheimia.*
13.   ,,   RHYNCHORIDÆ. G. *Ismenia, Delthyridæa, Rhynchora.*
14.   ,,   MAGASIDÆ. G. *Magas, Bouchardia.*
15.   ,,   THECIDÆIDÆ. G. *Thecidæa.*
16.   ,,   ARGIOPIDÆ. G. *Argiope.*

The chief objection to Professor King's arrangement lies in the sometimes great sub-division of the genera, and to which we will revert hereafter when treating the respective sections.[1]

Since the last-named publication two other schemes have been introduced by Dr. Van der Hoeven[2] and Professor Quenstedt.[3] The first of those authors admits of eight genera :—

---

[1] I take this occasion to express my warmest thanks to Professor King for the friendly and unreserved manner in which he has communicated to me his views, and for the liberal loan of all the specimens belonging to his collection, now forming part of the Museum of the Queen's College, Galway; and I hope that any difference of opinion I may have expressed, will be considered as having only been dictated by a desire similar to his own, viz., the advancement of science. I am happy to repeat the same, relative to Professor M'Coy.

[2] Handbuch der Zoologie, 1850.   [3] Handbuch der Petrefactunde, 1851.

Lingula, Orbicula, Crania, Calceola, Thecidea, Terebratula, Spirifer, and *Productus.* The second six, viz. : *Terebratula, Spirifer, Orthis, Productus, Lingula,* and *Thecidea.*

In 1852, Professor M'Coy divides the Palliobranchiata into two orders :[1]

1st Order. *Rudistes* (Lam.) contains three families :—1, *Thecideæ;* 2, *Caprinidæ;* 3, *Radiolidæ.*

2d Order into eleven families: —1, *Craniadæ;* 2, *Orbiculidæ;* 3, *Terebratulidæ;* 4, *Magasidæ;* 5, *Spiriferidæ;* 6, *Uncitidæ;* 7, *Rhynchonellidæ;* 8, *Orthisidæ;* 9, *Productidæ;* 10, *Calceolidæ;* 11, *Lingulidæ.*

Professor M'Coy has fallen into the same mistake as M. d'Orbigny, viz. by placing *Thecidium* among the Rudistes. In the 2d Order his 4th and 6th families may be dispensed with, the 4th belonging to the *Terebratulidæ,* the 6th to the *Spiriferidæ.*

From the above brief summary it is evident that the tendencies of some have been to multiply, others to reduce, the number of genera; this fluctuation of opinion proceeds from the different relative value or importance attributed to certain characters which are by some considered *generic,* by others only *specific.* Thus, for example, Prof. Quenstedt divides his genus *Terebratula* into 1, *Bicornes;* 2, *Calcispiræ;* 3, *Annuliferæ;* 4, *Loricatæ;* 5, *Cinctæ;* 6, *Impressæ;* 7, *Nucleatæ;* 8, *Triplicatæ;* 9, *Spiriferinæ.* Now, in my opinion, as well as in that of many other Palæontologists, his TEREBRATULÆ CALCISPIRÆ (*T. prisca* and *prunum*) and his TEREBRATULÆ SPIRIFERINÆ (*T. concentrica, tumida, ferita,* &c.) from possessing internal spires for the support of the cirrated arms, similar in essential or family character to those of *Spirifer,* would be much more naturally and properly classed under the last-named genus, and the objection to placing in the same section or family such shells or animals as *Ter. vitrea, R. psittacea, A. prisca* and *S. concentrica,* is too obvious to demand any lengthened explanations; for both the dispositions of the oral processes, and other details of the animal economy, are essentially dissimilar, and therefore could not possibly be confounded under a single denomination; otherwise the drawing up of a uniform and clear diagnosis would be impossible, and thus lead to very serious practical inconveniences. Those who have devoted serious attention to the external and internal characters of these shells, have come to the conclusion that the class is now sufficiently investigated to be susceptible of subdivision into natural groups or families, and these again into genera. Thus, for example, we are acquainted with upwards of a hundred species of *Terebratulæ,* presenting exactly the same internal dispositions, namely, *a short* simply-attached loop supporting similarly disposed cirrated arms; to such a group the same generic appellation is natural and proper; but not if likewise made to include another equally numerous series of species, in which, besides differences in the external and shell structure, we find free extensile spiral arms (*Rhynchonella*), supported only at their origin by two slender curved lamellæ. Still, according

---

[1] British Palæozoic Fossils, part 2, p. 186, 1852.

to Prof. Quenstedt and some others, this last group would also take place among the *Terebratulæ*, an assumption surely in opposition to zoological law.

The great difficulty seems how to establish a definite and APPLICABLE arrangement; and this question will probably continue unsolved, until our knowledge of the internal structure of the many at present ambiguous forms has been cleared up, which fortunate circumstances, time, and persevering researches can alone accomplish.[1]

The many years I have devoted to the study of the species composing this class, has tempted me to propose certain amendments and changes in the schemes already published by Prof. King, M. d'Orbigny, and others; and I do so in the hope that any steps, *however small*, towards the future settlement of a so-much controverted question, may be considered worthy of the attention of those who feel interested in the subject.

---

[1] A celebrated French conchologist stated some years ago, that Palæontologists could never hope to arrive at a knowledge of the interior of all the species; but since that period the numberless discoveries and well-conducted researches make us believe, on the contrary, that ere long we may become acquainted with the most delicate internal details of all those species which have hitherto resisted our endeavours. And no better example can be selected of what patience and diligent exertions can achieve, than the splendid work recently published on the 'Silurian Fossils of Bohemia,' by that eminent and justly celebrated geologist and palæontologist, M. Barrande; the author mentions, among other examples, that to obtain a perfect example of one of the most common Trilobides (*Dalmanites socialis*), it required a number of years of the most assiduous researches. Among the Brachiopoda, we may likewise quote *Stringocephalus* and *Uncites*, whose internal arrangement baffled all researches for a long period.

SYNOPTICAL TABLE, or CLASSIFICATION OF THE BRACHIOPODA INTO FAMILIES AND GENERA.[1]

Families and Sub-Families.	Genera and Sub-Genera.	Author.	Date.	A few Types.	Silurian.	Devonian.	Carboniferous.	Permian.	Triasic.	Jurassic.	Cretaceous.	Tertiary.	Present.
TEREBRATULIDÆ	TEREBRATULA .	Llhwyd	1699	vitrea, maxillata, pervoalis, diphya	?	*	*	*	*	*	*	*	*
	Sec. A. Terebratulina .	D'Orbigny	1847	capat-serpentis, sub-striata, gracilis	:	:	:	:	.	*	*	*	*
	" B. Waldheimia	King	1849	australis, lagenalis, digona, numismalis	:	:	:	:	*	*	*	*	*
	TEREBRATELLA .	D'Orbigny	1847	chilensis, dorsala, Coreanica	:	:	:	:	*	?	*	:	*
	Sec. A. ? Trigonosemus	Kœnig	1825	elegans .	:	:	:	:	.	.	*	.	.
	" B. ? Terebrirostra	D'Orbigny	1847	lyra .	:	:	:	:	.	.	*	.	.
	" C. Megerlia .	King	1849	truncata, lima	:	:	:	:	.	.	*	.	*
	KRAUSSIA .	Davidson	1852	rubra, cognata, pisum, Lamarckiana	:	:	:	:	.	.	*	.	*
	MAGAS .	Sowerby	1818	pumilus, orthiformis	:	:	:	:	.	.	*	.	.
	BOUCHARDIA .	Davidson	1849	tulipa or (rosea)	:	:	:	:	.	.	.	.	.
	MORRISIA .	Davidson	1852	anomoides = (appressa)	:	:	:	:	.	.	.	:	*
	ARGIOPE .	Deslong-	1842	decollata, decemcostata, cuneata	:	:	:	:	.	.	*	:	*
? STRINGOCEPHALIDÆ	STRINGOCEPHALUS	Defrance	1827	Burtini, giganteus	.	*	.	.	.	.	.	.	.
THECIDEIDÆ	THECIDIUM	Defrance	1828	radiatum, digitatum, dorsatum, Klipst. sp.	.	:	:	:	*	*	*	:	.
	SPIRIFER .	Sowerby	1815	striatus, alatus, Archiaci, glaber	*	*	*	:	*	.	.	.	.
	Sec. A. ? Spiriferina	D'Orbigny	1847	rostrata, cristata, Tessoni, Münsteri	:	*	*	*	*	*	.	.	.
	" B. ? Cyrtia	Dalman	1827	trapezoidalis .	*	*	.	.	.	.	.	.	.
SPIRIFERIDÆ	ATHYRIS .	M'Coy	1844	tumida, Herculea, scalprum	*	*	*	:	.	.	.	.	.
	SPIRIGERA .	D'Orbigny	1847	concentrica, pectinifera, Roissyi	*	*	*	*	.	.	.	.	.
	Sec. A. ? Retzia .	King	1849	Adrieni ? ferita	*	?	*	.	.	.	.	.	.
	UNCITES .	Defrance	1848	gryphus	.	*	.	.	.	.	.	.	.
	ATRYPA .	Dalman	1827	reticularis, marginalis, prunum	*	*	*	.	.	.	.	.	.
? KONINCKINIDÆ	KONINCKINA .	Suess	1853	Leonhardi	.	.	.	.	*	.	.	.	.
RHYNCHONELLIDÆ	RHYNCHONELLA .	Fischer	1809	loxia, acuminata, psittacea, octoplicata	*	*	*	*	*	*	*	*	*
	CAMAROPHORIA .	King	1844	Schlotheimi, multiplicata, globulina	.	.	*	*	.	.	.	.	.
	PENTAMERUS .	Sowerby	1813	Knightii, galeatus, conchidium	*	*	.	.	.	.	.	.	.
? PORAMBONITIDÆ	PORAMBONITES .	Pander	1830	æquirostris, reticulatus, Ribeiro	*	.	.	.	.	.	.	.	.
	ORTHIS .	Dalman	1827	calligramma, ? rustica, elegantula	*	*	*	.	.	.	.	.	.
STROPHOMENIDÆ	ORTHISINA .	D'Orbigny	1849	adscendens, anomala, hemipronites	*	*	.	.	.	.	.	.	.
	STROPHOMENA .	Rafinesque	1825 ?	planumbona, alternata, analoga	*	*	*	.	.	.	.	.	.
? DAVIDSONIDÆ	LEPTÆNA .	Dalman	1827	transversalis, Davidsoni, oblonga	*	*	*	.	.	.	.	.	.
	DAVIDSONIA .	Bouchard	1849	Verneuili, Bouchardiana	.	*	.	.	.	.	.	.	.
	CHONETES .	Fischer	1837	sarcinulata, lata, concentrica	*	*	*	*	.	.	.	.	.
PRODUCTIDÆ	STROPHALOSIA .	King	1844	excavata, Morrisiana, Gerardi	.	*	*	*	.	.	.	.	.
	Sec. A. ? Aulostages	Helmersen	1847	Wangenheimi = (variabilis)	.	.	*	*	.	.	.	.	.
	PRODUCTUS .	Sowerby	1814	semireticulatus, Martini, horridus	:	*	*	*	.	.	.	.	.
CALCEOLIDÆ.	CALCEOLA .	Lamarck	1809	sandalina	.	*	.	.	.	.	.	.	.
CRANIADÆ .	CRANIA .	Retzius	1781	Brattenburgensis, anomala, costata	*	*	*	*	*	*	*	*	*
	DISCINA .	Lamarck	1819	lamellosa, striata, Caminegii	*	*	*	*	.	.	*	*	*
DISCINIDÆ.	Sec. A. ? Orbiculoidea	D'Orbigny	1847	elliptica, Forbesii	*	*	*	.	.	.	.	.	.
	" B. ? Trematis	Sharpe	1847	terminalis, cancellata	*	.	.	.	.	.	.	.	.
	SIPHONOTRETA	Verneuil	1845	unguiculata, verrucosa, anglica	*	.	.	.	.	.	.	.	.
LINGULIDÆ	Sec. A. ? Acrostreta	Kutorga	1848	subconica, disparirugata, recurva	*	*	.	.	.	.	.	.	.
	LINGULA .	Bruguiero	1789	anatina, Dumortieri, Beanii	*	*	*	*	*	*	*	*	*
	OBOLUS .	Eichwald	1829	Apollinis, sculptus, transversus	*	*	*	*	.	.	.	.	*

All naturalists are aware that linear arrangements do not occur in nature, therefore the present as well as all similar attempts must be viewed as artificial, and more or less conventional; our endeavours have, however, tended to group together those animals which in the present state of our knowledge seem to possess essential similarity of character, and have, therefore, provisionally divided the class into ten families and four sub-families; but it must be observed that when Palæontologists are better acquainted with the internal characters of the genera and species, the limits, as well as the number of the groups, will most probably require modification, and the more so since those forms located for the present in *sub-families* have not been sufficiently worked out to warrant their admission in the groups already defined.   No small amount of investigation has taken place in the endeavour to find out *what principle* or *what characters* should be considered of sufficient importance to authorise the separation of families and *genera;* the difficulty will be easily understood when it is remembered that out of forty *sections* or *sub-sections* here introduced, fourteen alone possess living representatives, and the animal of several of these last has not hitherto been completely anatomically examined.   The difficulties, therefore, naturally increase in those cases, unfortunately by far the most numerous, where all that can be gathered of the inhabitant of the shell consists in a few impressions existing in the interior of the valves.   The study of the animal of *Terebratula, Crania, Discina, Lingula,* and a few other recent types, has to a certain extent opened the way to a knowledge of portions of the animal of many extinct genera, and it is by analogy and careful comparisons that modern Palæontologists have perceived the absolute necessity of separating certain things which early naturalists had confounded together, from not being able to interpret or understand the use or importance of the calcified processes or impressions visible in the interior of the shell.

All Brachiopoda are possessed of two valves *articulated* by the means of teeth and sockets, or *unarticulated,* and kept together entirely by muscular action.   Out of the forty-five sections or sub-sections enumerated, thirty-three are distinctly articulated, but in one of the most natural groups, viz. that of the *Productidæ,* we find genera presenting both conditions.   The animal of the *Brachiopoda* lived entirely free, or was fixed to marine bottoms during a part or the whole of its existence either by the substance of a portion of the ventral valve, or by the means of a muscular peduncle passing between the valves, or through an aperture existing in the ventral or dental valve, it is probable that the majority of the species were fixed, at least in an early stage of their existence, however much they may have dispensed with that temporary attachment at a more advanced period.   In the same family and even genus, however, we discover forms which had been attached, while others present no traces of such a condition.   It would, therefore, be impossible to separate the *attached* and *unattached* species, genera, and families, from this character alone.   The shape and position of the *foramen,* or its absence, when combined with other peculiarities, may help to define and circumscribe certain groups.   The value to be attached to the *shell structure* has already been ably discussed by Dr. Carpenter, that we need only remind our readers of the fact that certain families are entirely composed of *punctated*, others of *unpunctated* species, while a few are made up of genera, which seem to assume both conditions. *Spinose* and *foliaceous* expansions cannot be considered of generic importance, since they occur in some species of most genera.

Coloration cannot be made use of as a generic character, and its value to the Palæontologist is small; but when occurring on fossil forms it should always be noted. Prof. Forbes has kindly informed me, " that his observations on the distribution in depth of recent species have led him to the conclusion that definite patterns, *i. e.*, stripes, bands, and waves of colour vividly marked, do not occur, except in rare instances, on shells living beyond moderate depths, as below fifty fathoms or thereabouts, and that thus we may be enabled to come to approximate conclusions respecting depths of ancient seas from the patterns preserved to us on fossil shells." The coloration is of some use in distinguishing the recent forms of Brachiopoda, green, yellow, red, and blueish-black being the prevailing colours; several forms are striped or spotted with red. Among the fossil species some examples have preserved traces of their colours, as already mentioned in Part III, p. 6, and several other examples will be hereafter noticed, so that in all probability the species now extinct, when alive, presented all the rich varieties of tint observable in the present inhabitants of our seas.

MUSCULAR IMPRESSIONS.—We will now briefly notice those impressions left by the muscles in the interior of the shell in the different genera. The character of several of these in *Lingula*, *Discina*, *Crania*, *Terebratula*, and *Rhynchonella*, had been pointed out in 1802 and 1833 by Cuvier and Prof. Owen, but since that period considerable progress has been made, and we are already sufficiently acquainted with the bearing of the majority of these in the forms now alive to venture to interpret their homologues in the fossil, *extinct genera;* but further researches on the function of certain of these, especially in *Lingula*, *Discina*, and *Crania* will be necessary before the subject can be considered completely and finally settled.

Professor Owen has made known his most recent views on the muscular system of *Terebratula* in the first chapter of this Introduction.[1]

Three kinds of muscular impressions particularly interest the Palæontologist.

I. The ADDUCTOR (*valvular muscles* of King, *Adductor longus*, *anticus*, and *posticus*, of Professor Owen) leaves in the *dorsal* or *socket valve* of all the articulated genera four

---

[1] The account which I have given of the muscular system of the recent Brachiopoda is also derived from a personal examination of the animals of all the recent genera, aided by the important Memoir of Professor Owen, in the 'Zool. Trans.' for 1833, and that of Professor King, in his work on the 'English Permian Fossils,' published by the Palæontographical Society in 1849. These examinations, and their application to the muscular impressions in the *extinct* genera, were made in conjunction with Mr. Woodward, (as stated in the 'An. Nat. Hist.' for May, 1852).

Since then, Professor Owen kindly offered to publish, in my Introduction, the valuable materials accumulated in his researches during the last twenty years; but after comparing the learned professor's printed proof with my own, I find that he has given to some of the muscles names differing from those

impressions at times divided by a *dorsal mesial* plate; and a glance at the figures introduced in Plates VI, VII, VIII, and IX, will explain their position and modifications in the different genera better than could be done by a lengthened description.

The muscular fibres that produce this quadruple impression, after converging to near the centre of the *ventral* or *dental valve,* are there fixed, and produce a *single impression* (at times divided); but in some species of *Spirifer,* such as *S. rostratus, Munsteri,* &c., where there exists a largely developed ventral septum, the base of which occupies the whole space generally devoted to that muscle, its attachment is then shifted to the upper sides of the plate, as represented in Pl. VI, fig. 60.

which I employed, and which, as our plates are printed, it is now impossible to harmonize except by the following comparison here added to assist the reader:

PROF. OWEN.		DAVIDSON.
Adductor longus anticus	} =	*Adductor muscle.*
Adductor longus posticus		
Adductor brevis	=	*Cardinal muscle.*
Cardinalis	=	*Accessory cardinal.*
Retractor superior	} =	*Pedicle muscles* { *Dorsal.*
Retractor inferior (of the Peduncle)		*Ventral.*

From the above, it will be seen that I used Professor King's term *cardinal muscles* for Professor Owen's *adductor brevis* and *cardinalis.* The term *pedicle muscles* (which I divide into *dorsal* and *ventral*) appeared to be a name which it was very desirable to retain, and I find that the distinguished Hunterian Professor admits, in p. 8, that his *retractor superior* and *inferior* are essentially *peduncle muscles.*

In the 'Ann. and Mag. of Nat. Hist.' for July, 1846, Professor King enters into the following details relative to the muscular system of *Terebratula,* "in order that the use of certain parts to be mentioned hereafter may be properly understood. From a specimen of *T. dorsata,* at present before me, containing the entire muscular system desiccated, and freed of the visceral mass, I have drawn up the following details:—The rostral or umbonal cavity is occupied with a dense fibrous cylindrical body called the pedicle; considering the convexity of the foraminal valve, as the upper side of the shell, the inferior end of the pedicle fits into the foramen; while its superior end, which is somewhat flattened or dilated in the transverse direction of the shell, is situated at the entrance or anterior part of the rostral cavity, to the surface of which it appears to be attached by means of tendinous or membranous cords: the truncated extremity of the pedicle itself not being adherent. A little in advance of the upper extremity of the peduncle, three pairs of muscles pass off to different parts, the outermost pair (which consist of those muscles implanted nearest the lateral margins of the valve), passes at a slight angle into the upper part of the pedicle: within these muscles, and somewhat in front of them, another pair passes downwards (slightly converging at the same time), and becomes attached to a flattened prominency, situated in the centre of the hinge of the lower or imperforate valve. To distinguish these pairs of muscles from each other, it will be necessary to name the former the *superior pedicle muscles,* and the latter the *cardinal muscles.* In close proximity to the superior end of the pedicle, and a little behind, and within the cardinal muscles, and therefore near the medio-longitudinal line of the shell, is situated the origin of the remaining pair, which passes directly down to a little behind the centre of the opposite valve, each muscle at the same time becoming dichotomous in its inferior half; these may be termed the *valvular muscles.* Besides supporting the cardinals and the valvulars, the imperforate valve affords attachment to other two muscles which pass upwards from the *crural base* (where each one is divided), and become inserted in the upper part of the pedicle: it is proposed to name these, the *inferior pedicle muscles.* With one exception, the foregoing description agrees with that

II. The position of the *Cardinal muscle* of King (*Adductor brevis* of Prof. Owen) is always recognisable in the interior of the articulated genera. In the *ventral* or *dental valve* it leaves two large pyriform scars always clearly defined, and extending somewhat in advance of the confluent insertion of the *Adductor*. In passing to the opposite valve the two divisions of this muscle converge and are fixed in the *dorsal* or *socket valve* to a *Cardinal process* (*boss* of King), which varies to a very remarkable extent in certain *Terebratulidæ, Stringocephalus, Camarophoria, Orthis, Strophomena productus* &c., while in some forms, from the non-development of this process, the muscular attachment to that valve is less clearly defined. In the un-articulated genera, *Discina* and *Crania*, the equivalent of the cardinal muscle has not been ascertained with certainty; it is perhaps the *sliding muscle* of Professor Owen.

given by Professor Owen, in his 'Memoir on the Anatomy of the Brachiopoda' (1833), in which it is stated that the muscles which have been termed valvulars pass into the upper part of the pedicle—a statement which I am led to suspect may have arisen simply from the superior termination of these muscles, in the specimens examined by this distinguished anatomist, being so close to the upper part of the pedicle as to appear as if attached to it." In 1849 Professor King diagramatically represented the *muscular* system of *Ter. australis*, and the position of our *adductor* and *cardinal* muscles (*adductor longus* and *brevis* of Professor Owen) is clearly exposed ('English Permian Fossils,' pl. xx, figs. 10—12). Professor Owen likewise states, that in the year 1846 he left for publication at Naples most of the details on the muscular system he has introduced in the first chapter of this Introduction. In a very useful and interesting manual on the 'Mollusca,' published by Mr. S. P. Woodward (1851), a figure of *Ter. australis* is given (p. 8) in which the *retractors* (cardinal muscles) are there termed for the first time (to my knowledge) "muscles by which the valves are opened." However, one year after (1852), Professor M'Coy states in his 'British Pal. Fossils in the Cambridge Mus.,' p. vii of the advertisement, and p. 191 of the letterpress, that he had discovered "that the valves of Terebratula were opened by the action of a pair of muscles, which arise from the middle of the perforated or receiving valve, and the tendons of which are inserted into the internally prolonged entering beak of the entering valve, which thus forms a lever moving on the hinge-teeth as a fulcrum." I have not, however, been able to find where the learned author had published these discoveries, which to claim priority must have been made public before 1846.

In the short zoological account of the Brachiopoda, published by the learned author of the 'Paléont. Française,' in the 'Annals des Sciences Nat.,' vol. viii, 1847, some of the muscles seem to have been incorrectly described.

To conclude this account of the views hitherto published on the muscular system of the genus *Terebratula*, we may mention, that in Dr. Gray's 'Catalogue of the Terebratulæ in the British Mus.' (1853), two diagrams by Mr. Woodward illustrate the muscular impressions in the *Ter. australis,* and which we have reproduced with Dr. Gray's permission, in p. 64, figs. 6, 7. The woodcut here appended (fig. 1), was likewise drawn for Dr. Gray by Mr. Albany Hancock, to which were attached the following explanations : *a, adductors; r, retractors; x, accessory retractors; p,p, pedicle muscles; z, function uncertain; o, mouth; v, vent; l, loop; t, dental socket.*

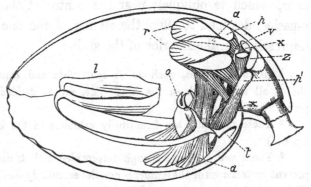

Fig. 1. *Waldheimia flavescens.*

Our *Accessory cardinal muscles* (*Cardinalis* or *hinge muscles* of Prof. Owen) produce very slight impressions on the internal surface of the shell; their position is represented at (*x*) in figs. 7, p. 64, and 1, p. 55, but seems to have been omitted in Prof. King's diagram, illustrating the muscular system in *Ter. australis*. ('English Permian Fossils,' Pl. XX.)

III. *The pedicle* muscles (*retractor superior* and *inferior* of Prof. Owen) have often but not always left impressions in the interior of the valves; they are wanting in those forms which appear to have lived free, or to have been but slightly attached in the early part of their existence. It is likewise worthy of remark that commonly the peduncle muscular impressions are more clearly defined in those species where the peduncle and the foramen for its passage is comparatively large. The *pedicle* muscles may be designated as follows :—

1. *Ventral pedicle muscles* (*retractor inferior* of Prof. Owen) form two scars exterior of the *adductor*, and can be traced in the interior of the *ventral valve* of all the *Terebratulidæ* in *Athyris, Spirigera, Atrypa, Rhynchonella,* &c.

2. *Dorsal pedicle muscles* (*retractor superior* of Prof. Owen) produce two or four recognisable impressions in the *dorsal valve*, especially in those species provided with a *hinge plate*, such as *Waldheimia, Terebratella,*[1] *Magas, Spirigera,*[2] &c. They are likewise observable in a similar position in *Rhynchonella, Atrypa, Strophomena,* and some others, where the *hinge plate* is less developed, and in which it is divided into two portions by a furrow.

In the unarticulated forms (*Productus* and *Aulosteges* excepted[3]) the muscular impressions appear more complicated, and by them *Lingula, Crania, Discina,* and others may be distinguished; but from the want of sufficiently well-preserved animals, the function and character of all these have not been as yet completely or clearly made out, nor can we add anything to what was published in 1802 and 1833 by Cuvier and Prof. Owen.

In *Discina* for example, eight muscular impressions can be distinctly traced in each valve; the posterior pair in the *perforated* or *ventral valve* lie on either side of the inner disk near the foramen, while the anterior two are situated further down towards the centre of the shell. In the upper or *dorsal valve* the posterior scars are placed nearly horizontally, at a short distance from the margin, and seem much smaller than the anterior ones, which lie obliquely near the centre of the valve; two other thin pair of diverging muscles, destined to affect the sliding of the one valve over the other, have likewise left impressions in the interior of the shell.

[1] In some species, such as *T. spathulata* and *Davidsoniana* (De Kon.), the hinge plate occupies almost all the breadth of the hinge line, and is divided into four large concave spaces for the dorsal peduncle muscles.

[2] These four pits are beautifully exhibited in the hinge plate of *Spirigera concentrica*. (See Pl. VI, fig. 66.)

[3] These two genera, although unprovided with teeth and sockets, possess a straight hinge line, and present impressions of the *adductor* muscles essentially similar to those seen in all the other articulated genera; the *cardinal process* never performs the function of a *hinge-tooth*, as was erroneously supposed by several authors.

Very remarkable differences in the *vascular impressions* left by the mantle on the inner surface of the shell have been observed, and this subject deserves much more minute investigation, in order to be used as a substantial character in the distinction of families.

It now remains for us to examine what results may be obtained from the characters afforded by the labial appendages or cirrated arms, the most important portions of the animal in the Brachiopoda; and I have little doubt that in them will be found the best character for distinguishing the great families as well as most of the genera;[1] but in order to better appreciate some of these dissimilarities, we will successively and briefly pass in review the principal modifications they assume in the different families, reserving all details connected with the genera for that portion of the introduction which treats of them in particular.

1. In the TEREBRATULIDÆ the cirrated arms[2] are variously twisted or folded upon themselves, and connected, to a greater or lesser extent, by a membrane, becoming free only at their spiral ends.[3] These labial appendages are always more or less supported by a calcified lamellar process, which assumes various shapes, but more commonly that of a loop, which is either simply fixed to the hinge plate, or likewise so to a medial septum in the dorsal valve. From differences observable in the shape and character of this kind of internal skeleton, the family may be readily and conveniently sub-divided into the following sections or sub-sections, *Terebratula, Terebratulina, Waldheimia, Terebratella, Megerlia, Kraussia, Magas, Bouchardia, Morrisia,* and *Argiope.*

2. STRINGOCEPHALUS has been provisionally placed in a sub-family, of which it is the only member, on account of its peculiar organisation; the exact character of its oral appendages has not hitherto been completely elucidated, but from the discoveries made by

[1] M. d'Orbigny justly attaches great importance to characters derived from the arms in the classification of the ORDER.

[2] Commonly described as *"ciliated"* arms; but the term *cilia* ought to be restricted to the microscopic vibratile organs, described by Dr. Grant and Prof. Sharpey.

[3] Baron von Buch mentions that Otto Frederic Müller having dredged from the lake of Dræbach, in Norway, a number of *Terebratulæ* (*probably belonging to Rhynchonella psittacea*), and placed them in a glass of water, he observed that they gracefully extended their spirally coiled arms. Professor Owen and M. Phillipi appear not to have watched the actions of any Brachiopoda in the living state; but from the examination of preserved specimens, they have arrived at the conclusion, that they were able to unfold the spiral portion of their arms. Professor Forbes, who has frequently dredged living *Terebratulæ* on the coasts of Scotland, and also in the Mediterranean, when describing the animal of *Terebratulina caput-serpentis* ('History of British Mollusca'), states that the arms themselves *cannot be protruded.* The exact resemblance of these arms to the calcareous spiral appendages of *Atrypa* and *Spirifer,* led me to doubt the accuracy of the observations made nearly a century ago, by the excellent naturalist quoted by Baron v. Buch, and to suppose that the canal and muscular fibres of the oral arms might be intended to inject the *cirri,* and not to protrude the arms themselves; but I may have been mistaken in my suppositions. Through the kindness of Mr. Cuming, Mr. Woodward and myself were enabled to examine these beautiful appendages in the *R. nigricans,* dredged alive by Mr. F. J. Evans, R.N., in Foveaux Straits, about five miles N.E. of Ruapuke Island, New Zealand (19 fathoms); and after having forcibly extended the arms, we were able to count the coils, which numbered eight or nine.

Prof. King and M. Suess it seems evident that the arms were largely supported by a remarkable process somewhat in the shape of a loop, but from the inner edge of which numerous lamellæ branched off. The place this genus should occupy in the classification must therefore remain unsettled until a complete knowledge of the calcified process has been attained.

3. In the THECIDEIDÆ the oral arms have not been as yet perhaps completely understood, and it will be very desirable to examine fresh examples of the animal of the only species we now find alive. In the greater number of fossil forms there exists a more or less complicated disk variably grooved for the reception of a peculiar loop or apophysary ridge which seems to become free near the visceral cavity; but other important peculiarities in the general habit of the genus *Thecidium* has necessitated its separation from the *Terebratulidæ*, to which it appears allied in many respects. Important discoveries recently made by M. Suess, of Vienna, may cast additional light on this group, which is so particularly worthy of the attention of the Palæontologist.

4. THE SPIRIFERIDÆ may be easily and clearly characterised by the largely-developed spiral lamellæ which form two horizontal or vertical cones filling the greater portion of the shell, and no doubt supported spiral cirrated arms: but we may likewise infer from the extent of these processes that the oral appendages were not extensile, as may possibly have been the case in *Rhynchonella*. Two principal modifications in the position and direction of the spires have been ascertained, that common to *Spirifer*, and the other to *Atrypa* of Dalman (*reticularis, marginalis, prunum*, &c.,) various other details of internal arrangement and shell-structure have rendered desirable further subdivisions of the group, but in all the character of the *spire* prevails. Some of the species, as we have already noticed, bear a close *outward* resemblance to certain *Terebratulidæ*, but active researches have hitherto brought to light but rare instances of true *Terebratulæ* with loops in the Palæozoic epoch, and on the contrary the generality of these ancient *terebratuliform shells* were provided with calcified spires.

5. The only other genus hitherto discovered which possessed spiral processes for the support of the labial appendages, not here included among the *Spiriferidæ*, is that proposed by M. Suess under the name of KONINCKINA (Pl. VIII, figs. 194—198), for a shell confounded with *Productus*, but which the Viennese author found to possess spiral lamellæ somewhat similar to that of *Atrypa*, but we have not ventured to include this genus for the present in the *Spiriferidæ*, although its place cannot be far removed.

6. In the great family RHYNCHONELLIDÆ, the labial appendages seem to have been free, spiral, fleshy, perhaps extensile, and supported only at their origin by two short calcified curved lamellæ. The family includes the genera, viz. *Rhynchonella, Camarophoria*, and *Pentamerus*.

7. But little is known of the labial appendages in the families STROPHOMENIDÆ and PRODUCTIDÆ; and in the sub-families *Poramobnitidæ* and *Davidsonidæ*, the arms were in all probability fleshy and spirally rolled, evidence of which may be observed in the interior

of several species, but no developed calcified process existed for their support, and the arms were perhaps supported in a somewhat similar manner to what we observe in the *Craniadæ* and *Discinidæ*.

8. Nothing is known regarding the arms in *Calceola*, a genus completely anomalous in the class.

9. The labial appendages in the *Craniadæ*, *Discinidæ*, and *Lingulidæ*, are fleshy and spirally coiled. From this rapid examination it seems evident that the best characters for the subdivision of the class may be derived from the shape of the arms and their calcified supports; and by taking into account other peculiarities which accompany the modifications above described, one may hope to arrive in time at a rational arrangement of the numberless forms comprised in the Palliobranchiata.

The geological and vertical range of some of the genera and species has to a certain extent been alluded to by several authors, but M. d'Orbigny claims the merit of having first endeavoured to arrange the general distribution in a tabular shape.[1] He divides the sedimentary strata into 27 epochs,[2] and therein locates the genera, taking care to illustrate his views at the same time as to the numerical development of species by peculiar signs. This, like all similar attempts, must be considered as merely approximative, since fresh discoveries are every day liable to modify the limits and numerical importance we may assign to the genera, since Palæontologists are not yet able to place *all* the Palæozoic species with *certainty* in their proper places. So much confusion likewise exists in the synonyms that it would require much time and investigation to settle, if even possible, the true amount of known species. We differ from the learned author of the 'Paléontologie Francaise' in certain details regarding the first appearance and disappearance of some of the genera; and in the Table accompanying this introduction, I have endeavoured to mark the extent of development at *present* known derived from positive examination, and under each genus have only mentioned a few examples which I had thoroughly examined and known to belong to the group; the principal defect in the 'Prodrome' lately published by the French author above mentioned, as well as in all similar catalogues, consisting in the necessity of forcing into sections a number of species with whose internal character the compiler is necessarily unacquainted.

Thus, for example, M. d'Orbigny characterises his genus *Atrypa* (not of Dalman), as a shell *deprived of perforation* for the passage of a muscular pedicle, *with free extensile fleshy arms*, supported *only* at their *origin* by *two short curved lamellæ*, in every respect similar to those of *Rhynchonella*, from which the author states it to be distinguished *simply* by the (*often very questionable*) want of a foramen. He gives a list of 188 species which are stated to agree with his diagnosis, and consequently members of *his* section. Now let any one examine this list with care, and he will soon be convinced of the dissimilarity of the internal organisations there assembled; and I may boldly assert that a number of the

---

[1] Cours Elémentaire de Paléontologie et de Géologie Stratigraphiques, tableau ix, 1852.

[2] The exactness of such a system requires still to be proved.

species do not possess the character assigned by the author to his genus, viz.:—A great number possessed *spiral calcified* processes for the support of the arms, similar to those seen in *Tumida* and *Herculea*, and were not consequently free, such as *pisum*, Sow., *Nitida*, Hall, and a number of the species taken from M. Barrande, and which partly belong to the genus *Athyris*, M'Coy, (*Spirigera* of D'Orbigny,) while a few, such as *prunum*, from possessing vertical spires, require to be classed with the true *Atrypæ* of Dalman, or should, according to the French author's views, have been admitted into his genus *Spirigerina*. M. d'Orbigny's genus *Atrypa* is, therefore, made up of species belonging to the genus *Rhynchonella* (which he admits), as well as of others partaking of the character of his *Spirigera* and *Spirigerina*, &c. Numerous examples could be adduced to show the difficulty at present existing in the classing of many species into their true genus or place from *mere external appearances*, for, if we are to characterise a group by the presence of free extensile arms simply fixed at their origin to two slender curved processes, departing from the hinge-plates, and another by the arms being supported throughout in a peculiar manner by spiral lamellæ, forming vertical and horizontal cones, it is evident that species with the first character cannot take place along with those of the second; and experience has proved that often two forms, somewhat similar *exteriorly*, may have been inhabited by a very different animal, the one with free, probably extensile arms, the second in which that appendage was unextensile, and entirely supported by a largely-developed calcified process, &c. The only way, therefore, to be able with time to arrive at a proper and natural catalogue of *all the species*, is to examine and develop the internal dispositions of those forms as yet uninvestigated, and to gradually and carefully illustrate their characters. A great difference exists likewise on the probable periods of the first apparition and extinction of the different genera, and to this point geologists as well as Palæontologists justly attach great importance; some few genera, viz., *Lingula*, *Discina*, *Crania*, and *Rhynchonella*, appear to have traversed the whole geological vertical range; they appear in the older Silurian deposits, and with similar or but slight modifications in character, are still represented in our seas by a limited number of species. The *Terebratulæ* with short loops appear, from the present state of our information, to have first appeared during the Devonian era, since no example has as yet been pointed out in older rocks; but it is well to observe, that the genus *Terebratula* seems to have been of very rare occurrence in the Palæozoic period, and to have become truly developed only from the epoch of the Oolites to the present day. *Terebratulæ* with long loops (*Waldheimia*), as well as the other sections in the Family TEREBRATULIDÆ, appear, during the Oolitic period, and upwards; thus, for example, *Waldheimia* and *Terebratulina* are found in the *Oolitic Cretaceous, Tertiary,* and *present* epoch, *Megerlia* and *Argiope* only in the last three, while *Kraussia, Bouchardia,* and *Morrisia,* are only positively known in the present period.

Thecidium has not yet been discovered under the Oolites, and there also it seems to have attained its maximum of specific development, although not much less numerous in the Cretaceous strata; it suddenly diminished in the Tertiary period, since only one species, likewise now alive, has hitherto been discovered. The Family SPIRIFERIDÆ, on the contrary, presents its greatest development in the early periods of the habitable world, and no genus or species with spiral lamellæ has been as yet noticed above the *Lias;* and it seems as if the spiral forms had been suddenly replaced by those provided with loops, which certainly implies a most important modification in the form and condition of the labial appendages; and it seems much more natural, and I may say proper, to place in one family all the species provided with loops, and into another all those possessed of spires. The same may be stated as regards the *Strophomenidæ* and *Productidæ,* no example of which has been found higher up than the Lias.

Viewed, therefore, in a general way, the number of sections, according to our present information, would be distributed somewhat as follows:

In the *Silurian* 22, in the *Devonian* 23, in the *Carboniferous* 16, in the *Permian* 13, in the *Triasic* 8 or 9, in the *Oolitic* 11, in the *Cretaceous* 14, in the *Tertiary* 10, in the present period 14 : or, in other words, 33 genera lived during the *Palæozoic epochs,* or up to Trias included; from thence upwards to the present period, the number seems to have been reduced to about 20, of which 8 only are found in both the two great divisions; but, as above observed, every day may bring fresh changes in our *statistics,* and the views here expressed can only be looked upon as approximative, besides which a few of the sections, such as *Trigonosemus, Terebrirostra, Spiriferina, Cyrtia, Retzia, Aulosteges, Orbiculoidea, Trematis, Acrotreta,* and perhaps one or two others, require further examination before being *finally* admitted.

We will now proceed to describe each genus in detail, as far as our present information will admit.

FAMILY—TEREBRATULIDÆ.

Animal fixed to submarine bottoms by a muscular peduncle issuing from a perforation in the beak of the larger valve; this aperture is partly surrounded by a deltidium in one or two pieces : oral appendages entirely or partially supported by calcified processes, which commonly assume the shape of a loop, variable in form and dimensions, but always fixed to the smaller or dorsal valve: shell structure always punctated.

Obs. In this natural family we have admitted seven genera and five sub-genera or sections; with the exception of two of these last, viz. *Terebrirostra* and *Trigonosemus,* which are still problematical, all other genera and sections are based on well-defined modifications in the form and position of the calcified supports of the arms.

Genus—TEREBRATULA, *Llhwyd,* 1696.[1]

Type—T. VITREA, *Linn.,* sp. Introduction, Pl. VI, figs. 1—4.

CONCHA ANOMIA, (part) *Columna,* 1606 and 1616.

ANOMIA, (part) of *Linnæus, Gmelin, Da Costa, &c.*

TEREBRATULA, (part) *Llhwyd,* and the generality of authors.

LAMPAS, (part) of *Humph.*

GRYPHUS, *Megerle,* 1811. *Philippi,* 1853.

TEREBRATULITES, (part) of *Schloth.*

PYGOPE (DIPHYA), *Link.*

TEREBRATULÆ, with short loops, *Dav.*

EPITHYRIS (ELONGATA), *King.*

ANTINOMYA (DIPHYA), *Catullo.*

Shell, oval, elongated, or transverse, externally smooth or plaited, valves more or less

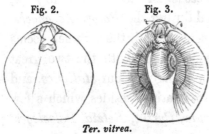

Fig. 2.   Fig. 3.

Ter. vitrea.

2. *Dorsal valve showing the loop.*

3. *Dorsal valve with the animal; the œsophagus passes through the opening of the loop.*

unequally convex, margin even or waved : hinge line curved, beak short, truncated by a foramen, variable in size, and partly margined by a deltidium in one or two pieces; loop short, confined to the posterior portion of the shell, and not exceeding much more than a third of the length of the valve, simply attached to the hinge plate. The two riband-shaped lamella are soon united by a transverse lamella, bent upwards in the middle. The cirrated arms are supported by the crura, and project considerably in front of the loop : no internal septum in the socket valve.

Obs. Some Palæontologists[2] seem disposed to consider the species with short loops as the types of the genus TEREBRATULA, and propose to place into separate *sections* or *subsections* those forms in which the calcified supports extended to near the margin before becoming reflected (*Waldheimia*), as well as those in which the crura unite in the form of a band behind the mouth of the animal (*Terebratulina*); but these sub-divisions, if of little essential value, are very convenient in the arrangement of the species, which may even be distinguished by external characters : thus in those forms with long loops, the mesial septum may easily be traced on the external surface of the smaller or dorsal valve,[3] while none such occurs in the short-looped species.

The genus *Terebratula,* as here limited, includes a vast number of species, more than a hundred having been described and figured. The dimension of the foramen is very variable, being sometimes so large as almost to admit the end of a finger, (*Ter. grandis*) while at other times it will hardly afford space for the passage of a hair. (*Ter. carnea.*)

[1] Llhwyd; Lith. Brit. Ichn., 1696, *Terebratula minor subrubra,* pl. ii, fig. 890 = *T. maxillata,* Sowerby.

[2] This view is adopted by Professor King, Dr. Gray, Mr. Woodward, and others.

[3] This occurrence was first observed by M. Deslongchamps; the length of the septum may easily be ascertained by the use of a little acid.

Geological range.—*Terebratulæ* with short loops are known to have lived from the Devonian epoch to the present day; they probably occurred likewise in the Silurian era, although no positively authenticated example has been ascertained.

Examples: T. *vitrea*, Gmelin, sp. *grandis*, Blum.; *ampulla*, Brocchi; *bisinuata*, Lamarck; *carnea*, Sow.; *obesa*, Sow.; *depressa*, Lamk.; *biplicata*, Brocchi, sp. *semiglobosa*, Sow.; *Harlani*, Morton, *sella*, Sow.; *insignis*, Schübler; *maxillata*, Llhwyd; *globata*, Sow.; *perovalis*, Sow.; *Kleinii*, Lamk.; *simplex*, Buck.; *sphæroidalis*, Sow.; *plicata*, Buck.; *fimbria*, Sow.; *diphya*, Columna, sp. *flabellum*, Def. *hastata*, Sow.; *elongata*, Schl.; *coarctata*, Park; *longirostris*, Walh.; &c. &c.

Section A, sub-genus—TEREBRATULINA, *D'Orb.*, 1847.[1]

Type—T. CAPUT-SERPENTIS, *Linn.* sp. 1767.   Introd., Pl. VI, figs. 7-8.

ANOMIA and ANOMITES, (part) of *Linnæus, Chemnitz, Wahlemberg,* &c.
TEREBRATULA, (part) of the generality of Authors.
TEREBRATULINA, *D'Orbigny.*

Shell, generally longer than wide, and more or less oval; beak obliquely truncated by a foramen, which generally extends to the umbo; deltidium small, and at times indistinct; socket or dorsal valve less convex than the perforated one, exhibiting two variably developed auricle expansions. Surface striated or costellated. Valves articulating by means of teeth and sockets; loop short, not exceeding one third of the length of the shell, and rendered annular by the union of the oral processes in the shape of a shelly band. The cirrated arms are supported by the crura and project considerably into the interior of the shell; structure punctated.

Obs. The species comprised in this sub-section are intimately allied to *Terebratula*, but differ in the union of the crural processes, which form a shelly band behind the mouth of the animal, whereas the reflected border of the loop is always in front (below in the figure) of the mouth. In young examples of the recent *T. caput-serpentis*, the crural processes are not completely joined, showing an intimate relation to *Terebratula* proper. The disposition of the cirrated arms appear the same as in *T. vitrea* and *W. australis;* they are united by a membrane, and assume the form of three lobes, the central lobe being rolled up like the

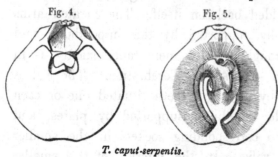

Fig. 4.   Fig. 5.

T. caput-serpentis.

4. *Dorsal valve of a young specimen, in which its oral processes are not yet completely developed.*

5. *Shell with the animal: the intestine is seen projecting above the oral aperture and fringe, the œsophagus passes through the annular part of the loop.*

[1] Considerations Zool. et Géol. sur les Brachiopodes, Comptes rendus de l'Acad. des Sciences, 1847; and Paléont. Franç. Ter. Crétacés, vol. iv, p. 56.

proboscis of a butterfly.   The animal in this sub-section was well figured by Godfroy Auguste Gründler, of Halle, in 1774,[1] as mentioned both by Cuvier and V. Buch.   Apart from the peculiarity of the loop, the *Terebratulinæ* form a small group so well characterised by form and sculpture as never to be confounded with any other.   The beak is truncated, and the foramen partly encircled by a deltidium united or disunited in different species. The ear-like expansions on the sides of the umbo are also characteristic of this sub-genus.

Geological range.—In our present knowledge the first species appeared in the Oolitic period, and continued under different forms to the present day.

Examples : T. caput-serpentis, Linn. sp.; *japonica,* Sow. sp.; *cancellata,* Koch, sp.; *Cumingii,* Dav.; *striatula,* Sow. sp.; *lacryma,* Morton, sp.; *striata,* Wahl. sp.; *Gisii,* Hag. sp.; *Martiniana,* D'Orb.; *gracilis,* Schl.; *echinulata,* D'Orb.; *substriata,* Schl. sp. ; &c.

Section B, sub-genus—WALDHEIMIA, King, 1849.[2]

Type—W. AUSTRALIS, *Quoy* and *Gaim.*, sp.   Int., Pl. VI, figs. 9, 10.

ANOMIA, (part) *Columna, Linnæus, &c.*
TEREBRATULA, (part) *Llhwyd, Lamarck, Brug., Dav.*, and the generality of Authors.
LAMPAS, (part) *Humph.*
WALDHEIMIA, *King, Gray, &c.*
TEREBRATULÆ, with long Loops, *Dav.*

Shell, variable in shape, more or less circular sub-quadrate, transverse or elongated, with both valves convex, or with the smaller or dorsal one depressed or concave ; margins straight or waved ; surface smooth or plaited ; beak truncated and perforated by a circular foramen of variable dimensions, partly completed by a deltidium in one or two pieces. Loop long, in general exceeding two thirds of the length of the valve, formed of slender shelly riband-shaped lamellæ simply attached by the crura to the hinge plate, and more or less folded back on itself.   The cirrated arms are partially supported by this appendage, and united throughout by a membrane, exactly as in the typical species of Terebratula.   The valves articulate by means of teeth situated one on each side of the deltidium, supported by plates, and fitting into corresponding sockets in the smaller or dorsal valve.   In the interior of the smaller valve, a prominent cardinal process or boss, and hinge plate, with four depressions, occupies the space between the socket ridges, under which originates

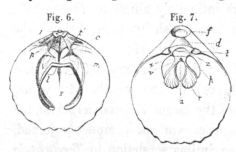

Fig. 6.   Fig. 7.

f. 6, *Dorsal valve;* j, *cardinal process or boss;* t', *dental sockets;* p, *hinge-plate;* s, *septum;* c, *crura of the loop;* l, *reflected portion of loop;* m, *quadruple adductor impressions.*

f. 7, *Ventral valve;* f, *foramen;* d, *deltidium;* t, *teeth;* a, *adductor-impression;* r, *cardinal muscular impressions;* p, *pedicle muscles;* x, *accessory cardinal muscles;* v, *vent;* z, *attachment of the peduncle-sheath* (capsularis, *Owen*).

---

[1] Naturforscher, i, 2d part, p. 86, tab. iii, figs. 1, 6.
[2] A Monograph of English Permian Fossils, p. 145, 1849.

a central mesial septum, which extends more or less into the interior of the shell. Structure punctated.

Obs. The animal has been admirably described and illustrated by Professor Owen in the first chapter of this Introduction; we need therefore only add, that the section was first proposed by Professor King, who at the same time represented the muscular system which he had described in 1846. As a section of *Terebratula*, I now believe the term *Waldheimia* may be advantageously preserved, especially as the number of species possessing a long loop is almost as great as that with a short one.

Geol. range.—The exact period at which *Terebratulæ* with long loops first appeared is as yet unascertained; the interior of many Palæozoic species not having been examined, some of which may possibly prove to belong to the section under consideration. *Waldheimias* were, however, certainly represented from the Trias up to the present period.

Examples: W. *flavescens*, Lamk., sp.; *lenticularis*, Desh., sp.; *californica*, Koch., sp.; *longa*, Roemer, sp.; *digona*, Sow., sp.; *lagenalis*, Schl., sp; *numismalis*, Lam., sp.; *cornuta*, Sow., sp.; *resupinata*, Sow., sp.; *Waterhousii*, Dav., sp.; *Grayii*, Dav., sp.; *cardium*, Lamk., sp.; *Moorei*, Dav., sp.; *cranium*,[1] &c.

Genus—TEREBRATELLA, *D'Orbigny*, 1847.[2]

Type—T. DORSATA, *Lamarck*, sp. CHILENSIS, *Brod.*, sp. Int., Pl. VI, figs. 11—14.

ANOMIA and TEREBRATULA, of Authors.
TEREBRATELLA, *D'Orbigny*.[3]
RHYNCHORA (COSTATA)? *Dalman.*

Shell elongated or transverse, variable in shape: both valves regularly and unequally convex or interrupted by a longitudinal depression in the smaller valve. Beak truncated by an oblique foramen of a circular or oval form, and partly margined by a deltidium in two pieces, at times disunited above the umbo: beak ridges more or less defined, and in some cases leaving between them and the hinge line a flat or concave cardinal area: external surface smooth or variously punctated. Hinge articulating by the means of teeth in the larger, and sockets in the smaller valve. In the interior of the dorsal valve, under the cardinal process and hinge plate, a more or less elevated medio-longitudinal crest or septum extends to about half the length of the valve: the loop is elongated and doubly attached, first to the hinge plate, and afterwards to the mesial septum, by processes given off at right angles near the centre of the valve, the remaining portion

Fig. 8.

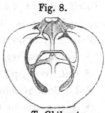

T. Chilensis.

---

[1] In part i, pl. i, fig. 8 *d*, I did not give sufficient development to the loop, which in this species is two-thirds as long as the shell.

[2] Considérations Zool. et Geol. sur les Brachiopodes.—Comptes rendus de l'Académie des Sciences Paris, 1847.—Annals des Sciences, vol. viii, 1847.—Paléont. Franç. Ter. Crétacés, vol. iv, p. 110, 1847.

[3] Professor M'Coy's genus *Delthyridea* may have been intended for this section; but on referring to

soon after becoming reflected.   The double impressions of the adductor are seen on either side near the mesial plate.

Obs. The animal has been ably described by Professor Owen, and the peculiar arrangement and double attachment of the loop had been observed, and more or less accurately figured, by several early naturalists, such as Grünther,[1] Davila,[2] Favanne,[3] Martini, Chemnitz,[4] Parkinson,[5] Fischer de Walderm,[6] and others ; but it was only in 1835[7] that the apophysary system and animal were correctly represented.   The form of the loop, as well as the development of the mesial septum, varies in different species of *Terebratella*, as may be seen in the annexed woodcuts.

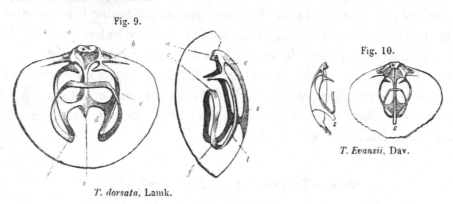

Fig. 9.

Fig. 10.

T. Evansii, Dav.

T. dorsata, Lamk.

Geol. range.—*Terebratella* is not positively known in the Oolitic period, but the genus is represented during the cretaceous epoch, and still continues to abound in our present seas, where from ten to twelve species have been obtained.[8]

Examples: T. menardi, Lam., sp.; *pectita*, Sow., sp.; *carantonensis*, D'Orb.; *canaliculata*, Roemer, sp.; *dorsata*, Lam., sp. ; *flexuosa*, King, sp. ; *chilensis*, Brod., sp. ; *coreanica*, Adam and Reeves, sp. ; *Bouchardii*, Dav. ; *rubicunda*, Sow., sp. ; *zelandica*, Desh., sp. ; *Evansii*, Dav.; *Sowerbii*, King, sp. ; *crenulata*, Sow., sp. ; &c.

---

the 'Synopsis of Carb. Foss. of Ireland,' p. 130, 1844, all I find there stated, is " *Delthyridea*, M'Coy, fig. 27), *or those species which have a long hinge line, and a distinct cardinal area (as in the Delthyridæ or Spirifers), the deltidium fissured in the middle, all the species are largely plaited.*"   No example is mentioned, nor is the figure sufficiently illustrative to justify the author's claim to priority over M. d'Orbigny, who clearly defined his genus in 1847.

[1] Naturforscher, 1774.
[2] Cat. Syst., pl. xx, 1767.
[3] La Conchyliologie, &c., pl. xli, fig. 13, 1780.
[4] Systematische Conchylien, pl. lxxviii, fig. 711, 1785.
[5] Organic Remains of a former World, tab. xvi, fig. 7, 1811.
[6] Notice sur le Système Apoph. des Ter., pl. i, fig. 34, 1829.
[7] Owen; Trans. of the Zool. Soc., vol. i, 2d part, pl. xxii, p. 141, 1835.
[8] For a list of the recent species, see Dav., 'Ann. and Mag. of Nat. Hist.,' May, 1852.

? *Section A.* TRIGONOSEMUS. *Kœnig*, 1825.[1]

Type—T. ELEGANS, *Kœnig*. Int., Pl. VI, figs. 15, 16.

TEREBRATULA (part), of Authors.
FISSURIROSTRA, *D'Orbigny.*
DELTHYRIDEA, *M'Coy.*

Shell inequivalve, oval, transverse or elongated : dental valve always convex, smaller one moderately so, flat or longitudinally depressed : beak produced, recurved and truncated by a small foramen originating at the summit of the beak and extending on the opposite side to the area. This last is large, margined and carinated exteriorly ; the deltidium occupying less than a third of the surface, a small portion only edging the foramen ; external surface striated or variously costated : hinge articulating by the means of teeth and sockets : between the socket ridges a remarkably large and produced cardinal process or boss fills a cavity in the beak of the larger valve ; the loop is doubly attached ; first near the base of the socket ridges and afterwards to a short mesial septum before becoming reflected : the adductor muscular impressions are deeply indented on either side close to the base of the septum.

Obs. The arrangements of the loop are very similar to those of *Terebratella* : the greatest difference resting in the form of the cardinal process, area, and position of the foramen. This sub-section has in reality very little importance, and cannot I think claim a generic rank.

Geol. range.—Hitherto no species has been observed out of the cretaceous period.

Examples : T. *elegans,* Kœnig ; *pulchellus,* Nils., sp. ; *incertus,* Dav. ; *pectiniformis,* Buch, sp. ; &c.

? *Section B.* TEREBRIROSTRA, *D'Orb.,* 1847.[2]

Type—T. LYRA, *Sow.* Int., Pl. VI, figs. 17, 18.

TEREBRATULA (part), of the generality of Authors.
TRIGONOSEMUS (part), *Koenig.*
TEREBRIROSTRA, *D'Orb., De Ryckholt,* &c.
WALDHEIMIA (part), *Gray* (not *King*).

Animal unknown. *Shell* more or less oval, inequivalve ; beak considerably elongated, almost straight, with a false area and narrow deltidium : the extremity of the beak is truncated by a circular foramen, partly margined by a deltidium ; valves articulating by the means of teeth and sockets ; internal arrangements of the calcified supports of the arms

---

[1] Icones Fossilium Sectiles, 1825.
[2] Cons. Zool. et Géol. sur les Brach.—Comptes rendu de l'Acad. des Sciences, 1847 ; and Paléont. Franç. Ter. Cretacés, vol. iv, p. 126, 1847.

unknown; a mesial septum exists in the interior of smaller valve; cardinal process large and produced; external surface plaited.

Obs. It is still very uncertain whether this sub-section should be retained, or placed under *Waldheimia* or *Terebratella*, from the mode of attachment of its loop not having been as yet ascertained: the elongation of the beak could not alone constitute a character of generic importance. M. d'Orbigny supposes that the loop was doubly attached, while Dr. Gray seems to be of a different opinion, since he places the shell in question with *Waldheimia*.[1] It will therefore be very desirable to examine the interior of *T. lyra*, which will alone decide whether the section should be admitted or cancelled.

Geol. range.—The shells placed by authors into this small section seem, in the present state of our knowledge, confined to the cretaceous period.

Examples: T. lyra, Sow., sp.; *Bargesana,* D'Orb.; *neocomiensis,* D'Orb.; &c.

Section C. MEGERLIA, *King*, 1849.[2]

Type—M. TRUNCATA, *Gmel.* sp., 1152. Int., Pl. VI, figs. 26, 27.

ANOMIA (part), *Linnæus, Pallas,* &c.
TEREBRATULA (part), of the generality of Authors.
ORTHIS, *Michelotti,*[3] *Philippi,*[4] (not ORTHIS, *Dalman.*)
TEREBRATELLA (part), *D'Orbigny.*
MEGERLIA, *King, Gray,* &c.
MEGATHYRIS (OBLITA), *D'Orbig.*
KINGENA, *Dav.* (part).

Shell inequivalve, suborbicular, transverse or longitudinally oval; beak short and truncated by a circular foramen; deltidium small; beak ridges well defined; external surface smooth, spinulose, or covered by fine radiating striæ; shell structure largely punctated; hinge line sometimes long and straight, with the articulating teeth and sockets widely separate. In the interior of the *socket* or *dorsal* valve a slightly elevated medio-longitudinal septum proceeds from under the cardinal process or boss, to less than half the length of the valve; the loop is three times attached, first to the base of the socket ridges, afterwards, by a horizontal process, to near the extremity of the septum, and

Fig. 11.

Fig. 12.

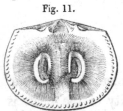
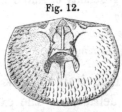

11, 12. *Megerlia truncata (interior of dorsal valve).*

---

[1] A Catalogue of the Brachiopoda in the British Museum, 1853.

[2] A Monograph of English Permian Fossils, p. 145, 1849.

[3] Desc. des Foss. des Terrains Miocènes de l'Italie, sup., p. 78, 1847. The *Orthis oblita,* Mich., is only a synonym of the *M. truncata,* Gmel. sp.

[4] Enumeration Molluscorum Siciliæ, vol. ii, 1844.

again by two additional processes, departing from the reflected portion of the loop to the central septum. The cirrated arms are large, their fringes extending to near the border of the shell. On either side of the mesial septum are seen the double scars produced by the adductors.

Obs. Prof. King proposed to distinguish those species possessing a triply attached loop from *Terebratula* and *Terebratella*, by the name of *Megerlia;* but after a minute examination of the three or four species at present known, and which seem to vary from one another to some extent in the details of their loop, it will be preferable for the present to admit *Megerlia* simply as a section of *Terebratella.* The *M. truncata* has been by some authors placed in the genus *Orthis,* but it in no way possesses the characters of Dalman's genus. The interior of the valves, especially in the last-named species, are covered with spinulose asperities which radiate from the base of the hinge line to the margin, a character likewise observable in some species of *Productus.*

Geol. range.—In the present state of our knowledge the first species appeared during the cretaceous period, and continued under different forms up to the present day.

Examples: M. *truncata,* Gmel.; M. *lima,* Def.; *pulchella,* Sow., sp.

Genus KRAUSSIA, *Davidson,* 1852.[1]

Type—K. RUBRA, *Pallas,* sp.[2] Int., Pl. VI, figs. 28, 29.

ANOMIA, *Chemnitz, Gmelin,* &c.
TEREBRATULA (part), of the generality of Authors.

Shell subcircular, with a nearly straight hinge line; beak truncated; foramen large, round; deltidial plates small, not united; beak ridges well defined, leaving a flat false area between them and the hinge margin; in most species a longitudinal depression exists in the smaller valve: external surface smooth or variously plaited, structure punctated: the dorsal pedicle muscles produce two wide eye-shaped impressions close to the hinge, and between the inner walls of the socket ridges, a small slightly elevated mesial ridge extends to about half the length of the valve, at the extremity of which arise two small forked diverging lamellæ expanded at their extremity. The cirrated arms are unusually small, their fringes not extending to more than half-way towards the border of the shell; in the first part of their course from the mouth forward, the cirri are few or wanting: the whole brachial apparatus is supported by the small forked process above described, no other part of the apophysary system being calcified: cardinal process or boss, small.

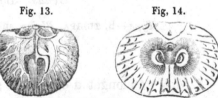

Fig. 13.   Fig, 14.

13. *Kraussia rubra* (dorsal valve).
14. *Kraussia Lamarckiana* (animal).

[1] Annals and Mag. of Nat. Hist., vol. ix, 2d series, p. 369.
[2] Mis. Zool., tab. xiv, figs. 2, 11. 1766.

Obs. The animal of this small genus or sub-genus (examined in two species), differs from *Megerlia* in the relative small size of the cirrated arms.

Geol. range.—Hitherto only known in the recent state.

Examples: *K. rubra*, Pallas, sp. ; *K. cognata*, Chemnitz, sp. ; *K. pisum*, Lamarck, sp.; *K. Lamarckiana*, Dav. ; *K. Deshayesii*, Dav.[1]

Genus—MAGAS, *Sowerby*, 1818.[2]

Type—MAGAS PUMILUS, *Sow.*   Int., Pl. VI, figs. 19, 21.

TEREBRATULA (part), *Lamarck, V. Buch, Deshayes*, &c.
MAGAS, of the generality of Naturalists.
ORTHIS (MILLEPUNCTATA), *De Koninck*, sp. (not ORTHIS, *Dalman.*)

Animal unknown; *shell* small, inequivalve, more or less regularly oval; beak truncated by a foramen extending to the umbo of the smaller valve; surface smooth or slightly striated; shell structure largely punctated; apophysary system in the *dorsal* or *socket valve*

Fig. 15.   Fig. 16.

Magas pumila.

15. *Interior of Dorsal Valve.*
16. *Section of both Valves.*

composed of an elevated longitudinal septum reaching from one valve to the other, and to which are affixed two pair of calcareous lamellæ; the lower ones are riband-shaped, attached first to the hinge-plate, they afterwards proceed by a gentle curve to near the anterior portion of the septum, to the sides of which they are affixed; the second pair originate on either side of the upper edge of the septum, extending in the form of two triangular anchor-shaped lamellæ.

Obs. Having fully described this genus in Part II, p. 19, it will not be necessary to add any other observations.

Geol. range.—Hitherto only known in the cretaceous period.

Examples: *Magas pumila*, Sow.; *M. orthiformis*, D'Archiac, sp.

Genus—BOUCHARDIA, *Davidson*, 1849.[3]

Type—B. TULIPA, *Blainv.* sp.=(ROSEA, *Mawe*, sp.)   Int., Pl. VI, figs. 22, 25.

TEREBRATULA (part), of Authors.

Shell of an elongated oval shape; valves thick and almost equally convex; beak nearly straight and truncated by a small circular foramen; no true area or deltidium; external

---

[1] The last two species are described by myself, in the 'Zool. Proc.,' for 1852.

[2] Min. Con., vol. ii, p. 40, tab. 119, figs. 1—5. 1818.—See likewise Bouchard and Davidson 'Bull. de la Soc. Géol. de France,' vol. v, 2d ser., p. 139, 1848.

[3] Bulletin de la Soc. Géol. de France, vol. vii, 2d ser., p. 62, pl. i, 1849.   [4] Dic. Sc. Nat.

surface smooth; structure punctated. Valves articulating by the means of teeth and sockets. In the interior of the smaller or dorsal valve the socket ridges are largely developed, extending to nearly a third of the length of the valve, and fitting into corresponding grooves in the dental valve; the hinge plate or platform is large and massive, filling up the space between the socket ridges; and from the extremity of the umbonal beak originates the cardinal process or boss, which assumes the shape of two deviating elongated crests, grooved along their upper surface, and probably serving as an attachment to the cardinal muscles; pedicle scars are visible on the outer sides of these; at the base of the platform above described a short mesial septum gradually arises, to the upper edge of which is affixed an anchor-like pair of lamellæ, and close to the base of the septum was attached the adductor muscle. In the *dental valve* a mesial ridge separates the large cardinal muscular scars placed on either side, and towards the centre of the valve a small ovate impression indicates the place of the adductor. The ventral pedicle-muscles occupy a deep groove, which extends from the last-named scar to the extremity of the foramen, which is excavated in the substance of the beak.

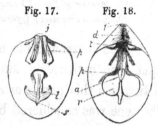

Fig. 17.   Fig. 18.

Bouchardia tulipa.

17. *Interior of dorsal valve*; j, *cardinal process*; l, *loop*; p, *hinge plate*; s, *septum.*
18. *Interior of ventral valve*; f, *foramen*; t, *teeth*; a, *adductor*; p, *pedicle*; r, *cardinal muscular scars.*

Obs. The remarkable organisation of this little genus is so different from that observable in the rest of the *Terebratulidæ* that I could not refrain applying to it a separate denomination; unfortunately the animal has not yet been examined, so that little can be said of the soft parts or arms, but the *cardinal process* presents a very unusual shape and character, and the whole internal details have their peculiarities.

Geol. range.—Hitherto only known in the recent state.

Example: Bouchardia tulipa, Blainv., sp.=(*rosea*, Mawe, sp.)

Genus—MORRISIA, *Dav.*, 1852.[1]

Type—M. ANOMIOIDES, *Scacchi*, sp. (=T. DEPRESSA, *Forbes*.) Int., Pl. VI, figs. 30, 31.

ORTHIS, *Scacchi and Philippi*.[2]
TEREBRATULA, *Forbes*.[3]

Shell minute, circular, depressed; foramen large, round, encroaching equally on both valves; larger or dental valve with a small straight hinge area: deltidium plates minute, widely separated; valves articulating by the means of teeth and sockets; smaller or socket valve deeply notched at the umbo; apophysary system consisting of two branches, originating at the base of the dental sockets, and united to a small elevated process arising from the centre of the valve.

---

[1] Annals and Mag. of Nat. Hist., vol. ix, 2d series, p. 361, 1852; Proc. of the Zool. Soc., 1852.

[2] Enumeratior Moluscorum Sicilie, vol. ii, pl. xviii, fig. 9, 1844.

[3] Report of the Mollusca of the Ægean Sea, 1843.

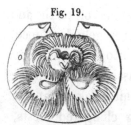
Fig. 19.

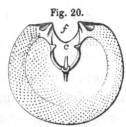
Fig. 20.

Morrisia anomioides (Scacchi, sp.)

19. *Interior of Dorsal Valve with the Animal.*
20. *c, oral processes; f, foramen.*

Animal furnished with two subspiral or *sigmoid* arms fringed with comparatively large cirri ; these arms originate beside the mouth (as shown in the figure) supported by the crural processes, and after passing forwards and converging in front of the mouth, they again turn outwards, each having the shape of the letter S : external surface smooth and largely punctated.

Obs. The remarkable arrangement of the oral processes, large foramen, encroaching equally on both valves, and form of the apophysary system, seems to me quite sufficient to distinguish this genus or *sub-section* from the other genera in the family of *Terebratulidæ.* The type was considered by Scacchi and Philippi to belong to Dalman's genus *orthis,* of which, however, it possesses *none* of the characters.

Geol. range.—I am inclined to believe that this little genus existed in the cretaceous period, possessing several examples of a small shell from the chalk of Gravesend, which so closely resembles the recent type that I can hardly distinguish them even specifically.

Example : M. *anomioides,* Scacchi, sp. (*recent.*)

Genus—ARGIOPE, *Deslongchamps,* 1842.[1]

*Type—*A. DECOLLATA, *Chem.,* sp., 1785. Int., Pl. VI, figs. 32, 34.

ANOMIA and TEREBRATULA, in part of Authors.
ORTHIS, of *Michelotti, Philippi, Hagenow, &c., but not* ORTHIS, *Dalman.*
MÉGATHYRIS, *D'Orbigny,* 1847.

Shell semi-orbicular, quadrate or transversely oval ; valves unequally convex, smooth or variously costated ; ventral or dental valve the deepest, beak produced, with a depressed triangular area ; foramen large, completed by the umbo of smaller valve ; hinge line straight ; valves articulating by means of teeth and sockets.   Interior of smaller valve furnished with a central septum, and sometimes with one or more lateral septa radiating from beneath the cardinal muscular prominence, and terminating at some distance from the margin in elevated processes ; apophysary system consisting of a distinct loop, originating at the base of the dental sockets and folded into two or more lobes occupying the interspaces of the radiating septa to which they adhere on their inner sides ; the labial processes or arms originate on the sides of the mouth, and diverge right and left parallel with the margin of the shell, but at some little distance from it ; when they arrive at the raised septa they turn inwards, forming one or more lobes on each side of the

[1] Mémoirs de la Soc. Linn. de Normandie, vol. vii, p. 9, 1842.

middle line, but on reaching the central septum, they become perhaps free at their extremities. The mantle adheres closely to the shell and is not seen except as a part of it.

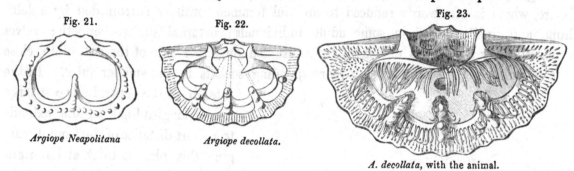

Fig. 21.   Fig. 22.   Fig. 23.

Argiope Neapolitana.   *Argiope decollata.*

A. decollata, with the animal.

Obs. Having described this Genus in part 1, p. 8, and in part 2, p. 16, it will not be necessary to add any further details.

Geol. range.—The genus in the present state of our knowledge seems confined to the cretaceous, tertiary, and recent periods.

Examples: *A. decemcostata,* Rœmer, sp.; *A. hirundo,* Hag., sp.; *A. decollata,* Chem., sp.; *A. cuneata,* Risso, sp.; *A. Neapolitana,* Scacchi, sp.; *A. cistellula,* S. Wood, sp.; &c.

Sub-Family—STRINGOCEPHALIDÆ.

Antmal unknown, attached during the greatest part of its existence by means of a muscular peduncle: labial appendages supported by a largely developed calcified process or loop: a mesial septum: small in the dorsal, large in the ventral valve. Cardinal process extending to the ventral septum: shell structure widely punctated.

Obs. The peculiar assemblage of remarkable characters possessed by *Stringocephalus*, has rendered the exact place the section should occupy somewhat difficult to determine. The presence of a loop and punctated structure connects the genus with the *Terebratulidæ*, but it cannot be properly admitted into that family, on account of its acute beak, position of the foramen, deltidium and largely developed septa, which last-named characters belong more particularly to the *Rhynchonellidæ*; it will therefore be preferable, at least for the present, to follow Professor King in keeping the section distinct, by placing it in a small sub-family, close to the *Terebratulidæ*.

Genus—Stringocephalus, *Defrance,* 1827.[1]

Type—S. Burtini, *Def.* Int., Pl. VII, figs. 95—98.

Terebratula (part), *V. Buch,* &c.
Strigocephalus, *Def.*, and of the generality of Authors.

---

[1] Dic. d'Hist. Nat., vol. li, p. 102, pl. lxxv, fig. 1, 1a, 1827.
[2] Most authors write the name *Strigocephalus,* but the proper spelling is *Stringocephalus.*

Shell transversely or longitudinally oval; valves unequally convex; beak acute, entire, and slightly incurved; area defined and divided in young individuals by a large triangular fissure, which is afterwards reduced to an oval foramen, entirely surrounded by a deltidium, and finally closed in some adult individuals; external surface smooth; valves articulating by the means of a large tooth situated on either side of the deltidium, close to the hinge line, and fitting into corresponding sockets in the smaller valve. In the interior of the ventral valve a large mesial longitudinal septum extends to a short distance of the frontal margin: this plate is thick at its origin and base, but gradually decreases in width while increasing in depth, as it recedes from near the extremity of the beak. In the smaller or dorsal valve, a massive curved cardinal process stretches to the opposite valve where it clasps the ventral mesial septum with its bifurcated extremity; near the base of this process in the dorsal valve a longitudinal septum arises (smaller than that of the ventral valve), and which

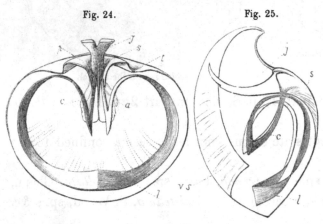

Fig. 24.   Fig. 25.

Stringocephalus Burtini.[1]

24. *Interior of the dorsal valve, partly restored, from a specimen in the Collection of Professor King.*

25. *A section of the valves; j, cardinal process or boss; s, septum in dorsal valve; vs, septum in ventral valve; t, sockets; c, crura of the loop; l, loop; a, adductor.*

divides the quadruple impressions of the adductor muscle; the socket walls are very much expanded, forming prominent hinge-plates curving inwards on each side of the cardinal process. The lower portion of the hinge-plate supports the crura of the loop in the shape of two flattened stems or lamellæ, which after proceeding with a slight upward curve to near the extremity of the septum, are suddenly reflected, and again approach the sockets before sweeping sub-marginally round in the shape of a large wide loop, from the inner edge of which a number of smaller lamellæ branch off and converge.

Obs. Great efforts have repeatedly been made to obtain a complete knowledge of the internal characters of this most remarkable shell, so solidly constructed, and so dissimilar from what we observe in the other sections of the class.[2] Some of the internal arrangements have been known for many years, since we find an attempt to illustrate the large

[1] I feel greatly indebted to Professor King for the communication of the original specimen from which his diagram ('English Permian Fossils,' pl. xix, fig. 1, 1849), had been drawn; figures 24 and 25 restored from the Professor's specimen, but want the branch lamellæ discovered by M. Suess. I am likewise greatly indebted to MM. de Koninck, Beyrich, Bouchard, and Suess for the communication of specimens and sketches of their best examples of the genus.

[2] Baron von Buch describes at great length the external appearance of this shell, but makes no allusions to any of its internal characters, he places it, with some doubts, among the *Terebratulæ.*

forked cardinal process in De Blainville's 'Conchyliologie et Malacologie;' but it is to Professor King and M. Suess that science is indebted for the knowledge we now possess of the calcified supports of the labial appendages. The first-named author published a restored figure of a loop, which he describes as a single piece, *having a tolerably close agreement with that of certain Terebratulæ*, but M. Suess subsequently discovered that Professor King's drawing, although correct, was incomplete, since a number of lamellæ, sometimes single, at other times bifurcated, branched off from the inner edge of the loop, as represented in the annexed woodcuts;[1] and it still remains to be known with certainty whether

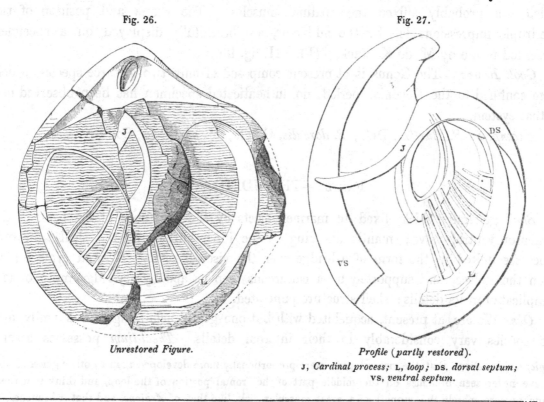

Fig. 26. Fig. 27.

Unrestored Figure. *Profile (partly restored).*

J, *Cardinal process;* L, *loop;* DS. *dorsal septum;* VS, *ventral septum.*

---

[1] Sketches of the diagrams here introduced were kindly forwarded by my Viennese friend, and are the result of his continued researches on the genus:—

Fig. (26) is an *unrestored* representation by M. Suess, of his best example, and exhibits the exact extent of our knowledge on the shape and character of the internal skeleton or loop.

Fig. (27) is a restored profile of both valves. M. Suess accompanies this diagram with a few observations which I have much pleasure in reproducing. "The profile view which has been added, although, perhaps incomplete, is correct, and contains all I know on the subject. The great lateral expansions of the loop (which may not exactly be termed crural plates, since the crura are fixed beneath them), have been omitted, because they would cover the upper part of the loop, the rim following exactly the contour of the shell, so that the *cirri* might surround the whole circumference of the shell, with the single exception of the cardinal region. A comparison of the loop and its branches with the corresponding process of the one-lobed *Argiope* (bearing one septum), has shown me much analogy between those forms: the flat, broad, widely developed loop is hardly recurved in front, the deflected portion so considerable in most *Terebratulæ* is barely indicated in *Stringocephalus* by a straightened frontal piece, similar to that in

those branches were attached or free at their outer end, although we have every reason at present to believe that they presented the last-named condition. It is therefore now quite certain that the oral supports did not assume characters similar to those of *Rhynchonella*, into which family the genus had been erroneously placed by the author of the 'Palæontologie Française.'[1] The hinge was well described by Professor King, as well as the expanded prominent hinge plates, which the author considers to constitute a divided crural base, and the largely developed cardinal process as strictly homologous, although a remarkable modification of the process seen in *Terebratula* and other genera,[2] and to which was probably affixed the cardinal muscles. The shape and position of the quadruple impressions left by the adductor, are beautifully displayed on a specimen presented to me by M. de Koninck. (Pl. VII, fig. 98.)

Geol. Range.—The Genus is at present composed of only two or three species, which seem confined to the Devonian period, no authenticated specimen has been observed out of that system.

Examples: S. Burtini, Def.; *S. dorsalis,* Goldf.;[3] *S. giganteus,* Sowerby.

FAMILY—THECIDEIDÆ.

Shell generally thick, fixed to marine objects by the substance of the beak of the larger or ventral valve; mantle adhering to the inner surface of the shell; the oral processes united in the form of a bridge over the visceral cavity; cirrated arms folded upon themselves, and supported by a calcareous loop or apophysary ridge more or less complicated in its details; shell structure punctated.

Obs. We are, at present, acquainted with but one genus belonging to this family, and the species vary considerably in their internal details. *Thecidium* possesses several

Argiope. I believe that the visceral parts were proportionally more developed than in other genera; and I have never seen branches on the middle part of the frontal portion of the loop, and think that these branches were merely the support of a great muscular disk, like that of *Argiope,* and that if branches, as long as the others, existed on the middle of the frontal piece, they would have necessarily traversed the mouth, I therefore believe, that they were shorter or did not exist at all in the centre, but do not yet know if the ends of the branches were fixed, although we have no reason to believe they were. The growth of the loop must have taken place in a very complicated manner; the measures of the profile view are quite correct, because a fracture in the specimen allowed me to remove the left half, so that I have a section of the entire profile; the great septum in the ventral valve begins at a short distance from the extremity of the beak." M. Suess has just published a detailed and very interesting account of his discoveries on the internal organisation of this remarkable genus, in the 'Verhandl. d. z. b. Vereines,' iii, 1853.

[1] Considérations Zool. et Géol. sur les Brachiopodes.—Comptes rendus de l'Académie des Sciences; Paris, 1847.

[2] Annals and Mag. of Nat. Hist., vol. xviii, p. 87, 1846.

[3] Beautiful illustrations of this species have been published by M. de Verneuil in the 'Trans. of the Geol. Soc. of London.' It is very doubtful whether three distinct species exist; they are all probably varieties of one common type. *S. brevirostris,* Phil., is a *Pentamerus.*

characters in common with *Argiope*, a point already observed by M. Deslongchamps,[1] D'Orbigny, Gray, and others, but although their relationship may on some grounds be admitted, still *Argiope* is much more essentially a member of the *Terebratulidæ* than *Thecidium*, which may be advantageously separated on account of its different mode of attachment, formation of the shell, and other internal and external details, we would therefore admit the THECIDEIDÆ to the rank of a family or sub-family.

Genus—THECIDIUM, *Defrance*, 1828.

Type—TH. RADIATUM, *Def.* Int., Pl. VI, fig. 35—46.

TEREBRATULITES and TEREBRATULA (part), of a few Authors.

THECIDEA,[2] *Def.*, *Goldfuss*, and of all modern Palæontologists.

Shell thick, subquadrate, triangular, oblong, or transversely oval; dental or ventral valve most convex, partially or entirely attached to submarine bottoms by the substance of the shell of its larger or ventral valve; beak straight or recurved with a more or less defined false area and pseudo-deltidium; smaller or dorsal valve slightly concave or convex, likewise at times exhibiting a narrow false area; external surface smooth or variously striated, the lines of growth passing uninterruptedly over the valves and false area; shell articulated by the means of teeth and sockets. Interiorly a thickened granulated margin generally encircles the valves, which, in the *dental* one, after reaching the front, extends inwards in the shape of a rounded mesial ridge; in the small free cavity of the beak, and beneath the deltidium, are seen three short elevated septa, the central one slightly exceeding the lateral ones; under these are situated two small contiguous elongated oval scars, probably due to the adductor, and on the outer side of these other two larger elongated impressions left by the cardinal muscles. The internal details of the *socket or dorsal valve* vary considerably both in species and age; the cardinal process or boss is more or less produced, the inner socket walls project considerably, and are united by a small bridge or arch-shaped process; the interior of the valve is more or less deeply and regularly furrowed to receive a testaceous ridge, folded into two or more lobes. This loop or apophysary ridge supports the brachial membrane, whose thickened and cirrated margin is apparently supported by the inner side of the sinuous grooves; *animal* small, with the mantle lobes disunited and adhering closely to the valves; structure punctated.

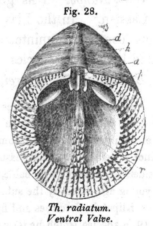

Fig. 28.

Th. radiatum.
Ventral Valve.

Fig. 29.

Th. Mediterraneum.
Dorsal Valve.

[1] Bulletin de la Soc. Géol. de France, t. vii, Dec. 1849.

[2] Defrance and most authors write the name *Thecidea*, but *Thecidium* is more correct.

Obs. We are indebted to Defrance for the institution of this excellent genus,[1] which has received the sanction of all naturalists; but perhaps no section contains an assemblage of shells more variable in their internal arrangements : these differences are chiefly observable in the interior of the smaller valve. In some species, such as the *Th. Moorei, Deslong-champsii, Wetherellii, rusticum, hippocrepis, sinuatum,* &c., the granulated margin on reaching the front extends inwards to a variable distance, assuming the shape of a narrow mesial ridge or expanded concave and sinuated dissepiment; in other forms, such as the *Th. radiatum, recurvirostris, Mediterraneum,* &c. the mesial dissepiment is irregular in shape, giving off at variable distances unequal lateral branches; while in some cases, *Th. digitatum, hieroglyphicum, rugosum,* (D'Orb.) &c., numerous dissepiments, variable in form and dimensions, arise from all round the margin, into which the apophysary ridge extends and follows the curves of the digitated grooves. Great dissimilarity likewise exists in the extent of the attached surface of the dental valve, which has a great influence on the regularity or shape of the species and valve; but however great or small the affixed portion is, it invariably originates from the extremity of the beak.[2]

The generality of species have both their valves convex, or the smaller one almost flat; but in *Thecidia Leptænoides,* (Deslong.) not only is the socket valve concave, but it is much smaller than the other, which protrudes considerably beyond the margin of the ventral one.

Geol. Range.—This genus, so far as we know, first appeared in the Salt-marls of St. Cassian.[3]   In the Lias, the species was both numerous and large, from that epoch the genus continued uninterruptedly to the present period, and the section is still represented by a single species.

Examples : *Th. Moorei,* Dav.; *Mayalis, Leptænoides, sinuatum, Koninckii,* E. Deslong. ;[4]

[1] Dictionnaire des Sciences Nat., vol. liii, 1828.

[2] M. D'Orbigny states in the fourth volume of his 'Paléont. Franç. Ter. Crétacés,' p. 151, that on examining the beak of some of the Terebratuliform species, he observed the remains of a foramen whence a pedicle must have issued; and in my second part, p. 13, I reproduced that view, but which I now consider to be very improbable, and feel much more disposed to believe that *Thecidium* was in all cases, in the young age, fixed by the substance of its beak to sub-marine bottoms.

[3] Klipstein describes and figures, in his 'Beitrage zur Geol. Kenntness der Astlichen Alpen,' plate xv, fig. 19, a species which he terms *Spirifer bidorsatus,* but on an examination of the original figured example now in the British Museum, Mr. Woodward pointed out to me that it most probably was a *Thecidium* approaching in character to the *Leptænoides,* Desl. It is certainly *not* a spirifer, and exhibits undeniable traces of its attachment by a portion of its ventral valve.

[4] In a very interesting and valuable paper on the *Leptænas* and *Thecidia* lately discovered in the Oolitic formations of Normandy, published by M. Eugene Deslongchamps, in the ninth volume of the 'Mémoires de la Soc. Lin. de Normandie (1853),' the learned author adds several new and interesting details relative to *Thecidium,* dividing the genus into two groups, from the presence or total absence of what we have termed " *the apophysary ridge,*" and I cannot do better than to add an extract in the author's own words—

" D'après l'inspection des intérieurs des différentes espèces de Thécidées de tous les terrains, je pense qu'on peut diviser facilement ce genre en deux sections. Si l'on regarde avec attention les intérieurs de deux espèces qui au premier coup-d'œil, semblent être assez voisines, la *Thecidea Mayalis* (espèce *triangularis*, D'Orb.; *Davidsoni*, Buv.; *Bouchardi*, Dav.; *rusticum*, Moore; *Deslong-champsii*, Dav.; *rugosum*, D'Orb.; *Wetherellii*, Morris; *recurvirostris*, Def.; *tetragonum*, Rœmer; *Hieroglyphicum*, Def.; *hippocrepis*, Goldf.; *Mediterraneum*, Risso; &c.

FAMILY—SPIRIFERIDÆ.

Animal free, or rarely attached by a muscular peduncle; oral appendages largely developed, and entirely supported by a thin shelly spirally rolled lamella; shell structure either punctated or unpunctated.

Obs. We have provisionally admitted five genera and two sub-sections; but it is probable that when the characters of many of the species are better known, the family will require further modifications.

Genus—SPIRIFER, *Sowerby*, 1815.

Type—SP. STRIATUS, *Martin*, sp. Int., Pl. VI, figs. 48—59.

ANOMIA, *Linn.* (part), *Martin*, &c.
TEREBRATULITES, *Schloth.* (part).
TEREBRATULA, *Lamarck* (part).
CHORISTITES, *Fischer*, 1825.
TRIGONOTRETA, *Kœnig*, 1825; *King*, 1849.
DELTHYRIS, *Dalman*, 1827; *V. Buch*, &c.
SPIRIFER, of the generality of Authors.
MARTINIA, SPIRIFERA, BRACHYTHYRIS, RETICULARIA (part), *M'Coy*.

Liasique) et la *Th. digitata* (espèce Crétacée), on verra dans la première un appareil unique naissant du bord frontal de la coquille, tandis que dans la seconde il y a, outre ce même appareil que j'appellerai *ascendant* un autre appareil suivant à peu près le contour du premier, mais dans une direction contraire; cet autre appareil, que j'appellerai *descendant* (*our apophysary ridge*), nait d'une bride transversale située au tiers à peu près de la coquille du côté de l'apophyse dentaire. Le double système se remarque surtout dans les espèces Crétacées et dans l'espèce vivante.

 "Je prendrai pour type de Thécidée à appareil double la *Th. digitata*, où il semble être le plus marqué et pour type de *Thécidée* à appareil simple les *Th. Mayalis, leptænoides, rustica*, &c. Les Thécidées à appareil simple semblent dominer dans le lias, tandis que c'est à la première section, ou à appareil double, qu'il faudrait rapporter les espèces crétacées. Les espèces jurassiques, que je rapporte à cette dernière section ne présentent même qu'incomplètement le système descendant; il y est représenté seulement par une petite bride demi-circulaire située de chaque côté de la côté centrale, comme on peut le voir dans la *Th. Deslongchampsii*, je ne pense pas qu'on puisse regarder les grosses granulations des *Th. Moorei, sinuata,* &c., comme suffisantes pour caractériser l'appareil descendant: je considère donc ces Thécidées comme des espèces à appareil simple."

Thus, according to M. Deslongchamps, in the first group, *with an apophysary ridge*, would be placed *Th. digitatum, radiatum, Mediterraneum, Wetherellii, recurvirostris, papillatum, rugosum, hieroglyphicum, hippocrepis, tetragonum*, &c. &c.

In the second group, where *no apophysary ridge* can be traced, the author places *Th. leptænoides, Bouchardii, Mayalis, sub-Mayalis, Perierii, Koninckii, rusticum, Moorei, sinuatum*, &c.; he likewise adds some details relative to the "*bridge-shaped process.*"

Animal unknown; shell somewhat trigonal, transverse or elongated, inequivalve, with or without a mesial fold and sinus; hinge line shorter or longer than the width of the shell; cardinal angles obtusely rounded, or extended into wing-shaped expansions; external surface smooth, or variously costated; shell structure unpunctated; valves articulating by means of teeth and sockets; beak straight or recurved; area in ventral valve large or narrow, and nearly parallel sided, bent backwards, flat or concave, and divided by a triangular fissure, always more or less closed by a pseudo-deltidium notched in the vicinity of the cardinal edge; area in the *dorsal valve* narrow, often linear, and likewise divided by a wide open fissure, partly or entirely occupied by the cardinal muscular process or boss.

In the interior of the smaller valve the labial appendages were supported by calcified lamellæ, assuming the shape of two large conical spires, which nearly fill the interior of the shell; the ends of the spires are directed outwards, towards the cardinal angles; the bases of the hollow conical spires nearly meet at the hinge side, but are wide apart in front; the hinge-plate is divided into two portions, to which the crura of the spires are attached; the crura are united, at some distance from the hinge, by an oral lamella; the divisions of the hinge-plate are hollowed, perhaps for the insertion of pedicle muscles; the cardinal muscles (adductor brevis of Owen), were no doubt fixed to the cardinal process or boss, which is small, and situated in the notch of the hinge-plate; under it are seen four large elongated scars left by the adductor.[1] In the interior of the *larger valve* a short hinge tooth is situated on either side at the base of the fissure, supported by vertical shelly plates, which extend from the beak to the bottom of the valve, thus forming the fissure-walls and assuming various forms and directions in different species; they are either small, regularly diverging, or converging to diverge again, and extending to a greater or less distance into the interior of the valve; between these a great portion of the free space at the bottom of the shell is filled up by muscles, which are generally divided by a blunt longitudinal crest; the adductor commonly producing a small longitudinal and mesial oval scar, and on either side of which are situated the cardinal muscles; pedicle impressions cannot be detected in this valve.

Obs. The shells placed under this head, have been sub-divided into many sections and genera, often established upon unimportant or superficial variations, partaking more of the specific than generic character. When the interior of a larger number of species shall have been carefully examined, it is possible that the genus may require to be subdivided into a few minor sub-sections, but I do not consider our present knowledge of the species sufficient to warrant the adoption of any other sub-sections than, perhaps, those of *Spiriferina*, D'Orb., and *Cyrtia*, Dalman; and even to these, I do not attach great value.

---

[1] On a remarkable cast of *Spirifer alatus*, figured by Professor King, in his work on 'English Permian Fossils,' the quadruple adductor scars appear to be situated in pairs one above the other, in a somewhat different manner to what we observe in the generality of Spirifers, as, for example, *Sp. striatus, Lonsdalii,* &c. On the same specimen, the vascular impressions are well preserved.

The *Anomites striatus* of Martin is the true type of SPIRIFER, since its remarkable spirals induced Sowerby to found the genus.[1]   Dalman, in 1827 or 1828,[2] from mere caprice, altered the names of several of Sowerby's genera : for Spirifer he proposed to substitute the term *Delthyris,* which was adopted by some few, but it is now almost entirely abandoned. Prior to Dalman, Kœnig had proposed the name *Trigonotreta,*[3] and Fischer de Waldheim that of *Choristites,*[4] for species agreeing with Sowerby's type; both are consequently synonyms.   Professor M'Coy's genus *Martinia* was proposed for those species *in which the hinge line was shorter than the width of the shell, and dorsal edges of the cardinal area obtusely rounded with a smooth surface, and small spiral appendages;*[5] but the length of the hinge line is so very variable, even in specimens of the same species, that on such a character it seems very unsafe to found a genus; the species are also both smooth and plicated and the disposition of the spiral appendages does not appear to be distinguishable from that of other *Spirifers.*   It will, therefore, be preferable to postpone the adoption of the section *Martinia* until some more tangible differences can be pointed out.

[1] A difference of opinion having arisen, as to the type of Sowerby's genus, it may be remarked, that in 1814 the author first called attention to the internal organisation of *An.* *striatus,* in a paper read before the Linn. Soc., and published in vol. xii, p. 514, of its transactions : the author likewise adds, "*I suspect* An. cuspidatus *figured since the reading of this paper as* Sp. cuspidatus '*Min. Con.,*' *tab.* 120 . . . . *may have a similar construction within, as well as* An. subconicus *of Martin, tab.* 47." This subsequent remark published in the same year as the generic description in the 'Min. Con.,' 1815, denotes that the author looked upon *A. striatus* as the type, and not *cuspidatus,* of whose internal character he was not *quite certain.*   Not being acquainted with these facts, Professor King in his ' Mon. of English Permian Fossils,' urged the adoption of *S. cuspidatus,* as the type of Sowerby's genus, and to employ the term TRIGONOTRETA, proposed by Kœnig in 1825, for shells similar to *S. striatus*; but the learned professor now abandons that view and admits *S. striatus* as the type of Sowerby's excellent genus *Spirifer.* Professor M'Coy has justly pointed out the fact above noticed, in his ' Synopsis of Carb. Foss. of Ireland,' p. 135, 1844, for when describing *Spirifer striatus,* he adds, "*This shell is very well known on the continent as the species in which Mr. Sowerby first discovered the spiral appendages.*"   While publishing my report on the Lamarckian species of *Terebratula,* I was enabled to prove that *Ter. spirifera* (Lamarck) was a synonym of *An. striata* of Martin, nor does the celebrated French naturalist omit to remark in the sixth volume of his 'An. sans Vert.,' p. 257, 1819, that "*Mr. Sowerby l'a distinguée comme genre.*"   The nature of the spiral processes has been much misunderstood : the celebrated author of the 'Mineral Conchology' believed them to be *cartilaginous tubes;* and, at a more recent period, Prof. M'Coy contended that they could not be the *labial arms,* and described them as " cardinal teeth," giving at the same time a woodcut, in which they are turned the wrong way in the shell.   Syn. Carb. Foss. Ireland, p. 127, 1844.

[2] Petref. suecana (Proc. R. Acad. Sc., Stockholm, 1828).

[3] Icones Fossilium Sectiles (cent. prima) 1825.

[4] Notice sur le Genre *Choristites*; Moscow, 1825.

[5] Synopsis of the Carb. Lime Foss. of Ireland, p. 139, 1844.   Professor M'Coy introduces a woodcut, No. 22, illustrating his idea of the proportions of the spirals, and which are stated to be *so small as only to occupy the rostral half of the shell;* and again, in another work on the ' British Pal. Foss. in the Camb. Museum,' p. 192, 1852, the same author states the spiral appendages to be *very small,* but it so happens that I have never been able to find a single specimen out of the great number of *Martinias* which have passed through my hands, exhibiting such small spirals as described by the learned author.

I cannot conclude these few remarks without mentioning how much science is indebted to the late Baron von Buch,[1] M. de Koninck,[2] De Verneuil,[3] Prof. M'Coy, and others, for their numerous investigations of the species composing this extensive genus.

Geol. range.—True *Spirifers* first made their appearance in the lower Silurian epoch, and continued to exist all through the Palæozoic formations, the last well authenticated examples being found in the Trias?

Examples: Sp. Striatus, Martin, sp.; *Cyrtæna*, Dal.; *Anossofi*, Vern.; *Verneuili, Archiaci*, and *Bouchardi*, Murch.; *Mosquensis*, Fischer, sp.; *Pailleti, Rojasi* and *Pellico*, de Verneuil; *alatus*, Schl.; *glaber*, Sow.; &c. &c.

Sub-section A.—SPIRIFERINA, *D'Orbigny*, 1847.

Type—S. ROSTRATA, *Schl.*, sp.   Int., Pl. VI, fig. 60.

SPIRIFER (part) of the generality of Authors.
SPIRIFERINA, *D'Orb.*, M'Coy, &c.

Shell generally transverse; valves unequally convex, with or without a mesial fold; surface smooth or costated; beak straight or recurved; area commonly largely developed and interrupted by a pseudo-deltidium, notched in the vicinity of the cardinal edge; shell structure punctated; surface spinose; hinge line commonly shorter than the width of the shell; valves articulating by means of a strong cardinal tooth situated on each side at the base of the fissure, supported by strong largely developed vertical shelly plates, and corresponding with sockets in the interior of the smaller valve; the space intervening between the dental plates in the interior of the ventral valve is occupied by the cardinal muscles, which are divided by an elevated mesial septum, wide and thick at its base, but gradually tapering into the shape of an acute blade, to the sides of which the adductor was, no doubt, fixed; in the interior of the dorsal valve the shelly lamellæ destined to support the cirrated arms assume the form of two large spiral horizontal cones.[4]

Obs. Besides the important difference in shell structure, a largely developed mesial septum exists in the punctated species, but no such plate is known to occur in the impunctate forms.   It may, perhaps, therefore be preferable to follow M. d'Orbigny in adopting *Spiriferina* as a sub-section of *Spirifer*, a view already embraced by Professor M'Coy.[5]

[1] Über Delthyris, 1837; and Mém. Soc. Geol. de France, vol. iv, 2d part, 1840.

[2] Desc. des An. Foss. de la Belgique, 1842-44.

[3] Geol. of Russia and the Ural Mountains, vol. ii, 1845, &c.

[4] Some of the species belonging to this sub-section have been fully described in Part iii, under the genus *Spirifer*.

[5] It is possible, that the *Sp. heteroclyta*, Def., *Demarlii*, Bouch., *Hispanica*, D'Orb. sp., and some other Palæozoic species, which are closely punctated, should be admitted into this sub-genus in preference to *Cyrtia*, which is impunctate.

Geol. range.—The exact period at which the punctated *Spirifers* first appear is still doubtful; they are known from the Devonian epoch to the upper Lias, above which no example has been as yet recorded.

Examples: *Spiriferina rostrata*, Schl., sp.; *Tessoni, Deslongchampsii,* and *Munsteri,* Dav.; *oxypterus* and *signensis,* Buv.; *cristata,* Schl. sp., &c.; *Haueri* and *Emerici,* Suess.

Section B.—Cyrtia, *Dalman,* 1827.[1]

Type—C. EXPORRECTA, *Wahl.* Int., Pl. VI, figs. 61, 62 (63 and 64 ?).

SPIRIFER, of the generality of Authors.
CYRTIA, *Dalman, D'Orbigny, M'Coy,* &c.

Shell somewhat trigonal; valves convex; hinge line nearly as long as the width of the shell, articulating by means of teeth and sockets; dental or ventral valve very deep, more or less pyramidal; beak straight, or slightly recurved; area wide and triangular; fissure entirely covered by a convex pseudo-deltidium in one piece, generally perforated extremely close to the beak by a circular foramen; sometimes this deltidium is longitudinally depressed along its centre, from the extremity of the beak to within a variable distance of the cardinal edge; at the extremity of this shallow groove is seen a small circular aperture, which up to a certain age afforded passage to pedicle muscular fibres; *the socket or dorsal valve* is very slightly convex, and is supposed to have been furnished internally with spiral cones; in the *larger valve* a mesial longitudinal septum extends from the extremity of the fissure to a short distance of the margin, and to the sides of which the dental plates converge, and are united after having formed the fissure walls.

Obs. Since the genus *Cyrtia* has been admitted by a number of Palæontologists, it may be desirable to investigate the grounds for such a conclusion. Dalman's diagnosis is unsatisfactory, and equally applicable to several species of his genus *Delthyris*: we are informed that in *Cyrtia* the shell is *inequivalve, larger valve raised into a semi-cone or half pyramid, with the hinge side vertically flat, foramen circular, hinge line straight.*—*Examples. C. exporrecta* and *trapezoidalis.*

The shells here mentioned are varieties of one species, well known in England, which must therefore be considered as the type of the section, and from it our diagnostic has been extended. It should also be remarked, that a microscopic examination by Dr. Carpenter, has proved its shell structure to be fibrous, impunctate, and similar in character to that of *Spirifer proper* (*Striatus*). The only differences at present known consist in a slight modification in the character of the deltidium, and presence of a circular perforation for the passage of a pedicle; likewise in the direction of the dental plates, and

[1] *Petrefacta suecana,* in 'Kongl. vet. Acad. Handl.,' 1827.
[2] This character is beautifully displayed in some Devonian species found in Belgium and China.

presence of a mesial septum, which may perhaps entitle those and similar species to be placed together as a small sub-section subordinate to *Spirifer*, as judiciously proposed by Professor M'Coy.[1] It now remains to be ascertained whether all the species classed under *Cyrtia* really belong to the sub-section,[2] thus, for instance, the *Sp. heteroclitus, Demarlii*, and a few others, which exteriorly appear to resemble the type of *Cyrtia*, possess entirely dissimilar *shell structure*, which Dr. Carpenter has proved to be largely punctated; and these last would, therefore, have quite as valid a claim to be separated from *Cyrtia*, as *Spiriferina* was, by its structure and plates, from *Spirifer proper*. It is therefore evident that much more will require to be known of the internal and structural character of the numerous species of *Spirifer* before the *genus* can be *properly* and *definitely* subdivided; and all we can say at present is, that a portion of the species composing the great genus SPIRIFER is punctated, while, perhaps, the greater number are not so, and it will be interesting to ascertain, by a minute investigation of the known species, how far the differences in the shell structure are accompanied by internal modifications.

Geol. range.—The exact period at which *Cyrtia* first appeared is uncertain; the types of the sub-section lived during the upper Silurian era, and the last representatives, at present known, are stated to have existed during the Trias.

Examples: C. trapezoidalis and *exporrecta*, Dalman, *Murchisoni*, De Kon.;[3] *cuspidata*, Martin, sp.; *Calceola*, Klipstein, sp.; &c.

Genus—ATHYRIS, *M'Coy*, 1852.[4]

Type—A. TUMIDA, *Dal.* or HERCULEA, *Barrande*. Int., Pl. VI, figs. 71—76.

ATRYPA (part) of *Dalman, King*, and some other Authors.
ATHYRIS (part) *M'Coy*, 1844.
TEREBRATULA, *Sow.* (part) 1815, and of the generality of Authors.
SPIRIGERA (part) *D'Orb.*, 1847.

Animal unknown; *shell* of a variable shape, circular, elongated, or transverse; valves more or less unequally convex, with or without a mesial fold and sinus; beak *apparently*

---

[1] British Pal. Fossils in the Camb. Mus., 1852.

[2] In the 'Tableau du Prodrome,' vol. iii, p. 56, we perceive that M. d'Orbigny has placed under CYRTIA the following species: *C. calceola, cristata*, Schl.; *dorsata*, M'Coy; *hispanica*, D'Orb.; *laminosa, mesogonia, subconica, exporrecta, heteroclita*, and *trapezoidalis*; but we feel surprised that the author should have admitted such an assemblage, since he positively states every time he has had occasion to mention *Cyrtia*, that the shell structure was fibrous and impunctate; *S. cristata*, Schl., in particular has been known some years to belong to M. d'Orbigny's section *Spiriferina*.

[3] This species is found in Belgium, and has lately been obtained from China, by Mr. Hanbury, and presented to the British Museum.

[4] Brit. Pal. Foss., 1852.

imperforated,[1] incurved, and in general overlying the umbo of the smaller valve; no area or defined beak ridges; valves articulating by teeth and sockets; external surface commonly smooth. In the interior of *larger or ventral valve* the dental plates are fixed to, and along the sides of a longitudinal prominence or convex, arch-shaped plate, which extends to less than a third of the length of the shell, with its narrow end fitting into the extremity of the beak, and its lateral diverging edges to the bottom of the valve. In the free medio-longitudinal region, between the gradually declining base, or prolongations of the condyle plates, were situated the cardinal and adductor muscles; this last has left a small, elongated, heart-shaped scar, under and along the outer sides of which are seen the much larger oval impressions of the cardinal muscles. The interior of the *smaller or dorsal valve* is partly divided by a large, deep, longitudinal septum, which extends from the extremity of the umbo to about two-thirds of the length of the shell, supporting at its origin the hinge-plate, which is divided into two portions by a narrow gradually widening channel; to the socket ridges are affixed the spiral cones, the extremities of which are directed towards the lateral margin of the shell; on either side of the septum are seen two muscular scars formed by the adductor.

Obs. I felt greatly embarrassed as to the course advisable to pursue relative to the name *Athyris* proposed by Professor M'Coy for a few species which do not always agree with the *derivation* of the term, or with the short undetailed diagnosis then published, viz., "*nearly orbicular, small; no cardinal area or hinge line; spiral appendages very large, filling the greater part of the shell.*" The author names several examples, of which some are decidedly perforated, *T. concentrica* (De Buch); others apparently not so, while a few belong to the genus *Spirifer* of Sow.; it is true the Professor had then the erroneous belief that his first-named type *when perfect*, was imperforated, and therefore gave to his section a name implying such a condition. In 1847, M. d'Orbigny objected to the denomination, stating it to be "*in complete contradiction with the zoological characters,*"[2] and proposed as a substitute the name SPIRIGERA—*T. concentrica* of Baron de Buch, being his type. Shortly before (1835 or 1836), Professor King having received from the Eifel a specimen labelled *T. concentrica*,[3] but which was really an example of *T. scalprum*, erroneously described the Baron's species "*with condyle plates attached to a process resembling* a shoe lifter;"[4] impressed with that idea, and justly perceiving certain important differences in a Permian species (*T. pectinifera*), he proposed for this last a separate generic

[1] Although no well-defined foramen is visible in the generality of half grown or adult individuals, I have clearly traced the existence of a small circular aperture in some young examples, which no doubt became cicatrized, and dispensed with at a more advanced period of the animal's life.

[2] Paléont. Franç. Ter. Cretacés, vol. iv, p. 357, 1847.

[3] Professor King having most obligingly forwarded for my inspection his so-called *T. concentrica*, I at once recognised in it a well-known Eifel species, *T. scalprum* of Rœmer.

[4] Ann. and Mag. of Nat. Hist., vol. xviii, p. 86, 1846, in that paper Professor King ably describes the remarkable process above mentioned.

appellation, selecting for that purpose CLEIOTHYRIS of Phillips,[1] but not exactly in the sense intended by that author. It however so happens that the *true T. concentrica* and *T. pectinifera*, although different species, are essentially similarly organised, and belong to the same group, while on the other hand, Professor King's so-called *Athyris concentrica* possesses the characters of another section, of which *A. tumida* (Dal.,) or *T. Herculea* of Barrande, may serve as types. If we preserve Professor M'Coy's misnomer, *Athyris*, for *T. concentrica*, another name would be required for those species possessing the shoe-lifter process described by Professor King, but which have been united into the same group, both by M. d'Orbigny, Professor M'Coy, and others. On perusing the last-named author's recent publication,[2] I find therein a slight but *important* modification in the description of *Athyris*, viz., "*a strong mesial septum exists in the rostral portion of the entering valve; dental lamellæ are moderate; no foramen.*" Example, *A. tumida*, Dal. Such a diagnosis could in no way answer to *T. concentrica*, while, on the contrary, it quite agrees with the state of things observable in the group characterized by *A. Tumida*[3] and *Herculea*. I therefore think that it would be preferable to retain for the last-named species and similar forms with an apparently imperforate beak or closed foramen, variously disposed septa, and largely developed dental plates, the term ATHYRIS, M'Coy; and to adopt for those forms similar to *T. concentrica, pectinifera, Roissyi*, &c., the name SPIRIGERA, D'Orb., by which means the palpable misnomer of Professor M'Coy is partially got rid of, and the name *Athyris* still retained for shells more closely partaking of the author's diagnosis and intentions.[4] The species belonging to this section are easily distinguished externally from *Spirigera* by a longitudinal line, which extends along the smaller valve from the umbo to half or more of its length, and which denotes the presence of the internal

[1] A Mon. of English Permian Fossils, p. 137, 1849. Professor Phillips proposed the name *Cleiothyris* as a substitute for Dalman's *Atrypa*, but did not employ it in his work, ' Fig. and Desc. of the Pal. Fossils of Cornwall, &c.,' p. 55, 1841.

[2] British Pal. Foss. in the Camb. Mus., p. 196, 1852.

[3] It is worthy of notice that Sowerby was the first author who described and figured the exterior and spires in this species, under the name of *T. obtusa*, and which denomination seems to hold priority over that of *tumida*, given to the same species by Dalman in 1827.—See Sowerby; Some Account of the Spiral Tubes or Ligament in the Genus *Terebratula* of Lamarck, &c., read in 1815, and published in vol. xii of the 'Trans. of the Linn. Soc of London,' p. 515, pl. xii, fig. 3, 4.

[4] Before coming to the above conclusion, I submitted my views to M. Deshayes, Mr. Salter, and others, who seemed to consider that this mode of compromising the difficulty could not reasonably be objected to by the two authors principally concerned, nor by the generality of Palæontologists. I have always been of opinion, that if a name implies a radical zoological error it should be expunged, as the great object of science is to improve the nomenclature, and not to endeavour to *perpetuate an error*, even on the grounds of priority.

I must not omit to observe, that there exists a slight difference in the interior of *A. tumida* and *herculea* from the shoe-lifter process, assuming only a rudimentary state in the first species; but it seems to possess all the other *essential* characters of the genus, and I therefore agree with Professor King who proposes to place them both for the present into the same section.

septa, and also by the two diverging lines which extend from the extremity of the beak to about two-thirds of the length of the valve, indicating the presence of the large condyle plates and shoe-lifter process.[1] Professor Quenstedt forms for *Athyris* and *Spirigera* a small sub-section, which he designates by the name of TEREBRATULA SPIRIFERINA ;[2] but it seems much more convenient, as well as desirable, to give to each group a distinct appellation, and to place these last in the family *Spiriferidæ*.

Geol. range. — This genus seems in our present knowledge restricted to the Palæozoic era; appearing first in the Silurian period, it is abundantly distributed in the Devonian age, but until the interior of many doubtful species has been examined, it will not be possible to point out its exact range.

Examples: A. *tumida,* Dal.; *Herculea,* Barr.; *pseudo-scalprum,* Barr.; *Scalprum,* Rœm.; *plebeia,* Ph., sp.; &c.

Genus—SPIRIGERA, *D'Orb.*, 1847.

Type—T. CONCENTRICA, *Buch.,* sp.   Int., Pl. VI, figs. 65—70 and 78.

ATHYRIS and ACTINOCONCHUS, *M'Coy* (part).
TEREBRATULA (part) of the generality of Authors.
SPIRIGERA, *D'Orbigny.*
CLEIOTHYRIS, *King* (non *Phillips*).
RETZIA? *King.*

Animal unknown, *shell* inequivalve and variable in shape; circular, subquadrate, elongated, transverse, globose, or depressed, with internal spires; valves articulating by teeth and sockets; beak short, more or less incurved and truncated by a small round aperture lying contiguous to the umbo of the socket valve, or separated by a deltidium in two pieces; no true area; beak-ridges more or less defined; valves convex and divided or not by a mesial fold and sinus; surface smooth, striated or variously costated, and marked by numerous concentric lines of growth, sometimes produced considerably beyond the margins of the shell in the shape of foliaceous expansions. In the interior of the *smaller* or *dorsal* valve, the hinge-plate presents four depressions or pits, which afforded attachment to pedicle muscles; and close to the extremity of the umbo, a small circular aperture, appears at times to communicate with a cylindrical tube[3] which, after

---

[1] Several of the species belonging to this group have been figured and described by M. Barrande, in the 'Ueber die Brachiopoden der Sil. Schichten von Bœhmen, 1847.' I am likewise indebted to M. Suess, of Vienna, for much valuable information regarding *A. herculea,* and for the loan of several of his unpublished figures.

[2] Handbuch der Petrefackunde, p. 474, 1851.

[3] The discovery of this remarkable appendage in *T. concentrica* is entirely due to M. Bouchard, of Boulogne, that distinguished author having many years ago beautifully worked out all the internal characters connected with the type of this genus which abounds in the Devonian Beds of Ferque. The tube has not, however, been hitherto observed in any other species, and therefore we cannot affirm that

originating under the platform, extends longitudinally and freely with a slight upward curve to about a third of the length of the valve; to the inner extremities of the socket ridges are affixed the spiral horizontal cones, with their extremities directed towards the lateral margins of the shell; the crura appear to have been united before the lamellæ commence their spiral coils:[1] no defined septum is visible along the bottom of the valve, but a minute rudimentary mesial ridge which divides the quadruple impressions of the adductor.

In *the larger or ventral valve* the dental plates are more or less developed, exhibiting on and close to their inner sides, impressions of the pedicle muscles; the adductor leaves, a small oval scar separated by a minute mesial elevation under and outside of which are seen two other larger impressions due to the cardinal muscles.

Obs. Under ATHYRIS I have mentioned all my reasons for the limitation here introduced, and shall therefore only add one or two other observations. Professor King separated from his genus *Cleiothyris*, certain shells which at least for the present I should wish either to include with *Spirigera*, or to place in a section of this last. The author proposes for these the term *Retzia*, with the following diagnosis:—" *A Spiriferidia; in general oval longitudinally, ribbed or striated, with large punctures; larger valve foraminated at or near the apex of the umbone, with a large triangular area, and a closed fissure.* Type—*T. Adrieni*, De Verneuil; *T. Baylii, Salteri, and Bouchardii*, Dav. *T. Oliviani. T. ferita*, is likewise mentioned.[2] I do not pretend that some of the shells here mentioned may not require to be separated and formed into a small sub-sectional group, but hitherto nothing has been seen or said regarding their interior; and in consequence its distinctive claims have not been sufficiently tested: much, indeed, has yet to be achieved before a multitude of Palæozoic species can be placed into their proper places, and which time alone can accomplish.

Geol. range.—The existence of this genus seems perhaps to have exceeded that of *Athyris;* appearing in the silurian epoch, it has continued up to the lower Lias, above which no examples have been discovered.

Examples: S. concentrica, V. Buch, sp.; *S. pectinifera*, J. Sow.; *S. Roissyi*, Def.; *S. lamellosa*, Def.; *plano-sulcata*, Phil., sp.; *serpentina*, Kon., sp.; *Hispanica, Ferronesensis; Esquerra*, Vern.; *axiocolpos*, Emmerich,[3] &c.

it always occurred. Professor King has figured the aperture in the hinge-plate of *T. pectinifera*, and I doubt not, that with time it will be recognised in many other species.

[1] This character has been well figured by J. De Carle Sowerby in *S. pectinifera*. Min. Con. T. 616.

[2] A Monograph of English Permian Fossils, p. 137, 1849.

[3] This species occurs in the lower or black Lias of Koessen (Tyrol), and has been clearly investigated by M. Suess, of Vienna, who kindly sent me his figures, and a description of this remarkable shell.

Genus—UNCITES, *Defrance*, 1828.

Type—U. GRYPHUS, *Schl.*, sp. Int., Pl. VII, figs. 79—86.

TEREBRATULITES (part), *Schloth.*
UNCITES, *Defrance*, and of the generality of Authors.
GYPIDIA (part), *Dalman*, 1828.[1]
TEREBRATULA (part), *V. Buch*, &c. (non *Llhwyd.*)

Animal unknown. *Shell* oval, elongated; valves nearly equally convex; *beak* long, produced, tapering and incurved at its extremity, hollow, and truncated in young specimens by a small oval foramen; no true area, a large concave deltidium partly surrounds the aperture and extends to near the cardinal edge; the umbo of the socket valve is considerably incurved and concealed by and under the deltidium of the other valve; the sides of the beak as well as of the umbo become in some examples considerably deflected inwards, producing deep, lateral, elongated, concave depressions, or pouches opening externally, and not communicating with the interior; valves articulating by means of teeth and sockets. In the interior of smaller valve, the calcified supports of the arms form two conical spires, attached close to the socket walls by crural processes; surface smooth or striated.

Obs. The genus *Uncites* was proposed in 1828 by Defrance,[2] and has been very generally admitted, although its true character had not been established. Several authors have endeavoured to define the section, but have evidently failed in their attempts. M. d'Orbigny's diagnostic[3] is essentially defective, since the beak was not, as he supposes, entire, but *hollow* and *truncated* (at least up to a certain period of the animal's existence), by a regular foramen, from which issued the pedicle of attachment: this point was proved beyond doubt by a series of examples of all ages I obtained from Nimes, near Couvin in Belgium.

The internal details are not yet completely known. The muscular impressions remain still to be ascertained; but we are indebted to Professor Beyrich, of Berlin, for the knowledge of the *spiral supports*, which he was so fortunate as to discover in a specimen from Paffrath, figured, with his permission, in Pl. VII, fig. 85.[4] The labial appendages were not therefore simply attached at their origin, as supposed by M. d'Orbigny, nor does the genus belong to the family of *Rhynchonellidæ*, but to that of the *Spiriferidæ*, and should be placed between *Atrypa* and *Athyris* (Retzia). The two singular lateral pouch-shaped

[1] Dalman proposed the term *Gypidia*, as a substitute for *Pentamerus*, Sow.! his first example was the *T. gryphus* of Schl., which is not a *Pentamerus*. Von Buch's synonymes of Uncites are very defective, as are also all his observations on the species.

[2] Dic. des Sciences Nat., *Uncites gryphoides.*

[3] Paléont. Franç., vol. iv, p. 347, 1847.

[4] I am indebted for the sketch of Professor Beyrich's specimen to my zealous friend M. Suess of Vienna.

depressions do not occur in all the specimens; they have been long known, and were figured in Blainville's 'Malacologie.'

Geol. range.—*Uncites* is at present known to occur only in the Devonian system, and seems to characterise a certain horizon.

Examples: U. gryphus,[1] Sch. sp. ?; *lævis,* M'Coy.[2]

Genus—ATRYPA, *Dalman,* 1827.

Type—A. RETICULARIS, *Linn.*, sp.   Int. Pl. VII, figs. 87—94.

> ANOMIA (part), of *Linnæus.*
> TEREBRATULITES (part), of *Schloth.*
> TEREBRATULA, of the generality of Authors.
> ATRYPA, of *Dalman,* (part), *King* and *Sow.*
> SPIRIGERINA, *D'Orb.*, 1847, *M'Coy.*
> TEREBRATULÆ CALCISPIRÆ, *Quenstedt.*
> HIPPARIONYX (part), *Vanuxem.*

Animal unknown. *Shell* circular, transverse or elongated, furnished with internal spires, valves articulating by teeth and sockets; beak produced or incurved, and slightly truncated by a small round opening separated or not from the hinge-line by means of a deltidium; false area at times well defined, dental valve convex, or almost flat, with a longitudinal depression or sinus; *socket valve* convex, with or without a mesial fold; surface smooth, striated or variously costated and imbricated by squamose lines of growth, often considerably produced beyond the margin, under the shape of tubular spines or foliaceous expansions; structure fibrous and impunctate; *spiral appendages* originating at the base of the socket walls, and forming two large hollow cones placed horizontally, with their apices directed inwards and towards the hollow of the same valve, which they almost fill; the inner sides of the spires are pressed together and flattened, with their terminations close to each other near the centre of the bottom of the shell.   In the interior of the *socket valve,* the quadruple impressions of the *adductor* muscle are separated by a medio-longitudinal ridge; the pedicle muscles were probably affixed to the two small cardinal plates.   In the *dental valve* at the base of the teeth a semicircular ridge curves on each side, forming a saucer-shaped depression open in front, and into which were implanted the shell and pedicle muscles; the *cardinal* muscles seem to occupy the largest portion of the depression, and to have been divided by an obscure mesial ridge; beyond these and a little higher up are placed the pedicle muscular impressions, and above the mesial ridge, nearer the beak, is seen the oval scar left by the adductor.

---

[1] This species was figured and described by Beuth, as a *Terebratula,* in a work entitled 'Juliæ et Montium Subterranea,' &c., p. 134, No. 74, 1776.   Professor Quenstedt likewise partly describes and figures the genus in his 'Handbuch der Petrefackunde,' 1851, pl. xxxvi, fig. 40, *a, b, c.*

[2] British Pal. Fossils of the Camb. Mus., pl. ii, A, fig. 6, 1852.

The vascular impressions in the *dental valve* consists of two principal trunks originating on each side between the cardinal and pedicle muscles: these soon divide into two primary branches which extend right and left almost parallel to the margin, and giving off at variable intervals smaller bifurcating veins, which are directed towards the edge of the shell.

Obs. Dalman founded the genus *atrypa*,[1] which he characterized as "*Inequivalve biconvex, hinge-line rounded, beak of larger valve covering the base of smaller valve, apex imperforate.*" Naming a number of examples,[2] the first and second being *A. reticularis* of Linn. and *aspera* of Schl.; but, unfortunately, these shells happen to have *perforation*, although the aperture is very often concealed by the curvature of the beak; the term *Atrypa* is therefore a misnomer, implying a zoological mistake, and should rightly be expunged; but many authors appear anxious to retain the name, *as a simple denomination*, casting aside its derivation, and considering alone the *first* or *typical species*. I have consented (against my own inclination) to adopt the term on the understanding that it will be restricted to those shells possessing the well-defined character of *A. reticularis*. M. d'Orbigny's name, *Spirigerina*, published in 1847, must therefore be considered as a synonyme.[3]

Many important characters distinguish and connect this genus with other sections; thus the beak of *Atrypa* bears resemblance to that of some *Terebratulæ*, being perforated by a circular opening partly surrounded by a deltidium; but this foramen is not visible in all examples of the same species, from the beak touching and overlying the umbo of the other valve, the animal was therefore probably unattached or free during a portion of its existence. To *Rhynchonella*, it seems allied by the arrangements of its muscular system, but not by those of its labial appendages, which in the last-named genus were free or affixed only at their origin, while those of *Atrypa* were supported by large spiral appendages,[4] so that the section may be said to combine characters belonging to *Spirifer*, *Rhynchonella*, and *Terebratula*, so as to constitute an organisation different from each in some particular. Out of the list of species placed in *Atrypa* by Dalman, three only can be

[1] *Petrefacta suecana* in the 'Konigl. Vet. Acad. Handl.' for 1827, published in 1828 (α, privative, and τρυπα, foramen).

[2] As observed by Professor King, in the 'Ann. and Mag. of Nat. Hist.,' vol. xviii, p. 29, 1846. Dalman included some very different shells in his genus, *A. gallata*, belonging to Sowerby's *Pentamerus*, and *A. nucella* to Fischer's *Rhynchonella*, &c.

[3] In his work on the 'British Pal. Foss. of the Camb Mus.,' Professor M'Coy adopts M. d'Orbigny's name *Spirigerina* in preference to *Atrypa*, but retains his own name *Athyris*, the derivation of which is quite as objectionable as that of Dalman.

[4] The spires of *A. reticularis* and *prunum* have been known for several years, and figured by Blainville, V. Buch, and Quenstedt, &c.; the last-named author, in his 'Handbuch der Petrefackunde,' 1851, proposes to place *A. reticularis* and *prunum* in a group termed "*Terebratulæ calcispiræ*," but the shells in question differ so essentially from *Terebratula* proper, as to require a separate denomination, and to be removed from that family.

retained. The author also forms two divisions in his genus, viz. *striatæ* and *lævis;* *A. reticularis,* and *aspera,* being the types of the first, *A. prunum* of the second.

Geol. range.—The genus first appeared during the Silurian period, and continued in the Devonian epoch, above which no authenticated example has been observed.

Examples: *A. reticularis,* Linn. sp.; *aspera,* Schl.; *spinosa,* D'Orb.; *marginalis,* Dalman, sp.; *comata,* Barr.; *prunum,* Dalman,[1] &c.

FAMILY—KONINCKINIDÆ.

Animal unknown; *shell* free; valves unarticulated? oral arms supported by two lamellæ spirally coiled.

Obs. Only one genus and species is at present known, which had formerly been classed among the *Productidæ,* but from which it has been since removed, on account of its peculiar internal organisation. I have placed it directly after the *Spiriferidæ,* on account of the spiral lamellæ observed in its interior, which bear some resemblance to those of Atrypa.

Genus—KONINCKINA, *Suess,* MS., 1853.

Type—K. LEONHARDI, *Wissmann,* sp., 1841. Int., Pl. VIII, figs. 194—198.

PRODUCTUS, *Wissman,*[2] *Münster,*[3] *Klipstein, De Koninck,*[4] &c. (part.)

Shell nearly circular, inequivalve, compressed; *larger or ventral* valve convex or gibbous, with a slight longitudinal depression; beak considerably incurved, with auricular expansions; *smaller* or dorsal valve concave, following the curves of the other; surface smooth; no area or deltidium; valves unarticulated? in the interior of *dorsal* valve a mesial ridge extends from the cardinal process to the frontal margin; arms or oral appendages supported by a spiral calcified lamella?

Obs. All the authors who have described this remarkable shell have placed it in *Productus,* although it possesses, in reality, none of the internal characters peculiar to that genus. Dr. Klipstein[5] has observed that it possessed internal traces of the spiral arms, and figures large spiral ridges. Having had an opportunity of examining, along with

---

[1] M. d'Orbigny places this species into *his* genus *Atrypa* (Prod., p. 37), and it must likewise be observed, that the French author's genus *Atrypa* is made up of a strange assemblage of dissimilarly organised forms, and which do not possess the characters assigned to them in vol. iv of the 'Paléont. Franç.;' most of M. d'Orbigny's *Atrypa* are shells with spirals, and belong to the same section into which he places *tumida,* viz., SPIRIGERA.

[2] Beitr. zur Petrefacten, Von Münster, vol. iv, p. 18, pl. vi, fig. 21, 1841.

[3] Ibid., p. 68, fig. 24, *P. dubia?* 1841.

[4] Recherches sur les Animaux Fossiles, 'Mon. des Productus,' p. 167, pl. xvii, fig. 4, 1847.

[5] Beitr. zur Geol. Kenntn. der Ostl. Alpen., p. 236, pl. xv, figs. 20 and 21*a, Prod. alpina,* 1845.

Mr. Woodward, the German author's typical specimens now belonging to the British Museum, I communicated on the subject with my Viennese friend, M. Suess, who had likewise examined a number of examples under more favorable circumstances, and I was informed by him that the arms were really supported by free calcareous spiral lamellæ, which he had succeeded in taking out in single bits, and kindly forwarded the interesting sketches reproduced in my plate, also requesting me to adopt his generic appellation, *Koninckina;* so compressed are the valves that scarcely any room existed for the animal, and slight impressions of the arms seem to have been reproduced by the mantle on the larger or ventral valve; traces of the vascular impressions have also been discovered by M. Suess, but there still remains a few internal points to be made out relative to the hinge line and muscular system.

Geol. range.—Hitherto only known in the triassic beds of St. Cassian.

Examples : The only species is the *K. Leonhardi*, Wissman, sp.

FAMILY—RHYNCHONELLIDÆ.

Animal free, or attached by a muscular pedicle issuing from an aperture situated under the extremity of the beak, in the larger or ventral valve; oral appendages spirally rolled, flexible, and supported only at their origin by a pair of short-curved shelly processes; structure fibrous, and impunctate.

Obs. This family is composed of the three genera, *Rhynchonella, Camarophoria*, and *Pentamerus.* It was represented from the first epoch of animal life, and has continued to exist through the whole geological scale : two species are still found alive in our seas.

Genus—RHYNCHONELLA, *Fischer,* 1809.[1]

Type—R. LOXIA, *Fischer.* Int., Pl. VII, figs. 99—107.

ANOMIA (part), *Columna,* 1616, *Linnæus,* and some other Authors.
TRIGONELLA, *Fischer,* 1809.
RHYNCHONELLA, *Fischer,* 1809, (part), of *D'Orbigny* and the generality of modern Authors.
TEREBRATULA (part), *Deshayes,* and of many Naturalists.
TEREBRATULITES (part), of *Schlotheim.*
CYCLOTHYRIS (*C. latissima*), *M'Coy.*
ATRYPA (part), of *Dalman,* &c.
HYPOTHYRIS, *Phillips,* 1841, *Morris, King,* &c.
HEMITHYRIS (part), *D'Orbigny* and *M'Coy.*
ACANTHOTHYRIS (part), *D'Orbigny.*

Animal free, or attached to submarine objects by means of a pedicle; visceral mass confined to a small space near the umbones, and separated from the general cavity of the shell by a strong *aponeurotic* membrane. The mouth is situated in the centre of this membrane, supported by the apophysary processes of the dorsal valve; the upper lip is

[1] Notice des Foss. du Gouv. de Moscow, 1809.

plain, the lower cirrated; both are united at the sides of the mouth, and form long appendages (or *labial arms*), coiled up spirally, with their ends directed inwards towards the cavity of the dorsal valve. The alimentary canal takes the same course as in *Ter. australis*, passing through the deeply-notched hinge-plate, and ending behind the point of insertion of the adductor muscle in the centre of the ventral valve. The pallial veins (like those of *Rh. acuminata*, p. 95, figs. 33, 34, and *Camarophoria*, p. 97, figs. 35, 36,) are much narrower than in *Terebratula*, and more angular in their mode of bifurcation; there are four principal branches in each lobe, opening into large sinuses similar to those of *Orthis* and *Strophomena*, but smaller in extent. The margin of the mantle is fringed with a few short horny *setæ*. The muscles are essentially like those of Terebratula; their impressions are referred to in the description of the shell. *Shell* inequivalve, variable in shape, transverse or elongated, circular or trigonal; valves more or less convex, with or without a mesial fold and sinus; beak entire, acute, prominent, or so much incurved as to leave no free space for the passage of pedicle muscles; foramen variable in its dimensions and form, placed under the beak, exposed or concealed, entirely or partially surrounded by a deltidium, the aperture being sometimes completed by a portion of the umbo of the smaller valve; *deltidium* in two pieces, at times either extending in the form of a tubular expansion, or rudimentary; surface striated or plaited, rarely smooth; structure fibrous, impunctate. Valves articulating by means of two teeth in the larger (ventral), and corresponding sockets in the imperforated (dorsal) valve; apophysary system, in smaller or dorsal valve, composed of two short, flattened, and grooved lamellæ, separate and moderately-curved upwards, attached to the deeply divided hinge-plate; in the *socket valve* the quadruple impression of the adductor muscle is clearly defined, and separated by a short medio-longitudinal ridge (*s*); the pedicle scars occupy the small cardinal plates, between which is the small and narrow cardinal process. In the perforated valve the two strong diverging cardinal teeth are supported by dental plates extending to the bottom of the valve, and at the base of these a semicircular ridge curves on each side, forming a more or less defined saucershaped depression, into which were affixed the shell and pedicle muscles; these last leave two narrow elongated scars, close to the inner base of the dental laminæ, the remaining and largest portion being chiefly occupied by the *cardinal* muscles, which are longitudinally divided by a small raised ridge; above these again is seen a small oval scar due to the adductor.

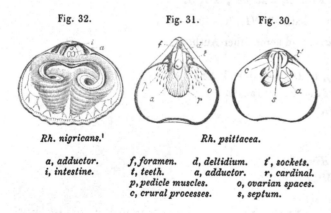

Fig. 32.   Fig. 31.   Fig. 30.

Rh. nigricans.[1]   *Rh. psittacea.*

a, adductor.	*f, foramen.*	*d, deltidium.*	*t', sockets.*
i, intestine.	*t, teeth.*	*a, adductor.*	*r, cardinal.*
	p, pedicle muscles.	*o, ovarian spaces.*	
	c, crural processes.	*s, septum.*	

[1] From Mr. Evans's specimen, see note to p. 57. The original was very unsymmetrical.

Obs. Having in Part III, p. 65, entered fully into the history and reasons for adopting the term *Rhynchonella*, in preference to other names subsequently proposed, I shall only add that I cannot coincide in Professor M'Coy's views on the supposed distinctions between this genus and that proposed by M. d'Orbigny, under the name of *Hemithyris*, since from an attentive examination of the interior of both *R. psittacea* (*Hemithyris*, D'Orb.) and *R. octoplicata* (*Rhynchonella*, D'Orb.), it will be easily perceived that both were provided with similar dental plates, and, in fact, with every essential external and internal character or detail of organisation; nor has M. d'Orbigny proved that his so-termed genus varied in any one important character. Professor M'Coy states that in *Hemithyris*, "the foramen is triangular, and not separated from the hinge," nor is it so in several admitted *Rhynchonellæ* (*R. concinna*, &c.), in which the foramen is partly completed by the umbo, the deltidial plates being lateral, and only rudimentary, as seen in the recent *R. psittacea* and *nigricans*; in the Palæozoic as well as Oolitic species we find some forms in which the beak is so much incurved, as to exhibit no aperture or passage for pedicle fibres, proving that the animal may have lived free during perhaps the whole, or a portion of its existence. Professor M'Coy is also mistaken while stating, in his 'Brit. Pal. Fossils,' that I refer *Hemithyris* to *Hypothyris* of Phillips; on the contrary, I clearly stated that I saw at that period no good reasons for adopting either of these denominations; besides which, Professor M'Coy gives examples of *Hemithyris*, which do not even possess the internal characters of his diagnosis.

Geol. range.—The genus *Rhynchonella* is one of the oldest types of animal life, having been repeated from the Silurian epoch up to the present period: two species are still found alive.

Examples: *R. acuta*, Sow., sp., *antidichotoma*, Buv., sp.; *concinna*, Sow., sp.; *Cuvieri*, D'Orb.; *decorata*, Schl., sp.; *furcillata*, Theodori, sp.; *plicatella*, Sow., sp.; *octoplicata*, Sow., sp.; *plicatilis*, Sow., sp.; *tetrahëdra*, Sow., sp.; *psittacea*, Chemnitz, sp.; *nigricans*, Sow., sp.; *Bouchardii*, Dav.; *acuminata*, Martin, sp.;[1] *pugnus*, Martin, sp.; *lacunosa*, Lin., sp.; *Wilsoni*, Sow., sp.; *spinosa*, Schl., sp.; &c. &c.

[1] In *R. acuminata*, the arms were attached to small curved lamellæ, similar to those of all the Rhynchonellæ; a beautiful example may be seen in the Museum of Practical Geology. I have thought that it might be interesting to introduce figures of two remarkable internal casts, on which the muscular scars, ovarian spaces, and vascular impressions are beautifully illustrated.

Fig. 33 is taken from a specimen belonging to Professor King: the view is drawn, as seen from the beaks, with the smaller or dorsal valve uppermost: A, *adductor;* R, *cardinal muscles;* O, *ovarian spaces;* V, *vascular impressions.*

Fig. 34 shows, in a beautiful manner, the impressions due to the vascular system in the dental or ventral valve, and has been drawn from a specimen in the collection of Mr. Morris.

Fig. 33. Fig. 34.

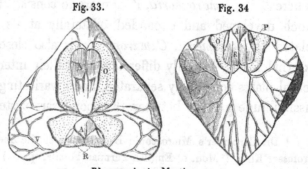

Rh. acuminata, Martin, sp.

Genus—CAMAROPHORIA, *King*, 1844.

Type—C. SCHLOTHEIMI, *V. Buch*, sp. Int., Pl. VII, figs. 108—113.

TEREBRATULA (part), of the generality of Authors.
RHYNCHONELLA (part), of *M. d'Orbigny*.

Animal unknown; *shell* of a subtrigonal shape, with convex valves longitudinally divided by a sinus and mesial fold; beak entire, acute, more or less incurved, under the extremity of which a small fissure is sometimes exposed; no area or deltidium; surface generally plaited, with or without marginal expansions; shell structure impunctate;[1] valves articulating by means of teeth and sockets.

In the *perforated valve*, the dental plates are conjoined at their dorsal margins, forming a trough-shaped process affixed to a low medio-longitudinal plate. In the *smaller valve*, the space between the sockets is occupied by a small cardinal muscular protuberance or boss, on either side of which two long slender processes curve upwards, to which were, no doubt, attached the free cirrated spiral fleshy arms; from beneath the cardinal process, a high vertical mesial septum extends to a little more than a third of the length of the valve, supporting along and close to its upper edge a spatula-shaped process, considerably dilated towards its free extremity, and projecting with a slight upward curve to nearly the centre of the shell.

Obs. We are greatly indebted to Professor King for the knowledge we possess of this genus.[2] Its affinities both to *Rhynchonella* and *Pentamerus* are obvious, and its natural place between the two did not escape the scrutiny of the learned Professor, who observed that the form of the trough and supporting plate of the larger or ventral valve in *Camarophoria* bears a striking resemblance to the same in *Pentamerus*, but that the apophysary system in the dorsal valve of the last-named genus "consists of two socket-plates largely developed, and passing to a considerable distance into the cavity of the shell. Whereas, in *Camarophoria*, it appears to consist of a medio-longitudinal plate, equally as much developed and expanded bilaterally at its free or upper margin. That although related to *Pentamerus*, *Camarophoria* is also closely allied to *Rhynchonella*,[3] especially in external form; the only difference is in their internal structure, *Rhynchonella* having the dental plates completely separated, and divaricating, as in *Orthis*, and the same amount of dissimilitude prevails between the apophysary system belonging to the small valve of both

[1] Dr. Carpenter's Microscopic Examination of the Shell Structure is different from that published by Professor King, ('Mon. of English Permian Fossils,' p. iv.)

[2] Refer for extended details to Professor King's description of the genus, in the 'Ann. and Mag. of Nat. Hist.,' July, 1846, and to the same authors 'Mon. of Permian Fossils of England,' p. 113, 1849.

[3] The term *Hypothyris* is here made use of by Professor King, but as that appellation is considered to be a synonyme of *Rhynchonella*, I have preferred this last to the name introduced by the Professor.

genera. In *Rhynchonella* it simply consists of the homologues of the crura of the loop starting from within the *socket plates,* which are rarely developed to any extent, offering, in this respect, a striking contrast with the large size of the corresponding structures of *Pentamerus."* But while agreeing so far with the founder of the genus, neither Mr. Woodward nor myself can entirely coincide with the views expressed on the muscular arrangement; the *adductor,* according to my learned friend, would have been fixed to the trough process of the larger valve, and to the spatula-shaped one of the smaller valve, while we feel disposed to consider that the second attachment took place at or close to the base of the septum in the smaller or dorsal valve;[1] the spatula-shaped process serving on the contrary as a visceral support; the cardinal muscles are supposed to have been fixed to the cardinal process or boss, and to near the base of the septum in the ventral valve. The vascular system is beautifully displayed on internal casts of *C. multiplicata,* as seen in the annexed diagrams, reduced from those published in the 'Monograph of English Permian Fossils.'

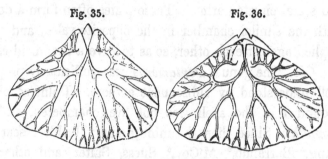

Fig. 35.   Fig. 36.

Camarophoria multiplicata.

Geol. range.—But few species of this genus have as yet been discovered, although I firmly believe their number will increase with future researches: for the present authenticated examples are only known to occur in the *Carboniferous?* and *Permian* deposits.

Examples: *C. Schlotheimi,* V. Buch, sp.; *C. globulina,* Phillips, sp.; *C. multiplicata,* King; and *C. crumena?* Martin, sp., according to Professor King, &c.

Genus—PENTAMERUS, *Sowerby,* 1813.

Type—P. KNIGHTII, *Sow.* Int., Pl. VII, figs. 114—119.

ANOMIA (part), of *Linnæus.*
PENTAMERUS, of the generality of Authors.
TEREBRATULA (part), of some Palæontologists.
ATRYPA (GALEATA), *Dalman.*
GYPIDIA (part), *Dalman.*

Animal unknown; *Shell* ovate, somewhat pentagonal, elongated or transverse, and articulated by means of teeth and sockets, rarely with any hinge line or area; *dental or ventral valve* in general the most convex, with or without a mesial fold or sinus; beak

[1] Professor King informs me, that he considers those impressions situated on each side of the septum in either valve of *Camarophoria* to be due to ovarian or genital organs, and refers to what is seen on a specimen of *R. acuminata,* of which we have given a woodcut (p. 95, fig. 33.) It appears to me that the rounded areas (fig. 35) on each side of the umbonal fissure, in the casts represented above, are more probably the ovarian spaces.

acute, entire and much incurved, concealing a triangular fissure rarely exposed except in the young state, and over which the umbo of the socket valve is greatly incurved; no area or deltidium; exterior smooth, striated, or ribbed. Inside the *larger valve*, two contiguous vertical septa coalesce into one median plate, extending from the beak to a greater or less distance; and then diverge and form the dental plates, inclosing a triangular chamber of much smaller dimensions than the lateral ones.

In the interior of the *smaller valve*, instead of a single median plate there are two distinct longitudinal septa of variable dimensions; (between which a small median ridge is occasionally found), to these the socket walls converge and join, forming two more or less developed and inclined plates, to the produced extremities of which were affixed the spiral cirrated arms. These plates often form a deep ∨ shaped chamber corresponding with the similar chamber in the opposite valve, and the edges of the two chambers are applied against each other, so as to leave a rhomboid cavity between them.

Obs. The genus, *Pentamerus* was judiciously proposed by Sowerby;[1] but not always well understood by subsequent authors. Dalman[2] objected to Sowerby's name, on the ground that the shell was not five chambered, and proposed as a substitute that of *Gypidia*. Since then the minute and valuable researches of M. de Verneuil,[3] Professor King,[4] Barrande,[5] M'Coy,[6] Suess, Salter, and others, have materially augmented our knowledge of the internal dispositions and affinities which *Pentamerus* bears to other genera. By the dispositions of the septa and dental plates in the ventral valve, it is closely related to *Camarophoria*. The position of the mesial plate and ∨ shaped process in the dental valve has been clearly shown, both by Baron v. Buch and Professor King, to be the equivalent of the mesial septum and dental plates of other genera; the dimensions of these varying almost in every species; they are most developed in *Pentamerus Knightii*, where the central septa extend nearly to the frontal margin, while in other forms, such as *P. lens*, the same plates are small and rudimentary;[7] affording, as justly remarked by Professor M'Coy, good specific characters. The same proportions and differences are likewise observable in the arrangements of the smaller valve; in *Pent. Knightii*, for example, the two sub-parallel longitudinal septa as well as the conjoined and inclined dental plates, are much extended and elevated, while in other forms they are considerably reduced, being almost rudimentary. Mr. Salter has observed that in *P. lens* the dental plates extend freely into the cavity of the shell, and that in (*P. liratus*)

[1] Min. Con., vol. i, 1813.   [2] *Petrefacta suecana*, in 'Kongl. Vet. Acad. Handl.,' 1827.

[3] Geology of Russia and the Ural Mountains, vol. ii, 1845; and Count Keyserling's Wissenschaftliche Beobachtungen, 1846.

[4] Annals and Mag. of Nat. Hist., vol. xviii, 1846; and Mon. of Brit. Permian Fossils, 1849.

[5] Uber die Brachiopoden der Silurischen Schichten, &c., 1847.

[6] Brit. Pal. Fossils in the Camb. Mus., 1852.

[7] See also the descriptions *P. lens, oblongus, undatus, microcamerus*, by M'Coy, 'Brit. Pal. Fossils in Camb. Mus.,' p. 209, &c.

they are produced in the form of free, long, and narrow lamellæ, to which the cirrated arms were of course attached. The exact position and form of the different muscles has not been yet completely made out; but in some species the quadruple impressions of the adductor are clearly defined.

Geol. range.—The genus first appeared in the lower Silurian epoch, and continued, uninterruptedly, to the carboniferous period, above which no authenticated example has been yet discovered.

Examples : P. Knightii, Sow.; *conchidium*, Linn., sp.; *borealis*, Eichw., sp.; *galeatus*, Dal., sp.; *Vogulicus*, Vern.; *linguiferus*, Sow.; *Sieberi*, Barr.; *oblongus*, Sow.; *oblatus*, Barr.; *pelagicus*, Barr.; *caducus*, Barr.; &c.

SUB-FAMILY? PORAMBONITIDÆ.

Animal unknown, but evidently attached, at least during a portion of its existence, by a muscular pedicle; no calcified process for the support of the oral appendages, which were probably fleshy and spirally coiled. In the interior of each valve two diverging septa originate from within the beaks; structure fibrous and impunctate.

Obs. As so little is known of the internal organisation of the curious shells composing the genus *Porambonites*, I have not ventured to follow M. d'Orbigny and Prof. King, in placing it in the family *Rhynchonellidæ*; but preferred, for the present, to leave it by itself in a small sub-family. Its place is, however, I believe between the *Rhynchonellidæ* and *Strophomenidæ*.

Genus—PORAMBONITES, *Pander*, 1830.[1]

Type—P. ÆQUIROSTRIS, *Schl.*, sp.   Int., Pl. VII, figs. 120—126.

TEREBRATULITES (ÆQUIROSTRIS), *Schl.*
TEREBRATULA (part), *V. Buch.*
SPIRIFER (part), of *Eichwald* and *V. Buch.*
PORAMBONITES, of *Pander, D'Orbigny, M'Coy*, and *Sharpe.*
SPIRIFERES ANORMAUX, A. ÆQUIROSTRES, *De Verneuil.*[2]
ISORHYNCHUS, *King*, 1849.[3]
ORTHIS (part), *Quenstedt.*

Animal unknown; *shell* circular, transverse or elongated, globose, sub-equivalve; valves articulating by teeth and sockets; beaks slightly unequal, rather more produced in dental valve, with a small area in each, generally rudimentary, and perforated by a

[1] Beiträge zur Geog. des Russischen Reiches, 1830.

[2] In the Geol. of Russia and the Ural Mountains, vol. ii, 1845. M. de Verneuil places the shells belonging to Pander's genus among his *abnormal Spirifers*, forming two subdivisions, which he distinguishes as *A. æquirostres (Porambonites)* and *B. Biforés*, these last belonging to the genus *Orthis*.

[3] From being unacquainted with Pander's type and priority, Professor King proposed, in 1849, the name *Isorhynchus*, which the author now admits to be a synonyme.

mesial elongated triangular fissure, which truncates the beak of the larger valve before extending to the hinge line; surface pitted, but the shell structure impunctate;[1] in the interior of *larger* or *ventral valve* the dental plates form two slightly elevated and diverging septa, extending to a greater or less distance along the bottom of the shell. In the other valve the socket plates form likewise diverging septa, extending also to a variable distance.

Obs. The genus *Porambonites* was proposed by Pander for a remarkable group of shells which abound in the neighbourhood of St. Petersburgh, but from the difficulty of obtaining interiors, they have been differently interpreted by several authors. M. de Verneuil seems uncertain as to the place these shells should occupy, since he states that he considers them to be nearer related to *Spirifer* than *Terebratula*, although removed by various characters, and principally by the septa of the socket valve, so that they might equally well be classed with *Orthis*. Since then M. d'Orbigny[2] has replaced the shells in Pander's genus, and admitted them into his family of *Rhynchonellidæ*, from a belief that the free spiral fleshy arms were probably fixed to the socket plates. From the beaks being in general so much incurved as almost to come in contact, the extent of the foramen or fissure is concealed, so that some authors, as for instance M. d'Orbigny and Professor King,[3] state only one of the beaks to be foraminated, while, on the contrary, M. de Verneuil,[4] Professor M'Coy,[5] Dr. Volborth,[6] and Professor Quenstedt,[7] believe that both possessed a foramen or fissure. Some additional light on the internal arrangements has been obtained through Mr. Sharpe's fortunate discovery of several internal casts of a Portuguese species (*P. Ribeiro*); we have figured, with that gentleman's permission, in Pl. VII, figs. 225-26, one which exhibits, in a beautiful manner, not only the dental and socket plates, but the position of the muscles and vascular impressions, not hitherto observed in any of the other forms; but the internal details of the different Russian species, require still further examination before the exact character of the genus will be completely known.

Geol. range.—The genus seems to be confined to the lower Silurian period.

Examples: P. *æquirostris*, Sch., sp.; *intercedens*,[8] Pander; *reticulatus*, V. Buch, sp.; *Ribeiro*, Sharpe, &c.

[1] The depressions or dots visible on the surface of the different species were found by Dr. Carpenter to be simply due to external sculpture, the shell structure being impunctate.

[2] Considérations Zool. et Geol. sur les Brach., in the 'Comptes rendus Heb. de l'Acad. des Sciences de Paris,' 1847; and Paléont. Franç. Ter. Crétaces, vol. iv.

[3] A Monograph of English Permian Fossils, p. 112, 1849.

[4] Geol. of Russia and the Ural Mountains, vol. ii, 1845.

[5] British Pal. Fossils in the Camb. Museum, p. 212, 1852.

[6] This gentleman has informed me, "that the *area* is in general so small as hardly to deserve the name, but that the triangular fissure is large and the teeth developed."

[7] Handbuch der Petref., p. 486, 1851. The author states that both beaks are perforated at their extremities: he likewise illustrates a fragment of the interior of one of the species, in pl. xxxix, fig. 5.

[8] Pander describes many species which have turned out to be mere varieties and synonymes. It is likewise doubtful, if the *Sp. Tcheffkini* of De Verneuil belongs to this genus.

Family—STROPHOMENIDÆ.

Animal unknown, some of the group appear to have lived free; others were fixed during the whole or a portion of their existence by means of a muscular pedicle; no calcified supports for the arms, which were no doubt fleshy and spirally coiled; shell with a straight hinge line, and a low triangular area in each valve; valves both convex, or with the dorsal or ventral one concave or convex; shell structure fibrous or punctated.

Obs. Much difference of opinion has been expressed as to what genera the present family should include, as well as to the name it should retain. Some authors, M. d'Orbigny, Professor M'Coy, &c. prefer that of *Orthisidæ*, (properly *Orthidæ*) while others, such as Professor King, advocate *Strophomenidæ*. I have admitted this last denomination more on account of its priority, than on any other consideration. As to the subdivision of the group, we have considered, that in the present state of our knowledge, it would be preferable to restrict the number of genera to four, viz., *Orthis*, *Orthisina*, *Strophomena*, and *Leptæna*. Professors M'Coy and King have admitted within its limits *Chonetes*, a genus, as I hope to prove, more nearly related to the *Productidæ*, and I am happy to be able to state, that the last-named author has expressed approval of my conclusions. *Porambonites* has likewise by some been placed in the present family; and although I am not prepared positively to deny such a view, I would defer its admission until more is known regarding its internal organisation. M. d'Orbigny is decidedly in error when placing *Leptæna* in one family, and *Strophomena* in another.

This is one of the oldest families, having originated in the lower Silurian epoch, and continued to exist to the end of the Liasic period, where hitherto the latest representatives have been found.

Genus—ORTHIS, *Dalman*, 1827.

Type—O. CALLIGRAMMA or O. ELEGANTULA, *Dal.*[1] Int., Pl. VII, figs. 127—135, and Pl. VIII, figs. 136—148.

HYSTEROLITHUS, of *Worm*, 1655; of *Linnæus*, 1753.
ANOMIA (part), of *Linnæus*.
ORTHIS, of *Dalman*, and the generality of Authors.
SPIRIFER of some Authors.
TEREBRATULA (part), of *Deshayes*.

---

[1] Dalman's first mentioned example is the *O. zonata*, which he figures with an *open* fissure. M. de Verneuil refers this species to *O. ascendens*, Pander; but in this last, the fissure is entirely covered by a deltidium, and does not agree in this respect with Dalman's figure. *O. callactis* or *O. calligramma* may be considered types.

PRODUCTUS (part), of *Pander* (not of *Sow*).
PLATYSTROPHIA, *King*.
DICŒLOSIA, *King*.
SCHIZOPHORIA, *King*.[1]

Shell variable in shape, subcircular or quadrate; valves equally or unequally convex; socket valve sometimes slightly concave, with or without a mesial fold or sinus; hinge line straight, generally shorter than the width of the shell; both valves furnished with an *area* divided by a triangular open fissure for the passage of the pedicle fibres; beaks more or less incurved, that of the larger valve generally more produced; surface smooth, striated, or ornamented by simple, bifurcated, or intercalated ribs; structure minutely or largely punctated; valves articulating by means of teeth and sockets. In the interior of the *larger* or *ventral valve* the vertical dental plates form the walls of the fissure, and extend from the beak to the bottom of the shell; between these a small rounded mesial ridge divides the muscular scars which extend over two elongated depressions margined on their outer side by the prolonged bases of the dental plates; the cardinal muscles appear to have occupied the greater portion of the anterior division of these two depressions, the pedicle muscles occupying the external and posterior part of the same space; the adductor was probably attached to each side and close to the mesial ridge. In the socket valve the fissure is partially or entirely occupied by a more or less produced simple shelly process,[2] to which were affixed the cardinal muscular fibres; the inner socket walls are considerably prolonged into the cavity of the shell, under the shape of projecting laminæ, to the extremity of which free fleshy spiral arms may, perhaps, have been affixed. Under this shelly process a longitudinal ridge separates the quadruple impressions of the adductor, which on each side forms two deep oval depressions, placed obliquely one above the other, and separated by lateral ridges branching from the central one; the pallial vessels, as well as their numerous minor bifurcations or veins, have often left impressions within the valves; the principal trunks seem both more numerous and better defined in the socket valve,—after extending in a somewhat radiate or sub-parallel direction from the muscular scars to near the anterior region, they sweep round sub-marginally on both sides of the valve, leaving wide ovarian spaces, and giving off a series of smaller branches.

[1] I am requested by Professor King to state that from not having been acquainted with Dalman's original work 'Petrefacta Suecana,' in the Proc. of the Royal Acad. Sc. of Stockholm, he had not perceived that the author had placed marks of doubt after his first two types of *Orthis*, otherwise he would not have proposed the term *Schizophoria*, which he now considers as a synonym of *Orthis*. This mistake he attributes to Hisinger having omitted Dalman's marks of doubt.

[2] Some authors term the bifid or trifid cardinal process a "*rostral tooth;*" but I strongly object to such a denomination, since the only true teeth are those existing in the dental valve, and the shell does not articulate by means of the bifid projection, which was in all probability destined, as in other genera where it exists, for the attachment of the cardinal muscles.

Obs. In 1827, Dalman proposed the genus Orthis for a group of shells which he compared to the *Scallop*, but placed a mark of doubt after his first two types, viz., *O ? pecten* and *O ? striatella*, his examples being in bad condition, so that he was not certain if they really belonged to the genus; but to his following types he placed no mark of doubt. *O. zonata, callactis, calligramma, elegantula*, &c., must therefore be considered the typical forms of *Orthis;* besides which, the first two species mentioned belong to different genera, neither of which agree with such shells as *O. zonata* or *O. calligramma.*

The genus *Orthis* forms a well-characterised natural group, abounding in species and easily recognised. In 1848,[1] I stated that the species described by various authors as *Spirifer biforatus* or *Lynx*, and *Sp. bilobus* or *cardiospermiformis*, did not belong to the genus *Spirifer*, but to that of *Orthis*, since they possess *all* the essential characters of Dalman's genus, and none of those peculiar to the *Delthyridæ* or *Spirifers*. My views were adopted by M. de Verneuil,[2] D'Orbigny,[3] Professor M'Coy,[4] and Mr. Salter;[5] but Professor King not being convinced of their correctness, proposed, in 1849, the appellations of Platystrophia for *O. biforata*, and Dicœlosia for *O. biloba*,[6] but to which sections the learned Professor does not I believe, at present, attach such great importance. I still, therefore, maintain my first view, which will be better understood by a glance at the respective figures (Pls. VII and VIII).

Geol. range.—*Orthis* is one of the oldest forms of animal life, having first appeared in the lower Silurian epoch, and continued uninterruptedly to the carboniferous period, above which we are not at present acquainted with any well-authenticated example, although some authors have most erroneously quoted it as occurring in the cretaceous, tertiary, and recent state; but an examination of the types of those so-called *Orthides* will soon convince any one that they do not possess the essential characters of the genus.

Examples: O. calligramma, Dal.; *testudinaria*, Dal.; *elegantula*, Dal.; *biloba*, Linn., sp.; *hybrida*, Sow.; *rustica*, Sow.; *Lewisii*, Dav.; *Bouchardii*, Dav.; *æquivalvis*, Dav.; *Michelini*, Lev.; *opercularis*, Vern.; *reversa*, Salter; *Gervillei*, Def., sp.; *palliata*, Barr.; *mulus*, Barr.; *Beaumontii*, Vern.; *obtusa*, Pander, sp.; *Dumontieri*, Vern.; *striatula*, Schl., sp.; *Lynx*, Eich.; *biforata*, Schl.; *resupinata*, Martin, sp.;[7] *Lyelliana*, De Koninck; &c.

---

[1] Bull. de la Soc. Géol. de France, vol. xxi, 2d ser., 1848.   [2] Ibid.

[3] Prodrome de Paléontologie Stratigraphique, vol. i.

[4] British Palæozoic Fossils of Mus. Camb., 1852.

[5] Mem. Geol. Surv., vol. ii, part 1, 1848.

[6] A Monograph of British Permian Fossils, 1849.

[7] Among the Carboniferous Brachiopoda in the Museum of the Geol. Soc., may be seen two very remarkable *ventral* or *dental* valves, which had been discovered by Mr. J. Yates, at Llanymynech, and which I am inclined to believe may have belonged to a species of *Orthis*, bearing great external resemblance to that form known by the name of *O. resupinata*, Martin, sp. These valves are most remarkable from their very large and deep *cardinal* and *adductor* muscular impressions, the last mentioned

Genus—ORTHISINA, *D'Orbigny*, 1849.

Type—O. ADSCENDENS, *Pander*, sp., 1830. Int., Pl. VIII. figs. 149—156.

ANOMITES, *Schloth.* (part).
ORTHIS (part), of the generality of Authors.
HEMIPRONITES and PRONITES, *Pander*, 1830.
ORTHISINA, *D'Orbigny, King, M'Coy.*
STREPTORHYNCHUS, *King*, 1849.

Animal unknown; *shell* semi-circular or subquadrate, with unequally convex valves, external surface variously striated and marked by numerous concentric lines of growth, sometimes projecting in the shape of pectinated lamellar expansions; shell structure impunctate; hinge-line straight, not quite so long as the width of the shell; area double, larger in the dental valve, straight or bent back at right angles to the valves; the fissure in both is entirely covered by a convex deltidium, which in the larger valve of a few species is perforated near its extremity by a circular foramen; the valves articulate by means of teeth and sockets. In the interior of the *dental* or *ventral valve*, the muscular impressions occupy an area bordered by the raised edges of the dental lamellæ, and divided by a short mesial ridge. In the interior of the *socket* or *dorsal valve*, a longitudinal median ridge occurs, and the adductors, in the shape of two pairs of transversely oval scars, are placed one above the other on each side of this ridge.

Obs. Palæontologists do not seem yet to have completely agreed as to the propriety of separating the shells above described from ORTHIS. M. de Verneuil[1] has placed them in several of his sub-divisions of Dalman's genus: they constitute his *Rectostriatæ*, with their fissures closed by a deltidium; but this last-named character, combined with other internal differences as well as the structure of the shell, renders it very desirable to separate the shells under notice from ORTHIS proper, which has always the fissure in both valves entirely open, with widely divergent and prominent inner socket ridges, a small cardinal process, and a slight difference in the muscular impressions. Pander[2] divided his species of *Orthis* into several sections, viz.: *Orthambonites, Gonambonites, Hemipronites,* and *Pronites,* the last two or three names referring to forms with a covered deltidium.

being divided by a more or less elevated mesial ridge or septum. The enormous thickness of the valve may be judged from the following measurements:—

Length of the ventral valve	29 lines.
Width	30 ,,
Thickness of the valve	8 ,,

---

[1] Geol. of Russia and Ural Mountains, vol. ii, 1845.
[2] Beiträge zur Geognosie des Russischen Reiches, St. Petersburg, 1830.

Seventeen years later M. d'Orbigny proposed the term ORTHISINA for exactly the same shells;[1] nor does he allude to the Russian author, whose name *Pronites* or *Hemipronites*, might have saved the necessity of a fresh appellation.[2] M. d'Orbigny states that he is only acquainted with three species belonging to his section, viz., *O. anomala*, *O. Verneuili*, and *O. adscendens*, and seems impressed with a false idea as to the value of the temporary foramen,[3] which in middle-aged shells is rarely visible. I feel disposed to extend the limits of *Orthisina*, by admitting therein a number of so-called *Orthis*, with closed fissures, such as *O. plana*, *O. pelargonata*, *crenistria*, and other similar forms, which would not, I think, be properly placed in *Orthis* or *Strophomena*. The section *Orthisina* seems to connect the two last-named genera, and would be found convenient in arranging and separating a number of species which could not be united without destroying the clearly-defined characters of both *Orthis* and *Strophomena*.

Geol. range.—*Orthisina* first appeared in the lower Silurian period, and continued to be represented in the Devonian, Carboniferous, and Permian epochs, above which no authenticated species has been hitherto recorded.

Examples: *O. adscendens*, Pander, sp. = *excelsa*, id. *O. Verneuili*, Eichw., sp.; *O. anomala*, Schl., sp.; *O. plana*, Pand., sp.; *O. hemipronites*, V. Buch., sp.; *O. crenistrias*, Phil., sp.? *O. pelargonatus*, Schl., sp., &c.

Genus—STROPHOMENA (*Rafinesque*), *Blainville*, 1825.[4]

Type—S. RUGOSA, *Raf.*?[5] = S. PLANUMBONA or S. ALTERNATA.   Int., Pl. VIII, figs. 157—175.

ANOMIA (part) of *Linnæus*.
PERIDIOLITHUS, *Hüpsch*.

---

[1] Considérations Zoologiques et Géologiques sur les Brachiopodes, Comptes rendus de l'Académie des Sciences, Paris, 1847.—Ann. des Sciences Nat., vol. viii, pl. vii, fig. 30.—Paléont. Franç. Ter. Crétacés, vol. iv, p. 339.—Prodrome, vol. i, p. 16.

[2] I felt embarrassed as to what name to preserve, and therefore consulted several Palæontologists, among others, Messrs. Salter, Woodward, and Morris, who agreed that as Pander had assembled a number of different things in his sections, and applied to the same shells several generic names, without pointing out their characters; and as M. d'Orbigny's name *Orthisina* had already been adopted by several authors, it was advisable to retain it in preference to those of Pander.

[3] The presence or absence of the foramen is of comparative little generic valve, that character being common to other sections, although in this case differing both from *Strophomena*, *Terebratula*, and *Rhynchonella* by its position, the aperture being entirely excavated in the substance, and near the extremity of the deltidium, and not in the beak itself, as in many other genera.

[4] Nobody has hitherto produced the American author's description or date; but Blainville describes the genus *Strophomena* in his 'Manuel de Malacologie,' in 1825, "*Coquille équilatérale, reguliere, subéquivalve; ayant une valve plate, l'autre un peu excavée, articulation droite, transverse, offrant à droite et à gauche d'une subéchancrure médiane, un bourrelet peu considérable crénelé ou denté transversalement, aucun indice de support.*"

[5] No one seems to have clearly pointed out or identified with certainty *S. rugosa* of Raf.

TEREBRATULITES (part), *Schloth.*
STROPHOMENA, *Raf., Blainville,* and other Authors.
LEPTÆNA, *Dalman,* and of the generality of Authors.
LEPTAGONIA, *M'Coy.*
ORTHIS (part) of *V. Buch.*
PRODUCTA (part), of *Phillips* and *Deshayes.*

Animal unknown; *shell* depressed, expanded, semicircular, sub-quadrate, transverse or elongated; external surface smooth, variously striated or costated. Hinge line straight, generally as long as the width of the shell; *dental* or *ventral* valve convex or concave; *socket* or *dorsal* valve following the curves of the other, the margin in some species being suddenly and abruptly bent down when the shell is half grown; *area* double, crenulated at its inner edges, and more developed in the dental valve; fissure in larger or ventral valve partly covered by a deltidium, the extremity of the beak entire, or perforated by a small circular aperture which became closed at a more advanced period of the animal's life. In the *socket valve*, the fissure is either partially covered by a deltidium, or entirely occupied by a projecting cardinal boss. In the interior of the *dental valve*, two widely-diverging teeth articulate with corresponding sockets in the other valves; the muscular impressions are more or less distinctly margined by a semicircular ridge continued from the base of the teeth, and curving on either side so as to produce a saucer-shaped depression of variable dimensions. The *cardinal* muscles fill on either side the anterior portion of this cavity; the pedicle muscles leave no definite scars, unless the external portions of the "cardinal muscular impressions" are due to them. The adductor lies close to a slightly elevated mesial ridge. In the interior of the *smaller* valve the cardinal process or boss is large and divided into two lobes, and not connate with the divergent socket ridges; to this process were no doubt affixed the cardinal muscles,. From this a slight mesial ridge runs down, and separates the two pairs of adductor scars, which are frequently bordered by prominent ridges; the vascular impressions consist of large primary vessels which run at once direct to the margin, following a somewhat radiate arrangement, or else the lateral trunks are greatly enlarged and reflected to surround the ovarian spaces.

Obs. Much has been written on the shells above described, and many have been the opinions entertained, and sub-divisions proposed, for Rafinesque's and Dalman's genera *Strophomena* and *Leptæna.*[1] Some authors have adopted one name in preference to the other for the same species, while a few would preserve both, but all are not agreed in what manner the division should be made. It will not be necessary to enter into the dis-

---

[1] *Petrefacta suecana,* in Kongl. Vet. Acad. Handl., 1828.

  1. *Leptæna,* "sub-equivalve, flattened, with the margins compressed and bent, hinge line straight, very wide; foramen O.; one valve with two blunt teeth. First example, *L. rugosa,* His. Second, *L. depressa,* Sow. Dalman speaks of Leptæna, as equal to part of Sowerby's genus *Productus,* which he abolishes. In the supplement to the *Mineral Conchology,* Mr. James De Carle Sowerby adopted *Leptæna* as a substitute for *Productus!*

cussions that have taken place on this subject,[1] a few observations, however, may be admissible. No author seems to have as yet discovered *substantial characters* by which the shells placed in *Strophomena* or *Leptæna* can be generically divided; those hitherto proposed, as justly observed by Professor M'Coy, "*do not seem worthy of generic rank.*"

After a careful examination of a great number of species I agreed, with Mr. Salter, to maintain both *Strophomena* and *Leptæna* subject to certain limitations, but we could not prefix to this last *many* very important distinctive characters. STROPHOMENA would include all species agreeing with *S. planumbona, alternata, grandis, filosa, euglypha, funiculata, antiquata, pecten, expansa, depressa,* &c. LEPTÆNA we would typify by such forms as *L. transversalis, sericea, transversa, oblonga, Davidsoni, tenuicincta,* &c. These two genera thus formed are well distinguished by habit, but the distinctive characters rest chiefly on the form of the muscular impressions and cardinal process.

Professor King, on the other hand, is desirous of dividing the above-named shells into three sections or genera, observing that the external and internal characters of *L. alternata, depressa,* and *transversa,* or *transversalis,* are sufficiently distinct to admit of the names *Strophomena, Leptæna,* and *Plectambonites,*[2] being advantageously applied: thus he proposes to typify the first by *S. alternata.* " *Valves regularly plano-convex; foramen, when existing, partly edged by the deltidium, pallial vessels running at once almost direct, or in a somewhat radiate manner from the muscular impressions to the front"* (see Pl. VIII, fig. 160). *Leptæna,* he typifies by *L. depressa,* Dal.; " *Valves geniculated, wrinkled; umbone of the dental valve foraminated; pallial vessels running parallel with each other till they have reached to near the anterior region, where they sweep round submarginally on both sides, giving off a series of smaller vessels."* But the learned Professor admits into his section *Strophomena,* shells essentially geniculated, such as *S. euglypha* and *imbrex;* and the

Fig. 37.   Fig. 38.

Strophomena analoga.[3]
Interior of the ventral and dorsal valves.
a, ovarian spaces.   e, foramen.   t, teeth.

---

[1] Refer to Professor M'Coy's Synopsis of the Carb. Limest. Fossils of Ireland, p. 104, &c., 1844; and Brit. Pal. Foss. in the Woodw. Mus., p. 232, 1852.—Professor King's paper on the Palliobranchiata, 'Ann. and Mag. of Nat. Hist.,' vol. xviii, p. 36, 1846; and the same author's Monog. of English Permian Fossils, p. 103, &c., 1849.—Mr. Salter's Palæontological Appendix, in the 'Memoirs of the Geol. Survey of Gr. Britain,' vol. ii, pt. 1, p. 377, 1847.—Mr. Sharpe; Quart. Journal of the Geol. Soc., vol. iv, p. 178, 1848.—M. d'Orbigny; Pal. Fr. Ter. Crétacés, vol. iv, p. 335, 1847; and Prodrome, vol. i, 1849.—De Verneuil; Bull. de la Soc. Géol. de France, vol. v, May, 1848, &c.

[2] This section was proposed by Pander, in his work entitled 'Beiträge zur Geognosie des Russischen Reiches,' 1830, his types being *P. planissima, transversa, lata, imbrex, oblonga,* &c., figured in his pl. xix.

[3] These figures are reductions from those published in Professor King's 'Monograph of the Permian Fossils of England,' pl. xx, figs. 6, 7.

value of the character to be derived from the pattern of the vascular impressions is considerably reduced, from the positive fact of a species, *S. expansa*, Sow.,[1] presenting along with the hinge and general habit of *S. alternata*, an arrangement of the vessels more like *L. depressa* or *analoga*, which, with a slight modification, are reflected, as in the last-named species, to surround the ovarian spaces, &c. (Pl. VIII, fig. 162.) Mr. Salter and myself were of opinion that we might, perhaps, in the present state of our knowledge, divide the genus *Strophomena* into three sub-sections, thus :—

Sub-section 1. *S. alternata, planumbona, grandis, filosa, euglypha, funiculata,* &c.

2. *S. expansa* (the only species with certainty known.)

3. *S. depressa, analoga, Bouei,* &c. (*Leptagonia,* M'Coy).

But I think it will be preferable not to introduce these divisions until more is known of the pattern of the vascular impressions and other characters of the hitherto unexamined species, as well as of their general habit. The area in some forms, such as *S. latissima,* Bouch.; *Naranjoana,* Vern., &c., has no deltidium or fissure, and yet the internal characters agree with other *Strophomenæ.*

Geol. range.—This genus contains some of the oldest forms of animal life, abounding in the lower and upper Silurian, Devonian, and Carboniferous periods, but it is not positively known above that epoch.

Examples: with the socket valve concave; *L. alternata, depressa, analoga, imbrex, Loveni;* &c.

with the socket valve convex; *L. euglypha; funiculata, antiquata, planoconvexa, planumbona, pecten, sulcata,* &c.

*Genus?—*Leptæna, *Dalman,* 1827, restricted.

*Type—*L. transversalis, *Dal.* Int. Pl. VIII, figs. 176—185.

Anomia, *Linnæus* (part).

Leptæna (part), of *Dalman,* 1827, as well as of the generality of Authors.

Plectambonites, *Pander,* 1830.

Animal unknown; *shell* involute, semicircular, transverse or elongated; surface smooth or variously striated; dental or ventral valve (always?) regularly convex; *socket valve* concave and following the curve of the other; area double, fissure in the dental valve partly covered by a deltidium; summit sometimes perforated by a small circular aperture; valves articulating by means of teeth and sockets; in the interior of the *dorsal valve* the socket ridges are largely developed, the cardinal process or boss small, multifid, and

---

[1] Mr. Salter called my attention to this remarkable species, figured by himself in the 'Silurian System,' pl. xx, fig. 14, and having borrowed the original specimen now in the Mus. of the Geol. Society, and compared it with other examples in the Mus. of Practical Geology, I was enabled to restore the entire impressions, incomplete from fracture in the original example.

connate with their bases; the muscular impressions left by the adductor are large, produced, and elongated, occupying more than two-thirds of the length of the valve, and bordered by well-defined ridges. In the interior of the *dental valve* the muscular scars are small, and not distinctly margined; the adductor lies close to a slight mesial ridge, while the cardinal muscles leave larger scars on either side; vascular impressions radiating.

Obs. On comparing with attention the shells above characterized with those forms we have admitted into *Strophomena*, such as *S. alternata* and *depressa*, the observer is struck by the great difference observable in the shape and position of the muscular impressions, especially in the interior of the socket valve: this character is constant, and essentially the same in *Lept. transversalis, sericea, oblonga, Davidsoni, tenuicincta, quinquecostata*, &c.; other differences likewise exist in the development of the socket ridges and cardinal process. It would, perhaps, therefore be advisable and convenient to retain for these species the term *Leptæna*, especially as one of Dalman's types is comprised in this section, and by so doing, the general opinion in favour of preserving the two names *Strophomena* and *Leptæna* is deferred to; in this view I am supported by Mr. Salter, who had long ago come to the same conclusion. Professor King admits the generic claim of this group, but would prefer the name *Plectambonites*, as he feels desirous of retaining that of *Leptæna* for such shells as *L. depressa*, and *L. analoga*, &c.; but we have already expressed under *Strophomena*, the objections to that arrangement. Some of the species in this group, (*L. Davidsoni, L. oblonga*, &c.) are likewise perforated by a small circular foramen, for the passage of a temporary pedicle, proving how unfounded were M. d'Orbigny's views when characterising *Strophomena* with a perforation, and *Leptæna* without one; but as the interior of a vast number of species of so called *Strophomena* and *Leptæna* are still unexamined, it would be premature to come to a *fixed* conclusion on the subject.

Geol. range.—Some of the shells composing this genus appeared for the first time in the oldest Silurian period, and continued with very similar shapes upwards and through the Lias, above which no examples have been hitherto observed.

Examples; *L. transversalis*, Dalman; *sericea*, J. Sow.; *oblonga*, Pand.; *Davidsoni*,[1] Ed. Deslongchamps; *quinquecostata*, M'Coy; *tenuicincta*, M'Coy; *transversa*, Pand.; *convexa*, Pand., &c.; *liasiana*, Bouch.; *Bouchardii*, Dav., &c.

SUB-FAMILY?—DAVIDSONIDÆ.

Animal unknown; shell fixed to marine bottoms (rocks, shells, and corals) by a portion of the dental or ventral valve; hinge line straight, with a more or less developed false area,

[1] The discovery of this remarkable species in the upper Lias of Normandy is due to Mr. Perier, of Caen, and has been described and illustrated in M. Eugene Deslongchamp's valuable paper on the Leptænæ and Thecideæ of the Lias. *L. Davidsoni* measures 10 lines in width, and bears the most striking resemblance in external shape to some Russian Silurian specimens of *L. transversa*.

and deltidium in the attached valve; no calcified supports for the arms, which were evidently fleshy and spirally coiled.

Obs. The place this genus should occupy in the classification, does not seem to have been quite satisfactorily established, I have therefore followed Professor King in constituting it for a small sub-family, whence it may be easily removed, should future discoveries require it to be admitted into any of those already established.

*Genus—*DAVIDSONIA, *Bouchard,* 1849.

*Type—*D. VERNEUILI, *Bouch.*   Int., Pl. VIII, figs. 186—193.

THECIDEA (PRISCA), *Goldfuss,* MS. Mus. of Bonn.

LEPTÆNA? *De Verneuil,* 1845.

DAVIDSONIA, *Bouchard, De Koninck, King, Schnurr,* &c.

Shell transversely oval, with thick unequal valves, fixed to marine bodies by a portion of the surface of the ventral valve, filling outward irregularities, but not reproducing the same internally; the unattached portion of the dental valve rises abruptly, especially in front; area more or less defined, and divided by a convex triangular pseudo-deltidium, considerably notched at its inner margin; upper or socket valve thick, slightly convex or concave; external surface smooth and marked by concentric lines of growth which extend uninterruptedly over the rudimentary area.   Valves articulating by means of strong teeth placed on either side at the base of the fissure close to the hinge line, and corresponding with sockets in the upper valve.   In the interior of the *attached valve* a hollow is visible under the pseudo-deltidium, affording space for the cardinal muscular process of the other valve; between and below the dental projections are situated the muscular scars, which extend to about a third of the length of the shell; these consist of two deep oval impressions, probably left by the cardinal muscles, and divided by a broad flattened and sinuated mesial elevation, to which the adductor was possibly affixed; the greater portion of the interior is occupied by two conical elevations projecting more or less beyond the level of the valve; the lateral and frontal portions of these solid cones exhibit five or six semicircular or spiral projecting ridges, diminishing in surface and width as they approach the summit of the cone, but not distinctly extending over the portion facing the cardinal edge, which is covered to its extremity with the same minute indentures or punctures which are seen on those portions of the internal surface unoccupied by muscular scars; the cones are separated by a longitudinal furrow, or space, which blends with the raised border surrounding the interior of the valve.   In the interior of the upper or dorsal valve, between the largely developed socket walls, and close to the hinge line, is situated a small cardinal process, under this the deep adductor impressions are seen lying in a semicircular depression, in the anterior half of the thickened valve are excavated two conical hollows, which correspond to the cones of the attached valve, and like them are separated by a rounded mesial ridge, united near the front to the raised border which surrounds the valve.

Obs. M. de Verneuil was the first to draw attention to the shell above described, by publishing a figure of the lower valve of a specimen from the Eifel,[1] which he classed with doubt in *Leptæna*, adding, "*that the spiral arms were coiled five or six times on each other, and placed in the attached valve, so as to rise perpendicularly against the smaller one;*" the author having kindly presented me with one of his four examples, M. Bouchard founded on it the genus DAVIDSONIA,[2] for which compliment I feel greatly obliged to my old and distinguished friend. M. Bouchard clearly proved that the shell was not a *Leptæna*, but it must likewise be observed, that at the period above-mentioned only the lower valve was known, so that the description was naturally incomplete; but this blank was some time after filled up by M. de Koninck's fortunate discovery of two upper valves.[3]

Much difference of opinion has been expressed as to the nature of the various impressions visible in the interior of the shell, as well as to the systematic position of the genus. M. de Verneuil's original view of the massive cones has already been noticed. M. Bouchard believes that they were produced *by secretions deposited by the posterior adductor muscle*, while M. de Koninck sees in them nothing more than the *common thickening of certain portions of the valves* in other Brachiopoda, such as *Bouchardia rosea*, &c.; and Professor King states, *that he is "still in favour of these cones having been produced by labial processes."*[4]

Some authors believe that the arms could not have influenced the form of the shell; but the absence of such impressions in recent forms, is no argument against the possibility of such occurring in some of those extinct species, where, as in *Davidsonia*, so small a space remains for the animal;[5] nor do I think it possible that the labial appendages could have occupied any other space than the small one existing between the cones; and that the mantle, by pressing on the free spiral arms, retained some impressions of their coils, which were transmitted to the shell it was secreting: and this view appears the more probable, since the vascular trunks and veins existing on the mantle, were likewise impressed on the outer sides of the cones as well as on other portions

Fig. 39.

Section of Davidsonia, *in the collection of M. de Koninck. The space left white between the valves is the only one occupied by the animal; f. is the portion attached.*

[1] Geol. of Russia and the Ural Mountains, vol. ii, p. 227, pl. xv, fig. 9, 1845.

[2] Mémoir sur un Nouveau genre de Brachiopode, &c., 'Annales des Sciences Nat.,' vol. xii, p. 84, 1849.

[3] Notice sur le genre *Davidsonia*, 'An. de la Soc. Royale des Sciences de Liege,' vol. viii, p. 149, 1852. These valves are now in the British Museum.   [4] Monograph of the Permian Fossils, p. 151, 1849.

[5] Similar impressions occur in certain examples of *Productus* and *Strophomena*, and which may be attributed to the same cause.

of the shell.[1]   M. Bouchard is of opinion that *Davidsonia* forms the passage from the articulated to the unarticulated genera.   M. de Koninck considers it to bear more analogy to *Thecidea* than to any other,[2] while Professor King and Mr. Woodward would place the genus in the family *Strophomenidæ*.

Geol. range.—In our present knowledge the genus is limited to the Devonian epoch. It occurs both at Chimay, and in the Eifel; in that horizon characterised by the *Calceola sandalina.*

Examples: *D. Verneuili*, Bouch.; *D. Bouchardiana*, De Koninck.

FAMILY—PRODUCTIDÆ.

Animal unknown; *Shell* entirely free, or attached to marine bottoms by the substance of the beak; valves either regularly articulated or kept in place by muscular action; no calcified supports for the oral appendages, which were no doubt fleshy and spirally rolled.

Obs.   We have admitted into this family the following genera: *Chonetes, Strophalosia, Aulosteges* and *Productus*, on account of the natural and intimate relation they bear to each other, as may be better understood by a glance at their respective figures (Pls. VIII and IX.)   In all we perceive the same dispositions of the muscular system,[3] the same reniform impressions (supposed to be vascular?) and the same elevation on each side of the cardinal ridge (marked g' in Pl. IX, fig. 214, 219, &c.)   In all, the surface of the valves is ornamented by tubular spines, which only vary in position and abundance. Professor M'Coy removes *Chonetes, Strophalosia,* and *Aulosteges* from the PRODUCTIDÆ,[4] and places them among the STROPHOMENIDÆ or LEPTÆNIDÆ; but we regret not being

[1] M. Deshayes, Professor King, and Mr. Woodward, also believe that the form of the cones is due to the influence of the free arms upon the mantle lining the inner surface of the shell.   M. de Koninck was correct in stating that the pallial vessels had left impressions on the outer sides of the cones and margins of the shell; but I was further able to trace the principal vascular trunks, from their origin between the muscles to near the front of the shell, where sweeping sub-marginally round both sides of the mantle they give off bifurcated veins, which have left impressions on the frontal and lateral margins of the cones, in a somewhat similar manner to.what we find so well illustrated in Professor King's figures of *Stroph. analoga* ('Mon. of English Permian Fossils,' pl. xx, figs. 6 and 7).

[2] I admit, with M. de Koninck, that the external shape of *Davidsonia* is very similar to that of certain *Thecidia*, such as *Th. Moorei* and *Bouchardii*, but as the Belgian author justly observes, the internal dispositions are very different from those seen in any species of Defrance's genus *Thecidium.*

[3] Professor M'Coy is mistaken when he states that no traces of the *dendritic* muscular impressions are visible in *Strophalosia* or *Aulosteges:*—I have distinctly observed them in both.   I must also confess I see a great difference between the reniform impressions of *Productus Strophalosia*, &c., and the vascular system of *Leptæna transversalis.*

[4] British Pal. Fossils in the Mus. of Cambridge, p. 387, 1852.

able to admit the conclusions arrived at by the learned Professor; and, on the contrary, firmly believe that the four above-mentioned genera are not only intimately related, but even lead me to *doubt* whether we are authorised in admitting as many as four genera.[1] None of the characters above described are found among the *Strophomenidæ*, where both the muscular and vascular system is essentially different; and the only similarity appears to consist in the presence in some of the PRODUCTIDÆ, of an area and pseudo-deltidium. The character of the area is essentially similar in *Chonetes, Strophalosia*, and *Aulosteges*, still the first two have a distinctly articulated hinge, while the last (*Aulosteges*), whose area and deltidium are far more developed than in the other two, is unarticulated, and thus intimately connect STROPHALOSIA with PRODUCTUS.

Genus—CHONETES, *Fischer*, 1837.

Type—C. SARCINULATA, *Schl.*, sp. Int., Pl. VIII, figs. 198—202.

PECTINITES, *Lister*.
PECTUNCULUS, *Volkmann*.
PECTUNCULITES, *Walch*.
PECTEN, *Ure*.
HYSTEROLITES ET TEREBRATULITES, sp., *V. Schloth*.
PRODUCTUS, sp., *J. de Sow., V. Buch*.
ORTHIS? (part), *Dalman, Goldfuss*, and of many Authors.
LEPTÆNA (part), *V. Buch, J. Sow., M'Coy*, &c.
SPIRIFERA (part), *Phillips*.
CHONETES, *Fischer, De Koninck, De Verneuil*, and the generality of Authors.
STROPHOMENA, sp., *J. Hall*, &c.
DELTHYRIS, sp., *Fahrenkohl*.

Shell inequivalve, compressed, transversely semicircular, with a straight hinge line, commonly as long as the width of the shell, or prolonged in the shape of auricular expansions; *dental valve* convex, depressed towards the cardinal edge; *socket valve* always concave, following the curves of the other; area distinct, almost equal in both valves, or larger and more produced in the dental one; the upper edge of the area in the larger valve is acute, and provided with a row of delicate spinose hollow tubes, varying in number in different species, and becoming gradually longer as they recede from the extremity of the beak, diverging obliquely from the hinge line; fissure covered by a pseudo-deltidium. In the *socket valve* the opening is entirely filled up by a projecting bifid or trifid cardinal process; surface ornamented with minute, longitudinal, dichotomised, or intercalated striæ, rarely largely plaited, but transversely marked by concentric lines of growth.

Internally, the valves articulate by means of teeth placed at the sides of the fissure of the

---

[1] M. de Koninck admits only two genera in the family: viz., 1, *Chonetes*; 2, *Productus*, which last includes *Strophalosia*, King, and *Aulosteges*, Helm.

dental valve, and corresponding sockets excavated on each side of the cardinal prominence already described. In the dental valve, a small longitudinal ridge divides the muscular impressions situated on either side, the cardinal muscles probably occupied the greatest space, the adductor lying on either side close to the mesial ridge. In the *socket valve* a blunt medio-longitudinal ridge divides the quadruple impressions of the adductor muscle, which forms on either side two oval scars from between which (in some specimens) two short *vascular impressions* proceed in an outward oblique direction, when turning backwards and inwards, they terminate at some distance from their origin. The interior of the valves is covered with minute granulous asperities, arranged in longitudinal lines; *animal* unknown, probably free or attached in the young by fibres issuing from the fissure.

Obs. In 1837, Fischer de Waldheim proposed the genus *chonetes*,[1] but did not characterise it sufficiently; and it was only after the publications of M. de Koninck[2] and De Vernueil,[3] that its value became known, and the appellation generally used. Several authors have remarked that the shells composing this section bear affinities, both to *Productus* and *Leptæna*. Professor M'Coy, in 1852,[4] makes *Chonetes* a sub-section of *Leptæna*. The outer resemblance it bears to many of the species of the last-named genus is obvious; but internally there exist differences which remove it from the *Strophomenidæ*, and unite it by family ties to the *Productidæ*, and in particular to *Strophalosia:* these are the disposition of its quadruple adductor scars and the *reniform vascular impressions*, which I was so fortunate as to discover in a specimen from Néhou (Pl. VIII, fig. 200), and although the latter character is unobservable or indistinctly marked in the generality of *Chonetes*, I have been able to trace its existence in several species. *Chonetes* also possesses the *double area, articulated hinge,* and *tubular spines* of the *Productidæ*, and although the last character cannot be claimed as of generic importance, it is at any rate common to all the *Productidæ*, and wanting in the *Strophomenidæ*.

Chonetes should therefore be placed first or last among the *Productidæ*, and near to the *Strophomenidæ*, as it is a genus or sub-genus which in a measure links the two families together.

We must refer our readers, for detailed descriptions, &c. to the excellent monograph on this genus, published in 1847, by M. de Koninck: the occurrence of spines only on the outer edge of the area of the dental valve, distinguishes the species of this section from those of the other *Productidæ*, a circumstance of great convenience in the present case.

Geol. range.—*Chonetes* first appeared in the lower Silurian age, and continued through

[1] Oryctographie du Gouv. de Moscow, p. 134.
[2] Description des Anim. Foss. du Terrain Carb. de Belgique, 1843.
[3] Russia and Ural Mountains, vol. ii, 1845.
[4] British Pal. Fossils, 1852. American authors have often erroneously placed *Chonetes* among their *Strophomenidæ*.

the Devonian period, attaining its specific maximum during the Carboniferous epoch, above which no authenticated example has been yet recorded.

Examples: *C. striatella*, sp., Dalman.; *comoides*, J. Sow.; *concentrica*, De Kon.; *papilionacea*, Phil.; *Dalmaniana*, De Kon.; *sulcata*, M'Coy; *variolata*, D'Orb.; *armata*, Bouch.; *convoluta*, Phil.; *Verneuili*, Bar.; *Buchiana*, De Kon.; &c.

Genus—STROPHALOSIA, *King*, 1844.

Type—S. EXCAVATA, *Geinitz*, sp.   Int., Pl. VIII, figs. 203—211.

SPONDYLUS (GOLDFUSSI), *Münster*, 1839.
ORTHIS (EXCAVATA), *Geinitz*, 1842.
STROPHALOSIA (=LEPTÆNALOSIA), *King*, 1844; *M'Coy*, &c.
PRODUCTUS, *De Verneuil, De Koninck*, and the generality of Authors.
ORTHOTHRIX, *Geinitz*, 1848.

Shell variable in shape, more or less circular, transverse or elongated, larger or dental valve convex, smaller one concave and following the curves of the other; surface ornamented by scattered tubular spines; beak often irregular, having been affixed by its substance to submarine objects; a small well-defined area in both valves, divided by a pseudo-deltidium: hinge-line straight; valves *articulated* by teeth and sockets, the former situated on each side of the base of the deltidium of the larger valve, and the latter on both sides of a cardinal muscular process of the socket valve. Internally, a longitudinal raised ridge extends from the cardinal process to about half the length of the valve, on either side of which is seen a small ovate raised muscular scar: the reniform impressions are rather large, their prominent outer edges issue from between the adductor scars, gradually arch forwards and outwards on each side, then turning backwards for about half their length, and finally run inwards horizontally to meet each other near the extremity of the mesial septum.

Obs. In 1844, Professor King proposed to separate from *Productus* certain shells possessing a well developed condyloid *hinge, area,* and *deltidium,* to which he applied the term *Strophalosia.*[1] This view has been sanctioned by some, but rejected by others who did not consider the distinctive characters sufficient to warrant the establishment of another genus.[2] About two or more years after the publication of Professor King's views, Dr. Geinitz proposed the generic appellation *Orthothrix*[3] for the same shells already known by that of *Strophalosia.* For a long time, I agreed with my friend M. de Koninck in resisting the establishment of this section, but after minute comparisons of the shells in question, aided by the kind assistance of Professor King, I must admit the existence of some grounds for establishing a separate group or sub-

[1] See Annals and Mag. of Nat. Hist., vol. xviii, p. 92, 1846; and Mon. of English Per. Foss., 1849.
[2] See De Koninck; Monographie des Genres Productus et Chonetes, p. 10, 1847.
[3] Die Vers. der Zechsteingebirges, &c., Dresden, 1848.

genus; the chief points of difference appear to consist—first, in the strongly articulated hinge of *Strophalosia*, while no *teeth* or *sockets* have yet been noticed in any true *Productus*, (such as *P. semireticulatus, longispinus, giganteus, horridus,* &c.,) where the valves are entirely kept in place by muscular action : the cardinal process or boss so often most improperly termed a tooth was simply a point of attachment for the cardinal muscle, a view clearly pointed out by Professor King, and which I have been able to verify in many recent forms of other genera. Secondly, by the presence of a well defined double area and pseudo-deltidium, of which scarcely any trace exists in true *Productæ*.[1] Internally, the muscular scars and reniform impressions seem also to offer some small differences, but preserve the essential characters of *Productus*.

Geol. range.—As far as our present knowledge goes, this genus first appeared in the Devonian period, was sparingly represented in the Carboniferous epoch, and attained its greatest numerical development in the Permian age, above which no authenticated example has been yet discovered.

Examples: St. Goldfussi, Munst., sp.; *S. Morrisiana,* and *Gerardi,* King; *productoides,* Murch., sp.; *Buchiana,* De Kon., sp.; *? parva,* King; *? subaculeatus,* Murch.;[2] &c.

Sub-Section—AULOSTEGES, *Helmersen,* 1847.

Type—A. WANGENHEIMI, *De Vern.,* sp. = A. VARIABILIS, *Helm.* Int., Pl. IX, figs. 212—216.

ORTHIS, sp., *De Verneuil,* 1845.

Animal unknown; *shell* of an irregular pentagonal shape, with its larger or ventral valve, by far the most convex; beak produced, and generally twisted either to the one or the other side, and furnished with a well defined triangular area, longitudinally interrupted by a pseudo-deltidium not quite reaching to the hinge line, which is straight and not provided with articulating teeth; smaller or dorsal valve slightly convex at the umbo, depressed or concave laterally; cardinal edge more or less developed; surface of valves ornamented by a multitude of short tubular spires. In the interior of smaller valve a large and produced trifid cardinal process extends under a portion of the beak and deltidium, filling up the uncovered portion of the fissure, and serving as a point of attachment to the cardinal muscle; under this process a longitudinal mesial ridge extends nearly to the margin, and on either side are placed the elongated

[1] M. de Koninck states that an *area* is at times observable in some examples of *Prod. giganteus* and *punctatus;* but Professor King considers this appearance more deceptive than real, and only occasioned by a strongly developed cardinal edge or margin; the presence of a largely developed area and unarticulated hinge in *Aulosteges* connects *Strophalosia* with *Productus*.

[2] This species is placed by Professor King in *Strophalosia;* and it is worthy of notice, that M. Bouchard had described the articulated *hinge, area,* and *cardinal process,* in a paper on this species, in the 'Ann. des Sciences Nat.,' 1842.

ramified adductor scars; the reniform impressions, after dividing the above-named muscle, extend by an outward oblique curve to near the margin, when turning abruptly backwards and inwards, terminate at some distance from their origin. Under the adductor, more towards the centre of the valve, exist two elevations (perhaps, brachial eminences).

Obs. In 1847, Colonel Helmersen proposed the genus AULOSTEGES[1] for a remarkable shell abundantly found in the Permian Limestones of Mount Grebeni, in Russia, and described by M. de Verneuil, under the name of *Orthis Wangenheimi*,[2] and which the Russian author considered distinct from both *Productus* and *Strophalosia* of King. In outward aspect it differs from *Productus* by its remarkable area and *pseudo-deltidium*, recalling certain shapes of *Orthisina*, but internally, the muscular scars, cardinal process, and reniform impressions are essentially like those of *Productus*, varying but little in their position and details. According to Professor King's present views, *Aulosteges* is a link connecting *Productus* and *Strophalosia*, but it seems to me, that if this shell is to be separated from either of those genera, it can only claim a *sub-generic* rank; its animal must have been very similar to that of *Productus*.

Geo. range.—Hitherto only one species has been discovered, which belongs to the Permian age.[3]

Genus—PRODUCTUS, *Sowerby*, 1814.

Type—P. MARTINI, *Sow.*, or SEMIRETICULATUS, *Martin*, sp.[4] Int., Pl. IX, figs. 217—223.

GRYPHITES, sp., *Hoppe, Walch., Schlotheim, &c.*

ANOMIA, *Da Costa, Chemnitz.*

PYXIS, sp., *Chemnitz.*

ANOMIÆ ECHINATÆ et CONCHÆ PILOSÆ, *Ure.*

ANOMITES, *Martin, Schloth.* (part).

ARCA, sp., *Bruguière.*

TRIGONIA, sp., *Parkinson.*

PRODUCTUS, *Sowerby, Fleming, Deshayes, De Verneuil, V. Buch, De Koninck, D'Orbigny, King*, and of the generality of Authors.[5]

TRIDACNA, sp., *Lamarck.*

TEREBRATULA (part), *De Blainville, Rang.*

PROTONIA, *Link.*

ARBUSCULITES, *Murray, Teste, Bronn.*

LEPTÆNA, *Dalman* (part), *Fischer de Waldheim, Goldfuss, J. de C. Sowerby, Phillips, &c.*

---

[1] Bull. Phys. Math. de St. Petersburgh, vol. vi, No. 9.

[2] Geol. of Russia and Ural Mountains, vol. ii, pl. xi, fig. 5, 1845.

[3] Professor King informs me that his *Productus umbonillatus* may require to be placed in the sub-genus *Aulosteges*.

[4] According to M. de Koninck, *Productus Martini*, Sow., would be a synonyme of *Anomites semireticulatus* of Martin.

[5] Pander places in PRODUCTUS shells we consider to belong to the genus *Leptæna* and *Orthis*.

MYTILUS, sp., *Fischer*.
PRODUCTA, *Conybeare, Phillips, Kœnig, J. de C. Sow., M'Coy*.
STROPHOMENA (part), *Bronn* (not *Rafinesque*).
PINNA, sp., *Phillips*.
PECTEN, sp., *Eichwald*.
LIMA, sp., *V. Buch*.

Shell variable in shape, inequivalve, transverse or elongated, furnished with lateral auricular expansions; *larger* or *ventral valve* regularly convex, geniculated, or perpendicularly incurved; hinge line straight, commonly shorter than the width of the shell; area very narrow, or no true area, but the cardinal edge considerably thickened; beak incurved. *Smaller* or *dorsal valve* concave, following the curves of the other valve. Exterior variously costated or striated, sometimes decussated by concentric lines of growth; from the striæ arise innumerable small closely packed tubular spines, especially abundant on the auricular expansions; in other species they are few in number, large, strong, and irregularly scattered over the surface, or arranged in regular or irregular rows near the cardinal edge; valves unarticulated, probably kept in place by the fibrous membranes and powerful muscular system of the animal.

In the interior of the *larger valve* a narrow mesial ridge separates the two elongated ramified muscular scars left by the adductors? (hepatic impressions of V. Buch); immediately under, but outside these, are seen other two very deep longitudinally striated, subquadrate impressions, probably formed by the cardinal muscles, and more or less widely separated by a mesial elevation or crest, and lower down more towards the centre of the shell two other deep concave sub-spiral depressions are visible,[1] all the remaining inner surface of the larger valve not occupied by the muscular scars is indented by a multitude of small depressions, increasing in number under the auricles; in the centre of the hinge line of the *smaller valve*, a prominent trifid cardinal process afforded attachment to the cardinal muscles, under this a narrow longitudinal ridge extends to more than half the length of the valve, and becoming more elevated towards its extremity; on either side, are seen the ramified scars left by the adductors, and corresponding to those situated in a similar position in the other valve; each scar is double, the two portions being separated by the commencement of the great vascular trunks (see Pl. IX, f. 219). Outside, and in front of the muscular scars, are two elongated or reniform impressions; their surface is smooth, and they are bounded by ridges, which, after dividing the adductors proceed in an outward oblique or almost horizontal direction, when turning abruptly backwards they terminate at a short distance from their origin; between the vascular impressions, and close to the muscular scars, exist two prominences, one on each side of the cardinal ridge;[2] the internal surface of the smaller valve is covered with innu-

[1] Supposed by V. Buch, De Koninck, King, and others, as intended to afford space for spiral or labial appendages.
[2] These are supposed to indicate the origins of the spiral arms; in which case *Productus* differs from

merable produced tubercles protruding often in the shape of short spines, especially visible near the marginal portions of the shell; no calcified supports for the arms, which were probably small and vertically coiled.

Obs. Although much has been published on this extensive genus by several learned Palæontologists, authors do not seem to coincide in the interpretation to be given to some of the impressions observable on its internal surface. Professor King has endeavoured to restore the muscular system; but from an attentive examination of the identical specimens on which the learned Professor conducted his examinations, as well as others in the Cambridge University Museum, the cabinet of Mr. Tate, &c., neither Mr. Woodward nor myself were able to recognise in a distinct manner more than the scars left by the cardinal and adductor muscle,[1] as seen in the annexed woodcut. A variety of opinions have been expressed, as to the manner in which the animal of this genus lived; some believe that the shell was affixed by its spines, others by the extremity of the beak of larger valve, or suspended by muscular fibres issuing from the margins of the valve![2] M. d'Orbigny supposes that the animal lived lying on soft bottoms, with its smaller valve uppermost, in a similar manner to *Oysters, Scallops,* and *Spondylus striatus,* in

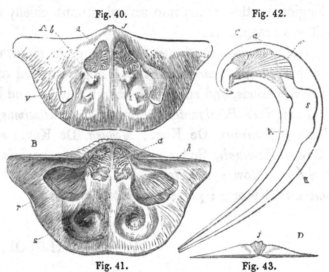

Fig. 40.　　　Fig. 42.

Fig. 41.　　　Fig. 43.
Productus giganteus (from Mr. Woodward's drawings).
Fig. 40. *Interior of dorsal valve.*—41. *Interior of ventral valve, with the umbo removed.*—42. *Ideal section of both valves.*—43. *Hinge-line of dorsal valve:* j, *cardinal process;* a, *adductor;* r, *cardinal muscles;* s, *hollows occupied by spiral arms;* v, *reniform impressions;* b, *brachial processes?;* h, *hinge-area.*

which the lamellæ and spines served to retain the animal in a fixed position; but M. de Koninck objects that these spines are often so long and so delicate, as to make one believe they would be fractured under such conditions. The remarkable tubular expansion of the larger valve in *Prod. proboscideus* led the last-named author to suppose that out of its extremity passed the muscular fibres of attachment; M. d'Orbigny, on the contrary, states it to be his belief that the prolongation was due to malformation produced by accidental circumstances, connected with the supposed constrained position in which the animal lived, having forced the mantle to

Terebratula, and agrees with *Crania* in *three* important particulars, viz. the absence of a hinge, the simplicity of the vascular impressions, and the shifting of the processes for the attachment of the arms, from the hinge-margin to the centre of the shell.

[1] Professor King states that he found impressions of pedicle muscles, and indeed that he was able to trace all the muscular scars observable in Terebratula. ('A Monograph of English Permian Fossils.')

[2] See Geinitz' 'Deutsch. Zechst.,' t. vi, f. 1.

prolong its edges so as to reach the surface of the sea bed : in the locality in which this form is found, a vast number of other species of the genus occur, which do not present this peculiarity, so that we must regard the structure as *normal*, although M. d'Orbigny's explanation of its function is probably correct. The thickness of the shell in many species is very considerable; but we must refer our reader for more ample details to the most valuable monograph of this genus, published in 1847, by M. de Koninck, as well as to several publications on the same subject, by Professor King, M. de Verneuil, and others, in which a multitude of details will be found which my limited space will not admit of mentioning here. Various authors have proposed to arrange the numerous species belonging to this genus into several groups, chiefly founded on the manner in which the shell was ornamented.

Geol. range.—*Productus* does not seem to have had so extensive a vertical range as either *Strophomena* or *Orthis*, being restricted to the *Devonian, Carboniferous*, and *Permian* periods, and specifically rare in the first and last of these.[1]

Examples: *P. giganteus*, Martin, sp.; *latissimus*, Sow.; *striatus*, Fischer, sp.; *cora*, D'Orb.; *Erminius*, De Kon.; *Medusa*, De Kon.; *semireticulatus*, Martin, sp.; *costatus*, J. Sow.; *Flemingii*, Sow.; *spinulosus*, Sow.; *scabriculus*, Martin, sp.; *fimbriatus*, Sow.; *horridus*, J. Sow.; *tessellatus*, De Kon.; *undatus*, Def.; *proboscideus*, De Vern.; *hemisphæricus*, Kutorga; *porrectus*, Kut., &c. &c.

FAMILY—CALCEOLIDÆ.

Animal unknown. *Shell* probably free, valves not articulated; ventral valve pyramidal, with a large, flat, triangular area; dorsal valve flat, semicircular, with a straight hinge-line, a small cardinal process, and two lateral groups of small apophysary (?) ridges; no foramen or muscular or vascular impressions.

Obs. Only one *genus* is known, and so anomalous in its characters that its true place in the class is far from being definitely settled.

Genus—CALCEOLA, *Lamarck*, 1809.[2]

Type—C. SANDALINA, *Linn., Gmel.,* sp., p. 3349. Int., Pl. IX, figs. 294—298.

ANOMIA (part), of *Gmelin*, &c.
SANDALIOLITHES, *Schröter.*
CONCHYTA JULIACENSIS, CREPITES, SANDALITES, CREPIDIOLITHUS, of *Hupsch.*
TURBINOLIÆ, sp., *Hisinger.*
CALCEOLA, *Lamarck,* and of the generality of Authors.

---

[1] Some doubt seems still to exist if the genus was truly represented in the Silurian period. Professor King believes, from the dispositions of the vascular system, that my so-called *P. Twamleyi* belongs to the genus *Leptæna.*

[2] Système des Anim. sans Vert., p. 139; Hist. des Anim. sans Vert., vol. vi, p. 234, 1819.

Shell of a triangular shape, with very unequal valves; *larger valve* sub-pyramidal; beak acute, bent backwards; hinge line straight; area large, triangular, slightly curved, and divided by a long narrow mesial elevation, extending from the apex to the cardinal edge, and longitudinally depressed along its centre; *smaller valve* semicircular, operculiform, slightly convex or almost flat, with a narrow sub-parallel triangular area. Exterior smooth, or encircled by concentric lines of growth, passing uninterruptedly over the area; hinge line crenulated. The cavity for the animal is small in proportion to the dimensions of the shell. The interior of the *smaller valve* is divided by a slightly elevated mesial ridge or septum, terminating at the hinge line in a small cardinal process, which is hollowed in the middle; from the whole extent of the hinge line a number of punctate striæ radiate to the margin; on each side, close to the hinge-line, in the vicinity of the cardinal angles, is a series of short parallel unequal elevated ridges. In the interior of the *larger valve* a corresponding series of punctate striæ proceeds from the whole margin, and converge towards the bottom or cavity of the shell, but becoming more elevated and produced at the cardinal edge, where they form unequal asperities; in the centre, and near the hinge line, a larger rounded projection (corresponding to the first described in the other valve) extends to the bottom of the shell, in the shape of a narrow mesial depression.

Obs. Of all the genera among the Brachiopoda, *Calceola* seems the most abnormal; no one has been able to point out the probable structure of the animal, nor has its true position in the class been yet satisfactorily ascertained. One would have imagined that such a ponderous shell would have been provided with a powerful muscular system, still no traces of it have been observed in the interior of the valves; and so great was the amount of calcareous matter deposited internally, that in many examples hardly any free space remained to lodge the animal. Some authors have expressed doubts whether this genus should be ranked among the Brachiopoda; Cuvier, in his ' Règne Animal,' classes it with the *Oysters;* Lamarck, and a few other Conchologists, place it in the family of *Rudistes,* whence it was removed by M. Deshayes and Goldfuss; it is now generally admitted to belong to the Palliobranchiata. The type of the genus, *Calceola sandalina,* has been long known under various appellations.[1]

Its *Geol. range* is confined to the Devonian period. Only one well-authenticated species has been discovered.[2]

Example: C. sandalina, Linn., Gmel., sp.

[1] Hüpsch described and figured it in 1768, and Beuth, in 1776. Goldfuss has likewise published excellent illustrations of the species, in his ' Petrefacten Musei Universalis,' pl. clxi.

[2] M. de Koninck has lately stated in the seventh volume of the ' Ann. des Sciences Royales of Liege,' that the so-called *Calceola Dumontiana* does not possess the characters of that genus, and even doubts whether the species belongs to the class; he terms it *Hypodema.*

Family—CRANIADÆ.

Animal fixed to marine bottoms by the substance of the shell of its lower or ventral valve; arms fleshy, and spirally coiled; no hinge or articulating processes; upper or dorsal valve patelliform.

Obs. This family contains only one genus, *Crania*.

Genus—CRANIA, *Retzius*, 1781.

Type—C. BRATTENBURGENSIS, *Stoboeus*, sp.   Int., Pl. IX, figs. 299—247.

NUMULUS, *Stoboeus*, 1732.
ANOMITES (part), of *Linnæus*, 1768.
OSTRACITES, *Beuth*, 1776.
PATELLA (ANOMALA), *Müller*, 1776.
CRANIA, *Retzius*, 1781, and of the generality of Authors.
ORBICULA (ANOMALA), *Cuvier*, 1798; *Lamarck*.
CRANIOLITES, *Schloth*.
CRIOPUS (for the Animal), *Poli*, 1791.
PSEUDO-CRANIA, *M'Coy*, 1852.
SPONDYLOBUS, *M'Coy*, idem.

Shell variable in shape, inequivalve, circular, subquadrate, transverse or elongated, partially or extensively attached by the substance of its lower or ventral valve; rarely free; *upper* or *dorsal valve* more or less conical, vertex subcentral; no articulated hinge or ligament, the valves being kept in place by the action of four muscles, which pass in a somewhat oblique manner from one valve to the other; the *attached* or *ventral valve* is generally the thickest, with or without a straight more or less produced beak, and false area.   External surface smooth, spiny, or variously ornamented by radiating costæ or foliaceous expansions; the concentric lines of growth passing uninterruptedly over the valves and area;[1] structure calcareous and tubular.   In the interior, four principal

---

[1] The *false area* of certain species of *Crania* is clearly described by M. Bouchard, in the twelfth volume of the 'An. des Sciences Nat.,' p. 87, 1849, as follows:—"Chez la *Crania antiqua* les espèces voisines (et la *Davidsonia*) . . . . . . ce que l'on nomme *area* n'en est pas une, ce n'est que le résultat d'un accroissement général, et pour ainsi dire oblique, de tout le corps de la valve, qui forme une sorte de *crochet* ou *talon*, sur lequel passent les mêmes lignes d'accroissement qui, se continuant, font tout le tour de la valve.   Cet accroissement est donc tout à fait circulaire et unique pour toutes les parties de la valve : mais il offre cette particularité, qui forme son obliquité, c'est que chaque lame est indiquée sur la face supérieure du talon par une simple strie qui, en s'en éloignant, s'élargit de chaque côté, de manière à atteindre une largeur de près de 1 millimètre vers le bord postérieur.   Ce *talon* ensuite n'est jamais creux : il est au contraire massif et calcaire : donc aucune partie de l'animal n'a pu s'y loger.   Enfin, la marche de cet accroissement est absolument semblable à celui des *Huitres*, et les *talons* des uns et des autres sont exactement de la même structure."

scars formed by the adductor muscles are observable. The anterior pair are approximate, and placed close to the centre, behind which a prominence is sometimes visible in the ventral valve. The posterior pair are situate near the cardinal edge, and are widely separated. The muscular impressions of the attached valve are sometimes slightly convex, at others deeply excavated; those of the dorsal valve are convex, the centre pair sometimes developing very prominent apophyses. (See Part II, Pl. I, f. 2 *b*.) The interior of the attached valve is surrounded by a raised and thickened border, exhibiting the tubular shell-structure in a conspicuous manner. The disk of each valve exhibits more or less distinct impressions of the vascular system, which is simply digitated.

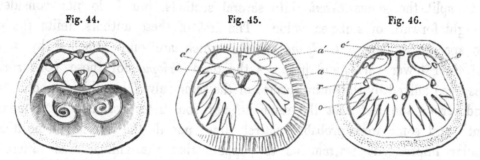

| Fig. 44. | Fig. 45. | Fig. 46. |

Crania anomala, Müll., from specimens dredged off Zetland, and communicated by J. S. Bowerbank, Esq. (From drawings made by Mr. Woodward.)

Fig. 44. *Upper valve with the animal; the mantle-lobe has been removed, to show the spiral arms and the muscles. The small central depression is caused by the process (c) of the lower valve. The mouth is concealed by the overhanging fringe of cirri.*

Fig. 45. *Lower (ventral) valve.*

Fig. 46. *Upper (dorsal) valve. Explanation of the letters: a, anterior adductor impression; a', posterior adductors; c, c, points of attachment of the protractor muscles; c', c', impressions of the cardinal muscle; r, r, points of attachment of the retractor muscles.*

The valves of *Crania* appear to have been opened by the action of sliding muscles, as described by Professor Owen, in *Discina;* those which advance the dorsal valve (*protractor* muscles), are aided by a single small muscle in the median line, answering exactly to the cardinal muscle of *Terebratula.* The pair of muscles which bring the free valve back to its place (*retractors*), are attached outside the anterior adductors of the free valve, and (apparently) to a point between the posterior adductors in the fixed valve (like their equivalents, in *Discina,* according to Owen). The two small impressions in the centre of the dorsal valve (fig. 46) are probably caused by the attachment of the muscles of the spiral arms. The labial arms are thick, fleshy, and spirally coiled; the volutions are few, and directed, vertically, towards the cavity of the dorsal valve, as in *Rhynchonella* and *Davidsonia,* and unlike *Producta* and *Discina.* The mantle lobes of *Crania* extend to the edges of the valves, and adhere closely, as in *Thecidium;* their margins are plain and thin.

Obs. Stoboeus appears to be the first author who figured and described a species belonging to this genus, under the names of *Numulus Brattenburgensis* and *minor,*[1] and

---

[1] Stroboeus; Diss. Epist. Lund., 1732, and *Opuscula*, &c.

shortly after (1768,) Linnæus changed the term to *Anomites craniolaris*. The genus was founded by Retzius,[1] in 1781, on the first of Stoboeus's types; Poli, in 1791, figured the animal of a recent species, which he termed *Criopus*;[2] but Müller had before him, in 1776 and 1781, described another recent form, picked up on the coast of Denmark, under the name of *Patella anomala*.[3] This shell has proved the subject of lamentable confusion; Cuvier, believing it distinct from the *Crania* of Retzius, proposed for it the generic appellation *Orbicula*,[4] a name which must unavoidably be expunged from the nomenclature and placed among the synonymes of *Crania*, since the so-called *Patella anomala* undoubtedly belongs to the last-mentioned genus.[5] Professors King[6] and M'Coy[7] have proposed to split the genus *Crania* into several sections, but I do not consider the points brought forward of sufficient value. The first of these authors limits the genus *Crania* to species similar to *C. Ignabergensis*,[8] which are only attached by a small portion of their lower valve, and proposes the term *Criopus* for those (*C. turbinatus, Parisiensis, &c.*) affixed by the entire surface of the same valve; but this character falls to the ground, from the fact that the same species sometimes assumes both modes of attachment,[9] and some were probably entirely free; nor do the internal dispositions of the muscular impressions warrant us to suppose that the animal was different in either case. Professor M'Coy proposes two other genera, *Pseudo-crania* and *Spondylobus*. The first is established on two Silurian species, the *C. antiquissima*, Eichw., sp., and *P. divaricata*, M'Coy, both are stated to have been unattached, but, as above observed, some well-known Cretaceous *Cranias* presented the same characters; the muscular scars in Professor M'Coy's species are very similar to those of *Crania*, the chief dissimilarities pointed out would appear to consist in the margin of the shell being smooth, or wanting the produced granulations, and in the vascular impressions; but these characters vary likewise to a great extent in different species of *Crania*. The Professor's

---

[1] *Crania oder*, &c., Berlin Gesellsch. Schrift, Brand ii, p. 66, 1781.

[2] Testacea utriusque Siciliæ, vol. ii, p. xxv, fig. 24.

[3] Zoologia Danica, 1777, tab. v, fig. 1—7.

[4] Cuvier; Tableau Élémentaire du Règne Animal, p. 435, 1798—" *Les orbicules (Orbicula)* . . . . . . On ne connait de ce genre qu'une seule espèce (*Patella anomala*)—Müller." Refer likewise to Cuvier's celebrated " Mémoir sur l'Animal de la Lingule," 'Mém. du Mus.,' vol. i, 1802, in which he again states that *P. anomala* is the type of his genus *Orbicula*.—Lamarck's views were exactly the same as those of Cuvier, as may be seen by referring to the 'Système des Anim. sans Vert.,' p. 140, 1801.—Bruguière does not appear to have been the first proposer of the genus *Crania*, as stated by Dr. J. E. Gray, in the 'An. of Phil.' for 1825.

[5] The species described by myself under the name of *Orbicula*, in Part I, p. 7, and III, p. 9 and 10, do not belong to the genus *Crania*, but to *Discina* of Lamarck.

[6] A Monograph of English Permian Fossils, p. 84, 1849.

[7] British Pal. Foss. in the Camb. Mus., pp. 187 and 255, 1852.

[8] The first-mentioned species by Retzius is the *C. Brattenburgensis* = *nummulus* of Lamarck.

[9] See Part II, p. 11.

genus *Spondylobus*[1] seems to me very doubtful, and although far removed by the author from *Crania*, and placed among the *Lingulidæ*, would appear to be more nearly related to Retzius's genus.

Professor M'Coy states that in *Spondylobus* the larger valve is "slightly longer, from the apex being perfectly marginal and slightly produced, channelled by a narrow groove below, the anterior end of which is flanked by two very *prominent, thick, conical shelly bosses, representing hinge teeth.*" In *Crania Hagenovii* (De Kon.) there are two similar bosses,[2] which did not perform the office of hinge teeth, but were probably points of attachment for one of the sliding muscles.

Numerous and varied forms of *Crania* have been described and illustrated by several authors, and in particular by Hœninghaus[3] and Goldfuss,[4] who published beautiful monographs of the species of the genus known to them.

Geol. range.—*Crania* is one of the oldest types of animal life, it first appeared in the lower Silurian period, and has lived throughout all subsequent periods up to the present time.

Examples: C. *Brattenburgensis*, Stoboeus, sp.; *Ignabergensis,*[5] Retzius; *abnormis*, Def.; *costata*, Sow.; *antiqua*, Def.; *spinulosa*, Nils.; *antiquior*, Jelley; *Moorei*, Dav.; *Parisiensis*, Def.; *tuberculata*, Nils.; *ringens*, Hœnig.; *rostrata, gracilis, armata, intermedia, bipartita*, of Munster; *obsoleta*, Goldfuss; *Cénomaniensis*, D'Orb.; *anomala*, Müller; *Hagenovii*, De Kon., &c.

FAMILY—DISCINIDÆ.

Animal attached by the means of a muscular peduncle, passing through the ventral or lower valve, a slit in the hinder portion, or a circular foramen excavated in the substance of the same valve. Arms fleshy; valves unarticulated.

In this family we have two or three *genera*, viz., *Discina, Trematis*, and *Siphonotreta*. The genus or sub-genus *Trematis*, as well as the sections *Orbiculoidea* and *Acrotreta*, will require further investigation before being definitely admitted.

---

[1] The species named as types, are *C. Sedgwickii*, Lewis, and *craniolaris*, M'Coy.

[2] My attention was called to this point by my friend Mr. Woodward.

[3] Beiträge zur Monog. der-Gattung Crania, 1828.

[4] Petrefacta Musei Universalis, 1840.

[5] This species was named after *Ignaberg*, a small village in Scania; Stoboeus, Retzius, and others have spelt it *Egnab*, an error I likewise committed in Part II, p. 11.

Genus—Discina, *Lamarck*, 1819.

Type—D. lamellosa, *Brod.*, sp.   Int., Pl. IX, figs. 248—252.

Orbicula, of *Owen, Sow.*, &c., but not Orbicula, *Cuvier* or *Lamarck*.
Discina, *Lamarck, Gray, King, M'Coy, Philippi*, &c.

Shell circular, transversely or longitudinally oval; upper or *dorsal* valve[1] conical, patel-liform, with the apex inclining towards the posterior margin; lower or *ventral* valve opercular, flat, or partly convex, and perforated by a narrow oval longitudinal slit, reaching to near the posterior margin and placed in the middle of an oval depressed disk.   Valves unarticulated, being kept in place by four thick adductor muscles, passing rather obliquely from one to the other; surface smooth, ornamented by numerous striæ radiating from the apex to the margin, or by concentric lines of growth, produced in the shape of foliaceous expansions; shell structure horny, and perforated by minute tubuli.   In the interior of the perforated valve the muscular disk is more or less prominent, and from the anterior extremity of the fissure a short longitudinal process arises.   Four pairs of muscular impressions are visible on the disk; the posterior, submarginal pair, are probably those of the sliding muscle; the central four are the adductor scars, of which two lie on either side of the foramen, while the anterior pair are situated more towards the centre of the shell; the fourth pair lie external to these.   In the upper valve, the posterior adductor scars are placed nearly horizontally, at a short distance from the margin, and are much smaller than the anterior pair, which lie obliquely near the centre of the valve; there are no projecting processes in the interior of this valve.

[1] In the letter addressed by Professor Owen to Professor M. Edwards, and published in the 'Ann. des Sc. Nat.,' vol. iii, 3d ser., p. 319, 1845; the learned Hunterian professor alludes to the valves of *Terebratula* and *Discina* (*orbicula* of Owen), as follows:—

"Pour en revenir aux *Térébratules*, j'ajouterai encore quelques mots relatifs à la manière dont j'envisage les rapports de position des parties molles et de la coquille.   Dans le *Terebratula flavescens*, le pharynx est entouré d'un collier nerveux simple, et les principaux nerfs naissent de petits renflements situés aux angles du côté de ce collier qui avoisine la base transversale des bras frangés.   Or, si le tube alimentaire était redressé par le tiraillement de la bouche et du pharynx en avant, cette base transversale des bras, et les points d'origine des nerfs analogues qui naissent ordinairement des ganglions sous-œsophagiens chez les Mollusques, plus élevés en organisation, seraient situés du côté de la grande valve perforée.   Je considère par conséquent cette valve comme étant la *valve inferieure* ou *ventrale*, et la position du cœur vient à l'appui de cette opinion, puisque ce viscère se trouve plus près de la petite valve ou valve dorsale que ne l'est l'intestine.   Jadis, j'ai décrit l'intestin comme se terminant du côté droit de la masse viscérale chez les Térébratules aussi bien que chez les Orbicules (*Discina*).   Je persiste encore dans cette manière de voir; mais, dans l'Orbicule (*Discina*), la valve dorsale et imperforée est la plus grande et la plus convexe."

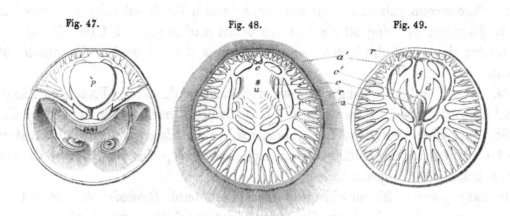

Fig. 47.   Fig. 48.   Fig. 49.

Discina lamellosa, Brod., Peru, Rev. W. Hennah.[1]

Fig. 47. *Animal, as seen on the removal of the lower valve, and part of the ventral mantle-lobe; the extremities of the labial arms are displaced forwards, in order to show their spiral terminations; p, is the expanded surface of the plug or pedicle; the mouth is concealed by the overhanging fringe.*

Fig. 48. *Dorsal mantle-lobe with its fringe of setæ.*

Fig. 49. *Ventral mantle-lobe (the setæ not represented); f, foramen, through which the fibres of the plug issue; a, a, anterior adductor muscles; a′, a′, posterior adductor muscles; c, c, protractor muscles; c′ c′ cardinal muscle; r, r, sliding muscle, or anterior visceral muscle of Owen.*

The *animal* of *Discina* is extremely delicate and transparent, allowing its internal organisation to be seen through the mantle. The mantle lobes do not adhere to the interior of the shell, which is smooth and polished; they are bordered with a dense fringe of long horny *setæ*, which are stiff and extremely brittle, resembling the bristles of certain Annelides (e. g., *Aphrodite*).[2] The relation of the animal to the perforate and imperforate valves is the same as in *Terebratula*, but as in *Crania*, the only process which can possibly have given any support to the oral arms at their origin, is developed from the centre of the *ventral* valve. The spiral extremities of the arms are directed towards the lower valve, and not dorsally, as in *Crania*. Professor Owen has shown in his anatomy of *Discina* (*Orbicula* 'Trans. Zool. Soc.,' 1833), that the valves must have been opened by a pair of oblique, or *sliding* muscles (*protractors, c, c,* fig. 48, 49); besides these there is a slender muscle (attached at the points *c′, c′*), similar to the cardinal muscle of

[1] From drawings made by Mr. Woodward: we are indebted to Dr. Gray, for permission to examine the specimens of *Discina*, preserved in spirits, in the British Museum.

[2] Professor Owen remarks, in p. 278, of his 'Lectures on the Comp. Anat. and Phil. of the Invertebrate Animals,' 1843: "That the margin is fringed with cilia, which are very long in Lingula and Orbicula (*Discina*). In the latter Brachiopod I found them beset with smaller cilia: these, with the brachial fringes and microscopic vibratile cilia; which, doubtless, beset the whole surface of the vascular mantle must be the chief agents in introducing the current of sea water within the cavity of the mantle for nutrition and excretion." It is, however, probable, that neither the brachial *cirri* or pallial *setae* are themselves vibratile organs. In certain species of *Terebratula* and *Spirifera*, the *cirri* had inflexible *calcareous supports*. (See Pt. III, Pl. II, fig. 18, and Pl. IV, fig. 2.)

Crania. The second pair of sliding muscles, by which the dorsal valve is restored to its place, is described as being attached at the points marked *r, r*. I failed in attempting to determine these, and also in ascertaining the points of attachment of the muscles of the peduncle.

Obs. Several authors, among whom we may mention Dr. Gray,[1] Professors King[2] and M'Coy,[3] have satisfactorily shown that the term *Orbicula*, having been applied by Cuvier, in 1798,[4] and subsequently likewise by Lamarck,[5] to *Patella anomala* of Muller, it ought not to be retained for the shells under description, as the species described by the Danish naturalist belongs to the genus *Crania.*[6]

For many years I felt unwilling to discard the term *Orbicula*[7] on account of its currency, occasioned by Professor Owen's description of the animal of *Discina* (*O.*) *lamellosa;*[8] but as the term *Orbicula* in the Cuvierian and Lamarckian sense is merely a synonym of *Crania*, we are obliged to adopt the genus *Discina*,[9] proposed and described by Lamarck, in 1819, and typified by *D. ostreoides*, a shell no doubt identical with *O. lamellosa* of Brodrip, but which never lived on the recent maritime shores of Great

---

[1] Annals of Philosophy, new series, vol. x, 1825; and An. and Mag. of Nat. Hist., vol. ii, 2d series, p. 439.

[2] A Monograph of Permian Fossils, p. 84, 1849.

[3] British Pal. Fossils in the Camb. Mus., 1852.

[4] Tableau Elémentaire du Règne Animal. "*On ne connait de ce genre qu'une seule espèce découverte par Müller en Danemarck, et rangée jusqu'ici parmi les Patelles ; Patella anomala.*" In 1802, the same author states that Lamarck had adopted his views.

[5] Annales du Muséum, vol. x, 1807. In that mémoir, Lamarck places *Crania* in the family of *Oysters; Orbicula, Terebratula*, and *Lingula*, in the Brachiopoda. In 1801, the same author describes the genus *Orbicula*, founded on *O. Norvegica = Patella anomala* of Müller, and in 1819, distinctly states that it was *not* attached by a pedicle, but by the *substance* of the lower valve.

[6] The Baron de Ryckholt figures in his ('Melanges Paléontologiques,' pl. iv, fig. 3) specimens of *Crania obsoleta*, under the appellation of *Orbiculoidea Cimacensis*, said to be from the Devonian of Chimay and Ferque, in which the shell is bordered by a fringe, which he terms "*impressions ciliaires*," and regards these as the remains of the horny *setæ* which surround the margin of the mantle of the recent animal, and which are also developed to a less extent in *Rhynchonella*. Not having seen the original specimens, I cannot offer an opinion, as to whether these beautiful examples really exhibit the fossilised border of the mantle, or only the fringed margin of the shell itself.

[7] In Parts I and III, I made use of the term *Orbicula* in the general sense, but not in that of Cuvier and Lamarck; the term *Discina*, will have to be there introduced, and *Orbicula* erased. Professor E. Forbes informs me, that "it would certainly be more convenient, under all circumstances, to prevent confusion, to use *Discina* instead of *Orbicula*, and to suppress altogether the latter name."

[8] Trans. of the Zool. Soc., vol. i, 2d part, 1833 (printed in 1835); and in the Annales des Sciences Nat., 1845.

[9] "*Discina; coquille inéquivalve, ovale-arrondie un peu déprimée; a valves de grandeur égale, ayant chacune un disque orbiculaire central très-distincte. Disque de la valve supérieure non percé, ayant au milieu une protubérance en mamelon ; celui de l'autre valve très-blanc, divisé par une fente transversale.*" Lamarck, 'Hist. Nat. des Animaux sans Vertebrès,' vol. vi, p. 236, 1819.

Britain,[1] as supposed by that author. The only point on which great doubt may remain is the propriety of admitting the term *Orbiculoidea*,[2] proposed by M. d'Orbigny for a vast number of fossil species, considered by the generality of authors to be true *Discinæ*, but which the author of the ' Paléont. Française' states to have been provided with a regular pedicle, as in *Terebratula;* I am, however, convinced that many of the species thus denominated were attached, as in the typical *Discinæ*, and have had a portion of their pedicle fibres extending over the external disk; therefore, while introducing the term *Orbiculoidea* as a sub-section, I do so with the greatest reserve and doubt, as no internal characters have as yet been brought to bear on the subject.

Geol. range.—It is highly probable that the genus *Discina* was represented during the whole series of Geological periods up to the present day; but according to M. d'Orbigny the Palæozoic and Secondary species are generically distinct, and *Discina* first appeared at the Tertiary epoch, becoming more abundant in the present seas.

Examples: D. lamellosa, Brod.; *lævis*, Sow., sp.; *nitida*, Phil.; *Townshendi*, Forbes, sp. ; *Cumingii*, Brod. ; *strigata*, Brod., sp. ; *striata*, Sow., sp. ; ? *Morrisia*, Dav., sp.

? *Section* A.—ORBICULOIDEA, *D'Orb.*, 1847.[3]

Type—O. ELLIPTICA, *Kutorga, sp.* Int., Pl. IX, fig. 253—255.

ORBICULA, of the generality of Authors.
ORBICULOIDEA, *D'Orbigny, De Ryckholt.*
SCHIZOTRETA, *Kutorga.*

Shell, sub-orbicular patelliform, longitudinally or transversely oval. Upper valve conical with the vertex close to, or removed from the posterior margin; *lower* valve conical or concave, no pedicle disk;? a narrow oval or circular aperture, more or less confined in its shape, is situated in a furrow or depression.

Obs.—M. d'Orbigny characterises his section as follows :—" *Coquille de contèxture cornée non perforée, dont la valve inférieure concave est pourvue d'une ouverture laterale au crochet pour le passage d'un pedicule simple;*"[4] and supposes that these shells were

[1] In the Linn. Trans., vol. xiii, 1822. Sowerby falls into the same mistake made by Cuvier and Lamarck; since he considers the *Patella anomala* to be generically distinct from *Crania*. His observations relative to Lamarck's genus *Discina* seem essentially erroneous, as it is *quite impossible* the French author could have traced his characteristic of *Discina* from a specimen of *O. Norvegica*, which is a *Crania*, and therefore could not have presented the fissure and disk alluded to by the celebrated author above noticed.

[2] Cons. Zool. et Geol. sur les Brachiopodes, 1847.

[3] Considérations Zool. et Geol. sur les Brachiopodes; Comptes rendus de l'Académie des Sciences, August, 1847.

[4] Cours Elémentaire de Pal. et Geol. Stratigraphiques, vol. ii, p. 90, 1852. The author states to be acquainted with 27 fossil species.

fixed to submarine bottoms by a pedicle, which afforded greater freedom of motion to the animal than that of *Discina*.   About the same period, Professor S. Kutorga likewise separated the above-mentioned shell from the Lamarckian genus, by the name of *Schizotreta*,[1] bringing forward similar reasons to those urged by the author of the ' Palæontologie Française;' but it is still very doubtful, if the section ought to be admitted.

Geol. range.—According to M. d'Orbigny, *Orbiculoidea* first appeared in the Silurian epoch, more numerous during the Carboniferous period, and disappeared in the Neocomian age.

Examples :   O. elliptica, Kütorga, &c.

Section B.   TREMATIS, *Sharpe*, 1847.

Type—T. TERMINALIS, *Emmons*, sp.   Int., Pl. IX, figs. 256—260.

ORBICULA, of the generality of Authors.
TREMATIS, *Sharpe* (June, 1847).
ORBICELLA, *D'Orbigny* (August, 1847).

Animal unknown ; *shell* somewhat depressed, sub-orbicular or transversely oval ; both valves slightly and unequally convex, giving a lenticular form to the shell ; umbo of the upper or imperforated valve submarginal, slightly projecting.   Lower or ventral valve with a sub-central umbo, behind which a narrow, oblong, oval slit reaches to near the posterior margin, and afforded passage to the pedicle fibres of attachment.   Internal dispositions unknown, a wide cardinal margin is visible in the imperforated valve.

Obs. Mr. Sharpe proposed to separate from *Discina* certain shells which in their general external appearance bore much resemblance to that genus, but appeared to differ by others, which he thus defined :[2] " *The valves are united by a hinge*, of which the details cannot be seen in the specimens ; *but it is probably formed* of two diverging lamellar processes in the dorsal valve, for where the shell of that valve has been worn away we can trace three calcareous plates diverging from the *hinge* of the dorsal valve, as in the Leptæ-noid species of *Orthis*, and in many of the Spirifers. Wherever these plates are found in the Brachiopoda, the outer pair appear to be continuations of the teeth or lamellar processes of the hinge ; so that the presence of such plates is sufficient to *show* that the valves played upon a hinge.   The third or mesial plate separates the two great adductor muscles." The above statements require, however, further examination before being admitted, nor have I been able to obtain examples adding much light to the subject ; but I am far from being convinced that *Trematis* was provided with an articulated hinge resembling that of *Terebratula;* on the contrary, I am inclined to believe the valves were un-

---

[1] Ueber dei Siphonotretaeæ aus den Verhandlungen der Kaiserlichen Min. Ges. für das Jahr 1847.
[2] Quart. Journal of the Geol. Soc., No. 13, vol. iv, p. 66.

articulated, and kept together by special muscles, as in *Discina*. Mr. Sharpe further observes, that "the shell consists of layers of two distinct structures; the outer layer is *punctated; the punctations* are so large as to be clearly visible to the naked eye, and are arranged quincuncially with great regularity. The inner layers of the shell are not punctated, and have a fibrous and slightly striated appearance and pearly lustre; these *impunctate* layers are thickest towards the middle of the shell, and do not quite reach the margin. It thus appears that the genus *Trematis* differs from *Orbicula* (*Discina*) in the punctated structure of the shell, and in having the valves united by a hinge, while it is distinguished from *Terebratula* and other hinged forms of Brachiopoda, by the ligament passing through the ventral valve." Having submitted a specimen of *Trematis terminalis* to Dr. Carpenter's microscopic examination, that gentleman informs me "*that it does not depart in any essential particular from the type of Discinidæ, the punctated surface being (as in Porambonites), a mere superficial conformation; and Mr. Sharpe's internal unpunctated layer being a succession of laminæ, just as in Discina, and that every thing now tallies satisfactorily in regard to the conformity between the general characters of the shell and its intimate structure.*"

In August 1847, M. d'Orbigny proposed for the same shells, the name *Orbicella*,[1] which must be considered as a synonym, Mr. Sharpe's paper having been published a month before that of the French author! M. d'Orbigny does not allude to a hinge or diverging plates, he mentions that the *structure of the shell* (which he erroneously states to be punctated), and *the convexity of the perforated valve* are the principal distinctions he had observed, the animal being fixed to submarine bottoms by a pedicle similar to that of *Terebratulæ*.[2] From all the above, it seems questionable whether the shells here described deserve to be erected into a distinct *genus*. They probably constitute simply a section of *Discina;* but as nothing is known of the internal impressions and characters, I shall for the present admit it with reserve and doubt.

Geol. range.—This section seems to have originated in the Silurian period, above which no examples have as yet been discovered.

Examples: *T. terminalis*, Emmons, sp ; ? *cancellata*, G. Sow.; ? *punctata*, Sow., sp.

Genus—Siphonotretra, *De Verneuil*, 1845.

Type—S. unguiculata, *Eichw.*, sp.   Int., Pl. IX, figs. 261—268.

Crania (unguiculata), *Eichwald*.
Terebratula (part), *Eichw.*
Siphonotretra, of all modern Authors.

---

[1] Considérations Zoologiques et Géologiques sur les Brachiopodes ; Comptes rendus des Sciences de l'Académie des Sciences; Paris, 5 Août, 1847.

[2] In the An. and Mag. of Nat. Hist., vol. iv, 2d ser., p. 316, 1849. Mr. Morris seems to doubt that *Trematis* was the same as *Orbicella*, D'Orb.; a reference to the types mentioned by both authors will convince any one of the correctness of my statement.

Animal unknown. *Shell* oblong oval, inequivalve; valves unarticulated; *larger* or *ventral valve* most convex, with a straight, thick, perforated, conical beak, more or less removed from the hinge line; foramen circular or elongated, opening on the back of the beak, and communicating with the interior of the shell by a cylindrical tube or siphon, destined to afford passage to the muscle of attachment; no area nor deltidium; each valve has a wide, crescent-shaped cardinal edge, covered by numerous horizontal lines of growth. *Smaller valve* slightly convex and depressed, the hinge-line forming an arch which merges imperceptibly into the lateral margins. Structure calcareo-corneous, with a distinctly punctured structure arranged in tubular layers; surface smooth, but presenting numerous lines of growth, and scattered or closely packed, slender, hollow spines, dilated at their base, and somewhat quincuncially arranged; no calcified supports for the arms. Muscular impressions unknown.

Obs. One of the species composing this remarkable genus was noticed in 1829,[1] but it was only in 1845,[2] that M. de Verneuil established the genus *Siphonotreta*, which has been recognised and adopted by all subsequent naturalists, on account of the clearly defined characters which separate it both from *Crania* and *Terebratula*, where it had been located, as well as from all the other genera.

Subsequent and most valuable information was thrown on the subject by the memoir of Dr. Kutorga,[3] in which beautiful illustrations and structural details had been delineated with the greatest care. Mr. Morris, who first determined its occurrence in England, (from a specimen in the cabinet of John Gray, Esq., of Dudley,) has likewise added some details.[4] M. de Verneuil pointed out that *Siphonotreta* should be placed near *Lingula* and *Discina*, but it is distinguished from the first by the pedicle not passing between the beaks, and from the second by the form of the perforated valve. I differ, however, from my learned friend in his belief that the perforated beak of *Siphonotreta* corresponds to the imperforate one of *Terebratula*. Mr. Morris considers the genus allied to *Crania;* and Professor King places it in his family *Craniadæ*. Professor Forbes includes *Orbicula* and *Crania* in the same family. My views coincide with those of M. de Verneuil, and I prefer placing it with *Discinidæ*, because the shell differs materially from *Crania* by its remarkable siphon or pedicle opening, no such aperture being visible in any species of the last named genus, while it would be nearly related to *Discina* and *Orbiculoidea* by being fixed by a pedicle issuing from an aperture. Its spines cannot be made use of as a generic character of much value, the same being common to other genera. Mr. Morris states that the spines are moniliform in the English species, a character not yet found in the Russian types.

---

[1] Professor Eichwald; Zoologia Specialis (*C. unguiculata*).

[2] Russia and Ural Mountains, vol. ii, 1845.

[3] Uber die *Siphonotretaeae*, by Dr. S. Kutorga, Verhandlungen der Kaiserlichen Mineralogischen Gesellschaft für das Jahr 1847.

[4] Annals and Mag. of Nat. Hist., vol. iv, 2d ser., 1849.

Geol. range.—The genus has not hitherto been discovered above the Silurian deposits.

Examples: *S. unguiculata*, Eichw., sp.; *verrucosa*, Vern.; *fornicata*, Kut.; *comoides*, Kut.; *Anglica*, Morris; *micula*, M'Coy.

Section A. *Sub-Genus*—?ACROTRETA, *Kutorga*, 1848.[1]

Type—A. SUBCONICA, *Kut.* Int., Pl. IX, figs. 271—275.

Animal unknown; *shell* triangular, larger valve conical; false area flat, and bent back at right angles to the margin of the valve, longitudinally grooved along the centre, and perforated at its extremity by a small circular aperture, the lines of growth encircle the shell and pass uninterruptedly over the false area; the smaller valve is almost flat and operculiform, smooth, and marked by concentric lines of growth. Valves unarticulated?

Obs. Never having been able to procure the inspection of any examples of the species composing this genus, I can offer no decided opinion on the subject. Professor Kutorga places this section in his family *Siphonotretaeae*; Mr. Morris, on the contrary, seems disposed to consider *Acrotreta* identical with *Cyrtia*,[2] in which opinion I cannot coincide, if the figures drawn by the distinguished Russian Palæontologist be correct, and of which I feel little doubt. The shape of the false area, the hollow groove, and the perforation of the summit of the truncated beak, are characters not observable in any species of *Cyrtia*.[3]

Geol. range.—Not hitherto observed above the lower Silurian beds of Russia.

Examples: *A. subconica, disparirugata,* and *recurva,* Kut.

FAMILY—LINGULIDÆ.

Animal fixed by a muscular peduncle passing out between the beaks of the valves; arms fleshy, unsupported by calcified processes. Shell unarticulated and sub-equivalve; texture horny.

Obs. The family is composed of two genera, *Lingula* and *Obolus*.

[1] Uber die *Siphonotretaeae* aus den Verhandlungen der Kaiserlichen Mineralogischen Gesellschaft für Jahr 1847, tab. vii, figs. 7—9.

[2] Ann. and Mag. of Nat. Hist., vol. iv, No. 23, p. 315, 1849.

[3] Having communicated on the subject with my friend Dr. Volborth, of St. Petersburgh, that gentleman states—"Sur aucun de mes exemplaires d'*Acrotreta* je n'ai pu constater de charnière ni d'apophyses dentaire; mais le nombre des echantillons est si petit que je n'oserai émettre une opinion positive à se sujet; cependant la physionomie générale de cette espèce la rapproche tellement des *Siphonotretaeae* que je leur assignerai des valves libres tant que je n'aurai les charnières et les dents devant moi."

Genus—LINGULA, *Bruguière*, 1789.

Type—L. ANATINA, *Lamarck.* Int., Pl. IX, figs. 276—279.

PATELLA, *Gmel.* (part).
PINNA (part), *Chemnitz.*
MYTILUS, *Dillwyn.*
LINGULA, of all modern Authors.

Shell thin, sub-equivalve, equilateral, more or less elongated, oval or sub-pentagonal, tapering at the beaks, widened at the pallial region; valves unarticulated, held together by the adductor muscles; both slightly convex, but depressed, beak of the valve more pointed and rather exceeding the other in length; surface smooth or concentrically striated, and covered by an epidermis. *Animal* attached to submarine bottoms by the means of a long

Fig. 50.

Lingula anatina.

peduncle passing out between the beaks of the two valves; no calcified supports; on each side of the mouth is situated an elongated subspiral arm, fringed exteriorly with numerous cirri. In the interior, several muscular impressions are visible; close to the beak are the conjoined depressions left by the pedicle muscle (A), under these, and laterally near the margin of the shell exist two large impressions due to the decussating muscle which produces the sliding action of the valves on each other (B), and towards the centre are situated two small oblique oval scars, caused by the posterior pair of adductor muscles (C), which in the larger valve are divided by a blunt rounded projection, and in the smaller by an elongated mesial elevated crest; under the last described impressions another triangular depression was occupied by the combined extremities of the anterior pair (D): shell almost entirely composed of laminæ of horny matter, perforated by minute tubuli.

Obs. Professor Owen having described the anatomical characters of the animal, I shall only notice that Cuvier was the first author who pointed out the internal dispositions of this remarkable genus,[1] and which prompted him to create a special class, to which he applied the term *Brachiopoda.*

Geol. range.—This genus is among the most ancient forms of animal life; it first appeared in the lowest Silurian epoch, and has continued uninterruptedly to the present period, where it is still represented by several forms living on the shores of tropical regions, and varying so little in external aspect through its long continued existence as to render specific distinctions often difficult to appreciate.

Examples: L. anatina, Lam.; *hians,* Swains; *Audebardi,* Brod.; *semen,* Brod., *albida,* Hinds; *Dumortieri,*[2] Nyst; *ovalis,* Sow. (non Reeves); *Lewisii,* Sow.; *Beanii,*

[1] Mémoir sur l'Animal de la Lingule (*L. anatina,* Lam.), Mém. du Muséum, vol. i, p. 69, pl. vi, 1802.

[2] In Part I, this species is every where erroneously spelt *Dumontieri.*

Sow.; *truncata*, Sow.; *quadrata*, Eichw.; *squamiformis*, Phil.; *spatulata*, Hall; *Credneri*, Geinitz; *tenuigranulata*, M'Coy; *Davisii*,[1] M'Coy, &c.

Genus—OBOLUS, *Eichwald*, 1829.[2]

Type—O. APOLLINIS, *Eichwald*. Int., Pl. IX, figs. 280—285.

OBOLUS, *Eichwald, De Verneuil, D'Orbigny*, and the generality of Authors.
UNGULA, *Pander*.[3]
ORTHIS, *V. Buch* (part).
AULONOTRETA, *Kutorga*.[4]

Animal unknown. *Shell* sub-equivalve, orbicular, equilateral, slightly transverse or elongated, depressed; valves unarticulated, being held together by muscular action. The *larger valve* (ventral of Owen) is most convex, with a short obtuse or pointed beak, and wide flattened cardinal edge or false area, over which the semicircular concentric lines of growth of the surface pass uninterruptedly; the flattened cardinal edge is longitudinally grooved by a semicylindrical furrow, destined to afford room for the passage of the muscular fibres of attachment; *smaller valve* rather shorter than the other, slightly convex or almost flat, without any prominent beak or vertex, the hinge line forming a regular arch, and passing imperceptibly into the lateral margins; the cardinal edge is likewise flattened and horizontally striated, but not interrupted by a mesial furrow; external surface smooth or irregularly sculptured by minute undulating wrinkles. Structure calcareo-corneous. In the interior of the larger valve, a narrow slightly elevated longitudinal ridge extends to about half the length of the valve, and four small oval muscular impressions are situated one on each side near the cardinal angles, and the other two towards the centre of the valve, a little beyond the base of the mesial ridge.

Obs. The genus *Obolus* was created by Professor Eichwald for the reception of two Russian shells, which could not be properly admitted into any of the sections hitherto established. MM. de Verneuil, Pander, D'Orbigny, Morris, &c., have most properly considered *Obolus* a member of the family *Lingulidæ*: the manner in which the pedicle passes between the beaks being essentially similar to that of *Lingula*, the structure is rather more solid than that of the last named genus, and from Dr. Carpenter's microscopic examination would appear to be impunctuate; the muscular impressions though different in their details bear also some analogy to those of *Lingula*. Professor Kutorga has ably described and illustrated this genus, in his paper on the *Siphonotretaeæ*, wherein

[1] This is the most ancient species at present known, see Professor M'Coy's paper, in the 'Ann. and Mag. of Nat. Hist.,' vol. viii, p. 405; and to Mr. Salter's very interesting observations on the *Lowest Fossiliferous beds of North Wales*, in the 'Reports of the Brit. Ass.,' for 1852.

[2] Zoologia Specialis, vol. i, p. 274.

[3] Beiträge zur Geognosie der Russischen Reich, 1830.

[4] Uber die *Siphonotretaeae*, Verhandlungen der Kaiserlichen Min. Gesellschaft für das Jahr 1847.

he changes the name to *Aulonotreta*, and objects to Professor Eichwald's denomination because the shell under consideration does not resemble the old Greek coin *Obolus*, and does not think far-fetched names befitting; but were we to reform all the absurd denominations existing in science, no material good would be procured, and imménse confusion, on the other hand, would ensue. If, therefore, a name is not essentially objectionable, it should not be changed, and we will therefore retain the old name *Obolus*, which has become current in science.[1] I have already stated the reasons for not admitting the genus under consideration into the same family with *Siphonotretaeae*, and for placing it, on the contrary, close to *Lingula*.[2]

Geol. range.—*Obolus* is one of the most ancient forms, abounding in the lower Silurian beds of Russia, and in the Wenlock Shales of England, above which no authenticated example has been as yet observed.[3]

Examples: O. *Apollinis*, Sch. (=*A. polita*, Kut.); *sculpta*, Kut., sp.; *Davidsoni* and *transversa*, Salter, sp.

[1] See also Mr. Morris's observations, in the An. and Mag. of Nat. Hist., vol. iv, 2d ser., 1849.

[2] I take this occasion to express my sincere thanks to Dr. Volborth and Professor Kutorga for having enabled me to examine, and kindly presented me with an extensive series of examples of this and other Russian forms.

[3] Having recently been requested by Mr. Salter to examine some curious impressions and internal casts derived from the Wenlock shales of Walsall, Parker's Hall, Dudley, and Woolhope, with a view of ascertaining the genus to which they belonged, I was not a little interested to find, after a minute comparison with an extensive series of *Obolus Apollinis* and *sculpta*, that our English Upper Silurian Fossil really belonged to the Russian genus, which had not hitherto been noticed above the Lower Silurian Rocks of Russia. Our British specimens appear to belong to two distinct species, and attained considerable dimensions, one of them is of a transversely oval shape, and appears to have been a thin shell, while the other is more circular, and appears from the depth of the muscular and other impressions to have been a very thick shell. Mr. Salter will shortly describe the two forms under the names of O. *Davidsoni* and O. *transversa*, and in order to facilitate the comparison of the different muscular impressions, I have accompanied these few observations by woodcuts illustrating both the Russian and British species.

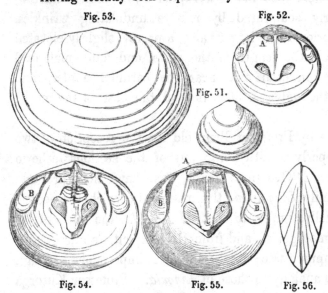

Fig. 53.   Fig. 52.
Fig. 51.
Fig. 54.   Fig. 55.   Fig. 56.

Fig. 51. Obolus Apollinis, *from Ruma.*
„   52. *Interior of the ventral valve.*
„   53. O. transversa (*Salter*).
„   54. O. Davidsoni (*Salter*), *Dudley, interior of ventral valve.*
„   55. *Interior of dorsal valve.*
„   56. *Side view of both valves.*

PLATE I.

Fig.

1. The anatomy of *Terebratula flavescens*, exposed by the removal of the right moieties of the valves, pallial lobes, and bent portions of the fringed arms. Magnified 4 diameters.

2. A dissection of the *Terebratula flavescens*, showing the muscular system, peduncle, and calcareous loop. The valves have been divaricated at the hinge. Magnified 6 diameters.

3. The visceral mass and cavity exposed in the cardinal part of the perforated valve of the *Terebratula flavescens*. Magnified 6 diameters.

4. The alimentary canal laid open, and part of the liver of the *Terebratula flavescens*. Magnified 20 diameters.

5. The anatomy of *Lingula anatina*, exposed by the removal of the valve and pallial lobe answering to the ventral ones in *Terebratula*, and principally to show the auricles, 1, 1, and the intestinal sinus, 8. Magnified 3 diameters.

6. The alimentary canal, liver, and intestinal venous sinus of *Lingula anatina*. Magnified 12 diameters.

7. Five embryos, removed from the ovarian cavity in the pallial sinus of *Lingula anatina*. Magnified 120 diameters.

3 *a*

PLATE. I.

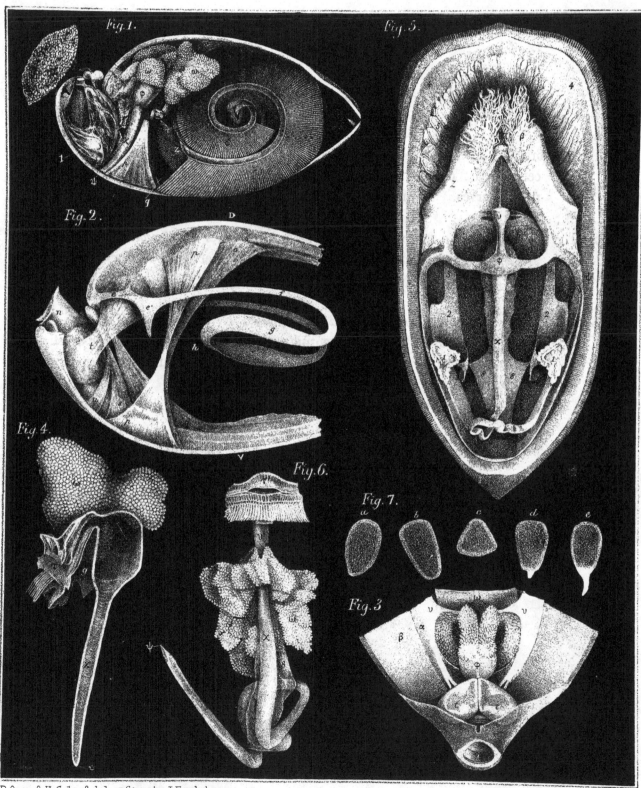

Fig. 1.

Fig. 5.

Fig. 2.

Fig. 4.

Fig. 6.

Fig. 7.

Fig. 3.

R. Owen & II Scharf, del. on Stone by J. Erxleben

Day & Son Lithrs. to The Queen.

PLATE II.

1. The brachial aponeurosis and spiral arms of *Terebratula flavescens*, showing the central part of the nervous system, with the brachial and beginning of the pallial nerves. Magnified 6 diameters.

2. The peduncle and a large proportion of the soft parts of *Terebratula flavescens,* showing the principal ramifications of the pallial nerves or the dorso-pallial fold of the mantle. Magnified 12 diameters.

3. Some of the soft parts of *Lingula anatina*, exposed by the removal of the valve and pallial lobe, answering to the ventral ones in *Terebratula*, and of the digestive organs, chiefly to show the trunks of the visceral and muscular nerves. Magnified 4 diameters.

Fig. III. *Fig. I.*

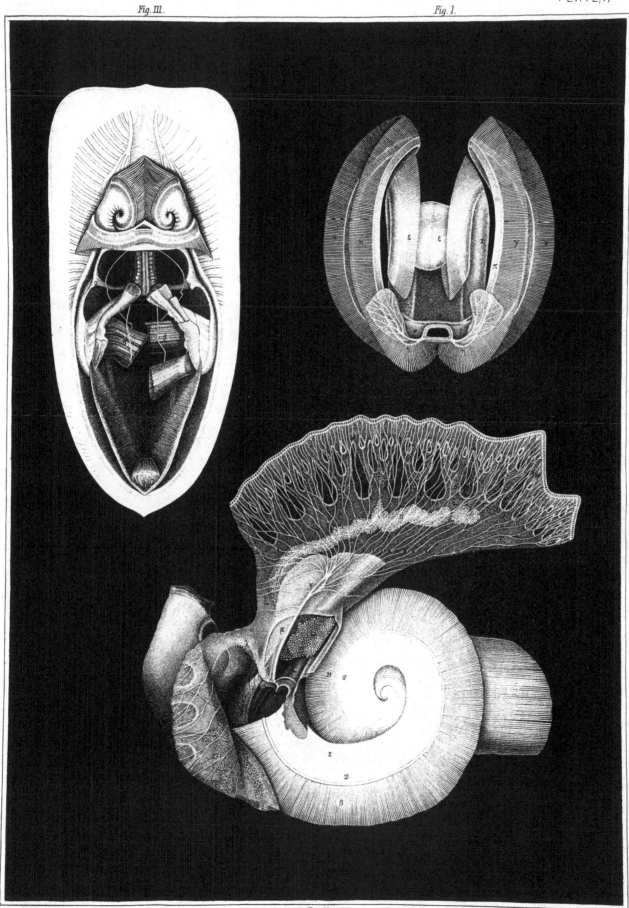

Fig. II.

PLATE III.

Terebratula flavescens.

Fig.

1. The visceral mass and portions of the mantle-lobes, dissected to show the hearts and principal venous sinuses.   Magnified 8 diameters.

2. The soft parts, as seen from the dorsal aspect, and exposed by the removal of the dorso-pallial fold and spiral extremity of the arms.   Magnified 6 diameters.

3. A similar view of the soft parts, with a minor degree of dissection.

The following are the letters or marks of reference used for the same parts in each of the figures.

D Dorsal, or imperforate valve.

     a Cardinal process.

     b Dental sockets.

     c Cardinal plate.

     d Median crest.

     e Crus, or basal process, of loop.

     e' Descending process of loop.

     f Produced plate of loop.

     g Reflected plate of loop.

     h Transverse, or confluent plate, of loop.

IMPRESSIONS OF

     o' Adductor longus anticus.

     p'   „   „   posticus.

     q' Adductor brevis.

v Ventral, or perforate valve.

     i Foramen.

     k Deltidium.

     l Cardinal teeth.

q'' Adductor brevis.

s' Retractor inferior.

o'' Adductores longi.

u' Cardinalis.

r' Capsularis.

ψ Position of vent.

Dp Dorso-pallial lobe.

       5 Sinuses.

Vp Ventro-pallial lobe.

       6 Sinuses.

       m Sheath of the peduncle.

       n Peduncle.

MUSCLES.

o Adductor longus anticus.

p ,,   ,,   posticus.

q Adductor brevis.

r Capsularis.

s Retractor inferior.

t Retractor superior.

u Cardinalis.

v Brachial aponeurosis.

w Basal fold of do.

x Lateral fold of do.

y Spiral fold of do.

z Muscular canal of arms.

α Bead, or border of insertion of.

β Brachial fringe.

γ Produced portion of fringed arm.

δ Reflected portion of do.

ϵ Spiral portion of do.

ζ Æsophageal ring.

η Pallial nerves.

θ Visceral nerves.

ι Subœsophageal ganglia.

κ Common trunk of pallial and brachial nerves.

λ Ventro-pallial nerve-trunk.

μ Dorso-pallial nerve-trunk.

ν Loop of do.

ξ Pallial nerves.

o Circumpallial nerve.

π Brachial nerve.

ρ Muscular nerves, lateral pair.

σ   „   „   median pair.

τ Mouth.

υ Œsophagus.

φ Stomach.

      „   cardiac portion.

      „   pyloric portion.

χ Intestine.

ψ Vent.

ω Liver.

1 Auricles.

2 Ventricles.

3 Arteries.

4 Circumpallial sinus.

5 Dorso-pallial sinus, (branchia in *Lingula*

6 Ventro-pallial sinus.

7 Common trunk of pallial sinuses.

8 Intestinal sinus.

9 Intervisceral sinus.

10 Testes.

11 Ovaria.

12 Ovarian loops of ventro-pallial ovaries.

Fig.1

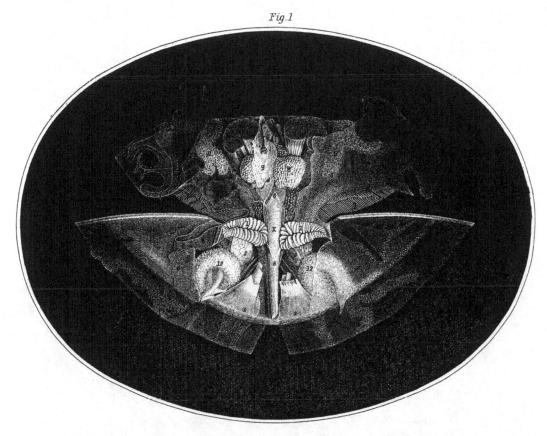

Fig. II

Fig.III

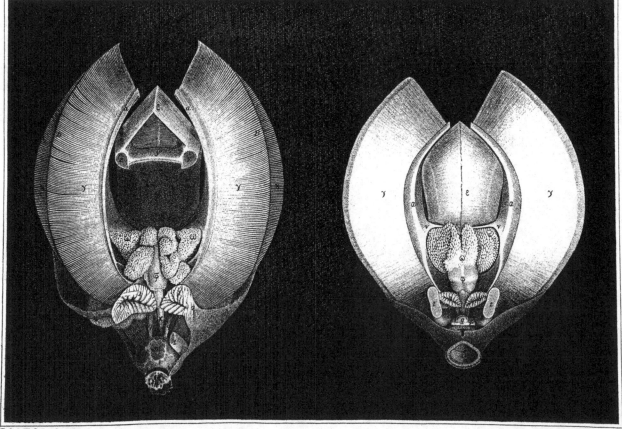

PLATE IV.

Fig.

1. Vertical section of the perforated valve of *Waldheimia* (*Terebratula*) *australis*, through the median plane, from the beak at A to the margin at B; showing that it mostly consists of but one layer, everywhere traversed by vertical passages. At *a a*, *b b*, and *c c*, are indications of the successive marginal additions which the shell has received. The layer, separated by the line *d d*, has the appearance of being a production subsequent to the general enlargement of the shell, apparently for the purpose of strengthening its older and thinner portion.—Magnified 6 diameters.

2. Portion of vertical section of the loop-bearing valve of *Terebratella rubicunda*, taken in the transverse direction through the origin of the calcareous appendages, *a a*. The higher magnifying power enables the vertical passages to be more distinctly seen; and it is shown that the calcareous appendages are not traversed by them.—Magnified 18 diameters.

3. Portion of fig. 1, near the extremity A, considerably enlarged, so as to show the ordinary trumpet-like form of the vertical passages, and the remarkable contraction which they exhibit in the inner and (probably) later-formed layer. The line *d d*, corresponds with the same line in fig. 1.—Magnified 100 diameters.

4. Another portion of fig. 1, at the part c, considerably enlarged, showing the occasional bifurcation of the vertical passages. Magnified 100 diameters.

5. Fragments of the shell of *Kraussia rubra*, separated by spontaneous disintegration.— Magnified 100 diameters.

6. Thin lamella of shell of *Waldheimia* (*Ter.*) *australis*, showing the parallel disposition of the flattened prisms, and the regular arrangement of the passages which intervene between them; and at *a* the outcrop of these prisms on the internal surface of the shell.—Magnified 100 diameters.

Fig.

7. Internal surface of shell of *Waldheimia* (*Ter.*) *australis*, showing the imbricated arrange ment of the extremities of the prisms, which are seen longitudinally at *a.*— Magnified 100 dimeaters.

8. External surface of shell of *Waldheimia* (*Ter.*) *australis*, showing the large trumpet-shaped ends of the vertical passages, covered-in by the opercular disks, which have radiating lines proceeding from them. Magnified 100 diameters.

9. Calcified membrane of the shell of *Waldheimia* (*Ter.*) *australis*, having attached to it some of the cæcal tubuli, crowded with cells. Magnified 150 diameters.

10. Portion of the surface of the calcareous loop of *Waldheimia* (*Ter.*) *australis*, showing the imbricated arrangement of the extremities of the flattened prisms, and the absence of perforations.—Magnified 100 diameters.

11. Portion of the internal surface of the shell of *Terebratulina caput-serpentis*, showing the imbricated arrangement, and the internal orifices of the vertical perforations, arranged in rows.—Magnified 100 diameters.

12. Portion of the external surface of the shell of *Terebratulina caput-serpentis*, showing the external orifices of the perforations, opening somewhat obliquely upon the ridges of the plications.—Magnified 100 diameters.

13. Portion of the shell of *Terebratulina caput-serpentis*, partially decalcified by maceration in acid, showing the cæcal tubuli, *in situ.*—Magnified 100 diameters.

14. Portion of the internal surface of the shell of *Megerlia truncata*, showing the usual imbricated arrangement and the internal orifices of the vertical perforations, with two of the peculiar elevations or 'bosses,' on which no passages open.— Magnified 100 diameters. [N.B. By an oversight on the part of the artist, the surface of these bosses is not represented as partaking of the general imbricated character of that of the rest of the shell, which is really the case.]

Pl. IV

PLATE V.

at *a* is seen a partial layer of prismatic substance arranged vertically, which was probably the seat of the muscular attachment in the young shell; at *b* a similar layer, in the situation of the muscular attachment in the adult shell.—Magnified 40 diameters.

9. Portion of the internal surface of the shell of *Crania Norvegica*, which affords the muscular attachment, highly magnified, showing the extremities of the component prisms, and the large orifices *a a* of the perforations.—Magnified 150 diameters.

10. External surface of the shell of *Thecidium pumila* (Recent), showing large orifices of perforations arranged in rows.—Magnified 100 diameters.

11. Portion of decalcified membrane of *Crania Norvegica*, with its attached cæca, these sometimes bifurcated.—Magnified 100 diameters.

12. External surface of the shell of *Crania Norvegica*, showing the radiating arrangement of the subdivisions of its canals.—Magnified 100 diameters.

13. Decalcified membrane of the prismatic portion of the shell of *Crania Norvegica*, showing at *a a* the cells in their longitudinal aspect, and at *b b* their truncated extremities.—Magnified 150 diameters.

14. Vertical section of shell of *Strophomena aculeata*, showing the shell *a a* to be traversed by perforations *b b*, round which the shell-structure is arranged in a funnel-like manner; *c c*, the matrix, the substance of which has penetrated the canals.—Magnified 34 diameters.

15. Section of shell of *Strophomena aculeata*, nearly parallel to the surface, showing the perforations divided transversely at *a a*, and obliquely at *b b*.—Magnified 34 diameters.

16. Section of shell of *Discina lamellosa*, showing its minutely tubular structure obliquely divided at *a a*, and transversely divided at *b b*, where the tubuli seem to lie in clusters, presenting an appearance of cellular arrangement; at *c* a more distinct indication of cellular structure presents itself.—Magnified 150 diameters.

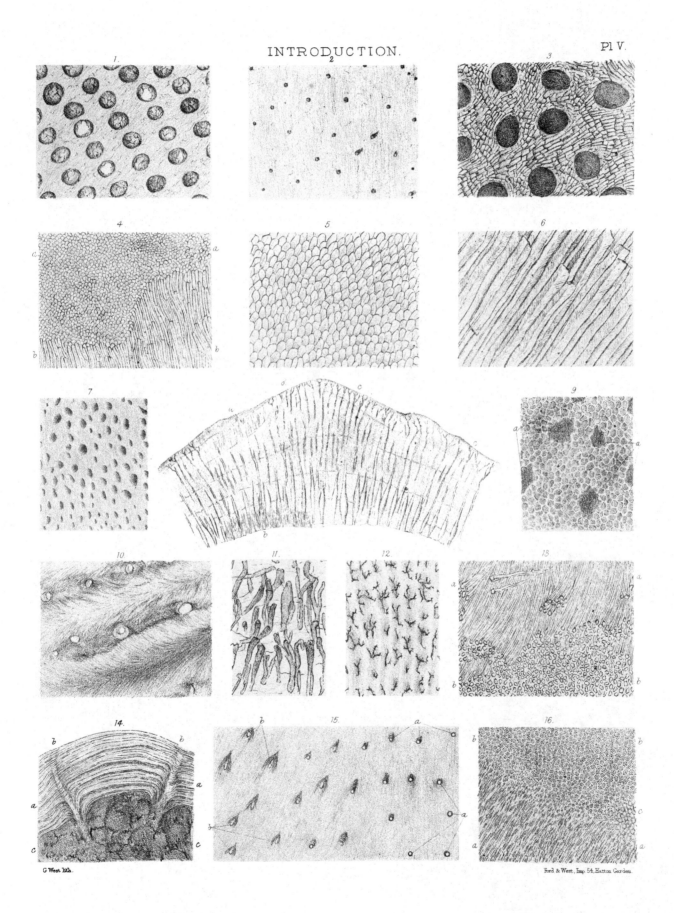

Ford & West, Imp 54, Hatton Garden.

PLATE VI.

The following Letters refer to the same parts and impressions in every figure.

A, *adductor muscle* (*add. longus*, Owen); R, *cardinal muscles* (*add. brevis* and *cardinalis*, Owen); P, *pedicle muscles* (*retractor sup.* and *inf.*, Owen); J, *cardinal process*; L, *loop*; T, *teeth*; B, *sockets*; D, *deltidium*; V, *vascular impressions*; O, *ovarian spaces*; S, *septum, &c.*

Fig.

1. TEREBRATULA VITREA. *Loop only once attached, and short.*
2.   „   „   Interior of dorsal or socket valve.
3.   „   *diphya.* Type of the genus *Pygope* of Link, *antinomia* of Catullo.
4, 5, 6.   „   *elongata* (5 loop), 6 is a portion of internal cast, shewing the place of the dental plate, x.
7. *Terebratulina caput serpentis.*
8.   „   „   Interior of dorsal valve.
9. *Waldheimia flavescens*, Lam. (*australis*, Quoy), type of Professor King's section.
10.   „   „   Interior of dorsal valve.
11. TEREBRATELLA CHILENSIS. Type of M. d'Obigny's section.
12.   „   „   Interior of dorsal valve, shewing the *double attachment of the loop.*
13.   „   „   Section to exhibit the septum, s.
14.   „   *Evansii.* Section to shew the largely developed septum, s, in this species.
15, 16. *Trigonosemus elegans*, Koenig = *Fissurirostra*, D'Orbigny.
17, 18. *Terebrirostra lyra.*
19. MAGAS PUMILA.
20.   „   „   Interior of dorsal valve.
21.   „   „   Section of both valves, to shew the largely developed septum, s, &c.
22. BOUCHARDIA TULIPA, Blainv. = *rosea*, Mawe, &c.
23.   „   „   Interior of ventral or dental valve.
24.   „   „   Interior of dorsal or socket valve.
25.   „   „   Section of both valves.
26. *Megerlia truncata.*
27.   „   „   Interior of dorsal valve.
28. KRAUSSIA RUBRA, Pallas, sp.
29.   „   „   Interior of dorsal valve.
30. MORRISIA ANOMIOIDES, Scacchi, sp. = *T. appressa*, Forbes. Exterior, o, ovarian impressions, v, labial appendages seen through the translucent shell.
31.   „   „   Interior of *dorsal* or *socket* valve, shewing the large notch for the passage of the peduncle in this valve.
32. ARGIOPE DECOLLATA.
33.   „   „   Interior of dorsal valve, shewing 5 ridges, z, and the position of the loop.
34.   „   *Neapolitana.* Interior of dorsal valve, as an example where the loop is only bilobed, one septum or ridge only existing.
35, 36. THECIDIUM RADIATUM, Def., type of the genus. This species was attached by a small portion only of the extremity of the beak.
37.   „   „   Interior of dorsal valve. z, dissepiments; L, apophysary ridge for the support of the arms and brachial membrane; w, bridge-shaped process.
38.   „   „   Interior of ventral valve.
39.   „   *hieroglyphicum.* Interior of dorsal valve.
40.   „   *digitatum.* Ditto.
41.   „   *hippocrepis.* Ditto.

Fig.

43 & 46. THECIDIUM *Deslongchampsii.* Interior of dorsal valve and profile, to shew the surface of attachment, *f.*

44.   „   *sinuata,* E. Desl. Exterior, *f,* point of attachment.

45.   „   „   Interior of dorsal valve.

47.   „   *Moorei.*   Profile to shew the extent of attachment in this species, *f.*

48. SPIRIFER STRIATUS, Martin, sp. Type of the genus. Interior of dorsal valve, shewing the internal
spires, from a specimen in the Cambridge Museum.

49.   „   „   Interior of *ventral* or *dental valve,* from a specimen in the British
Museum.

50.   „   *Lonsdalii.* Interior of *dorsal* or *socket valve,* from a specimen in the British Museum.

51.   „   *speciosus.* Viewed from the beaks, to shew the double areas and pseudo-deltidium.

52.   „   *Mosquensis.* Interior of *ventral* or *dental valve,* type of Fischer's genus *Choristites.*

53.   „   *Cyrtæna.* Interior of *ventral valve.*

54.   „   *Verneuilii.* = genus *Trigonotreta,* Kœnig.

55.   „   *Paillettii.* To show the large development of the mesial fold in this species.

56.   „   *cheiropteryx.*

57.   „   *glaber.* Type of M'Coy's genus *Martinia.*

58.   „   *reticulatus* = genus *Reticularia,* M'Coy.

59.   „   *hemisphæricus* = genus *Brachythyris,* M'Coy.

60. *Spiriferina rostrata.* Interior of *ventral* or *dental valve*, type of D'Orbigny's genus *Spiriferina.*

61. *Cyrtia trapezoidalis.* Type of Dalman's genus *Cyrtia*; F, foramen.

62.   „   „   A fragment of the area and deltidium, to shew that at a certain age the
foramen is closed; in a species from Belgium and China (*C. Murchisoni*),
the circular foramen exhibits a tubular expansion, as seen in some *Rhynchonellæ,* such as *R. octoplicata,* &c.

63. *Cyrtia? heteroclyta.* Exterior.

64.   „   „   Interior of *ventral* or *dental valve.*

65. SPIRIGERA CONCENTRICA. Exterior.

66.   „   „   Interior of *dorsal* valve, showing the largely developed hinge plates,
and four pits for the insertion of the pedicle muscles; *n'* visceral?
foramen, communicating with the tube.

67.   „   „   Interior of ventral valve.

68.   „   *pectinifera* = genus *Cleiothyris,* King (not Phillips).

69.   „   *plano-sulcata* = genus *Actinoconchus,* M'Coy, exhibiting the largely developed marginal
expansions.

70.   „   *phalæna,* Phil., sp. = *Hispanica,* Vern.

71. ATHYRIS TUMIDA. Interior of ventral valve.

72.   „   „   Fragment of the interior of dorsal valve, to show the mesial septum not occurring in *Spirigera.*

73.   „   *Herculea,* Barrande, sp.

74.   „   „   Specimen in which a small portion of the beak and umbo has been removed to
exhibit the shoe-lifter shaped process, *e',* on which repose the dental
plates, x; s, septum in dorsal valve.

75.   „   „   A specimen seen from the beaks, in which a portion of the shell has been
removed, so as to exhibit the shoe-lifter process, *e',* and dental plates, x.

76.   „   „   Section of both valves to shew the largely developed dental plates in the
ventral valve, and septum in the dorsal one. Fig. 74, 75, and 76, were
kindly communicated to me by my friend Mr. Suess, of Vienna.

77. *Retzia Adrieni,* Vern., sp., type of Professor King's genus *Retzia.*

78. *Spirigera? mucronata,* Vern., sp. D, deltidium.

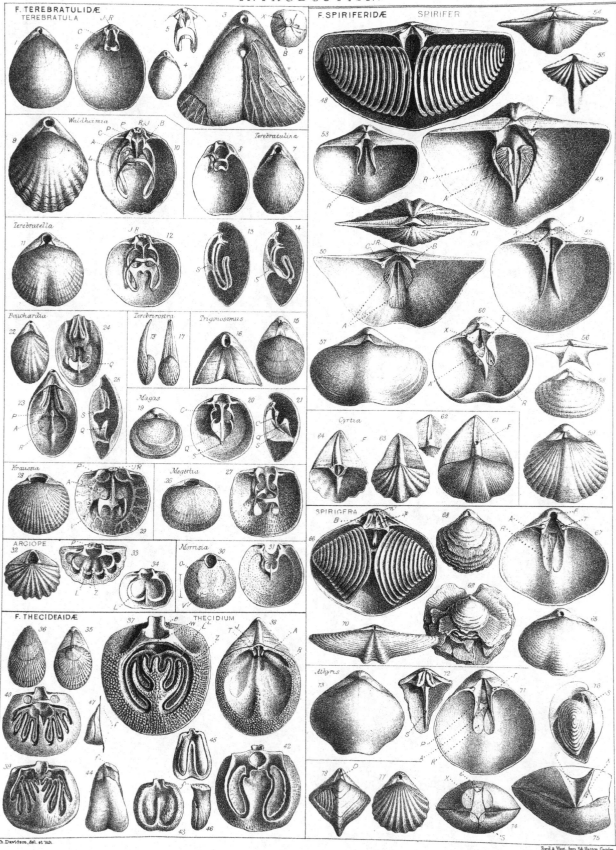

PLATE VII.

The following letters refer to the same parts and impressions in every figure.

A, *adductor muscle (add. longus, Owen)*; R, *cardinal muscles (add. brevis and cardinalis, Owen)*; P, *pedicle muscles (retractor sup. and inf., Owen)*; J, *cardinal process*; L, *loop*; T, *teeth*; B, *sockets*; D, *deltidium*; V, *vascular impressions*; O, *ovarian spaces*; S, *septum*, &c.

Fig.

79.   UNCITES GRYPHUS.   A Belgian specimen, shewing the depressions in the lateral portions of the beak, g''.

80.   „   „   Profile of the same, a portion of the shell having been removed, so as to show that the pouch-shaped depressions do not communicate with the interior of the shell; g', portion belonging to the dorsal valve; g'', that of the ventral valve.

81.   „   „   Profile of a specimen from Paffrath, in which the lateral depressions do not exist.

82.   „   „   A young individual from Chimay, considerably enlarged, to show that in the young state the shell was fixed by means of a peduncle, issuing from a perforation in the beak of the ventral valve.

83.   „   „   The extremity of the beak considerably enlarged, to shew the foramen and deltidium.

84.   „   „   A horizontal section of the hollow beak.

85.   „   „   Interior of a specimen from Paffrath, in the Collection of Professor Beyrich, of Berlin, exhibiting traces of the spiral lamellæ which supported the arms. This sketch was kindly communicated to me by M. Suess, of Vienna, with Professor Beyrich's permission.

86.   „   „   A specimen from Chimay, showing the position of the pouch-shaped expansions (g'', those belonging to the ventral valve), and the attachment of the crural lamellæ forming the spires.

87.   ATRYPA RETICULARIS, Linn., sp.   The first-mentioned type of Dalman's genus *Atrypa*, and of his first division *striatæ*; in this Silurian example no trace of the foraminal aperture is visible.

88.   „   „   from the Eifel, exhibiting in the most beautiful manner the foramen and deltidium.

89.   „   „   Interior of dorsal valve, showing the natural position of the spiral cones.

90.   „   „   Interior of *ventral valve*, from a specimen discovered by M. de Koninck (now in the British Mus.), and in which the vascular impressions are beautifully illustrated.

91.   „   „   Interior of *dorsal valve*, from which the spires have been removed, in order to exhibit the portion of the quadruple impressions left by the adductor muscle.

92.   „   „   Interior, from which the largest portion of the shell of the dorsal valve has been removed, to shew the position and shape of the spiral cones; from a beautiful example in the cabinet of Prof. Buckman.

93.   „   „   A transverse section, to show the extent of space occupied by the spiral cones; M, ventral; N, dorsal valve.

93 *bis*.   „   *marginalis*, Dalman, sp.

94.   „   *prunum*.   Type of Dalman's 2d section *læves*.

95.   STRINGOCEPHALUS BURTINI.   Exterior, showing foramen and deltidium.

96.   „   „   A young example from Belgium, with large triangular aperture.

97.   „   „   A fragment of the interior of both valves, from a specimen in the collection of M. de Koninck: J, cardinal forked process; c, portions of the stems or origin of the loop.   (See pp. 74 and 75, *for figures of the loop*.)

98.   „   „   Fragment of the *dorsal valve*, to illustrate the hinge plates, crura of the loop, cardinal process and quadruple impression of the adductor.

Fig.

99. RHYNCHONELLA LOXIA, Fischer, from Moscow, type of the genus *Rhynchonella* of Fischer.

100. „ *psittacea*, genus *Hemithyris* of M. d'Orbigny; D, deltidium.

101. „ „ Interior of *dorsal valve*: n', foramen for the passage of the intestine.

102. „ „ Interior of the *ventral valve*.

103. „ *octoplicata*. A fragment of the interior, to show the position of the teeth, and dental plates, x.

104 & 107. „ *scaldinensis*, D'Archiac, to show the tubular expansion of that portion of the deltidium which encircles the foramen.

105. „ *rimosa*. To show that from the great incurvature of the beak, there exists no visible foramen in some species.

106. „ *concinna*. Fragment of the beak, to show that the deltidium does not always surround the entire foramen, which is partly completed by the umbo of the dorsal valve.

108, 109. CAMAROPHORIA SCHLOTHERMI. Type of the genus.

110. „ „ A specimen with lamellose marginal expansions, from the collection of Professor King.

111. „ *multiplicata*, King. Internal cast.

112. „ „ Interior of a portion of both valves. In the *dorsal valve* is seen the disposition of the *septum*, s; *spatula shaped process*, G; curved lamellæ to which the oral arms were attached, c; cardinal process, J. In the *ventral valve*, the septum, s, and conjoined dental plates, x.

113. „ „ Section of the same, to shew the projection of the different parts; both these last figures are drawn from specimens kindly lent me by Professor King.

114. PENTAMERUS GALEATUS, Dal. The letter s, indicates the position of the septa in the *dorsal* and *ventral valves*.

115. „ „ Longitudinal section of the same.

116. „ *Knightii*. Type of the genus. Longitudinal section, shewing the position and shape of the largely developed septa and plates.

117. „ *galeatus*. Transverse section, to shew the relative position of the parts visible in the longitudinal section, fig. 115.

118. „ *liratus*. A fragment of the dorsal valve, from a specimen in the Mus. of Prac. Geol., shewing, besides the adductor impressions, the processes, c, to which the spiral arms were attached.

119. „ *linguiferus*. Exterior.

120. PORAMBONITES ÆQUIROSTRIS, Schl., sp. Viewed from the beaks; the letters B and x indicate the place and position of the plates.

121. „ „ Profile of the same.

122. „ „ Fragment exhibiting external sculpture (not punctation).

123, 124. „ *intercedens*, Pander.

125, 126. „ *Ribeiro*, Sharpe. Internal cast, showing the position of the plates, and muscular and vascular system, from a specimen in the collection of Mr. Sharpe.

127. ORTHIS CALLIGRAMMA.

128, 129. „ *striatula* or *sinuata*, showing the double area and fissures, = genus *Schizophoria*, King.

130. „ „ Interior *of ventral valve*, from Nehou.

131. „ „ Interior of *dorsal* or *socket valve*, from Nehou.

132. „ „ Fragment of the same valve, from a specimen from Chimay.

133. „ „ Internal cast of the *dorsal valve*, drawn from a beautiful specimen presented by M. de Koninck to the British Museum, shewing the quadruple impression of the adductor, ovarian spaces and vascular impressions.

134. „ *elegantula*, Dal. Interior of dorsal valve.

135. „ *resupinata*. Interior of dorsal valve, from a specimen in the British Museum.

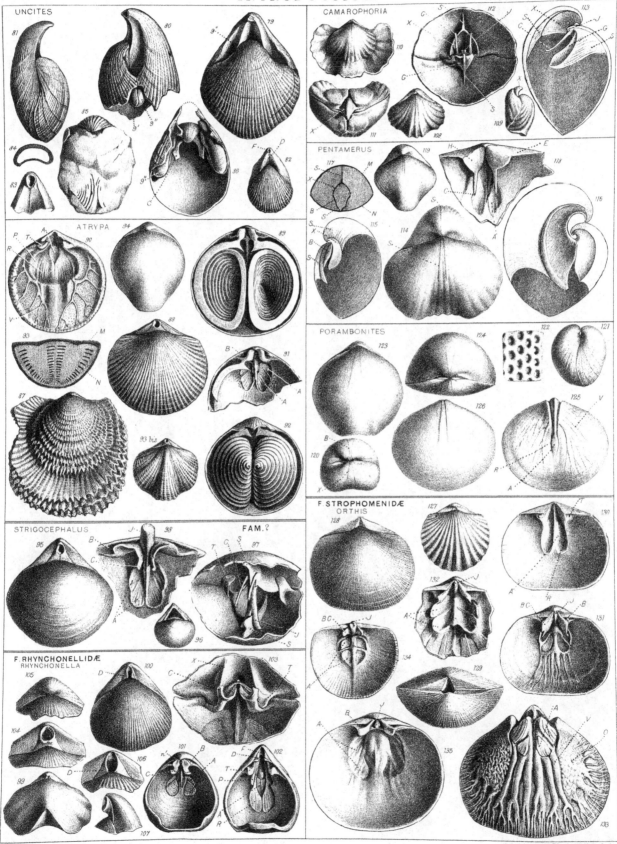

PLATE VIII.

The following Letters refer to the same parts and impressions in every figure.

A, *adductor muscles* (*add. longus*, Owen); R, *cardinal muscles* (*add. brevis* and *cardinalis*, Owen);
P, *pedicle muscles* (*retractor sup.* and *inf.*, Owen); J, *cardinal process*; L, *loop*; T, *teeth*;
B, *sockets*; D, *deltidium*; V, *vascular impressions*; O, *ovarian spaces*; S, *septum*, &c.

Fig.

136. ORTHIS RUSTICA, Sow.
137.   ,,   ,,   Interior of dorsal valve.
138.   ,,   ,,   Interior of ventral valve.
139.   ,,   ,,   The double area and fissures.
140.   ,, *elegantula*.
141—144 ,, *biloba*, Linn., sp. = genus *Dicælosia*, King.
145.   ,,   ,,   Interior of dorsal valve.
146.   ,, *biforata*, Sch., sp. = genus *Platystrophia*, King.
147.   ,,   ,,   Interior of *ventral* or *dental valve.*
148.   ,,   ,,   Interior of *dorsal* or *socket valve.*
149. ORTHISINA ADSCENDENS, Pander, sp. = genus *Pronites* and *Hemipronites*, Pander.
150.   ,, *anomala*, Schl., sp.
151.   ,,   ,,   Interior of dorsal valve, from a specimen in the collection of M. de Verneuil.
152.   ,, *hemipronites*. Viewed from the beaks, to illustrate the double area and closed fissures.
153.   ,, *plana*. Interior of ventral valve, from a specimen in the collection of M. de Verneuil.
154.   ,,   ,,   Interior of dorsal valve.
155.   ,, ? *senilis*, viewed from the beaks.
156.   ,, ? *pelargonatus* = genus *Streptorhynchus*, King.
157. STROPHOMENA PLANUMBONA, Hall. (*Dorsal valve convex.*)
158.   ,,   ,,   Interior of ventral valve.
159.   ,,   ,,   Interior of dorsal valve.
160.   ,, *alternata* (Conrad). Interior of *dorsal valve.*
161.   ,,   ,,   Fragment of the beak, to show the position of the foramen.
162.   ,, *expansa*, Sow., sp. Drawn partly from the original specimen, figured by Sir R. Murchison, in the 'Silurian System,' pl. xx, fig. 14; and from another example in the Mus. of Prac. Geol.
163, 164.   ,, *pecten*, Linn., sp.
165.   ,, *narangoana*, Vern., sp. No fissure or deltidium interrupts the area.
166.   ,, *latissima*, Bouchard. Ibid.
167, 168.   ,, *depressa* or *rhomboidalis*. No foramen seen in this specimen.
169.   ,,   ,,   Young example of the same, from Dudley, in which the perforation or temporary foramen is clearly seen.
170.   ,,   ,,   Interior of dorsal valve.
171.   ,,   ,,   Interior of ventral valve.
172.   ,,   ,,   Section of the valves.
173.   ,,   ,,   Interior of ventral valve, from a specimen from Gothland, in the collection of Professor King, showing curious sub-spiral labial impressions; a similar example has been figured by Professor Quenstedt. (Handbuch der Petref. tab. xxxix, fig. 20.)

Fig.

174.   STROPHOMENA *analoga.* Fragment of the interior of the ventral valve, to show the different muscular impressions and foramen.

175.   „   „   Fragment of the beak, to show the aperture.

176, 177. LEPTÆNA TRANSVERSALIS, Dal.   177 section.

178.   „   „   Interior of ventral valve.

179.   „   „   Interior of dorsal valve.

180.   „   „   View of the double area.

181, 182.   „   *transversa,* Pander = genus *Plectambonites,* Pander.

183.   „   *Davidsonii,* Eng. Deslongchamps. *Lias,* France.

184.   „   *oblonga,* Pander.

185.   „   „   Interior of dorsal valve.

186.   DAVIDSONIA VERNEUILI, Bouchard.

187.   „   „   Interior of ventral or dental valve, from the specimen figured by M. de Verneuil (Geol. of Russia, pl. xv, fig. 9).

188.   „   „   Interior of ventral valve, showing the vascular impressions, &c.

189.   „   „   Interior of dorsal valve (enlarged), from the specimen described by M. de Koninck, and now in the British Museum.

190, 191.   „   „   A specimen in the collection of M. de Koninck.

192.   „   *Bouchardiana,* de Kon.   Exterior, natural size.

193.   „   „   Interior of the dorsal valve, from the original specimen figured by M. de Koninck, and now in the British Museum.

194, 195. KONINCKINA LEONHARDI, Wissman, sp.

196.   „   „   Interior of dorsal valve, with a portion of the spiral lamella preserved.

197.   „   „   Translucid specimen made thinner by acid, and seen from the exterior of the ventral valve.

198.   „   „   Section of the valves. These last three enlarged illustrations were drawn and forwarded to me by Mr. Suess, of Vienna.

198 bis, 199. CHONETES LATA, Sow., sp., from Ludlow.

200.   „   n. sp. The interior of the dorsal valve, from the Devonian Limestone of Nehou, and showing besides the muscular scars, the reniform impressions, v, which prove that this section should be classed · among the *Productidæ.*

201.   „   „   „   ? Interior of *ventral valve,* from a species from the Eifel.

202.   „   „   Interior of *dorsal valve,* from a specimen from the Eifel.

203—205. STROPHALOSIA EXCAVATA = genus *Orthothrix,* Geinitz.

206.   „   *Goldfusii,* Munster, sp., showing the large area and deltidium.

207.   „   *Morrisiana,* King.   Interior of dorsal valve, enlarged ; M, fragment of the *ventral valve,* to show the articulation of the hinge.

208.   „   *Goldfusii.* Internal cast.

209, 210.   „   *Morrisiana,* King. *f,* point of attachment.

211.   „   *Gerardi,* King. (*Note.*—All the specimens from which the illustrations of *Strophalosia* have been drawn, were lent me by Professor King, and form part of the Collection of Queen's College, Galway.)

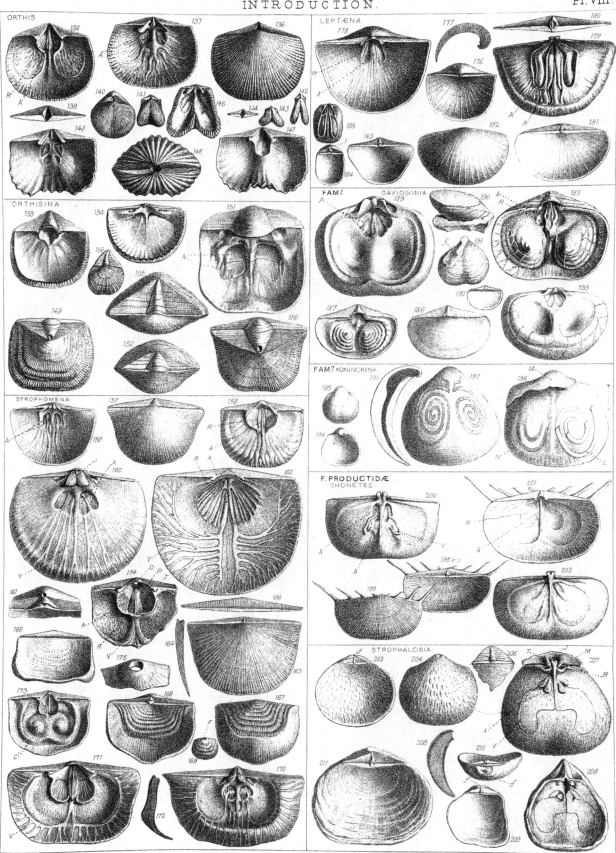

PLATE IX.

The following Letters refer to the same parts and impressions in every figure.

A, *adductor muscles* (*add. longus*, Owen); R, *cardinal muscles* (*add. brevis* and *cardinalis*, Owen); P, *pedicle muscles* (*retractor sup.* and *inf.*, Owen); J, *cardinal process;* L, *loop;* T, *teeth;* B, *sockets;* D, *deltidium;* V, *vascular impressions, &c.*

Fig.

212, 213. AULOSTEGES WANGENHEIMI, Vern., sp. = (*Variabilis*, Helmersen).

214.   „   „   Interior of *dorsal valve* (slightly enlarged).

215.   „   „   Cardinal process, J, from a very remarkable specimen, published by Col. Helmersen. (Bull. Phys. Math., tab. vi, fig. 12.)

216.   „   „   The same, from another specimen in my collection.

217. PRODUCTUS SEMIRETICULATUS. From a specimen in the collection of Professor King.

218.   „   *punctatus.* Ventral valve, with portions of the spines preserved.

219.   „   *horridus.* Interior of dorsal valve.

220.   „   „   Interior of ventral valve.

221.   „   *longispinus.* Interior of dorsal valve.

222, 223.   „   *proboscideus,* Vern. From specimens in the collection of M. de Koninck.

224—226. CALCEOLA SANDALINA, from the Eifel.

227.   „   „   Interior of *ventral valve.*

228.   „   „   Interior of *dorsal valve.*

229. CRANIA BRATTENBURGENSIS (NUMMULO, Stoboeus).

230—232.   „   *antiqua,* Def.

233.   „   „   Interior of the ventral valve.

234.   „   „   Interior of the upper or dorsal valve.

235, 236.   „   *Ignabergensis,* Retzius.

237, 238.   „   *anomala* (PATELLA *anomala,* Müller) = genus ORBICULA, Cuvier and Lamarck.

239.   „   *Hagenovi,* De Koninck, MS.

240.   „   „   Interior of lower or *ventral valve,* exhibiting two bosses, *h,* for attachment of sliding muscle?

241—243. *Spondylobus craniolaris,* M'Coy.

244, 245. *Pseudo-crania antiquissima,* Eichw. sp. Type of Professor M'Coy's genus.

246, 247.   „   *divaricata,* M'Coy, sp.

248—250. DISCINA LAMELLOSA, Brod., sp.

251.   „   „   Interior of lower or *ventral valve.*

252.   „   „   Interior of upper or *dorsal valve.*

253—255. ? *Orbiculoidea elliptica,* Kut., sp. Genus *Schizotreta,* Kutorga.

256—258. *Trematis terminalis,* Emmons, sp. (enlarged) = *Orbicella,* D'Orb.

Fig.

259.   *Trematis terminalis.*   Fragment, showing the outward sculpture (*not punctation*).

260.   „   „   A small specimen figured by Mr. Sharpe.

261—263. SIPHONOTRETA UNGUICULATA, Eichw., sp.

264.   „   „   Interior of dorsal valve.

265.   „   „   Interior of ventral valve.

266.   „   *conoides,* Kut.

267, 268.   „   *verrucosa,* Eichw.

269.   „   *unguiculata.*   Section from Kutorga's paper on the genus.

270.   „   „   A spine enlarged.

271—273. *Acrotreta subconica,* Kut.

274, 275.   „   „   Enlarged.

276, 277. LINGULA ANATINA.

278.   „   „   Interior of *ventral valve.*

279.   „   „   Interior of *dorsal valve.*

280.   OBOLUS APOLLINIS, Eich. (ventral valve), genus *Ungula,* Pander ;   *Aulonotreta,* Kutorga.

281.   „   „   *Dorsal valve.*

282.   „   „   Interior of *dorsal valve,* from a specimen sent to me by Professor Kutorga.

283 & 285.   „   „   Interior of the ventral valve, published in Professor Kutorga's 'Mémoir.'

284.   „   „   Fragment of the interior of dorsal valve, in Professor Kutorga's 'Mémoir.'

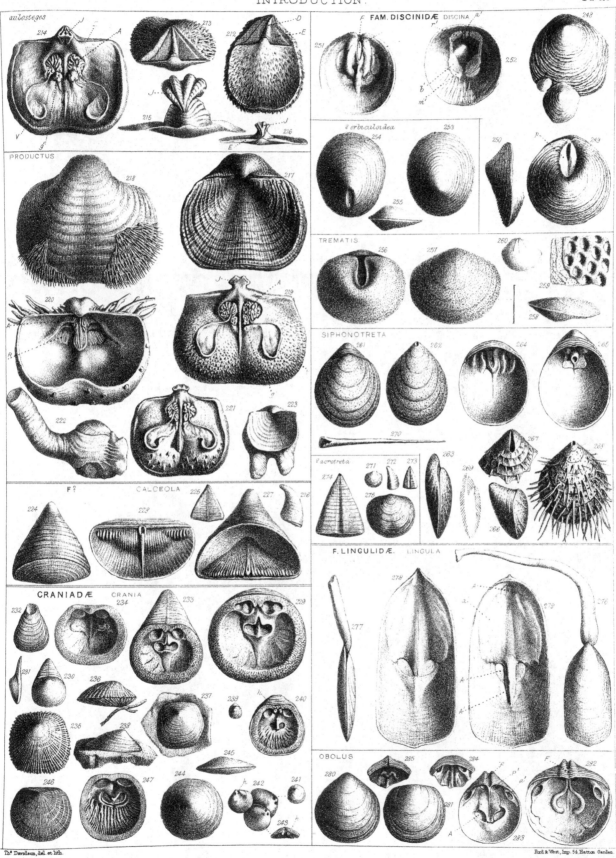

A MONOGRAPH

OF THE

BRITISH FOSSIL BRACHIOPODA.

PART I.
THE TERTIARY BRACHIOPODA.

BY

THOMAS DAVIDSON,
MEMBER OF THE GEOLOGICAL SOCIETY OF FRANCE.

LONDON:
PRINTED FOR THE PALÆONTOGRAPHICAL SOCIETY.
1852.

A MONOGRAPH

OF

BRITISH TERTIARY BRACHIOPODA.

PRELIMINARY REMARKS.

THE Tertiary Deposits, so rich in fossil remains of most classes of the Animal Kingdom, are remarkably poor in Brachiopoda, few species having lived at that period, especially when compared with the multitude of forms that filled the Cretaceous, Oolitic, and Palæozoic seas; the whole class having singularly diminished in number after the cretaceous period up to the present day; for, out of from fifty to sixty recent species, only five are found alive near our shores.

Our supercretaceous strata is principally made up of a vast assemblage of clays, sand and sandstones, gravel and limestones; succeeding and alternating with one another, and sometimes acquiring great thickness and extent. The division of these strata into distinct periods has been the study and aim of many of our most eminent Geologists, who generally seem disposed to admit three principal divisions, as follows:

Tertiary Formations
Upper	. . .	Supérieur	. . .	(Pliocene of Sir C. Lyell.)
Middle	. . .	Moyen	. . .	(Miocene „ „)
Lower	. . .	Inférieur	. . .	(Eocene „ „)

But the exact limits of these have not, in our opinion, been completely established. Some authors object to the use of the terms *Eocene, Miocene,* and *Pliocene,* and M. D'Orbigny in particular, in a small work he has lately published,[1] wherein he proposes to replace Sir Charles Lyell's names by those of—

---

[1] 'Cours élémentaire de Paléontologie et de Géologie Stratigraphiques,' Première Partie, p. 260, 1850; and 'Prodrome de Paléontologie Stratigraphique Universelle,' vol. i, Introduction, p. xliv, 1849.

Terrains Tertiares { 27. Subapennin,
                     26. Falunien,
                     25. Parisien,
                     24. Suessonien,

to avoid the *Cènes*, as he declares to have found the identifications of the recent Eocene species inexact, and not existing in the recent state. Without wishing to decide the question of who is right, and what names should be adopted in preference to others, we shall admit three great divisions, *Lower, Middle*, and *Upper*, each liable to subdivision, as shown by Sir C. Lyell, D'Archiac, and other authors. The important facts lately brought to light in the arrangement of the lower division, are mainly due to the zealous and indefatigable researches of Mr. Prestwich,[1] that author having established that the Barton clays and Braklesham sands were the equivalents of the French *Glauconic Grossiere*, or lowest beds of the Calcaire Grossier, while the London clay and Bognor rocks represent the French *Sables Inférieurs*, and *Lits Coquilliers*. The *Middle Division*, or *Miocene*, appears wanting in our island, and therefore we will not allude to it, but mention that the *Upper Division*, or *Pliocene*, is well represented in our island, and may, like the other great divisions, be subdivided into distinct periods. This upper division, comprising our newest Tertiary Deposits, is composed of sands, gravel, irregular beds of limestone, and layers of greenish marl, known by the name of Crag. These have been divided into three sub-divisions, viz., the *Coralline*, the *Red*, and the *Mammaliferous Crags*.

Some Geologists place the Coralline and Red Crag in the Miocene periods, and the upper crag in the Pliocene.[2] Others have separated the two lower crags, placing one in the Miocene and the other in the Pliocene;[3] and lately,[4] the two lower crags have been considered as belonging to one period (*older Pliocene*), and the upper or Norwich crag (*newer Pliocene*).

Mr. S. Wood objects, however, to this arrangement, now considering the crags to belong to three distinct periods; so that, by subdividing the newer Pliocene into lower (Coralline Crag), medial (Red Crag), and newer (Norwich or Mammaliferous Crag), we would be nearer the probable state of things as far as our island is concerned.

I have deemed it necessary to enter into these few details in order to explain the reasons why we arrange the different species contained in this Monograph more in one period than in another.

For more ample details on the Geology of this system, we would refer to the works

[1] 'Quarterly Journal of the Geol. Soc.,' Nov., 1847.

[2] See Viscount d'Archiac's 'Histoire de Progrès de la Géologie de 1834 à 1845,' vol. ii, 2de partie, p. 447, &c., 1849.

[3] See S. Wood, 'Monograph of the Crag Mollusca,' Introduction, 1848. (Mr. S. Wood is now, however, of a contrary opinion, placing all the Crags in the newer Tertiaries.)

Also formerly by Sir C. Lyell.

[4] Sir Charles Lyell, 'A Manual of Elementary Geology,' p. 362, 1851.

of our British and Foreign Geologists. Of the nine species discovered in the super-cretaceous deposits of our island, four exist still in the recent state, three of which are found near our coasts. It has therefore been thought advisable to include, in this Monograph, our few recent species so closely connected with the fossil ones, ably described and illustrated by Professor Forbes and Mr. S. Hanley in their excellent work on British Mollusca, which valuable information has assisted me so materially in drawing up the descriptions of our five recent species.[1]

It is stated by Professor Forbes, that "Brachiopods are so rare or so local in the British seas, that ordinary collectors are not likely to meet with any. Not long ago a British Brachiopod was one of the brightest gems in any collection so fortunate as to contain it. Three or four minute and undeveloped examples of *Ter. caput-serpentis* and a few *Crania* were all we were likely to meet with, after exploring the great majority of public and private cabinets: of late years a great number of that interesting Terebratula have been taken, and *Crania* has also been found in abundance, so that there is no difficulty in obtaining an indigenous type of the order."

The other three species are still great rarities, two of which, *Ter. cranium* and *Rhynchonella psittacea*, being only known by a few solitary specimens. In the upper Tertiaries, of the six known species, one only, *Ter. grandis*, may be called common, and in the lower Tertiaries of the three forms mentioned, one only has hitherto been found in an incomplete state.[2]

---

[1] Consult Sir Charles Lyell's works; those of Mr. Webster, in the 'Geol. Transactions,' vol. i; Mr. Prestwich's excellent papers in the 'Quarterly Journal of the Geol. Soc.;' those of Mr. Charlesworth, in the 'Mag. of Nat. Hist.,' 1837, and 'Phil. Mag.' for 1835; S. Wood, in the 'Annals of Nat. Hist.' and Palæontographical Society's works; Viscount d'Archiac's numerous papers, particularly his 'Essai sur la Coordination des Terrains Tertiares' ('Bull. Soc. Géol. de France,' vol. x, p. 168, 1839,) and 'Histoire des Progrès de la Géologie,' vol. ii, 2de partie, 1849; as well as the works of Cuvier and Brongniart, Constant Prevost, Deshayes, Sowerby, Hebert, D'Orbigny, F. Edwards, &c. &c.

[2] It gives me very much pleasure in here stating, that since the publication of last year's portion of my work, I have again received kind and zealous assistance from many of the gentlemen there named; and I have now the pleasure of adding those of Messrs. S. Wood, Fitch of Norwich, Harris of Charing, Catt of Brighton, Image of Whepstead, Wood of Richmond (Yorkshire), Falkner of Devizes, Ferguson of Redcar, Prof. Sedgwick, Mr. Carter of Cambridge, Mackey of Folkstone, R. Jones, Griffith of Dublin, Dr. Lewis, and MM. D'Orbigny and Schnurr, who have liberally lent me the specimens contained in their valuable local collections.

The different recent and supercretaceous species of Great Britain may be thus arranged:

RECENT		Found in deep water near our shores		*Crania anomala.* *Argyope cistellula.* *Terebratulina caput-serpentis.* *Terebratula cranium.* *Rhynchonella psittacea.*
TERTIARY OR SUPERCRETACEOUS EPOCH.	*Upper division, Pliocene.*	*Upper or Newest div., Pliocene.*	*Fluvio-Marine Crag of Norwich,* composed of sand and loam, with numerous remains of mammalia, a few land and fresh water; and many marine shells.	*Rhynchonella psittacea.*
		Middle div., Pliocene.	*Red Crag,* composed of sands, gravel, and loam, stained by oxide of iron, abounding in shells, often much rolled and waterworn.	*Terebratula grandis.*
		Lower div., Pliocene.	*Coralline Crag,* made up of calcareous sand, flaggy beds of limestone, and small layers of greenish marl, abounding in mollusca, corals, Zoophytes, &c.	*Orbicula lamellosa?* *Lingula Dumontieri.* *Argiope cistellula.* *Terebratulina caput-serpentis.* *Terebratula grandis.*
	Middle division, Miocene.		Wanting	
	Lower division, Eocene.	*Upper div., Eocene.*	Wanting	
		Middle div., Eocene.	Fresh water and fluvio-marine beds of Headon Hill, Barton clay, limestone, clays and sands (no Brachiopoda), Bagshot and Bracklesham beds of clay, grey and green sands, and sandy beds.	*Ter. bisinuata!*
		Lower div., Eocene.	London clay and Bognor beds, clays and limestones, plastic clay, &c.	*Lingula tenuis.* *Terebratulina striatula.*

CHALK.

Genus—LINGULA, *Bruguière*. 1789.

Shell inequivalved, one valve more convex than the other, more or less oval, elongated, tapering, and pointed at the beaks, widened at its palleal region, without hinge, valves held together by the adductor muscles; attached to submarine bodies by a long muscular peduncle issuing from between the beaks, a groove existing for its passage in that of the larger valve, without any shelly support; structure horny, covered by an epidermis; two muscular impressions on the one, four on the other valve.[1]

Obs. We are not acquainted with any recent British Lingula. Two species are found in the Tertiary strata.

1. LINGULA DUMONTIERI, *Nyst.* Plate I, figs. 10, 10 *ᵃ ᵇ*, 11.

LINGULA DUMONTIERI, *Nyst.* Coq. et Poly. Test. de la Belgique, p. 337, pl. xxxiv, fig. 4 *ᵃ ᵇ ᶜ*, 1843.

— MYTILOIDES, *Nyst.* 1835. Rech. sur les Coq. Fos. d'Anvers, p. 21, pl. iv, fig. 80 (non Sow.)

— FUSCA, *S. Wood.* Mag. of Nat. Hist., p. 253, 1840, (not figured or described.)

— — *Morris.* Cat. of Brit. Fossils, p. 122, 1843.

— — *Bronn.* Index Pal., p. 655, 1848.

— — *Tennant.* A Stratigraphical List of British Fos., p. 17, 1847.

Diagnosis. Shell almost inequivalve, of a lengthened oblong form, valves convex, slightly compressed, and rounded anteriorly; beak acute, not much produced; shell thin, and brittle; surface smooth, shining, of a ferruginous brown colour, and marked by numerous concentric lines of growth. Muscular impressions strongly marked in the interior of both valves, arranged in pairs, as in all Lingulas. Length 12; width 5 lines.

Obs. This species seems to have been first noticed in the Crag of Antwerp by M. Nyst, under the erroneous name of *L. mytiloides*, Sow., which error was afterwards acknowledged by the same author in another work. In England, it was first discovered in the Coralline Crag of Sutton by Mr. S. Wood, who published it under the appellation of *L. fusca*, unfortunately without figure or description, in the 'Annals of Nat. Hist.,' 1840; later in 1843 it was described and figured by M. Nyst, under the name of *L. Dumontieri*, which denomination we feel bound to accept in preference to that of Mr. S. Wood, as a species published without description or figure cannot claim priority. M. Nyst mentions it as abounding in the Crag of Antwerp, where it has not, however, been found perfect, the anterior portion being always broken, and this is generally the case with most of our English specimens, no doubt owing to the extreme thinness of its shell. *L. Dumontieri* somewhat approaches in form to *Lingula Hians* (Swains), from Port Essington, but it is perhaps smaller and more acute posteriorly. However, it is difficult to distinguish the different

---

[1] For more ample details, see General Introduction.

species of Lingula, as these generally vary but little from one another.   Sir C. Lyell remarks, in his 'Elementary Geology,' that the presence of species of Lingula in the Crag is worthy of notice, as these Brachiopoda seem now confined to more equatorial seas.

Plate I: figs. 10 and 11, are figured from specimens found in the Crag of Sutton, and kindly lent to me by Mr. S. Wood.

Fig. 10^a.   Interior of the smaller valve considerably magnified.

Fig. 10^b.   Interior of the larger valve likewise magnified.

2. LINGULA TENUIS, *Sow.*   Plate I, fig. 12.

LINGULA TENUIS, *Sow.*   M. C., tab. xix, fig. 3, p. 55, vol. i, 1812.
—   —   *Morris.*   Catalogue, 1843.
—   —   *Tennant.*   A Stratigraphical List of Br. Fossils, p. 32, 1847.
—   —   *Bronn.*   Index Pal., vol. i, p. 656, 1849.

Diagnosis.   Shell of a lengthened, lanceolate, oval form, flattish, the anterior edge short and straight; surface smooth, bright and shining, marked by numerous concentric lines of growth.   Length 5; width 1½ lines.

Obs.   This small species is described by Sowerby as not unfrequent in the sandy limestone of Bognor.   It has also, I believe, been found near Highgate, in the London Clay; it is easily distinguished from *Lingula Dumontieri*, Nyst, (*L. fusca*, S. Wood,) by its dimensions and more lanceolate shape.

Plate I, fig. 12, from the original specimens in the Min. Con.; we regret having been unable to procure better specimens for illustration.

Genus—ORBICULA, *Cuvier.*   1808.

Shell inequivalved, more or less orbicular, upper valve conical, with apex inclining towards the posterior margin, lower valve depressed, pierced by a longitudinal fissure, from which issues a tendinous pedicle spreading over a small disk placed near the posterior part of the lower valve, and externally adhering to rocks, corals, and other substances; valves smooth, concentrically lamellose or longitudinally striated; structure almost entirely horny; animal symmetrical, mantle free all round, with numerous long, horny and unequal cilia, body small; no calcareous supports; arms fleshy, ciliated, and united at their origin above the mouth, free only at their short spiral portion; muscular system composed of eight distinct muscles, leaving two oval impressions in upper or unattached valve, near the posterior margin, and two others near the palleal region.[1]

Obs.   We are not acquainted with any British recent orbicula, one only is found in the supercretaceous deposits.

[1] For more ample details, consult Professor Owen's excellent description of the animal of this genus, 'Zool. Trans.,' vol. i, 2d part.

3. ORBICULA LAMELLOSA? *Brod.* Plate I, fig. 9, 9$^{a\,b}$.

> ORBICULA LAMELLOSA, *Brod.* Zool. Proc., 1833, p. 124.
> DISCINA NORVEGICA? *S. Wood.* Mag. Nat. Hist., 1840.
> ORBICULA NORVEGICA, *Tennant.* A Stratigraphical List of British Fossils, p. 17, 1847.

Diagnosis. Shell inequivalved, nearly orbicular, longer than wide; upper valve of a flattened conical form, much depressed, vertex acute, prominent, situated at a third of the length of the valve from the posterior margin; surface ornamented only by minute concentric lamella or lines of growth; colour a ferruginous yellow. Structure horny. Length 2, width 1$\frac{3}{4}$ lines.

Obs. The discovery of this small orbicula is due to Mr. S. Wood, who found it in the Coralline Crag of Sutton. Unfortunately only one imperfect specimen of the upper valve has been procured, so that its determination is difficult and uncertain. We have referred it for the present to the recent *Orbicula lamellosa*, which it resembles, until a perfect specimen comes to light, on which a more accurate determination may be arrived at; it has also something of the appearance of *O. lævis.* Mr. S. Wood, in his 'Catalogue of Crag Mollusca,' attributes it to the *Discina norvegica*, which seems a mistake, that species being *Crania* (Patella) *anomala* of Müller. *O. lamellosa* is at present found recent in various parts of the coast of Peru; and is, as well as all the species of the genus, an inhabitant of tropical latitudes; and we may here state, also, that it is found in company with a Lingula *L. Dumontieri*, a genus likewise peculiar to much warmer seas than those which wash our shores.

Fig. 9. A specimen, natural size, from the Collection of Mr. S. Wood.

Fig. 9$^{a\,b}$. Enlarged representations.[1]

---

[1] It has been thought advisable to introduce a reference to the only recent Cranium, *C. anomala*, that occurs in the British Seas, in order to render the sequence of genera referred to in the table complete.

Otho Frederic Müller appears to have been the first to bring this species into notice, styling it "*Vermis singularissimus*," and placing it as an anomalous form of *Patella:* it has been well described by Professor Forbes, in his work on 'British Mollusca.' He states: "The arms are extended horizontally, each forming a rather short, graceful, plume-like curve; the fringes are long, rather stiff, and can be extended slightly beyond the shell; they are of a fleshy white colour; when the upper valve is removed, the fringed arms are seen lodged in it; the ramifying ovaries, which are of a tawny hue, remain on the under valve."

This species has hitherto been only found in the recent state; and it appears common near some of our shores, especially on the West Coast of Scotland. It was first found adhering to stones in deep water in Zetland by Dr. Fleming. Professor Forbes adds many interesting details relative to the localities and depth of water in which it has been collected. Thus, he adds, it is found "off Arran in twenty fathoms (Smith); Loch Fyne in thirty to eighty fathoms; plentiful on stones off Mull in twenty and ninety fathoms; off Lismore in from twenty to thirty fathoms; off Armadale in eighteen fathoms; off Copenhaw Head, Skye, in forty fathoms; on the Ling Banks off Zetland in fifty fathoms (M'Andrew and E. F.); Loch Alsh, Loch Carron, Ullapool, East of Lerwick, in forty fathoms (Jeffreys). In Ireland it has been taken off Youghal by R. Ball, and off Cork by Humphreys. It ranges throughout the Scandinavian seas."

Plate I, fig. 1. Specimens, natural size.

„ fig. 1^a. Interior of attached valve, considerably enlarged.

„ fig. ^{b}. Interior of upper valve, enlarged.

Genus—Argiope, *Deslongchamps.* 1842.

Mégathiris, *D'Orb.* 1847.

Shell inequivalved, variable in shape, semi-orbicular, quadrate, or transversely oval. Valves unequally convex, smooth, or variously ribbed. The larger valve deep, beak produced with a large depressed triangular area; foramen large, completed by the umbo of smaller valve, which generally becomes indented from the shortness of the peduncule, forcing the beak and umbo to lie close to the rock coral or other objects to which it is attached, and thus wearing by friction that portion of the shell; structure strongly perforated; margin thickened and granulated. Hinge-line straight; valves articulating by means of two single teeth in the larger valve, and corresponding sockets in the smaller one. Interior of the smaller valve furnished with a central septum, and sometimes with one or more lateral septa, radiating from beneath the muscular fulcrum, and terminating at some distance from the margin in elevated processes. Apophysary system consisting of a distinct loop originating at the base of the dental sockets, and furnished with converging processes; the loop is folded into two or more lobes, occupying the interspaces of the radiating septa, to which they adhere on their inner sides.

Obs. The shells composing this genus were first separated from Terebratula by M. Deslongchamps,[1] who pointed out its principal differences and affinities to Thecidea, the recent *Anomia decollata*, Chemnitz, or *detruncata*, Gmelin, being named as the type; later M. D'Orbigny (probably unacquainted with M. Deslongchamps' claims of priority) proposed likewise to separate the shells in question from the Terebratulæ under the generic name of Megathiris;[2] since that period the genus has been re-described by Professor Forbes,[3] who, unacquainted with M. Deslongchamps' priority, adopted M. D'Orbigny's name. It is well figured in several works;[4] but on the most important character of the genus authors have not yet agreed, namely, *if the shell was provided with fleshy arms or not.* M. D'Orbigny and Dr. Philippi are stated to have examined the animal anatomically, and to have found none, while Professor Forbes, who has had the same advantages, affirms the animal to be possessed of contorted spiral arms fixed to the margin of the apophysary septa above described, and to the cardinal teeth.[5]   Mr. J. E. Gray[6] places this

---

[1] 'Mém. de la Soc. Linn. de Normandie,' vol. vii, p. 9, 1842. M. Deslongchamps' detailed paper appeared in the 'Bull. de la Soc. Géol. de France,' vol. vii, 2d ser., p. 65.

[2] 'Comptes Rendus Hebdomadaires de l'Académie des Sciences,' August, 1847. Also in the 'Annales des Sciences Nat. Zool.,' tom. viii, p. 241, 3d sec., 1847; and 'Palæontologie Française Terrains Crétacés,' vol. iv, p. 146, 1847.

[3] Forbes and Hanley, 'British Mollusca,' vol. ii, 1849.

[4] In 1785, by Chemnitz, pl. lxxviii, fig. 705; by Sowerby, 'Thesaurus Conchyliorum,' pl. lxxi, fig. 70, &c.

[5] Since writing the above, Professor Forbes informs me that it must have been a small *T. seminulum* he examined, and not an Argiope.   [6] J. E. Gray, 'Annals of Nat. Hist.,' vol. xiv, pp. 271—9.

genus in his second order, or *Cryptobranchia*, stating the oral arms to be "entirely attached in the form of two or more lobed processes sunk into grooves in the disk of the ventral valve." Again Professor King[1] demurs to the order *Cryptobranchia*, and sides with Professor Forbes, believing Argiope to be a true "brachiferous Palliobranch."[2]

The absence of any notice of the loop of Argiope, in the descriptions above referred to, is probably owing to the imperfection of the specimens examined; from its extreme delicacy it is often broken away, both from recent and fossil specimens. The character of the loop and septa approximates this genus to Thecidea. Much variation appears to exist in the raised septa or ribs in the interior of the smaller valve, which in the recent *A. decollata* are of almost equal height,[3] but in all the fossil forms, attributed to the chalk by M. D'Orbigny, and in the tertiary *A. cistellula*, the central rib alone is prominent, while the others are faintly marked, and even imperceptible in most specimens. Dr. Philippi placed the type of this genus, and some supposed similar recent forms, in the genus Orthis, but, as remarked by Professor Forbes, they possess none of the characters of that genus.

[1] King, 'A Monograph of Permian Fossils,' Pal. Soc., p. 81, 1850.

[2] Since writing the above, my attention has been called by Mr. Woodward to the circumstance, that

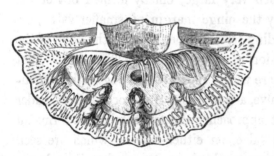

some of the minutest specimens of *A. decollata*, brought by Professor Forbes from the Egean Sea, contains the dried remains of the animal: we have examined two specimens in this condition, one of which is represented in the accompanying woodcut. The mantle adheres closely to the shell as in *Terebratula proper*, and is not seen, except as part of the shell; its margin is simple and *not ciliated:* the oral arms and their connecting membrane are very distinct, owing to the colour, which, is darker and redder than that of the shell. The dried cilia (or *cirri*) present the same glistening appearance as in *Thecidea*. The arms originate as in Terebratula, on the anterior side of the mouth, and diverge right and left, parallel with the margin of the shell, but at some little distance from it; when they arrive at the raised septa they turn inwards, forming two lobes on each side of the middle line: the outline of the arms is therefore four lobed, whilst in other recent species, *A. cistellula*, and in the Cretaceous *A. decemcostata*, which has only one septum, there is probably only one lobe to each arm. The cilia are few and thick. The arms are relatively connected, as in Terebratula, by a membrane filling up the whole interior space, thus forming an apparatus which forcibly reminds us of the Freshwater Polype *Plumatella*, figured by Mr. Hancock.

The distinguished Malacologist, M. D'Orbigny, appears to have mistaken the brachial disk of *Argiope* and the analogous structure in *Thecidea* for the *mantle*, and has founded upon these genera his order *Cirrhidæ*, into which he has admitted, in his most recent publications, the genera *Hippurites* and *Sphaerulites*, and those species of *Diceras which have one valve smaller than the other.*

From the peculiarity of structure above described, we regard *Argiope* as generically distinct from *Terebratula*, but belonging essentially to the same family.

[3] For the sake of reference, we have given figures of the interior of the smaller valve of the recent *Argiope decollata* in Part II, pl. iii, figs. 15, 16, but in which the loop is not introduced.

Ter. lunifera and *seminulum* of Philippi[1] have been referred by some authors to this genus, but I have been able to convince myself, from a perfect specimen of the last-named species, that its internal arrangements are completely dissimilar to those seen in the smaller valve of Argiope.

The genus Argiope seems to have originated, as far as our present knowledge goes, in the cretaceous period, and has continued to our day, one of the species, *A. cistellula*, from the crag being likewise found recent.

4. ARGIOPE CISTELLULA, *S. Wood*.   Plate I, fig. 13[a b].

TEREBRATULA CISTELLULA, *S. Wood*. 1840.   Catal. of Crag Shells, Ann. and Mag. of
Nat. Hist., vol. v, p. 253.
MEGATHIRIS CISTELLULA, *Forbes* and *Hanley*.   History of British Mollusca, pl. lvii,
fig. 9, 1849.

Diagnosis. Shell inequivalved, variable in shape, wider than long or otherwise, in contour hemispherical or transversely suborbicular, more or less truncated above.   Larger valve more convex than the smaller one, in which a longitudinal, central depression is visible, beginning at a short distance from the umbo, and extending to the front.   Beak produced, with depressed triangular area; foramen very large, chiefly formed out of the beak and area of larger valve, and completed by the hinge margin of smaller valve; no visible deltideal plates, a small groove only extending from the beak, along the edge of the foramen.   Hinge line straight, the central retrusion of the opposite margin shallow, but distinct.   The teeth and condyles on either valve are widely separated, owing to the dimensions of the foramen.   In the interior of smaller valve, a central septum proceeds from under the crura, becoming gradually more elevated as it approaches the front, where it forms an elevated central, longitudinal plate, dividing the valve, on either side of which are seen two slightly curved lateral elevations, not projecting much above the surface.   In larger valve, a slightly elevated longitudinal ridge is visible, extending from under the beak to within a third of the length of the valve; the interior and exterior of valves are strongly punctuated; often having the appearance of raised tubercles, the inner edge being more or less thickened and radiatingly scabrous.   Surface smooth, with only a few concentric lines of growth.   The colour is of a light, tawny yellow.   Length 1, width 1, depth a little more than half a line.   *Recent* and *Fossil*.

Obs. The first discovery of this curious little Brachiopod is due to Mr. S. Wood, who mentioned it under the name of *Ter. cistellula*, in his 'Catalogue of the Crag Mollusca,' published in 1840, several specimens of which he had found in the Coralline Crag of Sutton.   Professor Forbes gives a good description of this shell, as found in the recent state, in his valuable work on 'British Mollusca.'   We there find stated, that a few specimens had been found in forty fathoms of water by Mr. Jeffreys and Mr. Barlee, while

---

[1] 'Enumeratio Moluscorum Siciliæ,' pl. vi, figs. 15, 16, 1836.

dredging off Skye; also in thirty fathoms off Croulin Island, near Skye, by Mr. M'Andrew;[1] and lately at the Haaf, or deep-water fishing-grounds of Zetland, by Mr. Barlee. It may be said to be one of our rarest little Brachiopoda, both in the recent and fossil states.

Plate I, fig. 2, one of the recent shapes of this shell.

    „   fig. 2ᵃ, enlarged figure.

    „   fig. 13, nat. size of the Crag specimens, from Mr. J. Wood's collection.

    „   fig. 13ᵃᵇᶜᵈ, enlarged illustrations.

Genus—TEREBRATULINA, *D'Orbigny*. 1847.

Shell inequivalve, ovate, circular or irregularly pentagonal, variable at different stages of growth. Valves convex or depressed; beak more or less produced, and obliquely truncated by the foramen which is excavated out of the substance of the beak, completed by the umbo and by two small rudimentary, lateral, obsolete, deltideal projections, no area or distinct lateral ridges; the smaller valve deepest at the umbo, with two more or less developed auricle expansions. Structure punctuated; surface either minutely striated, plaited, or costellated, articulating by the means of two teeth in the larger, and corresponding sockets in the smaller valve. Apophysary skeleton short, not exceeding a third of the length of the shell, and formed of two short stems, simply attached to the extremity of the socket ridges, which, after converging, are united by a lamella, in the shape of a small, square, tubular ring, bent upwards in front, to the sides of which are fixed the free fleshy arms of the animal, these extending to near the front margin, bent back in the shape of a loop, the outer edges being covered by long cirri; body of the animal small, the edges of the mantle free, pedicule muscles very short; dimensions rarely exceeding one inch and a half.

Obs. The separation of the shells forming this genus from *Terebratula* is due to M. D'Orbigny, who pointed out the difference of their respective apophysary skeletons. We differ, however, from that author, when stating that this genus is deprived of deltidium; it generally is rudimentary, small, and lateral, but in some species, as in *T. substriata* of Schlotheim, the deltidium is large and almost complete. M. D'Orbigny is likewise in error in the statement, in his ' Pal. Française Ter. Crétacés,' vol. iv, p. 58, that the genus *Terebratulina* appeared for the first time in the Cretaceous period. I am perfectly well acquainted with two forms of the genus in the Oolitic period, one of which is the *Terebratulina substriata* of Schlotheim, placed by M. D'Orbigny[2] among the *Terebratulæ*.

---

[1] These specimens have been presented by Mr. M'Andrew to the Museum of Practical Geology, where they were pointed out to me by Professor Forbes; in size and shape they quite agree with those from the Crag.

[2] 'Prodrome,' vol. ii, p. 24, 1850. (This species belongs to the Cor. Rag: has been figured by Zieten in 1832, under the name of *T. striatula*. It is not the same as Sowerby's or Dr. Mantell's species, bearing that appellation.)

T. substriata has all the characters of the genus under notice, its apophysis being annular, and entirely similar to that of the type species *Terebratulina caput serpentis.* In the Cretaceous period this genus appears to have most abounded, and has lived through the Tertiary period up to the present day. It is found recent in many localities.

5. TEREBRATULINA CAPUT-SERPENTIS, *Linnæus,* Sp.   Plate I, figs. 3, 4, 5, 6 (recent); 14, 15 (fossil).

ANOMIA CAPUT-SERPENTIS, *Linn.*   Syst. Nat., 12 (non ed. 10), p. 1153; Fauna Suecica, ed. 2, p. 521; Acta Upsaliens, 1773, vol. i, p. 41, pl. 5, fig. 3.

— — *Born.*   Mus. Cæs. Vind., p. 119.

— — *Chimn.* Conch. Cab., vol.viii, p. 103, pl. lxxviii, fig. 712, 1785.

— — *Dillw.*   Recent Shells, vol. i, p. 293; Index Testaceolog., pl. ii, fig. 22.

— RETUSA, *Linn.*   Syst. Nat., ed. 12, p. 1151; Fauna Suecica, ed. 2, p. 521.

— — *Dillw.*   Recent Shells, vol. i, p. 292.

— PUBESCENS, *Linn.*   Syst. Nat., ed. 12, p. 1153.

ANOMIA PUBESCENS, *Schröter.*   Eintest. Conch., vol. iii, p. 397, pl. ix, fig. 10.

— — *Dillw.*   Recent Shells, vol. i, p.293; Index Testaceolog.,pl.ii,fig. 20.

TEREBRATULA PUBESCENS, *Müller.*   Zool. Danic Prodromus, p. 449, No. 3007.

— CAPUT-SERPENTIS, *Lam.* (non *Retzius.*)   Anim. sans Vert. (ed. Desh.) vol. vii, p. 332.

— — *Sow.*   Gen. of Shells, Tereb., fig. 2.

— — *Philippi.*   Moll. Sicil., vol. i, p. 95, pl. vi, fig. 5; and vol. ii, p. 66.

— — *Reeve.*   Conch. Systemat., pl. cxxvi, fig. 2.

— — *Sow.*   Thesaurus Conch., vol. i, pl. lxviii, figs. 1—4; and pl. lxxii, fig. 116.

— — *Forbes* and *Hanley.*   British Mollusca, pl. lvi, figs. 1—4.

— COSTATA, *Lowe.*   Zoologic. Journal, vol. ii, p. 105, pl. i, fig. 89.

— AURITA, *Fleming.*   Phil. Zool., vol. ii, p. 498, pl. iv, fig. 5; British Animals, p. 369; British Marine Conch., p. 127.

— — *Brown.*   Illust. Conch. G. B., p. 68.

— GREVILLEI, *S. Wood.*   Cat. of Crag Shells, Mag. Nat. Hist., vol. v, p. 253, Dec., 1840.

TEREBRATULINA CAPUT-SERPENTIS, *D'Orbigny.*   1847.   Comptes Rendus de l'Académie des Sciences et Pal. Franç. Ter. Crétacés, vol. iv, p. 58.

Diagnosis.   Shell ovate, rounded or pentagonal, longer than wide, variable in appearance at different ages; valves almost equally convex, beak produced and obliquely truncated by a moderately sized foramen, principally excavated out of the substance of the beak, completed by the umbo and by two small, rudimentary, lateral, deltideal plates; no distinct cardinal area or beak ridges; smaller valve convex, with two more or less developed auricles, much expanded in the young, smaller and more oblique in the adult state, where they sometimes almost disappear from the regular convexity of the valve; margin flexuous; in the young the valves are less convex. Imperforated valve deepest near the umbo, forming at times a rounded, elevated ridge, extending to the front, a corresponding depression or shallow sinus existing in the larger valve; while, in other specimens, a mesial longitudinal depression exists in the smaller valve.

The surface is ornamented by a great number of radiating little ribs or costæ, fewer, coarser, and simple in the young, augmenting rapidly in number at a more advanced period by bifurcation, and by the intercalation of a number of small plaits, intersected by regular concentric striæ, strongly produced in young specimens, giving the shell a granulated appearance, these becoming less distinct as the animal advances in age. The internal calcareous supports in smaller valve short and anneliform; the muscular impressions are marked; no mesial longitudinal septum is perceptible on either valve; their internal edges minutely crenulated; structure punctuated; colour squalid white. Dimensions variable: length 11, width 7, depth 5 lines.

Recent and Fossil.

Obs. This species of shell and its animal has been ably described by Professor Forbes in his work on ' British Mollusca:' "The arms or buccal appendages occupy the greater part of the cavity of the shell. They are fixed to and follow the course of the apophysary skeleton, and appear when the shell is forcibly opened in the form of a pair of brilliant orange or crimson fringed loops, lodged in each half of the cavity of the imperforated valve; the outer margins of each loop bear long cirrhi, also of a brilliant orange or crimson hue, and though the arms themselves cannot be protruded, their cirrhi are very extensile: when the animal is lively, the two valves separate and gape, for no very great distance from each other in front, and from their sides are seen the long crimson cirrhi, extended like a pair of double fringes, and borne somewhat stiffly, and with a slight curve outwards. Towards the edge of the strongly adherent mantle attached to each valve, are placed at regular intervals about forty small cirrhi of a softer texture, which do not appear to be protruded, at least conspicuously, beyond the edges of the shell: these cirrhi are tinged with crimson; also at their bases are seen, when a high magnifying power is used, coloured dots and cavities with vibrating corpuscules, which may be regarded as ocelli, and otilitic capsules. The whole surface of the mantle is studded with vibratile cilia. On each side of the inner surface of the perforated valve is seen an ovarium of an oblong shape and brilliant vermilion colour; and extending beyond these ovaria, in radiating fashion, are the glandular masses of the liver."

In the recent state it is one of our commonest Brachiopoda, first found by Dr. Fleming at Ullapool in Loch Broom, afterwards dredged by different collectors at Oban; on the West Coast of Scotland, Loch Fyne; at Lismore, near Oban, off Armadale, in the Sound of Skye, forty miles West of Zetland, &c., in depths varying from ten to fifty fathoms. It is likewise found in many spots throughout European seas, presenting slight variations in different localities. In the fossil state, it occurs in the Coralline Crag of Sutton, where

Mr. S. Wood discovered a few young specimens, noticed in his Catalogue under the name of *T. gervillei?* The question now is, whether *T. caput serpentis* really occurs lower down in the series, as some persons seem disposed to believe, from similarly shaped shells being found in almost all the series of tertiary deposits, as well as in those of the cretaceous period. We have compared with great care a number of these, and must confess that, in many cases, the variations are so trifling, that we find them in general reproduced on various specimens of the living type. It is only on adult shells, where all the characters are developed, that a positive determination as to specific difference can be established, great similarity existing in the young state of this genus. We believe, however, that there are some differences, which, though slight, may allow us perhaps to establish a few species and varieties, although we do not consider the subject as yet satisfactorily decided. Professor Forbes, in whose judgment I have great confidence, inclines to believe, that the recent type is really found lower down, both in the tertiary and cretaceous period, but for the present I have been unable to satisfy myself that such is truly the case, although it may be so. It is, however, certain, that too many species have been proposed in the genus, which will be referred to in their proper places.

6. TEREBRATULINA STRIATULA, *Sow.*, Sp.   Plate I, figs. 16, 16 ᵃ ᵇ.

TEREBRATULA STRIATULA, *Sow.* (Partim non *T. striatula*, Mantell,) 1829, vol. vi,
p. 69, tab. 536, fig. 5, non figs. 3, 4.
— — *Morris.* (Partim) 1843. Catalogue.

Diagnosis. Shell of a rounded oval or irregularly pentagonal shape, longer than wide; valves convex and compressed; beak not much produced, truncated by a moderately sized foramen, principally excavated out of the substance of the beak, and completed by the umbo and two small lateral obsolete deltideal plates; no distinct area or beak ridges; hinge lines circular; auricles small, often almost indistinct; valves ornamented by a great variety of minute striæ or costæ of unequal width, sometimes bifurcating, but more often augmenting by the intercallation of smaller costæ appearing at different distances from the umbo and beak, and extending to the front. Margin line slightly flexuous. Loop short, anneliform. Length 10, width 8, depth 4 lines.

Obs. Although the distinctions between this shell and the recent caput serpentis are not very great, still we think them sufficient to authorise its separation. Both young and adult specimens of the recent species just mentioned seem in general more convex, and tapering at the beak than in the London clay species, which is wider and more circular at the beak, the valves being likewise much more compressed, and the margin line less sinuous, nor do we find that mesial longitudinal depression so often visible in the cretaceous *T. striata;* the striæ which ornament the valves is so variable in number, as well as in dimensions, that they cannot serve as a distinguishing character; they increase in number rapidly at a short distance from the beak and umbo, much more by intercalation than bifurcation, smaller plaits appearing between the principal costæ, which lying often close to the original ones at their origin, give a false appearance of bifurcation; and draw so near to each other at the margin, that seven or eight may be counted in the breadth of a line, eighty or ninety ornamenting each valve.

This shell is found in the London clay; the largest specimens I have seen, 10 lines in length, were procured by Mr. Bowerbank from the London clay of Sheppey; it has likewise been met with in other localities.[1]

I have felt embarrassed to decide which name this species should retain, being aware that the term *striatula* had been applied by Dr. Mantell, in 1822, to a cretaceous species,[2] but which name was however only a synonym, the shell having received that of *striata*, in 1821, from Wahlemberg.[3] Subsequently Sowerby, as well as many other authors, indiscriminately adopted Dr. Mantell's name, both for the cretaceous and the species under consideration; I have therefore thought it advisable, to avoid a new name, to adopt that of *striatula* exclusively for the Tertiary species, and retaining that of *striata* for the Chalk shell. We may likewise observe, that from the description and figures given by Mr. Morton,[4] of his *Ter. lacryma*, we should conclude our London clay species distinct from the American one.

Fig. 16. Specimen from the London clay of Sheppey, in the collection of Mr. Bowerbank.

16*a*. An enlarged illustration.

Genus—TEREBRATULA, *Lhwyd.* 1699.[5]

Shell inequivalve, equilateral, elongated, transverse or circular; exterior smooth, rarely striated or plaited; valves generally convex, with or without sinus, corresponding to a mesial fold in smaller valve. Front straight or sinuated: beak always truncated by an apicial, emarginate, or entire foramen; deltideum in one or two pieces; internal ribbon-shaped lamella (partly supporting the ciliated arms), attached only to the crura, short or elongated, and more or less folded back on itself; animal fixed to submarine bodies, by muscular fibres passing through the foramen; structure perforated.[6]

[1] M. Deshayes has, within the last few years, found, in the Calcaire Grossier of the neighbourhood of Paris, young specimens of a shell, which is probably the same as our British species; but, from the great similarity presented by specimens at that age, it is very difficult to decide as to identity.

[2] 'Geol. of Sussex,' pl. xxv, figs. 7, 8, 12.

[3] 'Wahlemberg Petref.,' pl. vi, 1821.

[4] J. S. Morton, 'Synopsis of the Organic Remains of the Cretaceous Group of the United States,' pl. xvi, fig. 6; and pl. x, fig. 11. According to Sir C. Lyell, Viscomte D'Archiac, and M. D'Orbigny, *T. lacryma* would belong to the age of the London Clay; and it is placed by M. D'Orbigny in his 'Terrain Parisien Prodrome,' vol. ii, p. 396.

[5] 'Lithophylaci Britannici Ichnographia.' London, 1699.

[6] See Introduction, and Part III, p. 26.

Obs. We are only acquainted with one species of British recent Terebratula, and two from the supercretaceous period.[1]

7. TEREBRATULA GRANDIS, *Blumenbach.*   Plate I, fig. 18, Plate II, figs. 1—8.

> TEREBRATULITES GRANDIS, *Blumenbach.* 1803. Specimen Archæologiæ Telluris Ter-
> rararumque Imprimis Hannoveranarum, Tab. 1, fig. 4;
> figured, but not named, likewise by Knorr, in 1755.
> Lapides Deluvii Universalis Testes, Tab. B iv, figs. 1 and 2;
> reproduced by Walsh and Knorr in 1768. Die Natur-
> geschichte der Verst, Tab. B iv, figs. 1, 2; also Ency.
> Meth., pl. 239, fig. 2.
> TEREBRATULITES GIGANTEUS, *Schl.* 1813. Beiträge Zur Natur. der vers in Leonhard's
> Mineral Tasch., vol. vii.
> —   —   *Schl.* 1820. Die Petrefactenkunde, p. 278, No. 48.
> TEREBRATULA VARIABILIS, *Sow.* M. C., vol. vi, p. 148, Tab. D lxxvi, figs. 2—5, 1829.
> —   GIGANTEA, *V. Buch.* 1838. Mém. de la Soc. Géol. de France, vol. iii,
> 1 ser., p. 222, (non pl. xx, fig. 3, which is *Ter. bisinuata,*
> Lamarck.)
> —   VARIABILIS, *Galeotti.* 1837. Mémoir sur la Const. Géol. du Brabant,
> p. 151.
> —   —   *Nyst* and *Westendorp,* 1839. Nouv. Rech. sur les Coquilles
> Fossil de la Province d'Anvers, p. 15, No. 37.
> —   MAXIMA, *Charlesworth.* 1837, Mag. Nat. Hist., p. 92, figs. 13, 14.
> —   SOWERBII, *Nyst.* 1843. Coq. et Polyp. de la Belgique, p. 335, pl. xxvii,
> fig. 3 *a, b.*
> —   VARIABILIS, *Morris.* 1843. Catalogue.
> —   —   *Tennant.* A Stratigraphical List of Brit. Fos., p. 17, 1847.
> —   GRANDIS, *Bronn.* 1848. Index Palæontologie, p. 1237.
> —   VARIABILIS, *Brown.* 1838. Illust. of Foss. Conch. of Great Britain, pl. liv,
> figs. 16, 19, 21, 22.
> WALDHEIMIA VARIABILIS, *King.* 1850. A Monograph of Permian Fossils, p. 60.

Diagnosis. Shell inequivalved, variable in form, oval, more or less orbicular, generally longer than wide; valves almost equally convex; beak produced, not much recurved, and obliquely truncated by a large circular foramen, separated from the umbo by a narrow, cicatrised, concave deltideum, disunited in the young age; beak ridges indistinct; smaller valve regularly convex, two undefined, slightly elevated plaits existing towards the frontal

---

[1] The discovery of this species, *Terebratula cranium,* as a British Shell, is due to Dr. Fleming, who obtained three specimens in deep water to the Eastward of Bressay, in Zetland. It is found also on the Coast of Norway and in the Northern Seas. It has been well described and figured by Professor Forbes, G. B. Sowerby, and by Colonel Montague, in the eleventh volume of the 'Linnean Transactions,' &c. It has not been noticed in the fossil state; and, as stated by Sowerby, is well distinguished from *Ter. vitrea* by the greater length of its loop. Plate I, figs. 8, 8[a b].

edge in adult specimens; larger valve convex, with two shallow sinuses corresponding to the biplications of the imperforated valve; margin line slightly sinuated; surface of valves smooth, marked only by a few concentric lines of growth; loop in smaller valve short, attached only to the crura, and extending to about a third of the length of the valve; teeth of larger valve very strong, the posterior portion of the valve near the beak very thick in adult shells; structure punctuated; length four inches two lines, width three inches, depth two inches.

Obs. On examining the figures given by Blumenbach in 1803, as well as those before him, by Knorr and Walsh in 1756 and 1768, there seems to be little doubt that the shell under notice belongs to the first named author's *Terebratula grandis*, these views being likewise admitted by Professor Bronn and others; subsequently Schlotheim (1813) gave to this shell the name of *giganteus*, while acknowledging it to be the same as Blumenbach's species, referring to his work and figure. Baron von Buch adopts this last author's name, giving as synonyms, *T. bisinuata*, Lamarck, and *T. variabilis*, Sow. (the figure of the specimens inserted in the Mém. de la Soc. Géol. de France are those of the Lamarckian species, and not *grandis* of Blumenbach), but both Baron von Buch and Professor Bronn place under the same name shells, in my opinion, quite distinct, such as *T. bisinuata*, Lam., *Fragilis* of Kœnig, which certainly do not belong to the type of *T. grandis;* the shell of *T. bisinuata* being very thin and brittle, as Kœnig's name expresses, while that of the crag is very thick and strong, besides differing by various other characters of shape. In 1829 it was described and figured by Sowerby in the Min. Conch., under the name of *T. variabilis*, which name has been in general use in England. Mr. Nyst (in 1843) rejects Mr. Sowerby's name, on account of a similar denomination having been given by Schlotheim in 1813 to a plaited Lias Terebratula,[1] and proposes in lieu that of *T. Sowerbii*. Mr. Charlesworth advocated likewise in 1837, the name of *T. maxima* as a substitute for *variabilis*, but as we have stated above, we cannot but believe it must be the same as that figured by Blumenbach in 1803; the figure representing a large Terebratula measuring two inches nine lines in length, and two inches one line in breadth, and in every respect identical in shape to many of our crag specimens, which are, as Sowerby's name expresses, very variable in form, some being almost circular, others oval, and even considerably elongated, convex, or depressed, regularly rounded, or with a slight biplication in front. Mr. S. Wood having been able to trace specimens from less than a line in length to the largest dimensions, much confusion has arisen from the desire of some authors to combine, under one name, some strongly biplicated forms, such as *Ter. ampulla* of Brocchi, *T. bisinuata*, and *Pedemontana* of Lamarck, thus extending beyond reasonable limits the characters assignable to the type form. The best figures of this species, as found in England, are those given by Mr. Charlesworth; that author, besides transcribing many interesting details on the species, adds that, " during the early

---

[1] Schlotheim's *Ter. variabilis* is a true *Rhynchonella*, and therefore does not belong to the Terebratulæ, properly so called.

stages of growth, the edges of the valves do not encroach upon one another, there being simple adaptation of the margins in an even line from the excessive thinness of the shell at the line of junction; when, however, the shell has attained the length of three inches, the front edge is rather suddenly produced with an abrupt termination, which is received in a notch in the opposite valve."[1] The foramen is also variable in dimensions, sometimes so large as almost to admit the tip of the small finger, being doubtless the largest *Terebratula* as yet come under our notice. The internal apophysary skeleton is short,[2] never exceeding a third of the length of the valve; we therefore are at a loss to make out why this shell is announced by Professor King[3] as an illustration of his genus *Waldhemia*, where, according to that author, the process extends to near the frontal margin.

In England this species is common to both the *red* and *coralline crags*, but larger and more abundant in the last, where it is found in great numbers at Sudbourn, near Orford, on the estate of the Marquis of Hertford, and of the road leading from Aldborough to Leiston. In the red crag the specimens rarely have the two valves united, and are in general much water-worn, the best localities being Sutton, Walton on the Naze, Ramsholt, Felixtow, &c. On the continent it has been met with in several localities by Mr. Nyst, at Pellenberg near Louvain, and in the Crag of Antwerp in Belgium. In France it is stated to have been found à la Gresille near Doué; and during a late journey to Valogne, M. de Gerville showed me a basketful he had obtained from Bohon in the Dep. de la Manche. Blumenbach's types were obtained from Osnabruck.[4]

Plate I, fig. 18. A specimen from the Red Crag, in the collection of Mr. S. Wood.

„ II, fig. 1. Illustrates a remarkably fine specimen in the Museum of Mr. Bowerbank.

„   fig. 2. Interior of the large valve.

„   fig. 3. Interior of smaller valve, showing the loop, which, though incomplete, did not extend to a greater length, from a specimen in the collection of M. Bouchard.

„   figs. 4, 5, 6, 7. Different ages.

„   fig. 8. Enlarged portion of the beak from a young shell, showing that at a certain age the deltidcum was only lateral.

---

[1] 'Mag. of Nat. Hist.,' vol. i, 1837.

[2] From the coarseness of the sand which fills the interior, it has been as yet impossible to clear the apophysis completely; but, from a specimen in M. Bouchard's collection, it evidently does not exceed the dimensions stated.

[3] 'A Monograph of Permian Fossils,' Pal. Soc., p. 69.

[4] There appears to be a difference of opinion as to the age of this deposit; De Münster places it in the *older Pliocene*, while Goldfuss considers them *Miocene*, or *middle Tertiary*. See M. le Viscomte D'Archiac's valuable notes on this subject, 'Histoire des Progrès de la Géologie,' vol. ii, 2d part, p. 849, 1849.

8. TEREBRATULA BISINUATA? *Lamarck.* Plate I, fig. 17.

TEREBRATULA BISINUATA, *Lamarck.* 1819. An. Sans Vert., vol. vi, p. 252, No. 32; and Davidson, Notes on an Examination of Lamarck's Species of Foss. Ter., Annals and Mag. of Nat. Hist., 2d ser., vol. v, No. 32, pl. xiii, fig. 32.

—   —   *Deshayes.* Coq. Foss. des Env. de Paris, tom. i, p. 65.

—   FRAGILIS, *Kœnig.* 1825. Icones Sect., No. 45.

—   GIGANTEA, *V. Buch.* 1838. Mém. de la Soc. Géol. de France, vol. iii, p. 222, pl. xx, fig. 3. (non *T. gigantea,* Schl.)

—   GRANDIS, *Bronn.* 1848. Index Pal., vol. ii, p. 1237. (non *T. grandis,* Blum.)

—   BISINUATA, *D'Orb.* Prodrome, 1849, vol. ii, p. 395.

—   —   *D'Archiac.* Hist. des Prog. de la Géol., vol. iii, p. 276.

Diagnosis. Shell ovate, longer than wide; large valve convex, longitudinally keeled, a slight lateral depression causing the valve to project more in that part; beak nearly straight, and obliquely truncated by a large foramen separated from the umbo by two small deltideal plates; surface smooth, marked only by a few concentric lines of growth: length twenty-two, breadth nineteen lines.

Obs. The discovery of this Terebratula in our lower tertiary deposits is due to Mr. Prestwich, who unfortunately found only the larger valve in the Bracklesham sands, now considered to be the equivalent of the lower beds of the Calcaire Grossier of France, where this same species has been found at Grignon, Parnes, Chaumont, Courtagnon, Mouchi, &c. We cannot, however, help stating that our determination of this shell is not as satisfactory as we might have wished from the imperfect state of the specimen; this seems a rather thicker shell than the French *T. bisinuata,* the foramen is larger, and the deltideum more concave, and it *much* resembles a shell found in more recent tertiary formations in Sicily. It is to be hoped that future researches in the Bracklesham sands may bring to light a more complete specimen.[1]

Plate I, fig. 17^{ab}. Specimen from the collection of Mr. Morris.

[1] Since writing the above, and after my plate was printed, Mr. Cunnington, of Devizes, was so fortunate as to discover another and more complete specimen than the one described above. Mr. Cunnington's specimen is smaller, much compressed by pressure, but with both valves; it is circular, rather longer than wide, very thin, not much convex, and presenting *scarcely* any trace of the bisinuation,—a characteristic of the Lamarckian type, but which is not always distinctly visible on the Grignon specimens; and I think we may consider our determination as probable, if not certain. Mr. Cunnington's specimen is from the London Clay of Barton or Hordwell Cliffs, on the Hampshire coast. It measures in length 13½; in width 12 lines.

Genus—RHYNCHONELLA, *Fischer.* 1809.

Shell more or less circular, elongated or transverse; valves convex; beak acute, slightly or greatly recurved; no true area; peduncular perforation variable in form; entirely or partially surrounded by a deltideum, either lateral and rudimentary, complete, or tubular. Surface of valves variously ornamented, rarely smooth, generally striated, plaited, or costellated; structure non-punctuated, divisible into laminæ of extreme tenuity; valves articulating by means of two teeth in larger, and corresponding sockets in the imperforated valve; apophyses in smaller valve consisting of two short lamellæ, separate and moderately curved upwards, flattened and grooved, to which are attached the free, fleshy, spiral arms.

Obs. We consider M. D'Orbigny's lately proposed genus *Hemithiris* as synonymous with that established by Mr. Fischer in 1809, under the name of *Rhynchonella*, both genera being made up of shells of the same form, structure, apophysis, muscular impressions, &c.; the only distinction, according to M. D'Orbigny, consisting in the erroneous statement, that the foramen in *Hemithiris* was deprived of deltideum; while, in *Rhynchonella*, it was complete and tubular.

If we now examine with care some of M. D'Orbigny's types of *Hemithiris*, such as *H. psittacea* and *spinosa*,[1] we find that, in the first, the socket-walls do not form simply the sides of foramen, but that there exist two narrow plates, gradually widening as they proceed from the extremity of the beak and overlaying the socket walls, as we see in many species, considered to be possessed of deltideum. In the second, *H. spinosa*, D'Orb., we find regular lateral, deltideal plates, which do not completely surround the foramen, and this becomes evident in specimens where the beak is not so much recurved as to conceal the aperture. If, on the other hand, we cast our eyes on some of the shells admitted by M. D'Orbigny as true Rhynchonellas, the deltideum is found disposed in the following three ways:—

1. Small and lateral, as in *R. concinna*, the foramen not entirely surrounded, a small portion being completed by the umbo.
2. Surrounding the foramen, without tubular expansions, as in *R. obsoleta*.
3. Surrounding the foramen, thickened and produced in the form of a tubular expansion, as in *R. compressa, scaldinensis, vespertilio*, &c.

From the above it will be clearly seen that there is no important distinction between the two genera; both, in our opinion, are referable to the genus *Rhynchonella*. These views as to the form and value of the deltideum were published in a paper by M. J. Deslongchamps,[2] and by Mr. Morris[3] some years back. The last-named author,

---

[1] M. D'Orbigny afterwards placed this shell into his genus *Acanthothiris*.
[2] Soc. Linnéenne de Normandie, 1837.
[3] 'Quarterly Journal of the Geol. Soc.,' vol. ii, Part I, pp. 382-9.

besides admitting the presence of a slightly developed deltideum in *R. psittacea*, places it in the same genus as *R. vespertilio*, only under the generic name of *Hypothyris*, Phillips, not being aware at the time of Fischer's priority of date, his type having been established on *Rhynchonella loxia*, a Russian Oolitic species, somewhat similar in form to our common Lias *R. acuta*, and possessing the essential characters.

9. RHYNCHONELLA PSITTACEA, *Chemnitz*, Sp.   Plate I, fig. 7^{ab} recent; fig. 19^{ab}, tertiary.

ANOMIA ROSTRUM-PSITTACI, *Chemnitz*. Conch. Cab., vol. viii, pl. 106, p. 78, fig. 713, 1785.

— PSITTACEA, *Gmelin*. Syst. Naturæ, p. 3348.

— — *Turt*. Conch. Dic., p. 5, figs. 42, 44.

— — *Dillw*. Recent Shells, vol. i, p. 296; Index Testaceolog., pl. ii, fig. 27.

— — *Mawe*. Linn. Conch., pl. xv, fig. 3.

LAMPAS PSITTACEA, *Humphrey*. Museum Calonnianum, p. 834, 1797.[1]

TEREBRATULA PSITTACEA, *Lamarck*. An. sans Vert., vol. vi, 1819.

— — *Turt*. Dithyra Brit., p. 336.

— — *Fleming*. Brit. Animals, p. 368.

— — *Thompson*. Ann. Nat. Hist., vol. xiii, p. 433; Brit. Marine Conch., p. 127.

— — *Brown*. Illust. Conch. G. B., pl. lxiii and pl. xlvi, figs. 2, 3, 4.

— — *Alder*. Cat. Moll. Northumberland and Durham, p. 74.

— — *Crouch*. Introd. Lam. Conch., pl. xiii, fig. 4.

— — *Sow*. Genera Shells, Terebratula, fig. 5; Thesaurus Conch., vol. i, p. 342, pl. lxxi, figs. 78—80.

— — *Sow*. (jun.) Conch. Manual, p. 202.

— — *Gould*. Invert. Massach., pl. 142. fig. 91.

— — *Reeve*. Conch. Systemat., pl. 126, fig. 5.

— — *Forbes* and *Hanley*. British Mollusca, pl. lvii, figs. 1—3.

— — *Forbes*. Mem. Geol. Survey, vol. i, p. 407.

— — *Morris*. Catal. of Br. Fossils. 1843.

— — *Lyell*. Geol. Proc., vol. iii, p. 119; Encyc. Meth., p. 224, fig. 3.

— — *Bronn*. Index Pal., vol. ii, p. 1247.

— — *Tennant*. Strat. List of Brit. Fossils, p. 17, 1847.

HYPOTHYRIS PSITTACEA, *King*. Annals Nat. Hist., vol. xviii, p. 238; and Monograph of the Permian Fossils of England, p. 65, 1850.

— — *Morris*. Quart. Journal of the Geol. Soc., vol. ii, pp. 382—9.

[1] In 1797, Humphrey proposed the Genus *Lampas*, in which he places the following species:— *L. columbina* (*Anomia terebratula*, Linn.), *L. truncata*, *L. pectiniformis*, *L. psittacea*, *L. caput-serpentis*, and *L. sanguinea*. This genus, which has been adopted by no one that I am aware of but Mr. Gray, in the Coll. of Br. Mus., is composed of a number of species, all of which have since been placed in separate genera. In 1767, Davila ('Catal. Syst. et Raisonné des Cur. de la Nature') gives two very good figures, pl. xx, fig. B, of *R. psittacea*, under the name of *Bec de Perroquet*. Lister likewise illustrates this species in Tab. 211, fig. 46, of his 'Hist. sive Synopsis Meth. Conchyliorum,' 1685.

HÉMITHIRIS PSITTACEA, *D'Orb.*   Considérations Zoologiques et Géologiques sur les Brach.; Comptes Rendus de l'Académie des Sciences, 1847; et Pal. Franç. Ter. Crétacés, vol. iv, atlas pl. 490, fig. 10.

RHYNCHONELLA PSITTACEA, *Woodward.* 1851.   Manual of the Mollusca, p. 8.

Diagnosis. Shell of a rounded, somewhat globosely triangular shape, compressed laterally; smaller valve gibbous, with a slight mesial fold often becoming indistinct by the regular convexity of the valve; larger valve convex, but less so than the smaller one, with a wide, shallow sinus, beginning at a little distance from the beak, and extending to the front; beak acuminated, acute and recurved, leaning considerably over the umbo, and perforated by an elongated foramen extending from under the extremity of the beak to the umbo, which completes the circumference, two narrow deltideal projections laterally edging the aperture. A slight flatness is visible on each side of the beak, the ridges of which are indistinct; the marginal line of larger valve indents considerably the hinge line of smaller valve near the umbo.

The surface of the valves are ornamented by a great number of closely disposed small striæ, sometimes dichotomising at a short distance from the beak and umbo, also augmenting by the intercallations of smaller costæ, from forty-five to fifty in number on each valve; besides these, numerous lines of growth intersect the longitudinal striæ. The internal calcareous appendages in the imperforated valve consist of two small curved lamellæ, not exceeding more than one third the length of the shell, to which, in the living state, two free fleshy arms are fixed, and, according to Professor Owen, forming six or seven spiral gyrations, decreasing towards their extremities, which, when completely unfolded, extended beyond the shell to twice its longitudinal diameter. In the interior of the smaller valve a rudimentary septum divides the muscular impressions visible on either side of it. In the larger valve the dental plates are strong, and extend to the bottom of the valve, leaving also between them the corresponding muscular and peduncular impressions. Structure unpunctuated, and formed of minute plates. Colour blackish, and slightly glossy. Length 11; breadth 10; depth 8 lines.

Obs. This remarkable Brachiopod has been long known, its animal having been well described by Professor Owen.[1] The structure of the shell has likewise been examined microscopically by Dr. Carpenter,[2] and has been considered a type for the division of the great family of *Terebratulæ* into punctuated and unpunctuated genera. This species is very interesting, inasmuch that it is found recent in our seas, and fossil in our upper tertiary strata. We are indebted to Professor Forbes[3] for valuable information relative to its habitat, as found recent on our coasts. He states, that Mr. Maclaurin procured it from the Berwickshire coast;[4] also by Laskey, off Aberlady Bay; and in deep

---

[1] See Professor Owen's Paper, 'Transactions of the Zool. Soc.,' vol. i, 2d part.

[2] See Dr. Carpenter's Report, British Association, p. 18, 1844.

[3] Forbes and Hanley, 'British Mollusca,' vol. ii, p. 339, &c., 1849.

[4] Berwickshire Nat. Club, vol. i, p. 213.

seas, by dredging, in the Firth of Forth. Professor King showed me, likewise, some years ago, a valve of this shell, brought up from the depth of thirty fathoms, at a distance of twenty-five miles from the northern coast of Northumberland. It is therefore evident that it occurs at a short distance from our coast in the recent state; nor is there anything extraordinary in this fact, as it is a well-known Boreal shell, found abundantly in the northern seas near Greenland, Norway, Labrador, Melville Islands, &c. In the fossil state, it unquestionably occurs in our newer Pliocene beds or Mammiferous Crag of Bramerton, near Norwich, whence it has been collected by Sir C. Lyell, S. Wood, &c.; and several specimens may be seen in the collections of these gentlemen, as well as in those of Mr. Fitch,[1] Professor Tennant, British Museum, &c. These last are from Postwick. It is stated as occurring in similar beds in Ayrshire.

Fig. 7, 7^{ab}, are recent specimens.

Fig. 19, 19^{ab}, are the Crag specimens.

---

[1] Mr. Fitch informs me, that *R. psittacea* is found in a Crag containing a mixture of land, fresh water, and marine shells; that Mr. Woodward called it "*T. plicatilis*, and rare," in his 'Geol. of Norfolk;' but that he does not believe it rare, although he has only found two perfect specimens with both valves. Single valves, in a very fragile, imperfect state are not uncommon: in one of Mr. Fitch's specimens, the *deltidium* is distinctly seen.

PLATE I.

Plate 1.

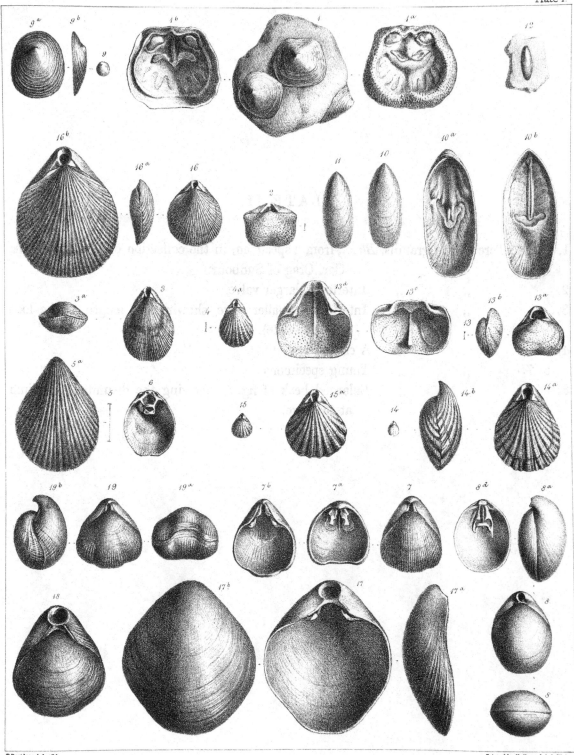

T. Davidson del et lith.

Printed by Hullmandel & Walton

PLATE II.

Fig.

1, 1ª, 1ᵇ. Terebratula grandis, *Blum.*, from a specimen, in the collection of Mr. Bowerbank, Cor. Crag of Sudbourn.

2. „ „ Interior of larger valve.

3. „ „ Interior of smaller valve, showing the length of the loop (not complete).

4. „ „ A circular variety.

5, 6, 7. „ „ Young specimens.

8. „ „ Enlarged beak of fig. 5, showing the disunited deltidium at that age.

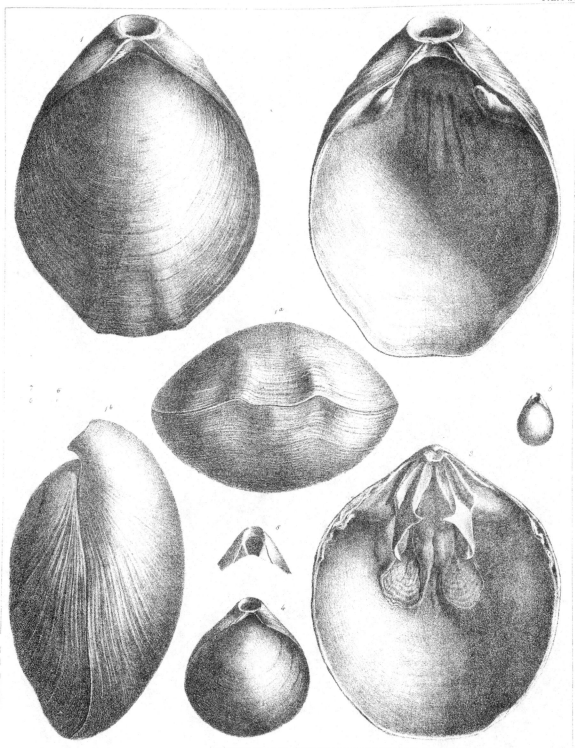

A MONOGRAPH

OF THE

BRITISH FOSSIL BRACHIOPODA.

PART II.
THE CRETACEOUS BRACHIOPODA.

BY

THOMAS DAVIDSON,

MEMBER OF THE GEOLOGICAL SOCIETY OF FRANCE.

LONDON:
PRINTED FOR THE PALÆONTOGRAPHICAL SOCIETY.
1852—1855.

A MONOGRAPH

OF

BRITISH CRETACEOUS BRACHIOPODA.

PRELIMINARY REMARKS.

In the present Monograph, our object is to describe those British species of Brachiopoda that lived during the deposition of the widely spread and remarkable accumulation of sediment known under the name of the CRETACEOUS SYSTEM, the stratigraphical position of which is between the lowest Tertiary and uppermost Oolitic deposits.

The great character of the period is well marked by its animal remains, which are abundant and often very perfect, owing to the nature of the sediment in which they were imbedded.

The active researches of a multitude of intelligent observers have shown, that the system in itself is susceptible of being subdivided with advantage into periods of secondary value, but far less important than those distinctions which separate the few great Geological systems. Indeed, in many cases, the lines of demarcation in the subdivisions of a system may be considered more or less arbitrary; forming links merely of a continuous series, as is proved by some of the same species being common to each. These secondary divisions are, however, good and useful as Geological horizons, and in most cases distinguished by a prevalence of certain forms peculiar to each in particular, and having a more or less prolonged existence.[1]

In Great Britain the Cretaceous System has a wide range. From Flamborough Head on the Coast of Yorkshire it extends in a South-Easterly direction to the Wash in Lincolnshire: commencing again on the North Coast of Norfolk, it proceeds in a South-West direction, and occupies a considerable portion of the Counties of Norfolk, Suffolk, Cambridge, Hertford, Oxford, Berks, Wilts, and Dorset. From the great central space of

---

[1] Consult Mr. Barrande's interesting paper on the "Migration of Species," 'Bull. de la Soc. Géol. de France,' vol. viii, 2d series, p. 150, 1851.

Salisbury Plain it takes an Easterly direction, by Andover, Alton, Guildford, Reigate, Wrotham, Rochester to Ramsgate, Deal and Dover, on the East Coast of Kent. In another direction, from Salisbury Plain, it proceeds by Winchester, Arundel, and Lewes, to Brighton, Newhaven, and Beachy Head on the Channel. It also passes through the centre of the Isle of Wight, from the Culver Cliffs and Needles at the Western extremity, occupying a large part of Hampshire, and greater or less portions of Surrey, Sussex, and Kent, by those two Southern divisions of its course. Chalk also occurs in the North of Ireland; but, for further details, we must refer to those excellent Geological works, in which every stratigraphical detail will be found admirably delineated.[1] Therefore we only mention here a few points connected with the distribution of our species.

In Great Britain the Cretaceous System is incomplete: some of the lower beds are wanting, such as the lower *Neocomien* of the French. The total thickness of our beds is supposed to be from 600 to 1000 feet, divided by the generality of British Geologists into six subdivisions, varying more or less in their mineralogical composition, but not all equally well defined by their organic remains; at least so far as Brachiopoda are concerned, as may be observed during the progress of this work. In the *red chalk* of Norfolk three species only have been noticed as yet, one common in the Lower Chalk, the second in the Upper Green Sand and Gault, and the third has not hitherto been discovered in any other British deposit, but is peculiar to the *Tourtia* of Belgium. Geologists seem to consider the *red* chalk to represent the Gault, from its being said to contain other species, such as *Am. dentatus, Bel. minimus, Inoceramus sulcatus*, and some other forms common to that strata. But few species are found in the *Speeton Clay*, these also occurring in the Upper Green Sand of Cambridge: most Geologists have considered the Speeton Clay as the equivalent of the Gault. We therefore believe it possible that the Upper Green Sand and the Gault are more intimately connected than is generally allowed. In the true Gault, few species are met with; those conjectured from the Gault of Cambridge turn out to be all from the Upper Green Sand.

The age of the Farringdon beds may yet afford a subject of discussion, although several distinguished geologists[2] state them to be *Lower Green Sand*. All I can say is,

---

[1] Consult the works of Messrs. Smith, Conybeare, and W. Phillips ('Outlines of the Geology of England and Wales,' 1822); the various papers and works of Dr. Fitton, in the 'Geol. Trans.' and 'Quart. Journ. of the Geol. Soc.;' Professor Phillips ('Geol. of Yorkshire'); Sir H. de la Beche ('On the Chalk and Green Sand of Lyme,' &c.), 'Geol. Trans.,' vol. ii; Dr. Mantell's several works on the 'Geology of Sussex,' in 1822, 1833, and 1844. Also the interesting papers of Professor Forbes, Messrs. E. Bennet, Lonsdale, Rose, Woodward ('Geol. of Norfolk'), Morris, Weaver, Clarke, Bowerbank, Lyell, Ibbetson, Austen, &c. Also the works of many distinguished foreign geologists, such as Viscount d'Archiac, D'Orbigny, Reuss, Rœmer, Cuvier, Brongniart, &c. &c.

[2] Refer to Mr. Austen's paper in the 'Quart. Journ. of the Geol. Soc.,' vol. vi, No. 24, p. 454, 1850; likewise, for the position of the Mans beds, to M. E. Guéranger's interesting section in the 'Bull. Soc. Géol. de France,' vol. vii, 2d ser., p. 800, 1850; also to Viscount d'Archiac's 'Memoir on the Tourtia,' in the 'Mém. de la Soc. Géol. de France,' vol. ii, 2d ser., p. 293, 1837.

that after a careful investigation of the locality, I was unable to convince myself of the real position of these beds. The shells do not appear to have lived on the spot in which they are found; some of the Brachiopoda are undoubtedly found in the *Tourtia*, the *Upper Green Sand*, and the *Lower Green Sand* of many localities. Among these we may mention *Ter. depressa* (Lamarck), a *Tourtia* shell occurring likewise in the *Hils. Cong.* of Essen, at least I have a specimen from the last-named locality, undistinguishable from the Tourtia species. *Ter. oblonga* is found at Farringdon, in the Lower Green Sand of Hythe, Maidstone, &c., in the Hils. Cong. of Essen, and in the French *Terrain Neocomien*, but Mr. Cunnington has found it also in the Upper Green Sand of Warminster. *Ter. sella* is abundantly distributed in the Lower Green Sand of many localities, and is met with at Warminster in the Upper Green Sand. *R. lata* (Sow.) is likewise common in the same conditions, &c.

It is therefore difficult to decide the question of the age of the Farringdon beds by the Brachiopoda; and I am convinced (notwithstanding M. D'Orbigny's efforts to prove the contrary) that several of the Cretaceous Brachiopoda lived in more than one of his divisions,[1] and consequently were possessed of a much greater vertical range. There is no reason why certain forms that lived while the Lower Green Sand was in progress of deposition, should not have existed also in the Upper Green Sand. All preconceived systematic views should be avoided, and it is advisable in the present state of Palæontology not to imagine that all species were restricted to such narrow limits. The Tourtia of Belgium reposes everywhere directly upon the Palæozoic rocks; some consider it a distinct formation in the Cretaceous System. M. D'Orbigny states it to belong to his *étage cénomanien*; M. Dumont supposes it *Neocomien*, and M. de Koninck, from its containing the *Ammonites varians*, and other fossils of the *Craie chloritée*, sup., refers it to that age; and it is perhaps represented in England by the *Chloritic Chalk with green grains at Chard*, the *Upper Green Sand*, and the *red chalk* of Norfolk.

---

[1] Consult a most interesting paper by M. J. Cornuel, bearing for title 'Catalogue des Coquilles de Mollusques Entomostracés et Foraminifères du Terrain Crétacé inf. de la Haute Marne, avec divers Observations relatives à ce Terrain,' ('Bull. de la Soc. Géol. de France,' vol. viii, 2d ser., p. 430, 1851.)

In page 446, that author states:—"Il est donc constant qu'il y a passage de quelques céphalopodes aussi bien que quelques gastéropodes et lamellibranches, des couches néocomiennes dans le Gault. Cette circonstance n'empêche pas la faune du Gault d'être, dans son ensemble, très distincte de celle du terrain néocomien.

"M. le Dr. Fitton a établi, par la comparaison des fossiles que toutes les couches du grès vert inf. du bassin de Paris y compris le terrain néocomien proprement dit, ne sont autre chose que le *lower Green Sand* d'Angleterre. Ce savant place la limite supérieur du *lower Green Sand* en Angleterre et en France au point ou commence le Gault. Il ne peut rester de doute chez nous qu'au sujet des sables et grès jaunatres et du sable vert (No. 14 et 15 ci-dessus), en ce sens seulement qu'ils paraissent former le passage entre le terrain néocomien et le Gault proprement dit," &c.

The following Table illustrates the principal characters of the Cretaceous System in England, as well as a few of the foreign synonymic appellations.

CRETACEOUS PERIOD OR SYSTEM.	Upper Chalk.	Nearly pure carbonate of lime, and minute fragments of shells and foraminifera, forming a white or yellowish-white, or light grey soft chalk, with horizontal layers of flinty nodules.	Lewisham, Grays, Northfleet, Norwich, Brighton, Dover, &c.	Corresponds to the *Craie blanche* of the French, the *Oberekreide* of the Germans, *Etage senonien* of M. D'Orbigny.
	Lower Chalk, and Chalk Marl.	Harder chalk than the former, almost without flinty nodules, under which a greyish marking chalk and sand, at times indurated.	Near Dover and Folkstone, Hinton, near Cambridge, near Swaftham, Lewes (Sussex), near Norwich, &c.	This is the *Untere kreide* and *Planer* of the Germans, the *Craie tufean* of the French, *Etage turonien* of M. D'Orbigny.
	Chloritic Marl, and Upper Green Sand.	Variable in its composition, a chloritic marl, containing most of the species of the Upper Green Sand, composed of a chalk with green particles, and minute grains of quartz. The Upper Green Sand is made up of a siliceous sand, or a marly calcareous sand, with green grains often consolidated into nodules of chert, and masses of limestone.	Chard, Chardstock, &c. near Warminster, Alton, Petersfield, Cambridge, &c.	This division seems to correspond to the *Glauconie crayeuse* of the French, the *Tourtia* of the Belgians, the *Green Sand* of the Germans, *Etage Cénomanien* of M. D'Orbigny.
	Red Chalk, Speeton Clay, Gault.	A thin bed of red chalk, coloured by oxide of iron, with minute siliceous grains. The Speeton Clay is also of a grey colour, containing a mixture of Upper Green Sand and Gault species. Dark blue tenacious clay, at times marly, with some concretions.	Hunstanton Cliff, Norfolk, Specton Cliffs, Yorkshire, Folkstone, Cambridge, Rigmer, &c.	The *Gault* of the French, *Galt* of the Germans, *Etage Albien* of M. D'Orbigny.
	Lower Green Sand.	Chiefly arenaceous deposits, sand with or without green grains, ferruginous sandstones, beds of clayey sand, clay, and bands of limestone known under the name of Kentish rags.	Folkstone, Hythe, Maidstone, Shanklin, and Atherfield, Isle of Wight.	This is said to represent the upper portion of the *Terrain neocomien*, or *Etage Aptien* of M. D'Orbigny.

Perhaps the fossils of the Cretaceous period have attracted more attention than those of any other system; a great many valuable works and memoirs are published on the subject by Sowerby, Rœmer, Geinitz, Reuss, Nilsson, Wahlenberg, D'Archiac, Brongniart, &c.; but the most complete is undeniably that of M. D'Orbigny's, 'On the Species of France,' although we regret not being always able to coincide in the determinations and observations communicated by that distinguished Palæontologist; our researches, both in the field and in collections at home and abroad, having led us to different results.

Notwithstanding that in some cases we may be mistaken, still our exertions have tended, we hope, to correct various errors prevalent, especially on the continent, where some of our British types are singularly misunderstood by authors who have not had the same advantages as ourselves of comparing their specimens with the original types preserved in this country. It may at the same time be remarked that, from the vast number of individuals obligingly forwarded from all quarters, we have succeeded in tracing some passages of form; and the great confusion in the nomenclature has arisen from describing species on the study of only one or two specimens; indeed, so perplexing are the minute shades that link certain shapes together, that we have often been embarrassed and uncertain where to draw a line of demarcation.

The internal arrangements of the calcareous appendages or muscular impressions are similar in all the individuals of the same species, and although diversified to some extent in different forms of the same genus, these last are sufficiently constant to warrant separation.[1] Our British Cretaceous period is very rich in species of Brachiopoda, but not so much so as in France, from being deficient in certain beds, which are there very prolific in a variety of forms. I am happy to say that among the numerous British specimens kindly lent to me, I have discovered a number of forms found on the continent, but unknown and unpublished in our English catalogues.[2]

We have admitted twelve genera among our Cretaceous Brachiopoda, viz., *Lingula, Crania, Thecidea, Argiope, Magas, Terebratella, Trigonosemus, Terebrirostra, Terebratulina, Kingina, Terebratula,* and *Rhynchonella,* and will elsewhere discuss their respective value and claims as sections in the class of Brachiopoda; they are based on *important internal*

[1] In page 7, Part III, I noticed the absolute necessity of abandoning the use of the terms *dorsal* and *ventral* in the descriptions of Brachiopoda, and proposed the use of other denominations. Since the publication of the above, Professor M'Coy, ('Annals and Mag. of Nat. Hist.,' Nov. 1851, vol. viii, p. 391,) following out these views, proposes to make use of the terms *receiving valve* for the imperforated valve of *Terebratulæ,* &c., and *entering valve* for the opposite one, of which the beak enters into the cavity of the receiving one. I have no objection to the use of these terms in the sense employed by Professor M'Coy, but should have preferred the name *dental valve* for the perforated one, and *socket valve* for the other. The use of two or three terms to designate them will be found of great convenience where the same technical designations must be often repeated.

[2] I am particularly obliged to Mr. S. Woodward for lending me a numerous suit of sketches of British Cretaceous Brachiopoda, preserved in various cabinets, which enabled me to procure the loan of the original specimens for illustration and description.

differences, proving modifications in the dispositions of the animal relative to the shell. Many persons from not having devoted sufficient attention to these most interesting variations, and from contenting themselves by the superficial observation of some external characters, would place together many animals essentially different in their details, while separating others whose internal organisation is similar; but those who have truly investigated the matter, after long and conscientious observations, all now admit the necessity of subdividing the few families composing the class into a certain number of genera or subgenera.

I regret not having always been able to examine the interior of the species, and therefore remain in some cases uncertain in which section such forms should be placed.[1]

Genus—LINGULA, *Bruguière.* 1789.

Shell inequivalved, one valve more convex than the other, more or less oval, elongated, tapering at the beaks, widened at its palleal region; without hinge, valves held together by the adductor muscles; attached to submarine bodies by a long muscular pedicle issuing between the beaks; a groove, existing for its passage in that of larger valve; arms fleshy, without any calcareous support; structure horny, covered by an epidermis; two muscular impressions in the one, and four in the other valve.

1. LINGULA TRUNCATA, *Sow.*   Plate I, figs. 27, 28, and 31.

> LINGULA TRUNCATA, *Sow. in Fitton.*   Observations on some of the Strata below the
> Chalk, read before the Geol. Soc. in 1827, printed in vol. iv,
> pl. xiv, fig. 15, of the Trans. of the Geol. Soc., 1836.
> —   —   *Morris.* Catalogue, 1843.
> —   —   *Forbes.* Catalogue of L. G. Sand Fossils, Quart. Journ. of the
> Geol. Soc., vol. i, p. 346, 1845.
> —   —   *Fitton.* Strat. Section, Quart. Journ. of the Geol. Soc., No. 11,
> p. 289, 1847.
> —   RAULINIANA, *D'Orb.* Pal. Franç. Ter. Crétacées, vol. iv, p. 80, pl. 490, 1847.
> —   TRUNCATA, *Bronn.* Index Pal., p. 656, 1848.
> —   —   *D'Orb.* Prodrome, vol. ii, p. 84, 1850.

---

[1] I would recommend to the study of all scientific observers the instructive and valuable work by Professor Milne-Edwards, bearing for title 'Introduction à la Zoologie Générale, ou Considérations sur les Tendances de la Nature dans la Constitution du Regne Animal.' Paris, 1851.

Diagnosis. Shell oblong, irregularly oval, slightly convex, but compressed and flattened longitudinally along the middle; valves almost equal; beaks not very acute; truncated or rounded in front, shell thin, shining, horny; surface wrinkled by numerous concentric elevated lines of growth; more numerous at the sides. Length 13, breadth 8, depth 3½ lines.

Obs. *Lingula truncata* was discovered by Dr. Fitton in the Lower Green Sand of Atherfield (Isle of Wight), Sandgate, and Peasemarch; it is the largest cretaceous *Lingula* with which I am acquainted, some specimens measuring full 13 lines in length. *L. truncata* is quite distinct from *Lingula ovalis,* Sow., M. C., tab. xix, fig. 4, as correctly stated by Professor Forbes in the 'Quarterly Journal,' vol. i, p. 346. I am now quite certain that Sowerby's specimens of *Lingula ovalis* were obtained from the Kimmeridge Clay, found in blocks at Pakefield in Suffolk, associated with other well characterised Oolitic species. *T. ovalis* was therefore never to my knowledge found in the Lower Green Sand of Sandgate, as stated by M. D'Orbigny, nor in the cretaceous period, where it is placed by several authors. M. D'Orbigny's *Lingula Rauliniana* is only a synonym of *L. truncata,* his description and figure quite agreeing with the specimens of that type, as any one may become convinced of on examination of the original specimens now deposited in the Museum of the Geological Society.

Plate I, figs. 27-28. A specimen, natural size, from the Lower Green Sand of Red Cliff, Isle of Wight, in my collection.

„ fig. 31. From a specimen found in the Lower Green Sand, of Sandgate; Dr. Fitton's figures were, unfortunately, not very well illustrated, which is perhaps the reason why some authors have not recognised the type.

2. Lingula subovalis, *Dav.* 1852. Plate I, figs. 29-30.

Diagnosis. Shell regularly oblong, oval, the beak and front presenting nearly the same shape; valves almost equal, slightly convex, compressed; shell thin; surface smooth, marked only by a few concentric lines of growth. Length 7, width 4, depth ¼ lines.

Obs. This small Lingula is found in the Upper Green Sand, near Warminster, where it does not seem much to exceed the dimensions above given; it appears distinct from *L. truncata* by its more regularly oval shape, and much smaller size, nor would it seem to be the *Lingula ovalis* of Sowerby, which is a Kimmeridge Clay Shell, but placed erroneously by most authors in the Lower Green Sand, which is the reason why I supposed it so, while writing on this genus in Part III; but this opinion I afterwards relinquished, having been able to compare specimens from the two formations. We are therefore at present acquainted with only two *British* Cretaceous Lingulas, viz., *L. truncata* and *L. subovalis;* and in the Oolitic series other two, viz., *Lingula Beanii* and *L. ovalis.* I have named our shell, *Subovalis,* to indicate its approximation in shape to *L. ovalis,* Sow.

2

Plate I, fig. 29. A specimen from the Upper Green Sand, near Warminster, in the Collection of the British Museum.

„   fig. 29ᵃ. A specimen, showing the interior and muscular impressions enlarged, likewise from the same locality, in the Collection of the British Museum.

„   fig. 30. Another specimen, from the Collection of Mr. Cunnington.

Genus—CRANIA, *Retzius.* 1781.

Shell inequivalved, circular, or subquadrate, more or less irregular, entirely or partially attached by the substance of its smaller valve to rocks, or other submarine bodies. Upper valve conical, with lateral or subcentral vertex, without hinge or ligament: four circular muscular impressions in each valve. Surface smooth, or longitudinally striated; arms fleshy, with spiral extremities; no calcareous supports. Structure strongly punctuated.

Some authors state to have found four or five species of British Cretaceous Cranias, but all our researches have only brought to light two, viz., *C. Parisiensis*, and *C. Egnabergensis*. Of those mentioned, some are only synonyms; others, such as *C. spinulosa*, Nilsson, and *C. costata*, Sow., seem founded on erroneous determinations.

3. CRANIA PARISIENSIS, *Defrance.* Plate I, figs. 1—7.

CRANIA PARISIENSIS, *Defrance.* Dic. des Sc. Nat. vol. ii, p. 313, No. 3, 1819.
—   —   *Lamarck.* An. Sans. Vert., vol. vi, p. 239, 1819.
—   —   *Brongniart.* Desc. Geol. des Env. de Paris, pl. iii, fig. 2, 1822.
—   —   *Sow.* Min. Con., vol. v, p. 3, tab. 408, 1825.
—   —   *Hœninghaus.* Mon. des Cranies, p. 9, fig. 8, 1828.
—   —   *Desh.* 1830. Ency. Meth., ii, p. 18, No. 8.
—   —   *Desh.* Nouv. Ed. de Lam., vol. vii, p. 300, No. 3, 1836.
—   —   *Dujardin.* 1837. Mém. de la Soc. Geol. de France, vol. ii, p. 222.
—   —   *Goldfuss.* Petref. Germ., p. 293, pl. 162, fig. 8, 1840.
—   —   *D'Orb.* Pal. Franç. Ter. Crétacées, vol. iv, p. 139, pl. 524, figs. 8—13, 1837.
—   —   *Rœmer.* Die vers Norddeutschen Kreidgeberges, 1840, p. 36, No. 3.
—   —   *V. Hagenow.* Jahrb., f. min., 1842.
—   —   *Morris.* Catalogue, 1843.
—   —   *Reuss.* Die Verstein. der Böhemischen Kreideformation, 1846, p. 53, No. 2.
—   —   *Tennant.* A Stratigraphical List of British Fossils, p. 47, 1847.
—   —   *Bronn.* Index Palæontologie, p. 342, 1848.
—   —   *D'Orb.* Prodrome, vol. i, p. 259, 1850.
—   —   *Dixon.* The Geology of the Tertiary and Cretaceous Formations of Sussex, p. 354, pl. xxvii, fig. 9, 1850.

Diagnosis. Shell irregular, inequivalve, transversely oval, entirely attached by the substance of the lower valve to rocks, corals, echinodermata, &c., modelling itself to the object of attachment, the irregularities of which it fills up to a greater or less extent; margin much elevated, especially in front, rising obliquely or even vertically all round, except on the posterior side; the central portion is irregularly hollowed out; structure largely cellular, spongeous or granular, particularly so round the margin; muscular impressions four in number, strongly produced in different examples; the two posterior ones variably circular, larger and more widely separated than the anterior ones; these last are usually in contact, being more elongated and depressed in the centre, between, and above which is seen a produced nose-shaped projection, somewhat variable in its details in different individuals; the digitated vascular impressions are likewise more or less marked. Upper or unattached valve thin, conical, patelliform, very convex, exteriorly covered by small granular asperities; vertex sub-central, with numerous concentric lines of growth. In the interior four deep muscular impressions; two irregularly circular or oval ones are situated near the posterior edge, widely separated, and corresponding to those seen in the same position on the attached valve; towards the centre, other two elongated muscular impressions are visibly in contact at their base, and forming long narrow uneven projections, detached, except at their origin, from the bottom of the shell, and directing themselves towards the lateral portion of the valve. Structure punctuated; dimensions variable. Length 7, width 11, height 4 lines.

Obs. This remarkable species has been fortunate, all authors having applied to it the same denomination,—an occurrence very rare among the Brachiopoda. It is frequently found in the Upper Chalk of many localities, but is a very irregular shell, few specimens bearing exactly the same shape; it is always attached by the whole surface of its lower valve, and, as stated above, modelling itself and filling up all the projections or depressions existing either on the rock or shell to which it is united, so that it cannot be detached from the place where it is fixed, provided this last is of a solid nature; but I have specimens that were joined during life to soft and perishable bodies, such as sponge, &c. One remarkable specimen, belonging to Mr. Catt of Brighton, presents an exact cast of the structure of a ventriculite to which it was fixed; it therefore remains very uneven and contorted where the object was rough and angular. In the Upper Chalk of Meudon this species is often met with, and I have picked up there specimens of *Ananchytes*, covered by more than fifteen or twenty individuals of all ages, illustrating in the clearest manner their formation. In the very young state the shell of the lower valve is so thin, except at its margin, that every accident of the object it sticks to is apparent and reproduced, the muscular impressions are faint and scarcely defined, the margin alone assuming a thickness ten or twelve times more considerable than that of the other portion of the valve; by degrees, however, as the shell acquires age, by successive layers of calcareous matter, it presents a greater and an unequal thickness, filling up and concealing the largest portion of the asperities existing on the object of attachment. Much difference may be noticed, likewise, in the shape of the muscular impressions visible in the interior of both valves, so well described by that conscientious observer, M. Bouchard, who has devoted much time to the study of this genus.[1] He states, that "The *Cranias* present two kinds of impressions left by the adductor muscles,—those which are often deeply excavated in the thickness of the valves are produced by muscles that do not deposit calcareous matter, and which, by their insertion on the valve, prevent the mantle from depositing its calcareous substance on the part occupied by their base; we then see this pallial secretion surround the base of the muscles, and circumscribing these impressions with a calcareous crest or rim. The others, on the contrary, possessing the faculty of depositing calcareous matter, form projections that assume all kinds of shapes; therefore, in the interior of the upper valves of *Crania Parisiensis* and *C. abnormis*, we perceive the muscular deposits under the form of lanceolate and pedicular laminæ, rising with age from the bottom of the valve, to the height of several millimetres." The upper valve, from its extreme thinness, is seldom preserved, while the attached valve is very abundant. In England, it is sometimes found in the Chalk of Kent and Sussex, but it cannot be said to be common. In Plate I, I have endeavoured to illustrate a few of the principal variations assumed by this shell from a beautiful series of specimens, for which I am indebted to the kindness of several friends.

Plate I, fig. 1. A specimen, enlarged, of the attached valve, very adult, from Gravesend, in the collection of Mr. Bowerbank.

,, fig. 2ᵃ. A very adult specimen, enlarged, from the same locality, in the collection of Mr. Bowerbank.[2]

,, fig. 2ᵇ. A very adult specimen of the upper valve, likewise from the above-mentioned collection.

,, fig. 3. A fragment, much enlarged, of the lower valve.

,, fig. 4. Profile view of the upper valve.

,, fig. 5. Profile view of the lower valve.

,, fig. 6. The exterior of the upper valve, from the collection of Mr. Bowerbank.

,, fig. 6ᵃ. Profile view of both valves, united, from the same collection.

,, fig. 7. A remarkably fine specimen, from the chalk of Meudon, in my collection, placed here to show how the shells sometimes clustered near each other in the young state; not finding room as they grew to develope themselves, they assumed the shape the space admitted, indenting and projecting one above another, one of the specimens preserving still its upper valve.

[1] 'Mémoire sur un Nouveau Genre de Brachiopode,' ('Annales des Sciences Nat.,' vol. xii, p. 84, 1849.)
[2] This valve has been rarely figured. Mr. Dixon gives a correct illustration in his work, and in a plate, entitled 'Collection de M. H. Michelin,' published some years back. Figs. 2 and 3 illustrate this valve, from Meudon, where it is extremely rare.

4. Crania Egnabergensis, *Retzius.* Plate I, figs. 8—14.

Numulo	. . . . ? *Stobæus.*	Dissert. de Numulo Brattenbergensis, 1732, pl. i, figs. 3-4; and Opuscula in quibus Petrefactorum, 1753, tab. 1, figs. 3-4.
Crania Egnabergensis,	*Retzius.*[1]	Crania Oder., &c., in Berlin Gesellsch Schaft, vol. ii, p. 75, tab. i, figs. 4, 7, 1781; Encyclop. Meth., pl. 171, figs. 6-7, 1789.
Crania striata,	*Defrance.*	Dict. des Sc. Nat., p. 315, No. 2, (non Nilsson, 1827,) 1818.
—	— *Lamarck.*	An. sans Vert., vol. vi, p. 239, No. 5, 1819.
Anomites craniolaris Egnabergensis,	*Wahlenberg.*	Petref. Telluris Svecanæ examinata in Nova Acta Societatis Scientiarum Upsaliensis, vol. viii, p. 60, 1821.
Crania striata,	*Hœninghaus.*	Beitrag zur Mondergattang Crania, No. 10, fig. A, 1828.
—	— *and* C. ovalis, *Woodward.*	An Outline of the Geol. of Norfolk, pl. vi, figs. 15-16, 1833.
—	striata, *Deshayes.*	Nouv. Ed. de Lamarck, vol. ii, p. 19, No. 9, 1836.
—	— *Hisenger.*	1837? Lethea Succ., p. 84, pl. xxiv, fig. 10.
Crania Egnabergensis,	*Bronn.*	Lethea Geog., vol. i, p. 665, tab. xxx, fig. 2, 1837.
—	— *Rœmer.*	Die Vers. Norddeutschen Kreidgeberges, p. 36, No. 6, 1840.
Crania striata,	*Goldfuss.*	Petref. Germ., p. 295, pl. 162, fig. 10, 1840.
—	ovalis, striata, spinulosa, *Morris.*	Catalogue, 1843.
—	Ignabergensis, *D'Orb.*	Pal. Franç. Ter. Crétacées, vol. iv, p. 142, pl. 526, figs. 1—6, 1847.
—	striata, spinulosa, *Tennant.*	A Stratigraphical List of British Fossils, 1847.
—	— *Dixon.*	The Geol. of the Ter. and Cret. Formations of England, pl. xxvii, 1851.
—	Ignabergensis, *D'Orb.*	Prodrome, vol. ii, p. 259, 1850.

Diagnosis. Shell irregular, inequivalve, circular, both valves presenting a more or less conical depressed shape; apex submarginal, nearest the posterior edge attached by a small or large portion of its lower valve to shells or corals, principally at or near the vertex of the lower valve. Surface ornamented by a variable number of costæ, radiating irregularly from the vertex in both valves, and after reaching the edge, pass it, forming a number of asperities all round, these likewise augmenting in number by the addition of smaller costæ appearing at a variable distance from the vertex, and irregularly intercalated between the larger ones; concentric lines of growth are also visible intersecting the striæ; the upper valve is most convex or conical; the interior of the lower valve is surrounded by a wide, rather flat, strongly granulated margin; four muscular impressions; the two pos-

---

[1] Retzius gives, in 1781, four good figures of his *Crania Egnabergensis,* and refers to the two illustrations published by Stobæus in 1732, under the name of *Numulo ;* but the figures of this last author are not sufficiently detailed to be identified with certainty. I am fully convinced, however, that his fig. 4 is the same as that described afterwards by Retzius.

terior ones are more or less circular, lying close, and partially surrounded by the inner edge of the granulated margin; these are separated from each other by an almost equally wide and depressed space; the anterior pair are situated towards the centre, united together, and diverging in the form of a V, being more or less oval and depressed in the centre; between and above them is a small, produced, nose-shaped protuberance, the same wide margin is likewise visible in the upper valve, which is deeper, with four corresponding muscular impressions, disposed as in the attached valve, with this difference, that the anterior pair are larger, and a hollow is seen between and above them, where the projection in the lower valve exists. Structure punctuated; vascular impressions well defined. Dimensions variable; length 3, width 3, depth $1\frac{1}{2}$ lines.

Obs. This small *Crania* has received several names from different authors; it seems, however, to have been first noticed by Retzius under that name of *Egnabergensis* in 1781, which denomination has been adopted by many Palæontologists, although commonly known also by that of *C. striata*; *C. ovalis* of Woodward is only an accidentally elongated specimen of Retzius' species, and many specimens attributed to *C. costata* in England, are only simple varieties of the form under examination with fewer costæ,[1] or where the intermediate elevated striæ are few in number, or even often wanting, especially in young specimens; thus have been found undoubted *C. Egnabergensis* with only sixteen costæ, while others have thirty-eight on each valve. I do not, however, pretend to dispute the existence of a distinct form under the name of *costata*, having well-characterised specimens of it from the Cretaceous beds near Valogne; but none of the British shells which have come into my hands authorised me to admit the two species, nor do the figures given of *C. costata* in Mr. Dixon's work entitle me to decide the question. The manner in which this species was attached is also very remarkable; in general, as in fig. 8, it seems to have been simply fixed by a very small portion of the summit of its apex to slender corals or bryozoa, or other branched bodies; sometimes, as in fig. 9, its attachment extended along the delicate coral from the vertex to the margin on one side only, interrupting and indenting the costæ; in other and rarer cases, when attached to rocks, flat objects, or shells, its fixed surface was much larger, and after minute examination of what takes place in many other species, as we have described under the head of *Th. Wetherellii*. I do not see reason for separating those specimens, figs. 13 and 14, attached by the greater portion of their surface on *Spatangus*, a part of the costated portion rising all round. The internal, muscular, and other impressions do not present differences of any value, from

[1] Sowerby is the first describer of *C. costata*, which he illustrates in No. 12, fig. 6 of his 'Genera of Shells,' stating that he found it at Orglandes, (near Valogne, in France,) but does not mention having discovered it in England. He characterises thus his species: "*Cr. valvula* superiore costis prominentibus radiantibus octo ad quindecem." In the same number and plate, fig. 3, he figures a small Crania, fixed by the greater portion of the attached valve to an echinus, stated to be from the Chalk of Norfolk, and which he supposes to be *C. Parisiensis*. This is, however, a mistake: it belongs to *C. Egnabergensis*, and is similar to the one I have figured, Pl. I, fig. 13.

what is seen in *C. Egnabergensis*. I have obtained from the chalk at Gravesend specimens of this species from the dimensions of a quarter of a line in diameter, to that of three lines; in these various specimens, the apex was likewise more or less sub-marginal, in some almost central.

C. Egnabergensis has been found in the upper and lower chalk of various localities, such as at Northfleet, Kent, by Messrs. Bowerbank, Morris, Woodward, myself, and many others; in that of Dover and Folkstone by Mr. Mackie, in the Sussex Chalk by Messrs. Catt and Dixon, in that of Norwich by Messrs. Woodward, C. B. Rose, Fitch, &c.; on the continent this species is found at Meudon, Fécamps, Vendôme (Loir et Cher), in France, but never at Hampton Cliff in England, as stated by M. D'Orbigny, whose figures of this species do not seem to have been drawn from very good specimens, the muscular impressions not being well characterised.

Plate I, figs. 8, 8ᶜ. From the collection of Mr. Bowerbank.

„ fig. 8ᶜ. Enlarged.

„ fig. 8ᵃ. Interior of attached valve, enlarged.

„ fig. 8ᵇ. „ upper valve „

„ fig. 8ᵈ. Enlarged figure, shewing both valves united and attached to a coral.

„ fig. 9. From the collection of Mr. Mackie.

„ fig. 9ᵃ. Enlarged.

„ figs. 10, 11. Two young specimens from Gravesend.

„ fig. 12. Elongated malformation, *C. ovalis*, Woodward.

„ fig. 13. A specimen attached by a great portion of its lower valve to a *Spatangus* from Gravesend.

„ fig. 13ᵃ. Enlarged figure.

„ fig. 14. Illustrating the profile of the same with both valves united, enlarged.

*Genus—*THECIDEA, *Defrance.* 1828.

Animal with the mantle-lobes disunited, and adhering closely to the valves. Shell free, or attached by the larger valve, oblong, or transversely oval, more or less irregular, thickened, especially round the margin, structure perforated. Dental, or largest valve, partially or entirely attached by its own substance; or when young, in some species, by a pedicle issuing from the extremity of the beak; upper or unattached valve always less convex than the dental one; surface smooth, or otherwise ornamented; hinge line more or less straight, with two strong teeth in the attached valve, adapting themselves into corresponding sockets in the smaller valve; beak more or less produced, with a well-defined area and deltidium. Interior of valves variable: in larger valve a longitudinal, central, and two lateral ridges are generally more or less visible, under which two deep muscular impressions are seen: upper valve furrowed more or less deeply and regularly, to receive the apophysary testaceous ridge, and leaving a small cavity in the upper portion of the valve for the body of the animal, these sinuous grooves varying in number, position, and extent, in different species; two strong lateral adductor muscles situated under the hinge.[1]

Obs. We are acquainted with only one British Cretaceous *Thecidea;* and although many species are found in the various beds of this system on the continent, they have not as yet been discovered in England.

5. THECIDEA WETHERELLII, *Morris.*   Plate I, figs. 15—26.

> THECIDEA WETHERELLII, *Morris,* 1851. Annals and Magazine of Natural History, pl. iv, figs. 1—3.

Diagnosis. Shell inequivalve, irregular, more or less circular or pentagonal, as wide as long, or longer than wide; triangular towards the beaks, rounded laterally, and slightly indented in front; larger valve convex, attached in various ways, either by nearly the whole surface of its valve, the edges only being slightly elevated, or by a small portion near the beak, which is more or less produced; area narrow, deltidium large, triangular, and elongated. A longitudinal depression is visible, extending along the centre of the larger valve to the front, which is repeated to a lesser extent on the smaller valve; this last is slightly convex, smooth, with many concentric lines of growth; hinge line straight, articulating by means of two teeth in larger, and corresponding sockets in smaller valve. In the interior of larger valve, beneath the deltidium, the short lamellar processes are seen to occupy about a fifth of the length of the valve; the central one being the longest and most elevated, the other two appearing at the base of the dental plates, converge gradually to near the central septum; the inner surface of this valve is covered with close granular longitudinal striæ, the interior of smaller or upper valve is divided on each side in a deep

---

[1] In connection with the *Argiope* described in Part I, p. 9, Mr. Woodward and myself have re-examined a suit of specimens of the recent *Thecidea Mediterranea,* one of which (in my cabinet) retains the animal in excellent preservation. *Thecidea* has a calcareous loop, folded into two or more lobes, and lying in hollows of corresponding form, and excavated in the substance of the smaller valve; this loop or apophysary ridge supports the brachial membrane, whose thickened and ciliated margin is

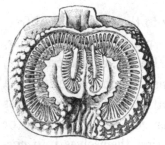

Th. Mediterranea.   Th. radiata.

apparently attached to the inner sides of the sinuous grooves. The *cilia* are very long, especially the outer series, which are directed inwards in the dried specimens. The loop exists in its most complete form in *Th. radiata* (Def.); each lobe of the loop adheres to part of the wall of the shell along its course, becoming free towards the visceral cavity.

arched or reniform sinus; the cardinal process is large, and the margin of the valve is minutely granulated.

Obs. This is the only British species of cretaceous *Thecidea* with which we are acquainted, and is known to us since 1847, at which period we had obtained specimens from Mr. Purdue, who found them while washing some chalk from Northfleet. It has likewise been discovered by Mr. Wetherell, and lately described and figured by Mr. Morris.[1] This species bears much external resemblance to *Th. Mediterranea*, and more especially to *Th. triangularis*, figured by me from the inferior Oolite and Lias in Part III, but quite distinct from both by the internal arrangements of the Apophysary system, which partakes most of the simplicity seen in *Th. hippocrepis*, but the dissepiment is not so wide in that species. It may likewise bear some resemblance to *Th. Broderipii*, of Michelotti;[2] but the figures of that shell are not complete enough to enable one to compare the internal disposition of the Apophysary system.

At the time Mr. Morris described this species, it was only known from the Upper Chalk of Gravesend or Northfleet, Kent, where it abounds; attached principally to *Echino-dermata*, *Inoceramus*, and other shells; the smaller valve being almost always wanting, but the larger one, from being attached by the greater portion of its surface, is found adhering to the object on which it lived; at times it is assembled in great numbers, nor is it rare to find *Ananchytes* with from twenty to thirty specimens fixed to them. During my examination of the rich and interesting cretaceous collections assembled with great care by Messrs. Cunnington and Faulkner, of Devizes, I at once recognised the species under consideration obtained by these gentlemen in the Chalk of Pewsey (Wilts.), where it is most beautifully preserved, with both valves united in the free state, owing no doubt to their having been only slightly attached during life to perishable bodies, such as sponges, &c.; it is also clearly seen, that they were not always fixed by the whole or greatest part of their larger valve, as in the Gravesend specimens, but only partially so, as is illustrated by our figures 20, 21, 22, 23, 24, 25, 26. I may also remark, that, as in *Crania*, in some cases the ornaments or markings on the object of attachment are reproduced on the valves of the *Thecidea*. It is also singular, that we should find in England no other species, so many being met with in the foreign cretaceous beds.

Plate I, fig. 15. Natural size. From the Upper Chalk of Gravesend.

„ fig. 16. Interior of unattached valve magnified.

---

[1] Mr. Morris's figures of this species are not very fortunate. From the artist not understanding the character of the genus, he overlooked some of the most important points of structure; thus, in the figure of the larger or attached valve, the area and deltideum are not sufficiently characterised; the teeth are wanting, and the three small lamellar processes, as well as the granulation, incorrectly illustrated. In the figure of the interior of the smaller valve, there exists in the specimens a small bridge-shaped process near the hinge, marked *b* in my fig. 16, but omitted in those given by Mr. Morris. We have endeavoured, in our numerous illustrations of this species, to make up for these deficiencies.

[2] Michelotti, ' Desc. des Foss. des Ter. Miocene d'Italie,' Sept. 1847, pl. i, fig. 26.

Plate I, fig. 17.   Interior of attached valve magnified.

„   figs. 18, 19.   Exterior and profile of the species magnified, from Gravesend.

„   figs. 20—26.   Different specimens from the Chalk of Pewsey, Wilts., showing various modes of attachment, from the Collection of Messrs. Cunnington and Faulkner, of Devizes.

Genus—ARGIOPE, *Deslongchamps*. 1842.   (*Mégathyris, D'Orb.*, 1847.)

Shell inequivalve, variable in shape, semi-orbicular, quadrate, or transversely oval. Valves unequally convex, smooth, or variously costated : the dental valve deepest, beak produced, with a large depressed triangular area, foramen large, completed by the umbo of smaller valve.   Structure strongly punctuated; margin thickened and granulated.   Hinge articulating by the means of two teeth in the larger and corresponding sockets in the smaller valve : apophysary system consisting of a distinct loop, originating at the base of the dental sockets, and furnished with converging processes : one or more radiating elevated ribs or dissepiments arising from the inner surface of the smaller valve, the central septum being more produced and larger than the lateral ones, which are often indistinct.

Obs. We are only acquainted with one British cretaceous *Argiope*. (See observations on this genus in Part I.)

6. ARGIOPE DECEMCOSTATA, *Rœmer*.   Sp. Plate III, figs. 1—13.

TEREBRATULA DECEMCOSTATA, *Rœmer*.   Dic. Vers. Norddeutschen Kreidgeberges, 1840, p. 41, tab. vii, fig. 13.

—   BRONNII, *V. Hagenow*.   Mon. der. Rügenschen Kreide Versteinerungen in Neuv. Jahb., fig. min. 1842, tab. ix, fig. 7.

—   BUCHII, *V. Hagenow*.   Ib., tab. ix, fig. 8.

—   DUVALII, *Dav.* 1847.   Lond. Geol. Journal, p. 113, pl. xviii, figs. 15—18.

MÉGATHIRIS CUNEIFORMIS, *D'Orb.* 1847.   Pal. Franç. Ter. Crétacées, vol. iv, p. 147, pl. 521, figs. 1—11.

TEREBRATULA DECEMCOSTATA, *Bronn.*   Index Palæont., p. 1234, 1848.

MÉGATHIRIS CUNEIFORMIS, *D'Orb.*   Prodrome, vol. ii, p. 259, 1850.

Diagnosis. Shell of a somewhat pentagonal transverse form, larger or dental valve convex, beak produced, not recurved, area large and triangular, variable, in its dimensions as wide as the shell; foramen irregularly circular or triangular, occupying often about one third of the area, and laterally distinctly margined by two narrow deltideal plates,[1] the anterior portion being completed by the umbo of smaller or socket valve, which is often

---

[1] I do not agree with M. D'Orbigny while stating that this genus is deprived of deltideum; to me it is quite apparent, although not very great.

worn away at the umbo by friction, presenting thus an accidental false area in the smaller valve; this valve is less convex and more compressed than the dental one. Surface ornamented by a variable number of large rounded costæ, between which a slightly concave space is seen; the two principal ones, or those nearer the centre, are the largest, and more or less separated by a mesial furrow existing in both valves; the costæ and furrows correspond to each other, and do not indent, as in the plaited Spirifers and Rhynchonellas, &c. Valves articulating by means of two teeth in the dental or larger valve, and two sockets in the smaller one; from under the crura in socket valve, a central triangular septum gradually arises, and attains its greatest elevation at about four fifths of the length of the shell, after which it descends by an almost perpendicular line to the bottom of the valve; margin wide all round, and strongly punctuated; in the interior of dental valve, besides the two strong teeth, a small slightly elevated longitudinal septum is seen to extend to about half the length of the shell. Structure punctuated; dimensions very variable. Sometimes length 2, width 1, and not quite 1 line deep; at other times, length 1, width 1, depth ½ line.

Obs. This little shell seems to have been first noticed in the Cretaceous Formations by Rœmer, under the denomination of *Terebratula decemcostata;* but the figures of that author not reminding one of the usual form, is probably the reason why the shell has received so many other names, owing likewise to its variable appearance, a character frequently lost sight of by most authors, who, out of varieties, have made many species; thus, *Ter. Bronnii* and *Buchii* of V. Hagenow seem to me, in all likelihood, only varieties with few or more costæ; we frequently find undoubted specimens of this species having either six, eight, or ten; it is therefore evident, that it is highly improper to give names from such a variable character. Some specimens do not present their greatest width at the hinge line, being rounded in that portion, while others, on the contrary, assume at the hinge the form of elongated wings, giving it a close resemblance to some Spirifers, &c. From my ignorance of Rœmer's and V. Hagenow's figures, I described this species in 1848 under the name of *Ter. Duvalii,* which must be considered a synonym; a little later, M. D'Orbigny fell into the same error, by proposing that of *Cuneiformis;* it is singular that author should have committed this mistake, since, in p. 147 of his 'Pal. Franç.,' vol. iv, while characterising his genus Mégathiris,[1] and after having enumerated several recent types, he states: "En espèces Fossiles le *T. decemcostata* of Rœmer;" and although, admitting this species in his 'Pal. Franç.,' he omits it completely in his 'Prodrome,' vol. ii, 1850; but allowing even that M. D'Orbigny considered *T. decemcostata* as specifically different from the chalk species here described, his name would have to give place to one of those published by V. Hagenow in 1842, but with which work M. D'Orbigny was probably unacquainted.

A. decemcostata bears the greatest outward resemblance to some recent forms of this

---

[1] The term *Mégathiris* is a synonym of M. Deslongchamps' genus *Argiope*, established long before.

genus found in the Mediterranean; but in the interior of these last, the three well-defined, elevated septa, figs. 14 and 16 *a b* of our Plate, are prominent, which is not the case in the interiors of the chalk species I have seen, and which, I must confess, I have never observed in the Meudon specimens, as represented in figs. 9 and 10 of M. D'Orbigny's Plate 521, the central one only being perceptible.

This remarkable variation presented by the fossil species above described, is likewise equally prevalent in the small recent type originally described under the appellation of *Anomia decollata*, by Chemnitz,[1] as I became convinced from a numerous suit of this species given to me by Professor Forbes, and dredged by himself in the Mediterranean, these variations having tempted Risso[2] to propose the following names :—*Ter. urna-antiqua, T. cardita, T. emarginata, T. cuneata*; some of which being likewise adopted later by Sig. Philippi[3] and others, proving how difficult it is often to pronounce with certainty the limits to be assigned to a species.

In many adult specimens of *A. decemcostata*, as already noticed, the umbo is much worn, no doubt from the shortness of the muscular pedicle of attachment forcing the shell to lie so close to the object to which it is fixed as to wear it by continual friction, caused by the opening and shutting of the valve; this is likewise the case with many specimens of the recent *A. decollata*, as is seen from the fig. 16 of our Plate; but in young specimens, and even in more aged shells, the pedicle fibres must have been longer, as both beak, area, and umbo are quite perfect, showing no traces of the friction above alluded to.

Argiope decemcostata is found in the Upper Chalk of Gravesend or Northfleet, and occurs more abundantly in the Chalk filling Echinodermata, in the washing of which Mr. Purdue procured many specimens. Mr. Harris has likewise met with it in the Chalk detritus of Charing (Kent); Mr. Cunnington discovered it at Pewsey, in Wiltshire, and is often found in company with *Th. Wetherellii*. On the Continent it is not very rare at Meudon and other places, although always troublesome to obtain, from its extreme minuteness. M. V. Hagenow found it also along with other interesting forms in the Chalk of the Prussian or Belgian dominions.

Plate III, figs. 1, 2, 3, 4, 5. A specimen from the Chalk detritus of Charing, Kent, in the Collection of Mr. Harris.

   „   fig. 6.   Interior of smaller valve magnified.

   „   fig. 6 *a*. Section of smaller valve magnified, *a* central septum.

   „   fig. 7.   Interior of larger or dental valve enlarged.

   „   figs. 8, 9.   A short square variation, with few costæ, from Gravesend.

   „   fig. 10.   Idem.

   „   figs. 11, 12.   A very transverse spirifer-shaped variation.

---

[1] 'Systematische Konchylien Habitat,' vol. viii, pl. lxxviii, fig. 705, 1785.

[2] 'Hist. Nat. des Principales Prod. de l'Europe Méridionale,' vol. iv, Nos. 175, 177, 179, 180, 1826.

[3] 'Enumeratio Moluscorum Siciliæ,' 1836.

Plate III, fig. 13.   An enlarged illustration of the beak and worn umbo; A. Septum in smaller valve seen through the foramen.

„   figs. 14, 15.   *Argiope decollata*—recent species placed here to illustrate the three elevated dissepiments *a* and *b*. The loop is not introduced, (see woodcut, Part I.)

„   fig. 16.   A profile view of the recent form, showing how close it lies to the object to which it is attached.

*Genus—*Magas*, Sowerby.* 1818.

Animal unknown; fixed to marine bodies by a pedicle issuing out of the foramen of the larger valve.   Shell small, inequivalve, generally more or less regularly oval; structure punctuated; beak truncated by a foramen, extending to the umbo of smaller valve; hinge articulating by means of two teeth in the larger and corresponding sockets in smaller valve; apophysary system in the imperforated valve composed of an elevated longitudinal septum, reaching from one valve to the other, to which are affixed two pairs of calcareous lamellæ, differently disposed; the lower pair are riband-shaped, attached first to the crural base : they direct themselves by a gentle curve to near the anterior portion of the septum, to the sides of which they are affixed; the second pair arise on either side of the upper edge of the septum, extending in the form of two triangular anchor-shaped lamellæ.

Obs. I have fully described this genus under the species *Magas pumilus*, and therefore must refer to it for further details.

7. Magas pumilus, *Sow.*   Plate II, figs. 1—12 and 33.

Figured by Faujas in 1789, 'Histoire de la Montagne de St. Pierre de Maestricht,' pl. xxvi, fig. 6 (but not named).

Magas pumilus, *Sow.*   Min. Conch., vol. ii, 1818, p. 40, tab. 119, figs. 1—5.

Terebratula concava? *Lamarck.* 1819.   An. sans Vert., vol. vi, p. 251, No. 26.
See likewise *Dav.*, Annals and Mag. of Nat. Hist., 2d ser., vol. v, June, 1850.

Magas pumilus, *Parkinson.* 1822.   An Introduction to the Study of Organic Remains, p. 227, pl. vii, fig. 14.

—   —   *Brongniart* and *Cuvier.* Descript. Géol. des Env. de Paris, 1822, pl. iv, fig. 9.

—   —   *Defrance.* 1823.   Dic. des Sci. Nat., vol. xxviii, p. 13, fig. 1.

—   —   *Blainville.* 1826.   Malacologie, pl. liv, fig. 1, (non *Magas pumilus, Sow.*)

—   truncata (*Rose*), *Woodward.* 1833.   Outlines of the Geol. of Norfolk, tab. vi, fig. 9.

TEREBRATULA PUMILA, *Von Buch.* Uber Ter., 1834; and Mém. de la Soc. Geol. de
    France, vol. iii, 1st ser., p. 206, pl. xix, fig. 5, 1838.

—   CONCAVA, *Deshayes.* 1836. Nouv. Ed. de Lamarck, vol. vii, No. 26.

MAGAS PUMILUS, *Bronn.* Leth. Géog., p. 662, pl. xxx, fig. 1, 1837.

—   —   *Brown.* Illustr. of Fossil. Conch., pl. xlix, figs. 5 and 13, 1838.

—   —   *Morris.* Catalogue, 1843.

—   —   *D'Orb.* 1845. Geol. of Russia and Oural, vol. ii, p. 495, pl. xliii,
        figs. 27—30.

—   —   *D'Orb.* 1847. Pal. Franç. Ter. Crétacés, vol. iv, p. 54, pl. 501.

—   —   *Bouchard* and *Dav.* Bulletin de la Soc. Géol. de France, vol. v,
        2d ser., p. 139, pl. ii, figs. 1—11, 1848; and *Dav.*, Bull. Soc.
        Géol. Fr., vol. vii, 2d ser., p. 62, pl. i, figs. 7—9, 1849.

—   —   *Bronn.* Index Pal., p. 699, 1848.

—   —   *Tennant.* Strat. List of Brit. Fos., p. 47, 1847.

—   —   *Dav.* Annals and Mag. of Nat. Hist., vol. v, pl. xv, fig. 2, 1850.

—   —   *D'Orb.* Prodrome, vol. ii, p. 258, 1850.

Diagnosis. Shell equilateral, inequivalve, oval or circular; larger valve convex; beak
more or less recurved, sometimes almost straight, slightly truncated by a triangular
foramen extending from the beak to the umbo; deltideum small and lateral, not surrounding the foramen; false area small or large; imperforated valve slightly concave or
moderately convex; surface smooth, marked by numerous concentric lines of growth;
structure largely or more or less closely punctuated in the form of lozenge-shaped punctures. The apophysary system is complicated, composed, in the smaller valve, of a mesial,
longitudinal, elevated triangular septum, extending to about two thirds of the length of
the valve, arising from under the crura by a gentle curve, reaches and touches the
larger valve near its centre, whence it descends by an almost perpendicular line to the
bottom of the valve; to this septum are attached two pairs of calcareous lamellæ,
differently shaped; they are fixed to the inner side of the strongly-developed socket walls,
first under the form of short, slender, converging stems, soon becoming delicate riband-
shaped lamellæ, directing themselves by a gentle elevating curve to near the anterior
portion of the septum, to the sides of which they are attached. The second or upper pair
arise on either side of the upper edge of the septum, and extend along its edge to more
than half its length, in the form of two triangular anchor-like lamellæ, wide at their base,
the sides converging to a point; each lamella forms an inward curve, this process being,
most probably, destined to lodge and protect the visceral parts of the animal, leaving on
either side a considerable space for the cirrated arms, attached to the under pair of lamellæ
above described.

The interior of the larger valve is simple; a central longitudinal, slightly elevated,
obtuse projection is seen to extend to about two thirds of the length of the shell, on either
side of which, well-defined, elongated muscular impressions are visible, laterally edged in
adult individuals by a thickness of the shell, probably occasioned by a superabundance of
calcareous matter deposited by a plait of the mantle at the basis of each adductor muscle,

the insertion of which at the bottom of the valve having prevented the calcareous deposition extending over the space occupied by the muscular fibres; on either side, at the base of the septum in smaller valve, are seen two muscular impressions, corresponding with those just described in the larger valve, but of much smaller dimensions; the valves articulate by means of two strong teeth in the larger, corresponding with sockets in smaller valve, their separation becoming impossible without fracture. Dimensions variable; length 4, depth 3 lines.

Obs. Sowerby wisely created this genus in 1818, by observing that its internal organisation differed materially from that of other Terebratuliform shells; he states: "In the middle of the shell rises a thin longitudinal septum, reaching from one valve to the other, the upper part of it is perpendicular; on each side are two shelf-like appendages, one over the other, the upper ones united by slender processes to the hinge. . . . . . The resemblance of the arched septum to the bridge of a violin has suggested the generic name; to which valve this septum is attached I have not been able to ascertain, because I could not open the shell without breaking it." In 1845, having received from my friend, the late Mr. Duval, a great number of perfect specimens of this species, derived from the Upper Chalk of Meudon, near Paris,—M. Bouchard and myself proceeded to develope and examine the internal structure, which we fully described and illustrated two years after in the 'Bull. de la Soc. Géol. de France,' vol. v, 2d ser., p. 139, pl. ii, principally on account of the imperfect figures given by M. D'Orbigny in his 'Pal. Française,' proving that that author had not been as fortunate as ourselves in finding out the exact disposition of the internal apophysary system; the upper anchor-like lamellæ having been omitted and unmentioned in the description which appeared long after the publication of our paper. From observation, we were enabled to clear the doubts expressed by Sowerby, showing that the septum was attached solely to the bottom of the smaller valve. Since 1818, this species has often been confounded and misunderstood by various authors, such as Baron von Buch, Lamarck, and others. It is, however, most beautifully and clearly distinguished from all other Brachiopoda by its internal organisation, approaching to the only two genera, *Bouchardia*[1] and *Waltonia*,[2] proposed by myself some time back, these forming, along with *Magas*, a small family, comprising but few species. From *Bouchardia* and *Waltonia*, *Magas* differs in the form of its beak, foramen, and teeth; in *Bouchardia* the foramen is completely surrounded by the substance of the shell, and separated from the umbo *without deltideum;* while in *Waltonia* the position of the foramen, deltideum, and umbo is similar to that of the genus *Terebratulina;* this last, however, approaching more to *Magas* by the simplicity of its hinge, which is remarkably complicated in the genus *Bouchardia*. Besides these particular family characters, we find the same central, elevated septum, but different in its details. In *Magas*, in addition to the two riband-

[1] 'Bull. de la Soc. Géol. de France,' vol. vii, 2d ser., p. 62, pl. i, figs. 1—6, 1849.
[2] Davidson, "On the Genus *Waltonia*," 'Annals and Mag. of Nat. Hist.,' vol. v, 2d ser., 1850, pl. xv, fig. 1.

shaped lamellæ described, we observe two upper anchor-like lamellæ, situated parallel to the under ones; while in *Bouchardia* the under ones are completely wanting. In *Waltonia*, on the contrary, we have hitherto observed only the lower ones.[1]

Since the period of Mr. Bouchard's and my own publications on the genus and species under notice, I have assembled a very extensive suit of British specimens, which enabled me to collect some additional facts regarding the external shape and variations sometimes assumed by this interesting little shell, which does not seem to have ever exceeded much more than 5 or 6 lines in length. The shape of the beak, as well as the form of the smaller valve, often producing great variations in its external aspect; in general, the smaller valve presents a slight longitudinal depression appearing at some distance from the umbo, and extending to the front, the beak being more or less incurved; but from several remarkable specimens discovered by Messrs. C. B. Rose, Fitch, Catt, Woodward, and myself, it appears that in some cases the smaller valve becomes more or less convex without depression, the beak being almost straight, and exhibiting a greater or less developed area and foramen, as may be seen from the series of illustrative figures in Plate II, which I have selected out of a vast number of specimens, to show the passages of form;[2] in figs. 1 and 2 the beak is much recurved, and the smaller valve more or less concave; in fig. 4 it is less so, the foramen being more visible; in 5 and 6 the smaller valve is slightly convex, the beak hardly recurved, and in the two extremes of 7 and 8, this last is almost straight, showing a large area and foramen. On closely examining the latter exceptional variations or malformations, I soon became convinced that they arose from local causes, and especially from the shortness of the pedicle muscular fibres, forcing the beak and area to lie so close to the object of attachment, as to prevent the curvature of the beak; this is likewise proved by the umbo of the smaller valve in these specimens being worn by friction, as we remark in many specimens of *Argiope*, &c.; in very young individuals the beak is also almost straight, becoming recurved at a more advanced period of growth. *Magas pumilis* is circular, oval, or more or less elongated, as may be perceived from figs. 9, 10, 11. Sowerby's illustration and type of *Magas* does not show a very recurved beak, which at the time we first described the genus we believed to be a mistake, owing to the Meudon specimens not

[1] While looking over the valuable collection of M. Deshayes in Paris, I was agreeably surprised to find in one of his drawers the interior of a Cretaceous species, from the *Tourtia* of Belgium, described by Viscount d'Archiac in the second volume of the 'Mém. de la Soc. Géol. de France,' p. 333, pl. xxii, fig. 4, under the name of *Terebratula orthiformis*, and ater, by M. de Koninck, under that of *Orthis millepunctata;* in this shell we find the same arrangements as in *Magas*, to which genus the species above noticed will have to be referred. Unfortunately, the upper pair of lamellæ are imperfect in M. Deshayes' specimen.

[2] Mr. Woodward, in his 'Synoptical Table of British Organic Remains,' 1830, p. 22, mentions three species of *Magas*, *M. pumilus*, Sow., *M. magna*, and *M. punctata*, all from the chalk of Norwich; but as these last two are MS. names, unaccompanied by figure or description, we cannot offer any positive opinion on the shells intended as types, but it is probable, if not certain, that they were made out of the extreme shapes of *Magas* above noticed and figured in our plate.

presenting this peculiarity; but I have since observed, as above stated, that specimens did really at times assume the aspect of Sowerby's figures, and varied even to a greater extent.

I cannot understand what can have induced M. D'Orbigny to place in the family of *Magasidæ* the genus *Terebratulina*, which is completely and widely separated from Magas by its internal arrangements, as may be seen by casting a glance to the interior of *Caput Serpentis*, or to that of any other species of the genus. If *Terebratulina* is admitted to belong to the same family as Magas, I see no reason why all the others with calcareous appendages should be separated, several having much more important family affinities; the only reason stated by M. D'Orbigny being based on the erroneous supposition, that *Terebratulina* was unprovided with a deltideum. I have elsewhere shown that it really exists to a greater or less extent in all the species of that genus, and particularly on *T. substriata* Schlotheim, an Oolitic form, possessing the true internal character of *Terebratulina*, but rejected and placed by M. D'Orbigny in that of *Terebratula*, from the well-developed deltideal plates, proving how dangerous it is to class certain shells from mere external appearances, without giving full weight to the far more important internal arrangements connected with the disposition and the organisation of the animal.

Magas pumilus, to my knowledge, has not been observed lower down than the Upper, and perhaps Lower Chalk; it was found at Maudesly by Mr. Sowerby, at Trimmingham by Mr. Bowerbank, at Norwich by Mr. Fitch and the Rev. Mr. Image, at Brighton by Mr. Catt, and at Letheringsett, West Norfolk, by Mr. C. B. Rose, &c. On the Continent it is abundant; at Meudon near Paris, at Sens, Fécamp, near Valogne, and St. Gervais near Epernay in France, at Vaclo in Belgium, Simbrisk in Russia, &c.

Plate II, fig. 1. From the Norwich Chalk, in my Collection, enlarged.
,, fig. 2. A specimen, in which the smaller valve is slightly concave.
,, fig. 3. From the Chalk of Norfolk, in the British Museum, showing a very thickened margin and adult shell.
,, fig. 4. From the Chalk of Trimmingham, in the Collection of Mr. Bowerbank.
,, fig. 5. From my Collection.[1]
,, fig. 6. In the Collection of British Museum.

[1] Some of my friends urged me to make two species out of all these forms of *Magas*, from the extraordinary difference presented between such specimens as fig. 1 and figs. 7, 8; but I should be puzzled where to draw a line of demarcation, when it is remembered that figs. 7, 8, and 33, as well as 10 and 11, are quite exceptional forms, arising, no doubt, from malformations and accidental causes; besides which, Mr. Sowerby's type of *Magas pumilus* was created on specimens such as figs. 2, 4, and 5, which are intermediate in shape between those extremes in which the beak is much incurved, and those that are not so disposed. I consider figs. 1, 2, 3, 4, and 5 the usual forms of *Magas pumilus*, the remaining figures being accidental malformations. The illustration of *Magas pumilus*, given by M. D'Orbigny, in the work on Russia and the Oural, pl. xliii, fig. 27, closely resembles the real type of this species as figured by Sowerby, and is similar to our fig. 4; in it, as may be observed, the smaller valve is slightly convex and not depressed; the beak is likewise not much recurved, showing the area and foramen in all its extent.

Plate II, figs. 7, 8. Two remarkable malformations, in the Collection of Mr. C. B. Rose, from Letheringsett, West Norfolk.

,, figs. 9, 10, 11. Elongated varieties, in the Collection of the Rev. T. Image.

,, fig. 12^b. Interior of smaller valve.

,,   a. The socket walls.

,,   i. The sockets where the teeth, j, in the larger valve articulate.

,,   f. Extremity of the septum.

,,   c. Lower pair of riband-shaped lamella.

,,   e—e'. Upper or anchor-shaped pair.

,, fig. 12^c. Illustrates a profile view or section of the interior of both valves, the septum is seen to touch the larger valve at the point k.

,, fig. 12^a. Interior of larger valve illustrated.

,, fig. 33. From a remarkable malformation from the Sussex Chalk, in the Collection of Mr. Catt; this unique specimen varying so much from the common type, exhibits three marked stoppages of growth.

Genus—Terebratella, *D'Orb.* 1847.

Animal attached to submarine bodies by the means of a pedicle issuing from the beak of larger valve. Shell inequivalve, elongated, or transverse, variable in shape; valves convex; structure punctuated, smooth, or variously striated and plaited; beak truncated by a foramen of an oval or irregularly triangular shape, placed more under than above the summit; deltideum in two pieces, disunited in many cases, the aperture being completed by a small portion of the umbo; cardinal area more or less defined; loop (to which are affixed the arms) doubly attached, proceeding from the crura, but before attaining its greatest length, it gives off a flat, wide, more or less horizontal process, likewise attached to a central longitudinal, more or less elevated septum, the principal lamella proceeding till it doubles itself in the shape of a loop, as in true Terebratulæ.

Obs. I am only acquainted with two certain species of British Cretaceous *Terebratella*; it is possible that there may exist one or two more, but which, until future discoveries, I have left under the genus *Terebratula*.

8. Terebratella Menardi, *Lamarck*, Sp. Plate III, figs. 34—42.

Terebratula Menardi, *Lamarck.* 1819. An. sans Vert., vol. vi, p. 256, No. 50.

—   —   *Parkinson.* An Introduction to the Study of Org. Remains, p. 227, 1822.

—   —   *Defrance.* 1828. Dic. des Sc. Nat., vol. liii, p. 160.

TEREBRATULA MENARDI, *V. Buch.* 1834. Uber Ter., and Mém. de la Soc. Géol. de France, vol. iii, p. 184, pl. xvii, fig. 6, 1838.

— — *Deshayes.* Nouv. Ed. de Lamarck, vol. ii, p. 344, No. 50, 1836.

— TRUNCATA, *Sow.* Min. Con., vol. vi, p. 71, tab. 537, fig. 3, 1829.

— *Morris.* Catalogue, p. 137, 1843.

— — *Forbes.* Quart. Journ. of the Geol. Soc., p. 346, No. 105, 1845.

— — *Tennant.* A Strat. List of British Fossils, p. 47, 1847.

TEREBRATELLA MENARDI, *D'Orb.* Pal. Franç. Ter. Cretacées, vol. iv, p. 118, pl. 517, figs. 1—15 (not *T. Astieriana*, D'Orb.), 1847.

TEREBRATULA — *Bronn.* Index Pal., p. 1241, 1848.

— TRUNCATA, *Austen.* Quart. Journ. of the Geol. Soc., vol. vi, p. 477, 1850.

— MENARDI, *Cunnington.* Quart. Journ. of the Geol. Soc., vol. vi, p. 454, 1850.

TEREBRATELLA — *D'Orb.* Prodrome, vol. ii, p. 172, 1850.

TEREBRATULA — *Guéranger.* Bull. de la Soc. Géol. de France, vol. vii, 2d ser., p. 803, 1850.

Diagnosis. Shell semicircular, generally transverse, as wide, or wider, than long; rarely longer than wide; hinge-line forming a very obtuse angle, sometimes nearly straight, and almost as wide as the shell; larger or dental valve most convex, with a longitudinal sinus, extending from the beak to the front; beak large, straight, presenting a well defined, oblique, triangular area, truncated by a large foramen, and completed by two indented deltideal plates, which separate it more or less from the umbo; smaller valve less convex than the dental one, with a well-defined mesial fold extending from the umbo to the front, and producing an elevated curve; surface of valves ornamented by a variable number of sharp plaits, rarely bifurcating, but more commonly augmenting by the intercalation of other costæ at different distances from the umbo and beak. From eighteen to thirty plaits may be counted on each valve, according to age; from six to seven ornamenting the mesial fold and sinus; these longitudinal plaits are more or less intersected by concentric wrinkled lines of growth, often so close as to give the costæ a somewhat granulated appearance. Structure punctuated; in the interior of smaller valve the boss is much produced, on either side of which two deep condyles are seen to receive the articulating teeth of the dental valve; the loop is doubly attached; a slightly elevated longitudinal ridge is visible in the interior of larger valve, extending to about half the length of the shell; dimensions variable: length $6\frac{1}{2}$, width 7, depth 4 lines.

Obs. Professor Forbes, Bronn, Morris, and others have long considered the *Ter. truncata*, Sow., as a synonym of *T. Menardi*, Lamarck, an opinion in which I quite coincide, from having compared with great care many specimens of our British shell with those from France, and collected by myself in both localities. M. D'Orbigny, however, believes that our views are erroneous, and I suppose without having been able to examine a series of

British specimens, has given it the name of *T. Astieriana*. It is necessary to remark that many of the shells found in the Green Sand of Farringdon, whence our British specimens were derived, are more or less rolled, and the valves disunited, the plaits being slightly blunted; but during a later visit to that locality, I was able to obtain a few examples quite as sharp and perfect as any of those from Mans, and corresponding exactly with the description of *T. Menardi*, by Lamarck and M. D'Orbigny, and which may be seen by comparing the two figures we have purposely given in Plate III of the French and British specimens: fig. 42 is a French *T. Menardi* from Mans; fig. 38 a British specimen from Farringdon; wherein these differ I am at a loss to perceive; the same number of plaits and mesial fold; the same character of area, foramen, and general aspect; they both possess the same wrinkled concentric lines of growth stated by M. D'Orbigny to occur in *T. Menardi*, but *not*, according to him, in *T. Astieriana*, in which the number of costæ are more numerous, of which I am convinced, from having two typical specimens now before me, given to me by M. D'Orbigny himself, differing by several characters from the Farringdon shell, being more oval, flatter, and possessed of a greater number of smaller plaits. Our British specimens are, it is true, generally smaller than the French ones, and rather thicker, but this no doubt arises from accidental and local causes, which favoured more the development of the Mans specimens, an occurrence so frequent among animals where local causes produce different races and varieties. *Ter. Menardi* is stated to occur in Lower Green Sand, near Devizes, by Mr. Cunnington;[1] on the Continent it is abundant in beds of Mans (Sarthe), &c.[2]

Plate III, figs. 34, 35, 36, 37, 38, 39. Specimens from the Green Sand of Farringdon[3] in my collection.

„ fig. 42. A French specimen from the Mans.

„ fig. 41. The smaller valve, showing the calcareous loop enlarged.

„ fig. 40. The dental valve enlarged.

9. TEREBRATELLA PECTITA, *Sow.* Sp. Plate III, figs. 29—33.

TEREBRATULA PECTITA, *Sow.* M. Con., vol. ii, p. 87, tab. 138, fig. 1, 1818. (Non *Pectita*, Nilsson Petrefacta Succana, pl. iv, fig. 9, 1827.)

— — *Lamarck.* An. sans Vert., vol. vi, p. 255, No. 46, 1819.

— — *Parkinson.* An Introduction to Org. Rem., p. 227, 1822.

TEREBRATELLA PECTITA, *Brongniart et Cuvier.* Descr. Geol. des Env. de Paris, pl. ix, fig. 3, 1822.

---

[1] 'Quart. Journ. of the Geol. Soc.,' vol. vi, p. 454, 1850.

[2] I believe Mr. Austen in error while stating, in the 'Quart. Journ. of the Geol. Soc.,' vol. vi, p. 477, that *T. truncata*, Sow., is likewise found in the Upper Oolitic beds of Germany, whence at least I have never seen any authentic specimen.

[3] The age of the Farringdon beds has not yet been satisfactorily settled. Mr. Austen and others state them to be Lower Green Sand.

TEREBRATELLA PECTINATA, *Smith.* Identified by Organised Fossils, fig. 4, 1826 or 1827.

—   PECTITA, *Defrance.* Dic. des Sc. Nat., vol. liii, p. 159, 1828.

—   —   *V. Buch.* Uber Ter., 1834, and Mém. de la Soc. Géol. de France, vol. iii, p. 168, pl. xvi, fig. 12, 1838.

—   —   *Deshayes.* Nouv. Ed. de Lamarck, vol. vii, p. 343, 1836.

—   —   *? Hisinger.* Leth. Succ., p. 79, pl. xxii, fig. 13, 1837.

—   —   *Rœmer.* Des Vers. Nordd. Kreid, 1840.

—   —   *Morris.* Catalogue, 1843.

TEREBRATELLA PECTITA, *D'Orb.* Palæont. Franç. Ter. Crétacée, vol. iv, p. 120, pl. 517, figs. 16—20.

TEREBRATULA PECTITA, *Bronn.* Index Palæont., p. 1244, 1848. (But not the fig. 3 of *Faujas*, as stated by Professor Bronn.

—   —   *Dav.* Notes on an Examination of Lamarck's Brachiopoda. (Annals and Magazine of Natural History, vol. v, 1850.)

TEREBRATELLA PECTITA, *D'Orb.* Prodrome, vol. ii, p. 173, 1850.

Diagnosis. Shell more or less circular, or irregularly pentagonal, generally a little longer than wide, sometimes the reverse; valves seldom equally convex, the smaller one being somewhat depressed, especially so longitudinally towards the centre, and proceeding to the front; no mesial fold or sinus; hinge lines often very wide, and nearly straight; beak more or less prominent, slightly recurved, and diagonally truncated by a large and entire foramen, partly surrounded by a deltideum in two pieces, which separates the aperture more or less from the hinge margin; beak ridges well defined, leaving between them and the hinge margin a flat triangular more or less wide area. Surface of both valves numerously and variably plicato-striated, augmenting rapidly in number from the intercalation of plaits at different distances from the beak and umbo; thus, in very young specimens, I have counted only 26 plaits on each valve, while in some adult shells their number at times exceeds 62, owing to intercalation at different distances; concentric lines of growth, often very strongly marked, intersect the longitudinal striæ. In the interior of smaller valve, the calcareous supports are doubly attached, first to the crural base, secondly to the mesial longitudinal septum, after which the riband-shaped lamella, again extending to a short distance, bend themselves back, forming the loop. Structure punctuated.

Dimensions variable; the largest specimen noticed in England measured length 11, width 13, depth 7 lines, but in general not exceeding, length 10, width 9, depth 6 lines.

Obs. All authors seem to have agreed in preserving Sowerby's name to this elegant shell. It is very variable in shape; the plaits in some specimens being delicate and numerous, while the reverse is observable in others. The hinge line is also sometimes as wide as the shell, but it does not in general exceed two thirds; these variations will be seen in the different illustrations we have given in Plate III, figs. 29 to 33, all the specimens being from the Upper Green Sand of Horningham, Hill Deveril, near

Warminster, where it is very abundant; but I am not aware of its positively having been found in any other locality, excepting perhaps in the Chloritic Marl of Chard, where the species is, however, very rare. *T. pectita* is easily distinguished from *T. Menardi*, from its want of a mesial fold.   On the Continent, *T. pectita* is found at Cap la Hève, near Havre, but we do not know it to occur at Swanage Bay, Dorsetshire, as stated by M. D'Orbigny.

Plate III, figs. 29-30.   Specimens in the Collection of Mr. Cunnington, of Devizes.

Sub-Genus—Trigonosemus, *Kœnig*. 1825.

Animal fixed to rocks, or other marine bodies, by a pedicle issuing from the extremity of the beak; shell inequivalved, irregularly oval, circular, or rhomboidal, as wide as long, or longer than wide; larger valve always convex; smaller valve moderately so, flat or longitudinally depressed, beak produced, more or less recurved, and truncated by a small oval elongated foramen, beginning at the summit of the beak, and directing itself on the opposite side to the area.   Area large, often nearly as wide as the shell, triangular, flat, edged and carinated exteriorly; deltideum occupying less than a third of the surface; a small portion only surrounding the foramen.   Surface ornamented by numerous small radiating costæ, augmenting by intercalation at variable distances from the beak and umbo. Hinge line very obtuse, sometimes straight; valves articulating by means of two teeth in larger and corresponding condyles in smaller valve; between these last, a remarkably developed boss, or cardinal muscular fulcrum is visible, extending in some species considerably beyond the hinge line, and filling a corresponding cavity in the beak of larger valve; a short, thick, elevated, longitudinal septum, occupies about half of the length of the valve, and on either side of which two deep oval muscular impressions are visible; apophysary system or loop doubly attached; the riband-shaped lamella are first fixed to the sides of cardinal muscular fulcrum, and after proceeding to a short distance, are again attached to the highest point of the septum, before bending back on themselves to form the loop.   Structure perforated, dimensions variable, the largest specimen known not exceeding 20 lines in length.[1]

Obs. This genus was proposed in 1825 by Kœnig (*Icones Fossilium Sectiles*) under the following characteristic, " *Trigonosemus nob*. (Mollusca Brachiopoda) Testa inæquivalvis ovato-rotunda, valvarum altera superne producta in rostrum acuminatum S. truncatum apicæ perforatum externe convexum, interne planum; facie plana signo subtriangulari aperturam clausam indicante notata; nomen generis ex rostri signo triangulari.   Fig. 73. Trigonosemus elegans."

---

[1] The largest specimen of this genus I have seen is preserved in the collection of Mr. Morris, obtained in Cretaceous beds near Valogne.   The dimensions of the smaller valve, the only one found, measures length 17, width 17 lines.

Kœnig, however, placed two other shells in his genus, which are now considered to belong to different sections or genera; *Trigonosemus elegans* must therefore be taken as the type; later, in 1847, M. D'Orbigny applied to the same shell another generic title, that of *Fissurirostra*,[1] but which cannot hold priority over the name proposed by Kœnig many years before.

The arrangements of the loop in this shell are those of *Terebratella*, to which it bears more affinity than to that of any other genus or subgenus, as admitted by M. D'Orbigny, but the shape and character of its area, beak, and foramen, as well as the remarkably-shaped boss and muscular impressions, seem to entitle it to sub-generic distinction, since differences, much less important, are very often made use of in the separation of genera in other classes of Mollusca. In some of the species, such as *Trigonosemus pulchella* and *pectiniformis*, the foramen is so very minute, as to be generally visible only by the lens, which makes M. D'Orbigny suppose, that in every adult specimen the aperture became obliterated, and that the animal lived with its smaller valve upwards, and the larger one under,—the reverse of what exists in *Terebratulæ*. I am not, however, convinced of the correctness of this observation, from having a number of specimens of both the above species before me, in all of which the foramen, although small, is perceptible; and, owing to the incurved beak, we cannot believe it was attached differently from what we observe in *Terebratulæ*. The sub-genus *Trigonosemus* seems at present peculiar to the Cretaceous period.

10. TRIGONOSEMUS ELEGANS, *Kœnig.* Plate IV, figs. 1—4.

TRIGONOSEMUS ELEGANS, *Kœnig.* Içones Fossilium Sectiles, p. 3, pl. vi, fig. 73, 1825.
TEREBRATULA ELEGANS, *Defrance.* Dic. d'Hist. Nat., vol. 53, p. 157, 1828.
— RECURVA, „ Dic., p. 161, 1828.
FISSURIROSTRA RECURVA, *D'Orb.* Pal. Franç. Ter. Crétacées, vol. iv, p. 133, pl. 520, figs. 1—8, 1847.
— ELEGANS. Ib., vol. iv. p. 135, pl. 520, figs. 9—13, 1847.
TEREBRATULA ELEGANS, *Bronn.* Index Pal., p. 1236, 1848.
— RECURVA. Ib., p. 1248.
FISSURIROSTRA ELEGANS and RECURVA, *D'Orb.* Prodrome, vol. ii, p. 259, 1850.

Diagnosis. Shell inequivalve, irregularly oval, or somewhat rhomboidal, generally longer than wide; valves unequally convex, the greatest width and depth towards the middle; dental valve very convex longitudinally, keeled; beak large, much produced, moderately recurved, and obtusely truncated by a small, narrow, elongated, oval foramen or fissure, through which the pedicle issued; area very large, triangular, wider than long,

---

[1] 'Considerations Zoologiques et Géologiques sur les Brach.,' 'Comptes Rendus Hebdomadaires de l'Académie des Sciences,' 1847, and 'Pal. Française Ter. Crétacées,' vol. iv, p. 132.

and almost flat, carinated and edged; deltideum occupying less than a third of the width of the area, and diminishing gradually till it reaches the foramen, a very small part of which it encircles; smaller valve slightly convex, or longitudinally depressed near the front: hinge-line very obtuse and long. Valves ornamented by a great number of small rounded, radiating costæ, rarely bifurcating, but augmenting rapidly at variable distances by the intercalation of additional plaits, thus in one specimen eighteen only are seen at the umbo, thirty-three towards the middle, and fifty-six near the margin; in another at the umbo, fifteen; near the middle, thirty; at the edge, forty-four, &c.; these are likewise intersected at various distances by well-defined concentric lines of growth. Boss or cardinal muscular fulcrum much produced; loop small, doubly attached; structure punctuated; shell very thick, especially at the beak and umbo. Dimensions variable; length 11, breadth 9, depth 5 lines.

Obs. This beautiful and elegant species seems to have been first figured and described by Kœnig in 1825, under the name of *Trigonosemus elegans;* about the same period it was likewise described by Defrance under that of *Terebratula elegans* and *recurva.* The typical specimens of both authors having been obtained from the Upper Chalk of the neighbourhood of Valogne. During a late visit to that locality, M. de Gerville kindly gave me many specimens of this shell of all ages, and on the examination of which I felt persuaded that M. Defrance, D'Orbigny, and Bronn are mistaken in proposing to split this type into two species, as we find every insensible gradation between those forms with a depressed smaller valve to those in which it is moderately convex; this is especially noticeable in younger shells, where the smaller valve is very often nearly flat; the form of the beak is also variable both in length and curvature. It is but lately that this species has been discovered as occurring in England, and the first knowledge of the fact I owe to Mr. Woodward, who had seen and sketched two specimens found in the Norwich Chalk by Mr. Fitch. Mr. Harris has also procured it from the chalk detritus of Charing, Kent: this species is very rare in England, only three examples having been as yet obtained; all of which will be found illustrated in our Plate IV, figs. 1, 2, 3. These specimens are beautifully preserved, and present those variations which tempted Defrance and others to propose two species. It likewise occurs in the chalk of Ciply in Belgium, also at Freville near Valognes (Dep. de la Manche).

Plate IV, figs. 1, 1 ^{ab}. Specimen, natural size; the largest as yet found from the Chalk of Norwich, in the collection of Mr. Fitch.

,,   fig. 1 ^{cd}. Enlarged illustrations of the same.

,,   figs. 2, 2 ^{a}. Another specimen likewise from the Chalk of Norwich, measuring length 9, breadth 8, depth 4 lines; also from the collection of Mr. Fitch.

,,   fig. 3. A young specimen from the Chalk detritus of Charing, Kent, measuring length 6, width 6, depth 2½ lines, from the collection of Mr. Harris.

Plate IV, figs. 4 and 4*a*. Interior of smaller valve, illustrating the arrangement of the apophysary system, from a specimen obtained near Valognes, and preserved in the collection of M. Deshayes in Paris, (enlarged.)

„ fig. 4*b*. The dental valve, also enlarged.

11. TRIGONOSEMUS INCERTA, *Dav.* Plate IV, fig. 5.

Diagnosis. Shell of an elongated oval shape, longer than wide; valves almost equally convex: beak produced, rounded, moderately recurved, and truncated by a small oval foramen; area triangular, nearly flat, short, deltideum edging a small portion of the foramen; surface of valves ornamented by about thirty-four rounded costæ; the greater number arising from intercalation; these are closely intersected by numerous small and close concentric lines of growth. Structure punctuated; length $4\frac{1}{2}$, width 4, depth $2\frac{1}{2}$ lines.

Obs. This beautiful little shell was discovered by Mr. Moore in the Chalk with green grains of chard, where it appears very rare. It is always unsafe to establish a species on the inspection of a single specimen, which is all we have been able to obtain of the present form; but the shell seems adult, and differs much in general aspect from *Trigonosemus elegans*, being more oval, and ornamented by fewer and stronger costæ, as well as by the almost equal convexity of its valves, and straight frontal line; the area and beak are shorter, and the foramen comparatively larger. I have therefore temporarily named it *T. incerta*, hoping that the discovery of other specimens may enable Palæontologists to fix its real place with greater certainty.

Plate IV, fig. 5. Specimen, nat. size, from the collection of Mr. Moore.

„ fig. 5*abc*. Enlarged illustration of the same.

Sub-Genus—TEREBRIROSTRA, *D'Orbigny.* 1847.

Shell generally elongated, more or less oval, inequivalve, the dental valve much longer than the smaller or socket one, extending in the shape of a long beak, with a flat, false area and narrow deltideum; the foramen truncating the extremity of the beak, and partially completed by the deltideum; hinge articulating by means of two teeth in larger and corresponding condyles in smaller valve; internal disposition of the calcareous supports unknown; a mesial longitudinal plate is seen in the interior of smaller valve, probably destined to support a doubly-attached loop; structure punctuated.

Obs. This sub-genus is at present characterised so very unsatisfactorily, from our ignorance as to its internal disposition, that it is with great difficulty I have been able to make up my mind even to its temporal admission in this Monograph. I do so in hope that the discovery of its complete interior will soon permit Palæontologists to ascertain its true position. It is possible that eventually it may turn out to be only a *Terebratella*,[1] or have a disposition of loop peculiar to itself, as we do not consider a beak being more or less elongated of sufficient generic value. Many authors, however, seem disposed to separate the shells in question from the true Terebratulæ, and they may probably prove distinct. Thus, Mr. Sowerby states that Mr. Cumberland had called it *Lyra Meadi*; but, finding the term *Lyra* so apt, he could not resist applying it to the specific name. Later (1825) Kœnig, in his 'Icones Fossilium Sectiles,' placed it in his *Trigonosemus*, naming first, *T. elegans*; secondly, *T. rustica*; and thirdly, *T. lyra*, as types and examples. All these, however, have subsequently been placed in distinct genera, and we have preserved the name of *Trigonosemus* to the first, which corresponds to M. D'Orbigny's genus *Fissuri-rostra*. Few authors, excepting Mr. Brown,[2] have applied Kœnig's generic appellation to *T. lyra*, and it was only in 1847 that M. D'Orbigny proposed his genus *Terebrirostra*.[3] In England, we are only acquainted with one species, the *Terebrirostra lyra*.

12. TEREBRIROSTRA LYRA, *Sow.* Sp. Plate III, figs. 17—28.

Encyclop. Méth. Pl. 243, fig. 1, 1789.

TEREBRATULA LYRA, *Sow.*		Min. Con., vol. ii, p. 87, tab. 138, fig. 2, 1818.
—	— *Lamarck.*	An. sans Vert., vol. vi, p. 255, No. 49, 1819.
—	— *Conybeare* and *Phillips.*	Outlines of the Geol. of England and Wales, p. 130, 1822.
—	— *Parkinson.*	An Introduction to the Study of Organic Remains, p. 234, 1822.
—	— *Kœnig.*	Icones Fossilium Sectiles, p. 4, pl. vi, fig. 77, 1825.
TRIGONOSEMUS LYRA, *Kœnig.*		Ibid., p. 4, pl. vi, fig. 76, 1825.
TEREBRATULA LYRA, *Smith.*		Strata identified by Organised Fossils, fig. 3, 1816—27. (As this work appeared at different epochs, I do not know the exact date of this species.)
—	— *Defrance.*	Dic. des Sc. Nat., vol. liii, p. 160, pl. lxii, fig. 7, 1828.

---

[1] In the 'Journal de Conchiliologie,' No. 11, p. 223, 1851, M. D'Orbigny states his acquaintance with five species of *Terebrirostra*, viz. *T. neocomiensis*, *T. arduennensis*, *T. lyra*, *T. Bargesana*, and *T. cana-liculata*. Another most beautiful and well-characterised form, from the Chalk of Ciply, is known to me. One of M. D'Orbigny's so-called species, the *T. canaliculata* of Rœmer, is a true *Terebratella*, and somewhat resembles such species as *T. Menardi* and *T. pectita*, the loop is doubly attached, and disposed exactly as we see in the types of the genus *Terebratella*. The position of the foramen relative to the extremity of the beak is different from *T. lyra* and its associates; and we cannot but feel surprised at M. D'Orbigny proposing to place Rœmer's species in his genus *Terebrirostra*.

[2] 'Illustrations of Fossil Conch. of Great Britain,' 1838.

[3] 'Considerations Zoologiques et Géologiques sur les Brachiopodes,' 'Comptes Rendus Hebdomadaires de l'Académie des Sciences,' Août, 1847.

TEREBRATULA LYRA, *Deshayes.* Ency. Méth., vol. iii, p. 1029, 1832.
— — *V. Buch.* Uber Ter., 1834, et Mém. de la Soc. Géol. de France, vol. iii, 1st ser., p. 173, pl. xvi, fig. 17, 1838.
— — *Deshayes.* Nouv. Ed. de Lamarck, vol. vii, p. 344, No. 49, 1836.
— — *Deslongchamps.* Soc. Linn. de Normandie, 1837.
— — *D'Archiac.* Obs. sur le Groupe Moyen de la Forme Crétacée, Mém. Soc. Geol. Fos., p. 295, vol. iii, 1839.
TRIGONOSEMUS LYRA, *Brown.* Illustrations of Foss. Conch. of Great Britain, pl. xlix, figs. 5, 13, 1838.
TEREBRATULA LYRA, *Morris.* Catalogue, 1843.
— — *Raulin.* Patria la France Ancienne et Moderne, p. 362, fig. 100, 1844.
TEREBRIROSTRA LYRA, *D'Orb.* Pal. Franç. Ter. Crétacées, vol. iv, p. 129, pl. 519, figs. 11, 19, 1847,
TEREBRATULA LYRA, *Dujardin.* Dic. Universelle d'Hist. Nat. Mollusques, pl. ix, figs. 5-6, 1848.
TEREBRATULA LYRA, *Bronn.* Index Pal., p. 1241, 1848. (But not his Synonyms, which are very defective.)
TEREBRIROSTRA LYRA, *D'Orb.* Prodrome, vol. ii, p. 173, 1850.
— — *D'Orb.* Journal de Conchyliologie, vol. ii, p. 224, 1851.

Diagnosis. Shell inequivalve, of an elongated, irregularly oval shape, moderately convex, rather compressed in the middle; beak much produced and elongated, very often exceeding in length the remaining portion of the shell, straight or gently curved, while tapering gradually to its extremity, which is truncated by a rather small transversely oval foramen, partly completed by a long and narrow deltideum, longitudinally depressed, and extending from the extremity of the beak to the hinge-margin; on either side a well-defined, almost flat, false area is perceptible.

Smaller valve oval or irregularly pentagonal, tapering at the umbo, widest near its middle, slightly curved or straight in front; valves ornamented by a great number of diverging, irregularly-disposed rounded costæ, sometimes bifurcating, more commonly augmenting by numerous intercalated plaits at different distances from the beak and umbo; these are intersected by concentric lines of growth, more numerous near the margins. The internal dispositions of the calcareous supports are as yet unknown; structure punctuated; dimensions variable. Length of the largest specimen 2 inches 3 lines, width 10, depth 7 lines.

Obs. This is one of the most beautiful among the Brachiopoda, and is much sought after by collectors, from its elegant shape as well as its rarity: the great length of its beak is one of its most striking features; as we have already stated, it often exceeds considerably the remainder of the shell, especially in young individuals; but, as judiciously observed by M. D'Orbigny, it seldom extends with age; on the contrary, the palleal portion becomes more elongated as well as augmented in width, so that in the generality of adult specimens the body of the shell exceeds the dimensions of the beak. The form and arrangement of the costæ are likewise very remarkable; they are rarely straight, except in the young, but are irregularly waving; the few central ones extending almost in a direct line to the front, but the lateral costæ, particularly in older shells, seem to lie more against the central ones, and augment rapidly by bifurcation and intercalation at different distances and periods of growth.   All we know of its interior is, that the valves articulate by the means of two teeth in the larger and corresponding sockets in the smaller valve; the beak seems to be hollow, and strengthened by two slender plates situated at the base of the teeth, and dividing it into three unequal partitions, as may be seen by the section I have given of the beak at half its length (Pl. III, fig. 25); these Plates, however, seem gradually to approach the lateral portions of the beak, leaving the foramen entirely surrounded by its substance and deltideum, as seen in fig. 28; in the interior of the smaller valve the boss or cardinal muscular fulcrum, fig. 27^h, is somewhat produced, and a short elevated longitudinal septum extends to less than half the length of the valve.   Whether the loop was simply attached to the crural base, as in true *Terebratulæ*, or doubly attached as in *Terebratella*, or otherwise disposed, is still a matter of conjecture.

Ter. lyra has hitherto been found but in a few localities; viz., in the Upper Green Sand at Chute Farm, near Horningham, where it is far from being common; also still more rarely in the Chalk, with green grains, of Chardstock, in which locality it was discovered by Mr. Bunbury, some years ago.   On the Continent it is found in the Upper Green Sand, or Chloritic Chalk of Cap la Hève, near Havre, whence no doubt the first specimen, figured in 1789 in the 'Ency. Méth.,' was obtained.

All our illustrations in Pl. III are from specimens belonging to the Upper Green Sand from the neighbourhood of Warminster, figs. 17, 18, 19, 20, belonging to the Collection of Mr. Falkner, of Devizes; fig. 26 from that of Mr. Cunnington.

Genus—TEREBRATULINA, *D'Orb.* 1847.

Animal fixed to submarine bodies by means of a pedicle passing through the foramen of the beak in larger valve, edges of the mantle free, body small.   Shell punctuated, inequivalve, variously shaped, generally longer than wide, and more or less oval; beak obliquely truncated by the foramen, which is large, and extending to the umbo deltideum often small and indistinct; smaller valve less convex than the perforated one, with two more or less developed auricle expansions.   Surface generally striated or costellated.   Valves articulating by the means of two teeth in the larger, and sockets in the smaller valve; apophysary system short, not exceeding one third of the length of the shell, and formed of two short stems simply attached to the extremity of the socket ridges, which, after converging, are united by a lamella in the shape of a small square tubular ring bent upwards in front; on the sides of these are fixed the free fleshy arms of the animal which extend to near the frontal margin, and bent back in the shape of a loop. The outer edges are covered by long cirri.

Obs. I have only found two species of British Cretaceous *Terebratulina;* the genus existed during the Oolitic, Cretaceous, and Tertiary periods, and is still abundant in the recent state.

13. TEREBRATULINA STRIATA, *Wahlenberg*, Sp.   Plate II, figs. 18, 28.

      Faujas St. Fond.   'Hist. de la Montagne de St. Pierre de Maestricht.' pl. xxvi, figs. 7 and 9, 1799.

      TEREBRATULITES CHRYSALIS, *Schlotheim.* Beiträge zur Nat. Vers., in Leonhard's Min. Tasch., vol. vii, p. 1813, (ref. to *Faujas* fig., pl. xxvi, fig. 7.)

      —   TENUISSIMA, *Schlotheim.*   Ib. (ref. *Faujas* fig., pl. xxvii, fig. 7.)

      —   CHRYSALIS, *Schlotheim.*   Die Petref., p. 39, 1820, (ref. *Faujas*, pl. xxvi, figs. 7, 9.)

      ANOMITES STRIATA, *Wahlenberg.*   Petrificata Telluris Suecanæ Examinata, in Nova Acta Regiæ Soc. Scien. Upsaliensis, vol. viii, p. 61, 1821, (no figure.)[1]

      TEREBRATULA DEFRANCII, *Brongniart et Cuvier.* Desc. Géol. des Env. de Paris, p. 383, pl. iii, fig. 6, 1822.

      —   STRIATULA, *Mantell.*   Geol. of Sussex, pl. xxv, figs. 7, 8, 12, 1822.

      —   PENTANGONALIS, *Phillips.*   Illustrations of the Geol. of Yorkshire, Part I, pl. i, fig. 17, 1825.

      —   STRIATULA, *Phillips.*   Ib., pl. ii, fig. 28.

      —   —   *Sowerby.*   Min. Con., vol. vi, p. 69, pl. 336, figs. 3, 4, 1829, (fig. 5 is the London Clay species, to which we preserve the name of *Striatula,* equally given to that shell by Sowerby.)

      —   DEFRANCII, *Nilsson.*   Petrefacta Suecana, p. 35, pl. iv, fig. 7, 1827.

      —   —   *Dalman.*   Acad. Handl., p. 52, 1827.

      —   DEFRANCII, *Defrance.*   Dic. des Sciences Nat., vol. liii, p. 163, 1828.

      —   GERVILLIANA, *Defrance.*   Ib., p. 157, 1828, (is only a young specimen.)

      —   GERVILLEI, *Woodward.*   An Outline of the Geol. of Norfolk, tab. vi, fig. 14, 1833.

      —   STRIATULA, *V. Buch.*   Uber. Ter., 1834, and Mém. de la Soc. Géol. de France, vol. iii, 1st ser., p. 164, pl. xvi, fig. 7, 1838.

      —   DEFRANCII, *V. Buch.*   Ib., p. 165, pl. xvi, fig. 8, 1838.

      —   CHRYSALIS, *V. Buch.*   Ib., p. 166, pl. xvi, fig. 9, 1838.

      —   STRIATULA, *Deshayes.*   Nouv. Ed. de Lamarck, vol. vii, p. 360, 1836.

      —   DEFRANCII, *Deshayes.*   Ib., p. 367.

      —   —   *Hisinger.*   Leth. Suecia, p. 78, tab. xxii, fig. 10, 1837.

      —   CHRYSALIS, *Bronn.*   Leth. Géog., p. 651, pl. xxx, fig. 6, 1837.

      —   STRIATULA, *Geinitz.*   Characteristik der Kreidegebirges, p. 14, 1839.

      —   DEFRANCII, *Rœmer.*   Die Vers. Nord. Kreid., p. 40, 1840.

---

[1] Described as follows: "*An. striata,* superficii ubique longitudinaliter striata conca subcuneiformis, nate brevi ampla—e Bahlsberg allatus longitudini sesquipollicari."

TEREBRATULA STRIATULA, *Rœmer.* Die Vers. Nord. Kreid., p. 39, 1840.

— FAUJASII, *Rœmer.* Ib., p. 40, tab. vii, fig. 8, 1840.

— AURICULATA, *Rœmer.* Ib., p. 39, tab. vii, fig. 9, 1840.

— CHRYSALIS, *Rœmer.* Ib., p. 40, 1840.

— STRIATULA, *Geinitz.* Char. Petref. Kreid., pl. xvi, fig. 12, 1840.

— CHRYSALIS, *Bronn.* Leth. Geog., p. 651, pl. xxx, fig. 6, 1837.

— STRIATULA, *Morris.* Catalogue, 1843.

— — *D'Orb.* Russia and Oural, vol. ii, p. 493, pl. xliii, figs. 18—20, 1845.

— — *Reuss.* Die Vers. der Böhemischen Kreideformation, p. 49, pl. xxvi, fig. 2, 1836.

— CHRYSALIS, *Reuss.* Ib., p. 49, pl. 26, fig. 3.

— FAUJASII, *Reuss.* Ib. p. 50, pl. xxvi, fig. 4.

TEREBRATULINA STRIATA, *D'Orb.* Pal. Franç. Ter. Cretacées, vol. iv, p. 65, pl. 504, figs. 9—17, 1847.

TEREBRATULA DEFRANCII, *Bronn.* Index. Pal., p. 1234, 1848.

— CHRYSALIS, *Bronn.* Ib., p. 1232.

— — *Dav.* Lond. Geol. Journal, pl. xviii, figs. 18, 20, 1847.

— STRIATULA, *Dixon.* The Geol. and Fossils of the Ter. and Cret. Form. of Sussex, pl. xxvii, fig. 21, 1850.

TEREBRATULINA STRIATA, *D'Orb.* Prodrome, vol. ii, p. 358, 1850.

Diagnosis. Shell variable in shape according to age; irregularly oval or pentangular, tapering at the beak, longer than wide; valves slightly and almost equally convex, sometimes presenting a slight longitudinal depression on each valve, beginning at about half the length of the shell, and extending to the front, which is at times more or less indented, or forms a regular outward curve, no traces of indentation being visible; beak short, obliquely truncated by a moderately sized foramen, formed partly out of the substance of the beak, and by two small deltideal plates on either side; the anterior portion being in general completed by the umbo. Beak ridges indistinct, sloping rapidly off, giving the beak a tapering appearance; auricles on either side of the umbo very large and straight in the young shells, small and oblique in the adult, and sometimes disappearing almost from the convexity of this portion of the valve. Surface ornamented by a variable number of radiated, small, elevated striæ; few in number, and granulated in the young; very numerous and smooth in the adult, augmenting rapidly at a short distance from the umbo and beak by bifurcation, but more generally by intercallation. Concentric lines of growth more or less marked; internal calcareous supports short and anneliform; structure punctuated; dimensions variable; our British specimens do not seem to exceed length 11, width 8, depth 4½ lines.

Obs. Many species have been made out of different ages of this variable shell, as may be perceived by a glance at the list of synonyms; having compared and studied a multitude of specimens of all ages of this form, from the size of a quarter of a line to near one inch in length, I feel satisfied that all those hitherto found in England belong to, and are mere variations of, one single type, scarcely distinguishable from the recent *Ter. caput-*

serpentis; this last seems, however, a deeper shell; but there is no difference that I can perceive in the striæ or internal details of the apophysary system. Most Palæontologists strongly object to the idea of a Cretaceous species being found recent, and although we are not convinced that such a thing is impossible, we have not been able to bring ourselves to a positive admission of the fact as certain. The tertiary species to which we have preserved the name of *Striatula* appears to us more distinct from the recent and Cretaceous form, than this last is from the recent. *T. striatula* is wider and more circular at the beak and umbo, and does not seem to taper away as in the other forms. I have experienced much difficulty in making up my mind as to what name this species should preserve, so many having been proposed by various authors.[1] I have at last decided to follow M. D'Orbigny, in adopting that of *T. striata,* described, but not figured, by Wahlenberg in 1821, although this name had been given by Brocchi in 1814[2] to another species illustrated by Soldani in 1780,[3] but on examining the very bad figure of this last-named author, it would be difficult to ascertain the type intended. It must also be remarked that several years before Wahlenberg, Schlotheim, in 1813, had given various names to young specimens and malformations, figured, but not named, by Faujas in 1799, such as *Ter. chrysalis.* This appears to be a malformation of a young specimen of *Ter. striata,* but not understood by most authors who have contented themselves to preserve and reproduce Faujas' figure, with all its defects. At the same period, Schlotheim likewise gave to another figure of Faujas the name of *T. tenuissima.* This is only another variation in age of the same form; and, as these do not illustrate in a satisfactory manner the shell in question, I think it preferable to adopt that of Wahlenberg. I am not astonished at early naturalists not having perceived the real difference presented by age in this species, but I am surprised that many modern authors should have quite lost sight of this character. I therefore believe the *Ter. striatula* of Mantell, *Defrancii* of Brongniart, *Gervilliana* of Defrance, *Pentagonalis* of Phillips, *Faujasii* and *Auriculata* of Rœmer, *Gervillei* of Woodward, and other names above mentioned, must all merge into one species, viz., *T. striata.* It is very probable, that one or two of the species lately proposed by M. D'Orbigny will require to be added to the synonyms. I have come to the above conclusion, but in which I may be mistaken, after a long and minute examination of several hundred specimens from various beds, localities, and countries. In England, we have not yet found any of those very adult specimens known under the appellation of *Defrancii,* which are rare everywhere; at Meudon, the chalk is full of young

[1] It is stated in M. Fischer's and Waldheim's 'Oryctographie du Gouvernement de Moscow,' 1830—37, that Dr. Mantell's *Ter. striatula* is a synonym of *Ter. scabra* (Fischer), described and figured by that author in 1809 ('Notice des Fossils du Gouv. de Moscow,' pl. ii, figs. 1 and 2); but, on examining the memoir just noticed, I must confess that no one could identify the shell figured with *T. striata.* *T. scabra* cannot, therefore, be taken as the type of the species under notice.

[2] Brocchi, 'Conchologia Fossile,' p. 466, No. 18, 1814, *Anomia striata.*

[3] Soldani, 'Saggio Orittografico,' Sienna, 1780, tab. xvi, fig. 82, o. p.

shells, but it is only now and then that a large adult specimen is discovered. With age the striæ do not seem to augment much in width; on the contrary, they are many times larger in proportion in the young.

T. striata is found in the Upper and Lower Chalk of Kent, Sussex, and Norfolk; in the Upper Green Sand near Warminster; and in the Speeton Clay of Knapton, on the Yorkshire Coast. At one time, I thought the Upper Green Sand specimens might belong to another species, from being generally more elongated and rarely notched or indented in front, this portion forming a regular outward curve; but, having found, in the Lower or Grey Chalk, near Dover, many examples quite identical with those from the Upper Green Sand, I determined to give up that idea. I will not offer so decided an opinion relative to the form found in the Speeton Clay, having been able to study only one specimen, lent me by Mr. Bean, and which apparently presents some variations; and, if constant, they would entitle it to a separate appellation. It much resembles the figure which M. D'Orbigny gives of *Ter. Companiensis*. This last-named shell is, however, always adult, with much smaller dimensions. On the Continent, *T. striata* is met with in the same stratigraphical range as in England, and is common to many localities. In Plate II, I have endeavoured to illustrate the principal variations that I have observed in the British specimens.

Fig. 18. From the chalk of Norwich, in the collection of Mr. Fitch. 18ᶜ. Enlarged illustration.

Fig. 19. Illustrates the interior and annular apophysary process, from a specimen in my collection.

Figs. 20, 21, 22, 23, 24. Young shells, from the Upper and Lower Chalk of Kent, principally from the collection of Mr. Harris.

Fig. 25. A specimen from the Gray Chalk, in the vicinity of Dover.

Fig. 26. From the Speeton Clay of Yorkshire, in the collection of Mr. Bean.

Figs. 27, 28. Two specimens from the Upper Green Sand near Warminster, from the cabinet of Mr. Cunnington.

14. TEREBRATULINA GRACILIS, *Schl.* Sp. Plate II, figs. 13—17.

TEREBRATULITES GRACILIS, *Schlotheim*. 1813. Beiträge zur Naturgeschichte der Versteinerungen in Leonhard's Min. Tasch., vol. vii, p. 112, tab. iii, fig. 3.

— — *Schloth.* Die Petrefactenkunde, p. 270, No. 35, 1820.

TEREBRATULA RIGIDA, *Sowerby.* Min. Conch., vol. vi, p. 69, tab. 536, fig. 2, 1829.

— GRACILIS, *Schlotheim.* Syst. Vers. der Petrefacten., 1832.

— — *Von Buch.* 1834. Uber Terebrateln, 1838; and Mém. Soc. Géol. de France, vol. iii, 1st ser., p. 167, pl. xvi, fig. 11.

— — *Geinitz.* Charak. der Schichten und Petrefac. der Säch. Kreidegebirges, 1839, p. 14, No. 4; and in 1840, pl. xvi, fig. 13.

TEREBRATULA GRACILIS, *Rœmer.* 1840. Die Vers. Norddeutschen Kreidgebirges, p. 40, No. 27.
—   —   *V. Hagenow.* Mon. der Rüg. Kreid. Vers., 1842.
—   RIGIDA and GRACILIS, *Morris.* Catalogue, 1843.
—   GRACILIS, *Geinitz.* Grundriss der Verst., pl. xxi, fig. 10, 1844.
TEREBRATULINA GRACILIS, *D'Orbigny.* Geol. of Russia and Ural, vol. ii, p. 499, pl. xliii, figs. 24—26, 1845.
TEREBRATULA GRACILIS, *Reuss.* Die Vers. der Böhemischen Kreideformation, p. 49, pl. xxvi, fig. 1, 1846.
TEREBRATULINA GRACILIS, *D'Orbigny.* Pal. Française Ter. Crétacées, vol. iv, p. 61, pl. 503, figs. 1—6, 1847.
TEREBRATULA GRACILIS, *Bronn.* Index Pal., p. 1237, 1848.
TEREBRATULINA GRACILIS, *D'Orb.* Prodrome, vol. ii, p. 198, 1850.
TEREBRATULA GRACILIS, *C. Puggaard.* (Géol. de l'Ile de Möen.) Bull. de la Soc. Géol. de Fr., vol. vii, p. 534, 1850.

Diagnosis. Shell inequivalve, orbicular, circular or elongated, either wider than long or longer than wide; larger or dental valve always convex; the smaller one either moderately so, or flat, never gibbous; beak not much produced; foramen small, formed partly out of the substance of the beak, and by two small lateral deltidial plates, the anterior portion being completed by the umbo; beak ridges well defined, leaving between them and the hinge line a flat space or false area; auricles on either side of the umbo rather small; valves ornamented by a variable number of radiating elevated striæ, often granulated, which, soon after leaving the beak and umbo, augment rapidly in number, rarely from bifurcation, but almost always by the intercalation of smaller costæ, at variable distances between the larger ones, and extending to the front and lateral margins; concentric lines of growth more or less prominent; structure punctuated; internal calcareous supports short and anneliform. Dimensions variable: length $5\frac{1}{4}$; width 5; depth 2 lines. The comparative depth in some specimens is greater than is here given.

Obs. Schlotheim, in 1813, appears to have been the first author who noticed and figured this species, stating it to occur in England; and it is the same as that described later, from the Chalk near Norwich, by Sowerby, under the appellation of *Ter. rigida.* It appears, likewise, to have been one of those fortunate forms, to which the generality of authors have applied the original name, notwithstanding its great variability in the convexity of its valves and the number of its plaits, the last varying from eighteen to fifty on each valve, round the edge. The costæ, also, are not of equal width and depth, as almost invariably, between two larger ones, one, two, or three smaller costæ intervene, appearing at different distances from the beaks, and extending to the front. It is, therefore, owing to the greater or less number of these intervening plaits that the shell becomes more closely or widely striated. It does not, however, present as much difference in the young state as we discover in *T. striata,* and from this last is always easily distinguished by its much more circular form and greater comparative width, as well as by the general aspect of the shell.

In England, *T. gracilis* is very abundant in the Upper and Lower Chalk of many localities; thus, at and between Dover and Folkstone, Mr. Mackie, myself, and others have found it in both the Upper and Lower Chalk; at Charing, Kent, it has been obtained from the chalk detritus by Mr. Harris; likewise in the Sussex Chalk by Mr. Catt. Mr. Bowerbank has it from the Chalk of Trimmingham; Messrs. Fitch, Woodward, and others, from that of Norwich; and Mr. C. B. Rose informs me, that he has found it in the Blue Gault of West Norfolk. I have not seen any specimens from the Gault of the South of England, nor from either the Upper or Lower Green Sand: its vertical range seems much smaller than that of *T. striata*.

On the Continent, *T. gracilis* is likewise abundantly distributed, and its principal varieties have been figured by Schlotheim, Sowerby, Rœmer, V. Buch, Geinitz, Reuss, and D'Orbigny.

Plate II, fig. 13. A specimen, with the smaller valve almost flat, very slightly convex, only at the umbo, from the Chalk of Trimmingham, in the collection of Mr. Bowerbank. Similar specimens are also preserved in the York Museum.

„ fig. 14. A remarkably fine specimen, the smaller valve being almost flat, from the Chalk near Norwich, in the collection of Mr. Fitch.

„ fig. 15. The interior of the smaller valve, from Dover, in my collection.

„ fig. 16. A variation, in which the smaller valve is slightly convex, from the Chalk of Dover.

„ fig. 17. A very convex variation, also from the Kentish Chalk, illustrating the greatest number of striæ I have noticed in the species.

Sub-genus—KINGENA, *Dav.* 1852.

Animal fixed to submarine bodies by means of a pedicle; shell inequivalve, more or less circular or ovate; valves convex; beak moderately produced, recurved; foramen circular, partly completed by a deltidium in two pieces, not always visible, from the foramen lying close to the umbo. Structure punctuated, surface variously ornamented by granular spinose or squamose unequal asperities irregularly disposed. Beak ridges well defined, leaving a false area between them and the hinge margin, valves articulating by means of two teeth in the larger and corresponding sockets in the other valve. In the interior of smaller valve, a deep hollow crura, or muscular fulcrum, widely separates the inner socket walls; no produced boss; a small deep depression lies along the centre of the crura, giving rise to a short elevated mesial longitudinal septum not extending further than half the length of the valve. On either side of the crural base two riband-shaped lamellæ are fixed; these, after proceeding a short distance, throw off a lateral process which attaches itself to the sides of the central septum, at a short distance from its origin;

the lamellæ extend again to about two thirds of the length of the valve; when, bending back upon themselves, they form a large wide loop, and, after leaving a very small free space just above the septum, are likewise fixed on either side to the central septum above the under pair.

Obs. The remarkable character and disposition of the lamellæ in this form, induces me to propose a subgeneric title, as such shells could not with propriety be placed in any of the sections now in use; and to afford greater facilities of comparison, I have introduced some illustrations of different sections of the family *Terebratulidæ:*

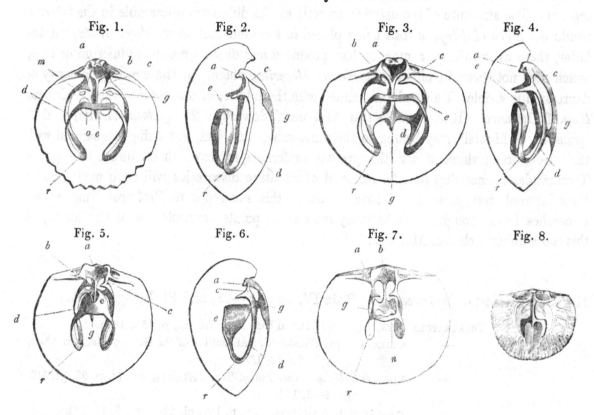

Figs. 1 and 2 represent the interior of Terebratula *Australis:* in this the loop is simply attached to the crural base, the riband-shaped lamellæ forming throughout a free and slender loop, extending into the cavity of the valve.

Figs. 3 and 4, Terebratella *Chilensis:* here the loop is doubly attached, first to the crura, then again to the central longitudinal septum, before proceeding forth, and bending back, to form a free slender loop, as in *Terebratula.*

Fig. 7, Megerlia *truncata:* a slightly elevated medio-longitudinal crest proceeds from under the muscular fulcrum (*a*) to less than half the length of the valve, near the extremity of which (*g*) two short central diverging branches arise (*n*) and support the calcareous loop, which consists of two riband-shaped lamella first attached to the inner side of the socket walls or crural base, afterwards extending to the extremity of the diverging branches (*n*) to which they are affixed before proceeding on both sides in the same direction to their extremity (*r*) under the shape of two nearly parallel longitudinal lamellæ; the inner ones are very wide, and joined by a narrow arched process; the loop is therefore three times attached on either side.

Fig. 8 illustrates the genus KRAUSSIA[1] *rubra*, (Pallas, Sp.)   Here the internal calcareous appendages consist only of two central diverging branches, somewhat spread out at their ends.

Figs. 5 and 6 illustrate KINGENA *lima*, as well as fig. 15, Pl. IV.   The last is viewed more in front, to show the connection of the four branches or lamellæ affixed to the septum.   The structure of the exterior, as well as the difference observable in the interior, would not allow of *Kingena lima* being placed in the same section as *Terebratella;* in the latter, there exists always a more or less produced and distinct muscular fulcrum or boss, which does not occur in either *Kingena* or *Megerlia,* where, on the contrary, a cavity or depression is visible.   I am only acquainted with the interior of one species, the *Terebratula lima* of Defrance.   It is probable that *Ter. ovata,* Sowerby; *T. rugulosa,* Morris; and *T. squamosa,* of Mantell, may belong to the same sub-genus; but, not being acquainted with their interiors, I think it for the present preferable to leave them under the genus Terebratula, whence they may be removed when future discoveries will have made known their internal arrangements.   I have dedicated this sub-genus to Professor King, whose researches have brought to light many interesting points connected with the history of this most difficult class of Mollusca.

15. KINGENA LIMA, *Defrance,* Sp.   Plate IV, figs. 15—28, and Pl. V, figs. 1—4.

TEREBRATULA LIMA, *Defrance.*   Dic. d'Hist. Nat., vol. liii, p. 156, 1828.
—   PENTANGULATA, *Woodward.*   An Outline of the Geol. of Norfolk, tab. vi, fig. 10, 1833.
—   LIMA, *D'Orbigny.*   Pal. Franç. Ter. Crétacées, vol. iv, p. 98, pl. 512, figs. 1—5, 1847.
—   HEBERTIANA, *D'Orbigny.*   Ib., p. 108, pl. 514, figs. 5, 11, 1847.
—   SPINULOSA, *Morris.*   Annals and Mag. of Nat. Hist., vol. xx, Oct., 1847, p. 253, pl. xviii, figs. 6, 6ª.
—   LIMA, *Bronn.*   Index. Pal., p. 1240, 1848.
—   SEXRADIATA, *Sow.* in *Dixon.*   The Geol. and Fossils of the Ter. and Cret. Form. of Sussex, pl. xxvii, fig. 10, 1850.
—   HEBERTIANA, *D'Orb.*   Prodrome, vol. ii, p. 258, 1850.
—   LIMA, *D'Orb.*   Ib., p. 172, 1850.
—   — *Guéranger.*   Bull. Soc. Géol. de France, vol. vii, p. 803, 1850.

Diagnosis.   Shell inequivalve, circular, oval, or irregularly pentagonal; as wide as long, or wider than long; valves more or less convex; larger or dental valve most so; beak

[1] Genus *Kraussia,* Dav. 1852; five recent species are known, *K. rubra, K. natalensis, K. pisum, K. cognata;* the genus has not been noticed in the fossil state.

short, much recurved, and truncated by a middle sized foramen, generally lying close on the umbo, so that little or none of the deltidium is seen; but when exposed, especially in young shells, it is composed of two plates, generally lateral and disunited; a small portion of the circumference being completed by the umbo; beak ridges well defined, exhibiting a flat or concave false area between them and the hinge margin; smaller valve less convex than the dental one, no mesial fold or sinus perceptible, but the margin line in front presents at times a slightly elevated curve. Structure punctuated and irregularly covered by a multitude of small granulations or short hollowed spines, often very large in comparison with the dimensions of the shell, and closely or more widely separated. The largest British specimen yet found, measures, length 12, width 12, depth 7 lines; others, length 9, width 6, depth 5 lines, &c.

Obs. This interesting shell has given me great trouble from the variability of its shape, and it was not before having minutely examined with much care a vast number of specimens, that I came to the conclusion, that this species existed in the Upper Green Sand, Gault, Lower and Upper Chalk; its stratigraphical and vertical range is therefore very great. Secondly, that *Ter. pentangulata* of Woodward, *Ter. Hebertiana* of D'Orbigny, *Ter. sexradiata* of Sowerby, *Ter. spinulosa* of Morris, are all variations of one and the same type, viz., *Ter. lima* of Defrance. The interior of all are exactly similar; the structure the same, as well as all the other characters, though these may be slightly modified in appearance in some beds and localities, owing to the matrix and other local causes; for example, if we take a small or young specimen of *Ter. lima* of Defrance, and compare it with a large specimen of *Ter. pentangulata* of Woodward, we perceive an apparent difference, but this is also seen in all extremes of any variable form; but if, however, we have a great number of specimens of the same species they will soon fill up all the links separating the two extremes by insensible passages, and this is just what I have found in *T. lima*, from the study of more than a hundred specimens obtained from the different deposits above alluded to; in the young, the shell is often somewhat compressed, but in all adult individuals it is more or less convex. To afford facility of comparison, I have given in Plates IV and V no fewer than forty-four figures from specimens found in all the beds mentioned, and from different localities; which I have carefully compared with the types found in the Green Sand of Havre, whence Defrance's original types were obtained. M. D'Orbigny does not appear to have noticed the fact, that on well-preserved shells of his so-called *Ter. Hebertiana*, the same granular spinose asperities noticed in *Ter. lima* do exist; all his specimens were worn on their surface, and only the punctuations visible, which he states to be strongly marked with checkerwise spaces between. Mr. Woodward first called my attention to the granulation in some from the Chalk of Norwich, and since then, I owe to the kindness of Mr. Fitch, a beautiful series, showing that it was as entirely and as closely covered, as any specimen from the Upper Green Sand. Let any observer compare for illustration, Pl. IV, figs. 21 and 22, from the Upper Green Sand, with figs. 19 and 20, the types of *Ter. pentangulata*, from the Norwich Chalk, and they will be at a loss to assign any differences; again Pl. II, fig. 3, from the Gault of Folkstone, is undistinguishable from the Chalk and Upper Green Sand forms; figs. 23 and 25 of Pl. IV are the same as found by Defrance, in the Green Sand of Havre, in France, and named by him *Ter. lima;* if any of these variable shapes had proved constant in any of the above sedimentary deposits, we might have supposed them distinct, but in one, as well as in all, the same variations are perceptible, and the only difference being that of colour, arising from the matrix wherein they have been imbedded, which is white in the Chalk, grey or blue in the Gault, and grey-yellowish or dirty-white in the Upper Green Sand. Such are my views; I may, however, be mistaken. The fortunate discovery of perfect interiors, both from the Chalk and Upper Green Sand, was a circumstance unknown to M. D'Orbigny, that author's figure in 'Pal. Fran.,' Pl. 514, fig. 8, being incomplete and incorrect.

Plate IV, figs. 21, 22, and 25. From the Upper Green Sand of the neighbourhood of Warminster, nat. size, in the collection of Mr. Cunnington; fig. 22 is the largest specimen I have seen.

„ V, fig. 4. Is a specimen from the Upper Green Sand of Cambridge, in the collection of Mr. Carter, in this locality, as well as at near Warminster; many varieties of shape are perceptible, of which our space does not permit of illustration.

„ IV, fig. 23. A specimen from the Chloritic Chalk of Chard, from the collection of Mr. Moore. Mr. Morris has similar specimens from Chaldon; and the same may be seen in the British Museum, from Havre, in France.

„ fig. 24. The original specimen of *Ter. spinulosa* (Morris), from the Gault of Folkstone, now in my collection.

„ fig. 24ᶜ. An enlarged illustration; we have also other specimens of this shell from the same Gault, proving the identity with those from the Upper Green Sand and Chalk.

„ V, fig. 3. Is likewise from the Gault of Folkstone.

„ fig. 3ᶜ. An enlarged illustration.

„ fig. 1. From the Grey Chalk, Maidstone, in the collection of Mr. Morris.

„ fig. 1ᵃ. An enlarged illustration. This is an exceptional form.

„ fig. 2. From the same bed, between Folkstone and Dover.

„ IV, fig. 15. From the interior of the smaller valve of a specimen from the Upper Chalk, near Norwich. Mr. Carter has a similar specimen from the Upper Green Sand of Cambridge. See also woodcut, figs. 5 and 6, in the Text.

„ fig. 16ᵃ. A very young specimen from the Upper Chalk of Dover.

„ fig. 16. Another young specimen from the Norwich Chalk, belonging to Mr. Woodward.

Plate IV, figs. 17 and 19. Two specimens from the Chalk of Letheringsett, Norfolk, in the collection of Mr. C. B. Rose.

„ fig. 18. Another variety from Norwich, a shape similar to M. D'Orbigny's type of *T. Hebertiana*.

„ fig. 20. A specimen from the Chalk of Norwich, in the collection of Mr. Fitch.

„ fig. 20ᶜ. An enlarged illustration; this is the shape on which Mr. Woodward established his *Ter. pentangulata*. Similar specimens exist in the collection of the Museum at Cambridge, and in that of the Rev. T. Image, &c.

„ fig. 26. The smallest specimen we have observed.

„ fig. 27. A young shell from the Chalk detritus of Charing (Kent), in the collection of Mr. Harris.

„ fig. 27ᵃᵇ. An enlarged illustration.

„ fig. 28. A young shell from the Upper Chalk of Pewsey (Wilts), in the collection of Mr. Cunnington.

„ fig. 28ᵃᵇ. Enlarged figures.

In many cases, I have given magnified illustrations of a small portion of the shell; these vary from different states of preservation, and from the spinose granulations being more or less separate; it will likewise be observed, that in most specimens a dark longitudinal line is seen to extend to a certain length along the smaller valve arising from the internal septum, but it is not visible on specimens in a fine state of preservation, and where the spinose granulation is perfect; it seems probable that this species, when alive, was slightly tinted with red, traces being discernible on many specimens in the Cambridge Museum. *K. lima* has also been found in the Chalk of Sussex, by Messrs. Dixon and Catt, and by myself in the Chalk Marl of Lewes. On the Continent, at Havre (Seine Inférieure), at Chavot and Césane, near Ay (Marne), &c.

Genus—TEREBRATULA, *Lhwyd*. 1698.

Animal fixed to submarine bodies by means of a pedicle issuing from a foramen in the beak of larger valve; edges of the mantle thin, entire, and fringed by short cilia, vascular system ramified, and situated on the mantle. Shell punctuated, variable in shape, inequivalve, elongated or transverse; convex or depressed, beak more or less produced, truncated by a foramen of variable dimensions, valves articulating by the means of two teeth in larger and corresponding sockets in smaller valve; internal riband-shaped lamella (partly supporting the ciliated arms), attached only to the crura, short or elongated, and more or less folded back on itself.

Obs. Nearly all our Cretaceous Terebratulæ had a short loop. *T. longa*, of Rœmer, is an exception, in it the apophysary system extended to near the frontal margin, as in *Ter. lagenalis*, &c.

16. TEREBRATULA CAPILLATA, *D'Archiac.* Plate V, fig. 12 [abcd].

> TEREBRATULA CAPILLATA, *D'Archiac*, 1846. Bull. de la Soc. Géologique de France, vol. iii, 2de ser., p. 336; and 1847, Mémoires de la Soc. Géol. de France, vol. ii, 2de ser., p. 323; pl. xx, figs. 1, 2, and 3.
> —   —   *Bronn.* Index Pal., p. 1232, 1848.

Diagnosis. Shell irregularly oval, or somewhat pentangular, longer than wide; valves unequally convex, the dental or larger one the most so, no mesial fold or sinus; valves regularly convex to the front; margin line presenting a slightly elevated curve, or biplication in front; beak short, rounded and recurved, diagonally truncated by a large circular foramen, a portion of which is completed by a small deltidium in two pieces; beak ridges not always well defined, soon turning in to meet the hinge margin; surface ornamented by a vast number of minute radiating capilliform elevated striæ close together, unequal, and more or less undulating, these being intersected at variable distances by concentric lines of growth. Structure punctuated, interior unknown, but probably possessing a short loop. Length 13, width 11, depth 8 lines.

Obs. Among the different specimens sent for my inspection by Messrs. Fitch and Carter, from the Red Chalk of Hunstanton Cliff, Norfolk, I was agreeably surprised to find two perfectly characterised specimens of the *Terebratula capillata*, D'Archiac, one of the most characteristic fossils of the Tourtia of Belgium, the small narrow capilliform waving striæ being perfectly preserved on our specimens, and distinguishes them from the other species peculiar to the Cretaceous period. I am surprised to find, in p. 172 of M. D'Orbigny's 'Prodrome,' that author does not perceive the marked differences between this and *Ter. depressa* of Lamarck, to which last he refers the present form, stating these differences to be only due to age and superficial wearing; this is, however, far from being the case; the very shape in both is quite distinct; the elongated beak and deltidium of *T. depressa* alone would distinguish it from *T. capillata;* and from an extensive and well-preserved series of both forms collected by myself in Belgium, I feel no doubt that at all ages *T. depressa* was smooth, while *T. capillata* is ornamented as described above. *T. capillata* has been well described, and beautifully illustrated, by Viscount D'Archiac, and is not very rare at Tournay, in Belgium.

Plate V, fig. 12 [abc]. Specimen from the Red Chalk of Hunstanton Cliff, Norfolk, in the collection of Mr. Fitch, of Norwich. A similar one is preserved in the collection of Mr. Carter, of Cambridge.

„ figs. 12[e] and 12. Enlarged illustrations.

17. TEREBRATULA OVATA, *Sowerby.* Plate IV, figs. 6—13.

TEREBRATULA OVATA, *Sow.* Min. Con., vol. i, p. 46, tab. xv, fig. iii, 1812.
— — *Parkinson.* An Int. to the Study of Org. Remains, 1822.
— — *Fleming.* A History of British An., vol. i, 1828.
— — *Woodward.* Synop. Table of Br. Org. Rem., p. 21, 1830.
— — *Brown.* Ill. of Foss. Conch. of Gr. Br., pl. lv, figs. 34, 35, 1838.
— — *Morris.* Catalogue of British Fossils, 1843.
— — *Tennant.* A Strat. List of British Fossils, p. 47, 1847.
— LACRYMOSA, *D'Orb.* Pal. Franc. Ter. Crét., vol. iv, p. 99, pl. 512, figs. 6—11, 1847; and Prodrome, vol. ii, p. 172, 1850.
— CARNEA, *Bronn.* Index Pal., p. 1232, 1848. (Non *T. carnea,* Sow.)

Diagnosis. Shell ovate, or oblong-ovate, depressed; beak produced, recurved, obliquely truncated by a rather large circular foramen, partly formed out of the substance of the beak, and completed by two small deltidial plates; beak ridges well defined, leaving a slightly concave false area between them and the hinge margin. Valves unequally convex, the smaller or imperforated one is less so than the other; uninterruptedly convex to the margin in the young state, but soon assuming a longitudinal depression, appearing at about the middle of the valve, extending and becoming deeper as it approaches the front. In the larger valve, a corresponding longitudinal elevation or keel occurs; marginal line wavy, and in front indented by the smaller valve. Surface covered by irregular elongated longitudinal elevated rugæ, little interrupted in the middle of the valves, but on the sides diverge and form innumerable small oblong tubercules, sometimes extending above the surface of the shell in the form of short spines. Concentric lines of growth numerous; structure distinctly punctuated, interior unknown; dimensions variable, the largest example yet found measuring, length 19, width 13, depth 11 lines; but the greater number of specimens do not attain these dimensions.

Obs. This is one of Sowerby's first-described species, which he obtained at "Chute, near Heytesbury, in Wiltshire." The species is perfectly known in England, and one of the most abundant of the tribe found in the Upper Green Sand of the locality above named, and discovered likewise by Messrs. Moore and Morris in the Chloritic Chalk of Chard and Chardstock; these specimens, however, rarely preserve the remarkable structure which characterises the species, from the nature of the sediment in which they were deposited having deteriorated the surface of most of them; this character was not observed by Sowerby; but, as the species is very common, it is not very difficult to procure specimens preserving their structure; and some, in the collections of Messrs. Cunnington, Moore, and Morris, show it to perfection. Many authors on the continent, among others Professor Bronn, have erroneously believed it a synonym of *Ter. Carnea,*[1] that species not having been

---

[1] Dr. Mantell, in his 'Geol. of the South Downs,' p. 209, 1822, erroneously places *T. ovata,* Sow., in the Upper and Lower Chalk, where the Sowerby type has never been as yet found. It is probable that this error is the cause of foreign authors believing *T. ovata,* Sow., to be a synonym of *Carnea.*

yet discovered in the Upper Green Sand; others, such as Nilsson,[1] Hisinger,[2] and Rœmer,[3] refer to Sowerby's species quite another shell from the Chalk, all three referring to the same figure of Nilsson. M. D'Orbigny admits, in his 'Pal. Tran.,' pp. 103-4, that those authors are mistaken, and places their *T. ovata* as a synonym of *T. Carnea*, but M. D'Orbigny figures and describes in page 99, under the new name of *Terebratula lacrymosa*, Sowerby's *Ter. ovata*, with which he was unacquainted. We regret being obliged to deprive that author of his supposed new form; but, as Sowerby's shell was figured and described in 1812, and as it is well known, we cannot give priority to that by M. D'Orbigny published in 1847.

More than two hundred specimens of this type have been minutely examined by myself from the dimensions of 2 to 19 lines; in young, and even in some more advanced ages, there are no traces of the longitudinal depression, sometimes so very deep in adult shells, as may be remarked in figs. 11 and 12, Pl. IV; the first appearance in the young is a slight depression, only quite near the margin.

Sometimes, as may be remarked in fig. 9 of the same Plate, up to a certain age, the valves were regularly convex, when a sudden stoppage in the growth occurred; and, on its being resumed, they at once presented a strongly marked depression extending to the front; this species is generally much depressed, but in some exceptional cases assumes great convexity, the width and depth being about the same as in fig. 12; but, as will be observed, the characters of the form are there as well preserved as on figs. 10 and 11, which may be taken as good types of the species. Sowerby's figure unfortunately does not exhibit this depression, and more resembles fig. 8. We regret not having been able to study the interior, our attempts having failed from the nature of the matrix; but we are disposed to believe, that the calcareous appendages were not arranged as in true *Terebratula*, perhaps more like what we find in *Kingena lima;* the external structure being somewhat similar, leads me to imagine them closely related; but, as it is of no use shifting species until we are certain of their true place, we will allow *T. ovata* to remain for the present in the genus Terebratula.

T. ovata has been discovered by M. D'Orbigny in the Lowest Green or Chloritic beds of Cap la Hève, near Havre; this bed and locality contains the same species we find at Chute, near Warminster.

Plate IV, fig. 6. A very young shell, from near Warminster.

„　　fig. 7, 8. Two specimens from the Chloritic Marl of Chardstock, in the collection of Mr. Morris.

„　　fig. 9. A specimen, showing a sudden stoppage of growth, &c., from the Chloritic Chalk of Chard, in the collection of Mr. Moore.

[1] 'Petrefacta Succana,' 1827, pl. iv, fig. 3.
[2] 'Leth. Succ.' 1837, pl. xxiv, fig. 3.
[3] 'Die Vers. Nord.,' 1840, p. 43.

Plate IV, fig. 10. Another more adult shell from the same bed and locality, also from the collection of Mr. Moore.

    ,,   fig. 11. A very adult specimen from the Upper Green Sand, near Warminster, in the collection of Mr. Cunnington.

    ,,   figs. 12, 13. Two exceptional shapes, likewise from the same locality and collection.

18. TEREBRATULA RUGULOSA, *Morris*. Plate IV, figs. 14, 14 *ᵃ*.

TEREBRATULA RUGULOSA, *Morris.* Annals and Mag. of Nat. Hist., vol. xx. p. 253, pl. xviii, figs. 5, 5 *ᵃᵃ*, 1847.

—   DISPARIALIS (pars), *D'Orb.* Pal. Franc. Ter. Cret., vol iv, p. 100, pl. 512, figs. 12, 13, (but not figs. 16, 17, which belong to *T. squamosa* of Mantell,) 1847.

Diagnosis. Shell ovate or somewhat irregularly pentagonal, longer than wide, straight or slightly indented in front; valves almost equally convex, dental one most so; a small depression existing near the front in some specimens, the margin of smaller valve slightly underlying the perforated one; beak rather produced, not much recurved, obliquely truncated by a large circular foramen; deltidium small, in two pieces, partly surrounding the foramen; beak ridges moderately distinct. Surface covered by minute rugæ, and in general the middle portion of each valve longitudinally, and but little interrupted. Those on the side diverge and have a tendency to break into small oblong tubercules, slightly projecting sometimes in the form of short spines. Structure distinctly and thickly punctuated. Interior unknown; dimensions variable; length 11, width 8, depth 7 lines.

Obs. This form was described and figured by Mr. Morris and myself, in 1847, under the name of *T. rugulosa;* a little later it received from M. D'Orbigny that of *T. disparialis*, but the last-named author includes in his species another form which we consider distinct and known in England under the name of *T. squamosa* (Mantell,) the surface being ornamented by wavy striæ and numerous squamose concentric lines of growth. Both *Ter. rugulosa* and *squamosa* occur in great numbers in the Chloritic Chalk of Rouen, and seem to me always easily distinguished: in England they are equally distinct, and separable from *T. ovata*, the last being much more convex, and deprived of that remarkable longitudinal depression in the smaller valve, and the corresponding keel-shaped projection in the larger one; in *T. rugulosa* the convexity extending in most cases regularly to the front, where sometimes a wide but slender depression is seen, quite different from that observable in *T. ovata*, and confined to the margin; by its exterior ornaments it approaches, however, to *Ter. ovata*, as justly observed by M. D'Orbigny, but the general aspect of the shell is otherwise quite different. Both species may, therefore, for the present, be conveniently retained under distinct appellations; the irregular manner in which the tubercules are disposed is very remarkable; some are shorter and wider than others, arising at different levels, and often, though close at their origin, diverge while others converge, so that notwithstanding that the greater number follow the same direction, they are not individually parallel to each other. We regret not having been able to examine the interior of this form, which we suppose different from that of true *Terebratula*, but with which it had better remain till further investigation.

T. rugulosa has been found in England by Messrs. Moore, Bunbury, and Pratt, in the Chalk with green grains at Chard Chadstock and Chaldon along with *T. ovata*. In France I have found it in great numbers at Rouen, in beds of the same age.

Plate IV, fig. 14. A specimen, natural size, from the collection of Mr. Moore.

    „   fig. 14 *d*. Another specimen, enlarged, in my collection.

19. TEREBRATULA SQUAMOSA, *Mantell.*   Plate V, figs. 5—11.

TEREBRATULA SQUAMOSA, *Mantell.* Geol. of Sussex, p. 132, 1822.
—   —   *Morris* and *Dav.* Annals and Mag. of Nat. Hist., vol. xx,
p. 254, pl. xviii, fig. 8 *a b*, 1847.
DISPARIALIS (pars), *D'Orb.* Pal. Franç. Ter. Crétacées, vol. iv, p. 100,
pl. 512, figs. 16, 17, (not figs. 12, 13, which belong to
T. rugulosa,) 1847.
SQUAMOSA, *Bronn.* Index Pal., p. 1251, 1848.

Diagnosis. Shell orbicular, or somewhat longitudinally ovate, valves nearly equally convex, sometimes gibbous; beak moderately produced and obliquely truncated by a circular entire foramen; deltidium small, rather dilated at the base, the convexity of the valves is at times equal to the front, while in other specimens there exists in the anterior portion of the imperforated valve a longitudinal depression and tendency to biplication. Surface marked with concentric squamose ridges, crossed by numerous radiating zig-zag raised striæ, giving to the squamæ an irregular serrated appearance; structure punctuated, punctæ rather widely separated; interior unknown. Dimensions variable, the largest specimen as yet discovered measuring, length 15, width 12, depth 9 lines, but the average size is, length 7, width 6, depth 5½ lines.

Obs. This species was described in 1822 by Dr. Mantell, who found it in the Gray Marl, of Hornsey. It is a form well known in England, but not having been illustrated by that celebrated geologist, Mr. Morris and myself thought it advisable to re-describe and figure the species in the 'Annals,' 1847; about this period or later M. D'Orbigny published the same shell under the name of *T. disparialis,* associating with it another form which we separate. Since that period the discovery of a great many British specimens in different localities allow us to notice several of its most important variations, and which are illustrated in Plate V. Dr. Mantell was provided only with small and young specimens, fig. 5 of our Plate; but Mr. Cunnington discovered at Pottern Butts, in the Upper Green Sand, specimens attaining much larger dimensions, as may be seen by figs. 8 and 9. In a few adult individuals there exists a tendency to biplication; a character which to that extent must be considered exceptional. It has likewise been lately discovered by Messrs. Moore and Morris in the Chalk with green grains or Chloritic Marl, of Chard, Chaldon, and Evershot; also in the Gray Chalk between Dover and Folkstone by Messrs. Mackey, Morris, and myself. Its vertical range is, therefore, from the Upper Green Sand to the Gray Chalk. I regret not having been able to examine the interior of this species, and suspect its organisation will prove to be distinct from that of true *Terebratula*, where however we must leave it for the present. *T. squamosa* is very common in the *Craie Chloritée* of Rouen, where it has been collected by M. D'Orbigny and myself.

Plate V, fig. 5. The original specimen on which Dr. Mantell founded his species from the Chalk Marl of Hornsey, in the collection of the British Museum.

„   fig. 6. 6*ab*. A specimen, natural size, from the Gray Chalk between Folkstone and Dover, in the collection of Mr. Morris.

„   6*d*. The same, enlarged.

„   6*e*. A fragment of the same, likewise considerably magnified.

„   fig. 5A. Another specimen, from the same locality, in the collection of Mr. Mackey.

„   figs. 7, 8, 9, 10. Specimens, natural size, from the Upper Green Sand of Pottern Butts, in the collection of Mr. Cunnington.

„   fig. 11. A specimen from the Chloritic Marl of Chard, in the collection of Mr. Moore.

„   fig. 11*abcd*. Enlarged illustrations of the same.

20. TEREBRATULA OBLONGA, *Sow.* Plate II, figs. 29—32.

TEREBRATULA OBLONGA,	*Sowerby.*	Min. Conch., vol. vi, p. 68, tab. 535, figs. 4, 5, 6, 1829.
—	OBLONGA, *V. Buch.*	Uber Ter., 1834, and Mém. Soc. Geol. de France, vol. iii, p. 359, pl. xvi, fig. 2, 1838.
—	— *Rœmer.*	Kreid., p. 39, No. 18, 1840.
—	QUADRATA, *Fitton.*	Trans. Geol. Soc., vol. iv, pl. xiv, fig. 9, 1836.
—	OBLONGA. *Morris.*	Catalogue, 1843.
—	— *Forbes.*	Quart. Journal of the Geol. Soc., vol. i, p. 346, No. 106, (non 105, as erroneously stated by M. D'Orbigny,) 1845.
TEREBRATELLA	— *D'Orb.*	Pal. Française Ter. Cretacées, vol. iv, p. 113, pl. 515, figs. 7—19, 1847.
TEREBRATULA	— *Cunnington.*	Quart. Journal of the Geol. Soc., vol. vi, p. 454, 1850.
TEREBRATELLA	— *D'Orb.*	Prodrome, vol. ii, p. 85, 1850.

Diagnosis. Shell oval, oblong, gibbous; beak nearly straight, rather produced; foramen entire, formed out of a portion of the truncated beak and the two deltidial plates which it indents, and is more or less separated from the umbo; beak ridges well defined, leaving between them and the hinge margin a flat, false area; the marginal line nearly straight all round; valves nearly equally convex; surface ornamented by a variable number of plaits, either simple or bifurcated, at irregular distances from the beaks and umbo. Structure punctuated; length 12, width 9, depth 7 lines.

Obs. When young, the shells of this species present less convexity than at a more advanced age, they are also very variable in the form and number of their plaits; in some specimens, and more especially in the young, the costæ are all simple from the beak and umbo to the margin, one or two only bifurcating on the sides; but in the generality of specimens, the bifurcation and intercalation of plaits at various distances from the beak and umbo is very remarkable, particularly in some French shells, these bifurcate several times before reaching the margin, presenting longitudinal undulations; the number therefore of the plaits is very variable; we have specimens with from twenty to thirty on each valve; others with from thirty to forty-six; sixteen to eighteen of which are due to bifurcation and intercalation within the anterior third of the length of the valves. This species is quite distinct from *Ter. cardium*, Lamarck; although mistaken by some authors, as Professor Bronn,[1] who believed it a simple synonym of the Lamarckian type. *T. oblonga* is a much more oblong and oval shell; the beak is more produced; the foramen completely different, being widely separated from the umbo, while in *T. cardium* it lies almost contiguous to it, and the plaits are fewer in number, generally only bifurcating in the young, while in the Cretaceous species this character is prevalent in all ages, and especially in the advanced state; I do not agree, therefore, with Mr. Austen,[2] when he states that this species is common both to the Oolitic and Cretaceous Formations; at least after a minute examination of a multitude of Foreign and British specimens of both, I have not come to that conclusion.

I am not quite certain as to the *genus* to which this species belongs, from not having been able to see the interior. M. D'Orbigny places it in his *Terebratella*, to which it may, perhaps, belong; but as he does not state it to have a doubly attached loop, I will leave it for the present among the *Terebratulæ*. *T. quadrata*, Sow., is only an exceptional variety of this species. *T. oblonga* is found in England, in the Lower Green Sand of Atherfield, Hythe, near Devizes, Maidstone, and Farringdon, whence it has been collected by Messrs. Walton, Harris, Morris, myself, and others; it has likewise been found in the Upper Green Sand of the neighbourhood of Warminster, by Mr. Cunnington, but the species is very rare in that locality. In France, it occurs at Wassy, St. Dizier, &c., and is well described and figured by M. D'Orbigny, in his 'Pal. Française;' in Switzerland, it is mentioned from Neufchâtel, and from several German localities, such as in the neighbourhood of Brunswick in the Hilsconglomerat of Essen, Schandelahe, Schoppenstedt, &c.

[1] Index Pal., p. 1243.

[2] 'On the Lower Green Sand of Farringdon, &c.,' Quart. Journal of the Geol. Society, vol. iv, p, 477, 1850.

Plate II, figs. 29, 29ᵃ.   A specimen from the Lower Green Sand of Hythe.

„   figs. 30, 30ᵃᵇᶜᵈ. From the Lower Green Sand of the same locality, in the collection of Mr. Morris.

„   figs. 31, 31ᵃᵇ. From the Green Sand of Farringdon, in the collection of Mr. Cunnington.

„   figs. 32, 32ᵃᵇ. From the Upper Green Sand of Warminster, in the collection of Mr. Cunnington.

21. TEREBRATULA OBESA, *Sow.*   Plate V, figs. 13—16.

TEREBRATULA OBESA, *Sow.*   Min. Con., vol. v, p. 54, tab. 438, fig. 1, 1825.
—   — *Fleming.* A Hist. of Brit. An., p. 371, 1828.
—   — *Brown.* Illust. of Foss. Conch., p. 133, pl. liv, figs. 28, 29, 1838.
—   — ? *Rœmer.* Die Vers. Nord. Kreid., p. 43, 1840.
—   — *Morris.* Catalogue, p. 134, 1843.
—   — *Tennant.* A Strat. list of Brit. Foss., p. 47, 1847.
—   — *D'Orbigny.* Pal. Franç. Ter. Cretacées, vol. iv, p. 101, pl. 513, figs. 1—4, 1847.

Diagnosis.   Shell irregularly ovate, oblong, very convex, straight or slightly indented in front; beak short, incurved, truncated by a large circular foramen lying close to the umbo, so that the deltidium is rarely exposed; beak ridges indistinct; margin line wavy; imperforated valve less convex than the dental one, presenting a very gentle longitudinal curve from the beak to the front; the central portion of the valve is nearly flat, with a slight longitudinal depression towards the front, giving rise to two lateral obtuse plaits; larger valve almost uninterruptedly convex; front elevated; surface smooth, marked with concentric lines of growth; structure punctuated; loop short, wide, anneliform, and confined to the posterior portion of the shell, not exceeding a third of the length of the valve; simply attached to the crural base, the two riband-shaped lamellæ are soon united by a transverse lamella bent upwards in the middle; dimensions variable; the largest specimen known measures length 2½ inches, width 2 inches, depth 1½ inch.

Obs. This is our largest British Cretaceous Brachiopoda, described in 1825 by Sowerby, under the name of *T. obesa,* and cannot be said to be common; indeed we believe it so closely allied to *Ter. Dutempliana,*[1] (D'Orb.), by all its most important characters, that it may probably be only the giants of that form, and I have no doubt that all the passages linking them together may be easily obtained; but Palæontologists seem anxious to retain both names, the present one for those very convex shells illustrated in our Plate V; we place them, therefore, under distinct heads while retaining our opinion as above.   One of the most marked characters of this species is its large and edged foramen, which

[1] *T. Dutempliana* is a shell known under the name of *Ter. biplicata,* Sow., but that name had been given to another species by Brocchi in 1814.

distinguishes it from *T. semiglobosa*, and I therefore cannot agree with Professors Bronn,[1] Reuss,[2] and V. Buch,[3] who place it as a synonym of that species.　M. D'Orbigny considers *T. Albensis*[4] a synonym of *Obesa*, but I am more inclined to place it as a variety of *Semiglobosa*.　*T. sulcifera*, Morris,[5] in my opinion more closely approaches *T. obesa* than any of the others mentioned, but from the remarkable, regular, and peculiar concentric lines of growth ornamenting its surface it may be conveniently retained under a distinct appellation.　The largest specimen of *T. obesa*, yet discovered, is from the Upper Green Sand; but almost as large a shell has likewise been obtained from the Chalk in the neighbourhood of Norwich.　Sowerby states his specimens to be from the Chalk at Norton Bevant, near Warminster.　In France it is rare at Rouen, and indeed everywhere, although common in the state of *T. Dutempliana*, which seems to confirm my views above mentioned.　On the sides of the valve may be remarked small raised striæ; this is also observable on *T. Dutempliana*.

Plate V, fig. 13, 14. The two largest known specimens, from the Chalk of Norwich, in the collection of Mr. Fitch.

　　„　fig. 15. A specimen from the same locality, in my collection, much resembling Sowerby's figure in the 'Min. Conch.'

　　„　fig. 16. The largest specimen yet discovered of this species, from the Upper Green Sand, near Warminster, in the collection of Mr. Cunnington.

[1] Index Pal., p. 1250, Prof. Bronn mentions as synonyms of *T. semiglobosa*, *T. subundata*, *undata*, *obesa*, and *albensis*, none of which we believe correctly placed with *T. obesa*.

[2] Die Vers. der Bohem. Kreid., 1846, p. 51.　The following are the synonyms here given of *T. semiglobosa*, *T. intermedia*, (a cornbrash shell,) *T. albensis*, *T. subundata*, *T. obesa*, *T. acuta*, none of Reuss' figures resemble Sowerby's *T. obesa*.

[3] Uber Ter. and Mém. de la Soc. Géol. de France, vol. iii, p. 225.　V. Buch's synonyms are still more defective; he considers *T. globata*, Sow., an inferior oolite shell, and *T. Sphœroidalis*, likewise from the same strata, as synonyms of *T. obesa*, this last with a point of interrogation.

[4] Leymerie Mém. de la Soc. Géol. de France, vol. xv, p. 11, pl. v, figs. 1—3.

[5] Morris and Dav. 'Ann. and Mag. of Nat. Hist.,' vol. xx, p. 254, pl. xviii, figs. 7, 7*. 1847.

22. TERERRATULA BIPLICATA, *Brocchi*, Sp.   Plate VI, figs. 1—49, and Plate IX, fig. 40?

ANOMIA BIPLICATA, *Brocchi*. Conchologia fossile, p. 469, pl. x, fig. 8, 1814.

TEREBRATULA BIPLICATA, *Sowerby*.   Min. Con., vol i, p. 201, tab. 90, 1815, and vol. v, p. 53, tab. 437, fig. 2, 3, 1825.

— — *Parkinson*. An Int. to the Study of Organic Remains, p. 227, 1822.

— OBTUSA, *Sow.* Min. Con., vol. v, p. 53, tab. 437, fig. 4, 1825.

— BIPLICATA, *Defrance*. Dic. des Sciences Nat., vol. liii, p. 151, 1828.

— — *Woodward*. A Synoptical Table of Brit. Organic Remains, p. 21, 1830.

— — and T. OBTUSA, *Brown*. Illustrations of Fossil Conch. of Great Britain, pl. lii, figs. 27, 28, and pl. liv, fig. 25, 1835.

— FABA, *Sowerby*. (*in Fitton*). Trans. Geol. Soc., vol. iv, p. 338, pl. xiv, tab. 10, 1836, (not *T. faba*, Sow. d'Orbigny, Pal. Tran., vol. iv.)

— BIPLICATA, (part) *Von Buch*. Class. des Térébratules, Mém. Soc. Géol. de France, vol. iii, p. 219, 1839.

— BIPLICATA and OBTUSA, *Morris*. Catalogue, 1843.

— DUTEMPLEANA, *D'Orbigny*. Pal. Franç., Terrains Crétacés, vol. iv, p.193, pl. 511, figs. 1—8, 1847.

— BIPLICATA, (part) *Bronn*. Index Pal., p. 1230, 1848.

— — and OBTUSA. Catalogue of the Terebratulæ, in the British Museum, pp. 23, 24, 1853.

— — *Sharpe*, Quart. Journ. Geol. Soc., vol. x, p. 191, 1843.

Diagnosis. Shell oblong, oval, somewhat pentagonal: ventral valve convex, and in general deeper than the dorsal one, which is more or less prominently biplicated; beak short, rounded, incurved, and obliquely truncated by a circular foramen of moderate dimensions. Deltidium very narrow, and generally inconspicuous from the aperture being contiguous to the umbo of the dorsal valve: lateral margins flexuous, bisinuated in front; surface smooth, marked by concentric lines of growth, and at times obscurely striated on the sides; loop simple, not exceeding one third of the length of the socket valve. Dimensions variable; length 23, width 16, depth 11 lines.

Obs. It was not until after much consideration and repeated comparisons of more than three thousand examples, that I could make up my mind to refer to a single species the *biplicated terebratulæ* illustrated in Plate VI. of the present monograph.

The term *Anomia biplicata* was first introduced by Brocchi for a cretaceous terebratula from San Quirico, in Tuscany; no description was appended, but the figure then published represents *a smooth, oblong oval, or somewhat pentagonal shell,* 14 *lines in length, by* 9 *in breadth, notched or biplicated towards the front, with a slightly incurved beak, circular foramen and inconspicuous deltidium.* Nothing more appears to be known of this Italian shell except that the same locality is mentioned for *Rh. vespertilio.*

One year after the publication of Brocchi's work, Sowerby described, under a similar denomination, a biplicated terebratula, common in the *Gault* of Folkstone, and *Upper Green*

Sand, near Warminster, Cambridge, and other localities, and these have been generally admitted to be the types of *T. biplicata*.[1]   M. D'Orbigny seems, however, to consider the Italian shell specifically distinct from our British one, not from a personal acquaintance with Brocchi's type, but from an assumption of his own; nor have I been more fortunate in my endeavours to obtain additional information regarding the San Quirico fossil, all we are therefore enabled to judge from, is the figure, and this so nearly agrees with many of the forms found in the *Upper Green Sand*, near Warminster, that strong doubts may be entertained whether the French author's opinion be really correct.[2]   The illustrations furnished by M. D'Orbigny, (Pal. Franc., Terrains Crétacés, vol. iv. pl. 511, figs. 9—15,) resemble, in my opinion, the Italian figure far less than some of those of which he forms his *Ter. Dutempleana* (same work, Pl. 511, figs. 1—8,) and to this last, he justly refers Sowerby's figures of *Ter. biplicata*.

No species varies to a wider extent than the one under consideration, but, at the same time, it does not seem difficult to trace the links which connect by insensible gradation the most extreme variations hitherto observed; nor am I yet prepared to surrender the opinion expressed in page 53, viz. that passages may not be found similarly connecting these last to the *Ter. obesa* (Sow.).

Local conditions have materially influenced the regular development of this, as well as that of other forms; producing varieties and races, not always easily referable to their original type.   Thus, when young, *Ter biplicata* is perfectly oval, uniformly convex, and without trace of biplication, but as the animal advances in its development, the dorsal valve becomes more or less prominently biplicated, with a mesial sinus of variable depth and width extending between the plicæ; but, in some examples, the regular convexity of the dental valve is but slightly influenced by the *biplication* of the dorsal one.   These differences may be observed in the numerous illustrations in Pl. VI, and especially in figures 5, 9, 17, 36, &c.   Much variation is likewise produced in the general contour, by the lesser or greater approximation of the two plaits and their ridges, as well as by the different degree of convexity presented by the valves.

Two principal varieties may be mentioned.

1. The one abundantly found near Warminster, and in the Isle of Wight, of which

---

[1] Sowerby mentions the *Gault*, at Cambridge, as containing his *Ter. biplicata*, but it would appear that the shell, although common in the *clay band* of that name at Folkstone and in other localities, at Cambridge is only found in the *Upper Green Sand*.   As justly observed by Mr. Deshayes, many other biplicated species from the oolitic and other formations have been confounded with the cretaceous type : thus Professor Bronn mentions as synonyms of *Ter. biplicata*, (Index Pal., p. 1230), the following forms which seem all specifically distinct.   *Ter. bicanaliculata*, *T. maxillata*, *T. sella*, *T. Harlani*, *T. bisuffarcinata*, and *T. perovalis*.   V. Buch has likewise erred in several of his supposed synonyms of this species, but was right in placing *Ter. faba* among them.

[2] In page 53 of this Monograph, published in 1852, I had so far sanctioned M. D'Orbigny's view by admitting the name *Ter. Dutempleana* ; but subsequent and more complete investigations disposed me to consider that conclusion both premature and uncertain.

figs. 33 to 44, are examples. This form approaches most to the Italian figure, and should be looked upon as the more typical shape. M. D'Orbigny's illustration, Pal. Franc., Terrains Crétacés, vol. iv, Pl. 511, figs. 5, 6, 7, is likewise referable to this form. *Ter. faba*, Sow. is simply a dwarf *Terebratula biplicata* from the Upper Green Sand of Warminster.[1]

2. Those wider and more flattened shapes commonly met in the Gault of Folkestone, and Upper Green Sand of Cambridge, (Pl. VI, figs. 1 to 9, 12, 29, &c.) in these the two plaits are much more widely separated, and to it, M. D'Orbigny's illustration, Pal. Franc., Pl. 511, fig. 1, may likewise belong. This is the variety I had intended to retain as *Ter. Dutempleana* (p. 53) and No. 1, as *Ter. biplicata*, (Brocchi,) a view I now abandon for the reasons already specified.

Ter. obtusa, Sow. (Pl. VI, figs. 10, 11, 13, &c.) is only a variety of *Ter. biplicata*, or *Dutempleana*, in which the shell has extended more in width than in length, and wherein the biplication has either entirely disappeared (figs. 10, 11,) or exists simply in a rudimentary state, (figs. 13, 23), but that this is simply an unusual condition of the species has been amply proved by a series of upwards of a thousand examples collected by Mr. Carter, at Cambridge, and in which every passage may be traced uniting such extreme forms as figs. 3, 6, and 10.

It seems difficult, in the actual state of our knowledge, to specify with certainty the precise period at which *Ter. biplicata* made its first appearance, but, if some shells (Pl. IX, fig. 40) lately discovered by Mr. Mackie in the highest bed of the *Lower Green Sand* series at Folkstone, really belong to this species, they would be the oldest examples with which we are acquainted.

Ter. biplicata occurs in the *Gault* of Folkstone, the *Speeton Clay* of Yorkshire; and the *red chalk* of Hunstanton; it abounds in the *Upper Green Sand* of Cambridge, in the neighbourhood of Warminster, at Farringdon; and although less common, still seems to be represented in the Chalk marl and Lower Chalk of Cambridgeshire:—thus having a wide vertical range, extending almost through the entire cretaceous system.

On the Continent, it occurs at Wissant, in the *Gault*, and in many of the *Upper Green Sand* localities of Europe.

Plate VI, figs. 1 and 2. *Ter. biplicata var. Dutempleana*, D'Orb. from the *red chalk* of Hunstanton, in the Collection of Mr. Fitch.

„ figs. 3, 4, 5. A large specimen from the *Upper Green Sand* of Cambridge, in the Cabinet of Mr. Carter.

„ fig. 6. A remarkable example from the same locality and collection, showing traces of coloration.

[1] Professor Forbes states, in his catalogue of the *Lower Green Sand* fossils, Quart. Journ. Geol. Soc., vol. i, p. 346, "*T. faba* (Sow. in Fitton, t. xiv, fig. 11). The original specimen is in the Geological Society's collection, and is from the Upper Green Sand of Warminster; it appears to be a young or starved state of *T. biplicata*, or some allied species."

Plate VI, fig. 7. Interior of the dorsal or socket valve with its loop, from the same collection and locality; figures 10 to 28, and 31, are all from Mr. Carter's Collection, and neighbourhood of Cambridge.

„ figs. 8, 9. Another Upper Green Sand specimen from Cambridge.

„ figs. 10 to 13. Different examples and ages of the variety (*Ter. obtusa*, Sow.) from Upper Green Sand, of Cambridge.

„ figs. 14 to 28. Other examples of different shapes and ages from the same bed and locality.

„ figs. 29, 30. *Ter. biplicata* from the Gault of Folkstone.

„ fig. 31. A specimen from the lower chalk near Cambridge.

„ fig. 32. Another from the Lower Chalk of Lewes, in the Cabinet of Mr. Catt.

„ fig. 33. An elongated example from Warminster, in the Museum of Practical Geology.

„ figs. 34 to 42. *Ter. biplicata*, a short inflated variety from near Warminster, in the Collection of Mr. Cunnington.

„ figs. 43, 44. id. (*Ter. faba*,) Sow.

„ figs. 45 to 49. *Ter. biplicata?* stated to have been found in Lower Green Sand, near Devizes, by Mr. Cunnington.

Plate IX, fig. 37. A specimen from the Upper Green Sand of Farringdon, in the Collection of Mr. Sharpe.

„ fig. 40. A specimen of *Ter. biplicata?* from the highest bed of the Lower Green Sand series in the vicinity of Folkstone, Collection of Mr. Mackie.

23. TEREBRATULA PRŒLONGA, *Sowerby*. Plate VII, figs. 1, 2.

TEREBRATULA PRŒLONGA, *Sowerby* (*in Fitton.*) Trans. of the Geol. Soc.,vol. iv, p. 338, pl. xiv, fig. 14, 1836.

— BIPLICATA, var. ACUTA, *V. Buch.* Class. des Térébratules, Mém. Soc. Géol. de France, vol. iii, p. 220, 1834. (Not *T. biplicata*, Brocchi, nor Sow.)

— PRŒLONGA, *Brown.* Illust. of Fossil Conch., pl. liv, figs. 8, 10, 1838.

— BIPLICATA, *Leymerie.* Mém. Soc. Géol. de France, vol. v, p. 29, (not Brocchi,) 1842.

— PRŒLONGA, *Morris.* Catalogue, 1843.

— — *Forbes.* Quart. Journ. Geol. Soc., vol. i, p. 345, 1845.

— — *D'Orbigny.* Pal. Franç., Terrains Crétacés, vol. iv, p. 75, pl. 506, figs. 1—7, 1847, and Prodrome, vol. ii, p. 85, 1850.

— — Catalogue of the *Terebratulæ* in the British Museum, p. 28, 1853.

Diagnosis. Shell of an elongated oval shape, with almost equally convex valves; beak prominent, not much incurved, obliquely truncated by a large circular foramen, partly surrounded by, and widely separated from the hinge line by a deltidium in two pieces; beak ridges undefined, lateral margins flexuous, slightly raised and bisinuated in front. The *dorsal* or socket valve is biplicated towards the front; surface smooth, with a few concentric lines of growth. Loop short, not exceeding a third of the length of the dorsal valve; shell structure minutely punctated. Dimensions variable; length 18, width 12, depth 9½ lines.

Obs. This species appears to be readily distinguished from *Ter. biplicata* by its very elongated oval shape; the cardinal half becoming considerably lengthened, gives the shell a somewhat scuttle-shaped appearance: while, on the contrary, *T. biplicata* is more pentagonal and rounded near the beak; the last becoming incurved, with the foramen situated so close to the hinge line, as to render the deltidium almost inconspicuous. Some exceptional examples might perhaps be obtained connecting the two forms, but these not having yet been procured, we must continue to consider this species as distinct.

Ter. prælonga appears to characterise the *Lower Green Sand* (*Néocomien* of the French), while the *Ter. biplicata* chiefly occurs in the *Gault* and *Upper Green Sand*.

T. prælonga was well figured by Sowerby in 1836 from a specimen stated to have been found in the *Lower Green Sand* of Sandgate (Kent); it occurs likewise near Maidstone (Kent), and in other localities; but it is one of our rarer British species.

On the Continent, M. D'Orbigny mentions it from the Lower Néocomien beds of Baudrecourt, Bettancourt-la-Ferrée (Haute Marne), Morteau (Doubs), the neighbourhood of Castellane and Marolles (Aube), at Neuchâtel in Switzerland, &c.

Plate VII, fig. 1. Type example from the *Lower Green Sand* of Sandgate, figured by Sowerby in Trans. Geol. Soc.

„ fig. 2. A very fine specimen from a similar bed at Maidstone, in the Collection of Mr. Morris.

24. TEREBRATULA SELLA, *Sowerby*. Plate VII, figs. 4—10.

TEREBRATULA SELLA, *Sowerby*. Min. Con., vol. v, p. 53, pl. 437, fig. 1, 1825.
— — *Fleming*. A Hist. of British Animals, vol. i, p. 371, 1828.
— — *Woodward*. A Synoptical Table of Brit. Org. Remains, p. 21, 1830, (marked Up. Oolite by mistake.)
— BIPLICATA (part), *V. Buch*. Classification des Térébratules, Mém. Soc. Géol. de France, vol. iii, p. 218, 1834.
— SELLA, *Brown*. Illust. of Fossil Conch. of Great Britain, pl. lii, figs. 31, 32, 1838.
— — *Rœmer*. Kreid., p. 43, No. 41, 1840.
— BIPLICATA, *Rœmer*. Ib., pl. vii, fig. 17, 1840.
— SELLA, *Leymerie*. Mém. Soc. Géol. de France, vol. v, p. 30, 1842.
— — *Morris*. Catalogue, 1843.
— — *Forbes*. Quart. Journ. Geol. Soc., vol. i, p. 345, 1845.
— BIPLICATA, (part) *Bronn*. Index Pal., p. 1230, 1848.

TEREBRATULA SELLA, *D'Orbigny*. Pal. Franç., Terrains Crétacés, vol. iv, p. 91, pl. 510, figs. 6—12, and Prodrome, vol. ii, p. 120, 1850.

— — Catalogue of the Terebratulæ in the British Museum, p. 28, 1853.

Diagnosis. Shell of a sub-quadrangular or somewhat pentagonal shape, rather longer than wide. Valves almost equally convex, slightly flattened; beak short, not much incurved, and obliquely truncated by a foramen of moderate dimensions, partly margined by a wide and short deltidium in two pieces. *Dorsal* or socket valve more or less prominently biplicated; the front is considerably elevated, narrow, regularly arched or bisinuated; lateral margins flexuous; surface smooth, with a few concentric lines of growth. Loop short, not exceeding a third of the length of the socket valve. Dimensions variable:

length 15, width 14, depth 8 lines.

„ 15, „ 12, „ 9 „

Obs. Sowerby observes that "when young, this shell is rather trigonal, in consequence of the length of the sides and roundness of the front; as it grows older, it becomes squarer, the front being more produced as well as more elevated; the beak is very slightly curved; the length and breadth are very nearly equal; the edges always sharp." But we may add, that the separation of the plaits, as well as the depth of the sinus between them, is very variable according to specimens, and is even at times almost entirely filled up.

Several authors, among whom we may quote Baron V. Buch and Professor Bronn, have considered the species under description to be only a variety or synonym of *Ter. biplicata*, but M. D'Orbigny justly remarks, that although individuals of *Ter. sella* are at times found bearing some of the external aspect of *T. biplicata*, they are commonly well distinguished by their beak, foramen, deltidium, and position of the plaits, the last being much more elevated and closer in *T. sella*, producing a biplicated mesial fold, and a deep sinus in the ventral or dental valve. It cannot be confounded with *Ter. prælonga*, which is a much more elongated or scuttle-shaped shell.

In England, *Ter. sella* abounds in the *Lower Green Sand* of Atherfield (Isle of Wight), at Reigate, Pluckley, Ashford, Sandgate, and Hythe; it is likewise (but rarely) found in the *Gault* of Maidstone, whence an undeniable example was obtained by Mr. Bowerbank; and it is probable that some individuals continued to exist in the lower beds of the *Upper Green Sand*, an opinion arrived at from the inspection of a few uncertain examples found at Warminster and Farringdon.

On the Continent, M. D'Orbigny mentions having found it in the Terrain Néocomien Supérieur and Aptien of Combles, Gargas, Renaud-du-Mont, aux Salles, Castellane, Marolles, &c.

Plate VII, fig. 4. A remarkably adult and characteristic example, from the Lower Green Sand of the Isle of Wight; British Museum.

„ fig. 5. From a bed of the same age, at Pluckley; in the Collection of Mr. Harris.

Plate VII, fig. 6. From the Isle of Wight.

    „   fig. 7. From Gault near Maidstone ; in Mr. Bowerbank's museum.

    „   fig. 8. A specimen from the Isle of Wight, in the Geol. Society's Museum.

    „   fig. 9. From the same locality, in the Collection of Mr. Morris.

    „   fig. 10. A young shell; Isle of Wight.

25. TEREBRATULA TORNACENCIS, Var. RŒMERI, *D'Archiac.* Plate VII, figs. 11—16, and Plate IX, figs. 1—8.

    TEREBRATULA TORNACENSIS, *D'Archiac.* Mém. Soc. Géol. de France, vol. ii, 2d series, p. 318, pl. xviii, figs. 2—5, 1847.

    —   RŒMERI, *D'Archiac.* Ib., pl. xviii, fig. 6.

    —   BOUEI, *D'Archiac.* Ib., pl. xviii, fig. 7.

    —   CRASSA, *D'Archiac.* Ib., pl. xviii, fig. 8.

    —   CRASSIFICATA, *D'Archiac.* Ib., pl. xix, fig. 1.

    —   RUSTICA, *D'Archiac.* Ib., pl. xix, fig. 2.

    —   BIPLICATA, *D'Orbigny* ? (not *Brocchi* nor *Sowerby*), Pal. Franç., Terrains Crétacés, vol. iv, pl. 511, figs. 9—15, 1847.

    —   RŒMERI, *Sharpe.* Quart. Journ. Geol. Soc., vol. x, p. 191, 1853.

    —   KEYSERLINGII, *Sharpe.* Ib.

    —   REVOLUTA, *Sharpe.* Ib. ?

Diagnosis. Shell variable in shape, somewhat pentagonal; as long as or longer than wide; valves almost equally convex or flattened. *Dorsal* valve more or less distinctly biplicated towards the front. *Ventral* valve with a mesial plait corresponding with the sinus of the opposite one. Beak produced, slightly incurved and truncated by a rather large circular foramen, partly edged by a deltidium of moderate dimensions; beak ridges tolerably defined, with a flattened space between them and the hinge line; lateral margins somewhat flexuous and bisinuated in front. External surface smooth, at times obscurely longitudinally striated and marked by numerous concentric lines of growth; shell structure largely punctated. Loop short and simple, not exceeding a third of the length of the smaller valve. Dimensions very variable:

    length 10,   width 8½,   depth 5 lines.

    „ 9,   „ 8,   „ 4 „

    „ 8,   „ 8,   „ 3 „ &c.

Obs. The most common species in the Sponge Gravel or *Upper Green Sand* of Farringdon is that represented in Pl. VII, figs. 11—16, and Pl. IX, figs. 1—8, of the present Monograph; it rarely exceeds 9 lines in length, with from 7 to 9 in breadth, is either elongated and moderately convex (Pl. VII, figs. 12, 13, 15, and Pl. IX, figs. 1, 2, 3, and 8), or as wide as long, and more or less compressed (Pl. VII, figs. 11, 13ᵃ, 14, and Pl. IX, figs. 4 to 7). These shells present every kind of variation, but they are connected by insensible passages, and unquestionably, in my opinion, belong to the same species.

The next point (and one which has given me, perhaps, more trouble than almost any other in this work) was the endeavour to find out to what species these shells really belonged, and it was not until after two excursions to Farringdon, and the minute examination of more than three hundred examples, collected in the same spot by Messrs. Sharpe, Lowe, Forbes, Waterhouse, Morris, myself, and others, that I at last determined to consider the shells above described as a *small race* or variety of *Terebratula Tornacensis*, D'Archiac; and, indeed, several of the forms found in that locality are only *dwarf races*; thus, *Terebratella Menardi* and *T. depressa*, Lamarck, have not there attained the full size of those species found in the *Upper Green Sand* of Mans (France), or the *Tourtia* of Belgium; and it appears evident that local conditions during the formation of the Farringdon deposits were unfavourable to the full development of many of the forms of Brachiopoda, and which is no doubt the true cause of the stinted growth observable in many of the species.

The next point to investigate is whether *T. Tornacensis* is the oldest denomination the species has received? and I am still uncertain if *Ter. phasiolina*, Lamarck[1] (published in 1819, but without figure), may not be a young state of Vicomte d'Archiac's species; but as doubts at present involve that question, we will give preference to the Vicomte's claims,[2] and endeavour to discuss the value of some of its probable synonyms.

M. D'Orbigny considers the following names to be synonyms or varieties of *Terebratula biplicata* (Brocchi):—*Ter. Tornacensis, Bouei, crassa, Robertoni, crassificata, rustica, Boubei, Virleti, revoluta, sub-pectoralis, Keyserlingii,* and *Tchiatcheffii* (D'Archiac),[3] but I differ from the author of the Palæontologie Française, (as elsewhere stated[4]) in the following particulars. I am of opinion that *Ter. biplicata* (Brocchi) is the same shell as that of Sowerby, which M. D'Orbigny considers distinct; the last-named author considers *Ter. Tornacensis* equivalent to *T. biplicata* (Brocchi), and of *T. biplicata,* Sow., forms a new species, to which he has applied the name *T. Dutempleana.* We, therefore, both so far agree in the opinion that *perhaps* there may exist two species, but disagree as to which Brocchi's figure should belong. I admit with M. D'Orbigny that *Ter. Rœmeri, Bouei, crassa, crassificata, rustica,* and perhaps *Ter. Murchisoni* and *Tchiatcheffii,* D'Archiac, may be only variations in age, &c., of *Ter. Tornacensis;* but I am not yet prepared to consider as such *Ter. Virleti* and *T. revoluta,* D'Archiac, and still less *Ter. Robertoni* and *T. sub-pectoralis* of the same author.

Mr. Sharpe and other palæontologists identify our Farringdon shell with *Ter. Rœmeri;* but after a lengthened examination, I was unable to find grounds of sufficient value for considering the last specifically different from *T. tornacensis.* It is true our British

---

[1] Animaux sans Vertèbres, vol. vi, No. 29, 1819.

[2] *Rapport sur les Fossiles du Tourtia.* Mém. Soc. Géol. de France, vol ii, 2d series, p. 316, &c., and plates xviii and xix, 1847.

[3] Prodrome, vol. ii, p. 172, 1850.

[4] See *Obs.* to our description of *Ter. biplicata.*

examples have not been hitherto obtained as large as the Belgian full-grown type, but this may be accounted for by the reasons already mentioned.

All our English examples of *T. biplicata* have the beak incurved, with the foramen close upon the umbo of the dorsal valve, so that the deltidium becomes inconspicuous and this seems likewise to have been the case with Brocchi's species, (if we are not misled by his figure),[1] while, on the contrary, in *T. tornacensis*, and in our Farringdon race, the deltidium is always more or less exposed, and the shell itself is also commonly wider in comparison to its length, than what we observe in the generality of our specimens of *T. biplicata*. Indeed, some of the Farringdon shells bear resemblance to some young examples of *Ter. maxillata*, Sow., so abundantly distributed in the Great Oolite of Hampton Cliff, near Bath. When quite young, *T. tornacensis* seems to be oval, with but a slight trace of biplication, in which condition it bears a great resemblance to many specimens of *T. biplicata* of a similar age. Viscount d'Archiac admits that his *Ter. Roemeri*, and *T. Bouei* may perhaps be only varieties of *T. tornacensis;* thus their close affinity had not escaped the scrutiny of that learned author.

The Sponge gravel of Farringdon is as yet our only British locality. On the Continent the species abounds in the *Tourtia* of Belgium, and in the *Upper Green Sand* of Mans, (France).

Plate VII, figs. 11 and 13. Wide examples from the *Upper Green Sand* of Farringdon, collection of Mr. Morris.

„ figs. 12 and 14. Other specimens, fig. 14 exhibiting a very thickened margin.

„ fig. 14ᵃ. A malformation in the cabinet of Mr. Lowe, in which a very exceptional tendency to triplication may be observed. Malformations of this kind may likewise be seen, though very rarely in *T. biplicata* and *T. sella*, (Pl. VII, fig. 7.)

„ figs. 15 and 16. Elongated variety.

Plate IX, figs. 1 to 8. A series of specimens from the same locality, in the collection of Mr. Sharpe, figs. 4, 5, and 7, are referred by that author to *T. revoluta*, (d'Archiac), and fig. 8 to *T. Keyserlingii*, (d'Archiac),[2] fig. 3 bears also some resemblance to *Ter. Tchiatcheffii*.

---

[1] This observation had not escaped the notice of the celebrated author of the "*Progrès de la Géologie*, Mém. Soc. Géol. de France, vol. 2, 2d series, p. 317, 1847.

[2] These specimens were obligingly lent to me by Mr. Sharpe, with those names inscribed on his tablets.

26. TEREBRATULA SULCIFERA, (*Morris*).   Plate VII, figs. 17—20.

TEREBRATULA SULCIFERA, *Morris and Dav.*   Annals and Mag. of Nat. Hist., No. 133,
vol. xx, p. 254, pl. xviii, fig. 7, 1847.
—   —   *D'Orbigny.*   Prodrome, vol. ii, p. 172, 1850.
—   —   A Catalogue of the Terebratulæ in the British Museum, p. 26,
1853.

Diagnosis.   Shell obovate, oval or somewhat pentagonal and ventricose; valves almost equally convex; beak short, thick rounded, moderately incurved, and truncated by a large circular foramen; deltidium partly concealed beneath the anterior portion of the aperture, which overlies, and nearly touches the umbo of the *dorsal* or socket valve; margins very sinuous, externally ornamented by numerous concentric ridges of growth, regularly disposed from the beak and umbo to the margin; loop short and simple, not exceeding a third of the length of the socket or dorsal valve.   Dimensions variable.   Length 22, width 16, depth 16 lines.

Obs.   This species varies greatly in external aspect; some examples (figs. 17 and 18) are much inflated, while others (fig. 19) are more or less compressed, and somewhat triangular in shape; at times, the frontal line is almost straight, but in the generality of specimens this portion of the ventral valve indents the opposite one; the sulci are likewise more or less produced, and regular.   It is readily distinguished from *Ter. semiglobosa*, by the large dimensions of its foramen, and from *T. obesa*, by its sulci, which are shorter and more inflated in appearance.

Ter. sulcifera occurs in the Lower Chalk of Cherry Hinton and Isleham, near Cambridge; at Hockwold, Norfolk, and in other localities.

Plate VII, fig. 17.   A very large, and fine example, discovered by Mr. Bunbury, figured in the 'Annals and Mag. of Nat. Hist.,' as the type of the species.
„   figs. 18, 19, and 20.   Other specimens from the Lower Chalk of the neighbourhood of Cambridge, in the collection of Mr. Morris.

27. TEREBRATULA SEMIGLOBOSA, *Sowerby*.   Plate VIII, figs 6—18.

TEREBRATULA SEMIGLOBOSA, *Sowerby.*   Min. Con., vol. i. p. 48, tab. xv, fig. 9, 1813.
—   SUBUNDATA, *Sowerby.*   Ib., p. 47, pl. xv, fig. 7, 1813.
—   SUBROTUNDA (part) *Sow.*   Ib., tab. xv, figs. 1-2, 1813.
—   SUBUNDATA, *W. Smith.*   Strata identified by organised fossils, p. 10,
Pl. iv, fig. 8, 1816.
—   —   *Lamarck.*   An. sans Vert., vol. vi, p. 248, No. 13, 1819.
—   SEMIGLOBOSA, *Lamarck.*   Ib., p. 251, No. 27.
—   —   and T. SUBUNDATA, *Parkinson.*   Trans. Geol. Soc., vol. v,
1821.
—   —   *Mantell.*   Geol. of Sussex, p. 209, 1822.
—   —   *Brong.*   Env. de Paris, pl. ix, fig. 1, 1822.
—   —   *Fleming.*   A Hist. of British Animals, vol. i. p. 369, 1828.

TEREBRATULA SEMIGLOBOSA, *Schlotheim.* System Vers., No. 80, 1832.

— — *V. Buch.* Classification des Terebratules Mém. Soc. Géol. de France, vol. iii, p. 205, pl. xix, fig. 9, 1834.

— SUBUNDATA, *Deshayes.* Nouv. Ed. de Lamarck, vol. vii, p. 333, No. 13, 1836.

— SEMIGLOBOSA, *Hisinger.* Leth. Suec., p. 82, pl. xxiv, fig. 2.

— — *Bronn.* Leth Geog., pl. xxx, fig. 11, p. 657, 1837, and Index Pal., p. 1250, 1848.

— — *D'Orbigny.* Pal. Franç., Terrains Crétacés, vol. iv, p. 105, pl. 514, fig. 1—4, 1837, and Prodrome, vol. ii, p. 258.

— — *Brown.* Illustrations of Fossil Conch., pl. liv, fig. 45-46, 1838.

— — *Geinitz.* Char. Kreid., p. 16, 1839.

— — *Rœmer.* Die Vers. Nord. Kreid., p. 43, No. 42, 1840.

— ALBENSIS, *Leymerie.* Mém. Soc. Géol. de France, vol. iv, pp. 288, 289, 1841; ii, p. 29, tab. xv, figs. 2—4, 1842.

— SEMIGLOBOSA, *Leymerie.* Mém. Soc. Géol. de France, vol. v, 1842.

— — T. SUBROTUNDA, and T. SUBUNDATA, *Morris.* Catalogue, 1843.

— — *Reuss.* Bohem. Kreid., pl. xxvi, figs. 5—8, 1846.

— CARNEA, *Reuss.* Ib., pl. xxvi, figs. 9—11.

— BULLA, *J. de Sow.* in *Dixon.* Geol. and Fossils of the Tertiary and Cretaceous Formations of Sussex, p. 346, tab. xxvii, p. 11, 1850.

— SEMIGLOBOSA and T. ALBENSIS. A Catalogue of the Fossil Terebratulæ in the British Museum, p. 25, 1853.

Diagnosis. Shell variable in shape, inflated, circular, or elongated oval; *ventral* or dental valve, commonly the deepest, and uniformly gibbous; beak short, more or less incurved, and perforated by a small circular foramen, contiguous to the umbone of the opposite valve; deltidium in two pieces, commonly inconspicuous; beak ridges undefined; margins flexuous, straight, or bisinuated in front; the *dorsal* valve is uniformly convex or biplicated towards the frontal margin. External surface smooth, marked by concentric lines of growth; shell structure minutely punctate; loop short and simple, not exceeding a third of the length of the dorsal valve.

Dimensions very variable: length 21, width 18, depth 15 lines;

„ 11, „ 11, „ 9 „

„ 13, „ 11, „ 7 „

„ 16, „ 11, „ 11 „ &c.

Obs. In the first volume of the 'Mineral Conchology,' the names *Ter. subundata* and *T. semiglobosa*, were proposed for varieties of a single form; but most authors have preferred the last denomination on account of its having been applied to the adult and common condition in which the species is obtained. *T. subundata* is only a less convex or more depressed variety, and it appears likewise, probable, that *T. subrotunda* of the same author, may have been founded (at least in part) on a variety of *T. semiglobosa*; but

Sowerby is certainly in error while supposing that the same shell was common to the Chalk and Cornbrash. Baron V. Buch, and others, have likewise erroneously added to their synonyms of *T. semiglobosa*, the *T. intermedia* (Sow.), a Jurassic species, distinguished by shape and character.

Sowerby mentions that, in the true type of *T. semiglobosa*, the frontal margin is slightly "*undulated with two risings*" or plaits; but after inspecting a series of several hundred specimens from the Lower Chalk of Lewes, Chardstock, and other localities, it appeared evident that the front was at times almost straight or arched, without any defined biplication, and it was from shells presenting this last condition, that Mr. Leymerie founded his *T. albensis*.[1] *T. semiglobosa* is also distinguished from *T. carnea* (Sow.), this last being a much more depressed shell, with a uniform straight margin, a character observable only in young examples of the species under consideration. And it may be worthy of notice, that we rarely find both forms associated in the same bed or locality; where the one abounds the other seems wanting: thus, in the Upper Chalk of Norwich, Brighton, Meudon, &c., where *T. carnea* is common, *T. semiglobosa* is absent; while the inverse takes place in the Lower Chalk of Lewes, Gravesend, Chardstock, &c.

Ter. bulla, Sow., figured in pl. xxxii, of Dixon's work, is also only an unusually large and more elongated form of *T. semiglobosa*, possessing no other valid distinguishing feature.

The vertical range of this species appears to be greater than that of *Ter. carnea*; we first find it in the *Red Chalk* of Hunstanton, believed by some geologists to represent the age of the Gault: it abounds in the *Chalk Marl* and *Lower Chalk* of Lewes, Charing, Gravesend, Tytherleigh, Chardstock, and other localities. On the Continent, it is very common near Rouen, in the Dep. de l'Aube, &c.

Plate VIII, fig. 6. *Ter. semiglobosa*, from the Lower Chalk of the neighbourhood of Lewes (Sussex).

	fig. 7.	„	from Gravesend.
„	fig. 8.	„	a remarkable variety, from Lewisham (Kent).
„	fig. 9.	„	Sowerby's original figure of *T. subrotunda;* specimens very similar to this may be collected near Lewes.
„	fig. 10.	„	A large example from Lewes, in the collection of Mr. Catt.
„	fig. 11.	„	another specimen (*T. bulla*, Sow.), in the collection of Mr. Wetherell.
„	fig. 12.	„	a specimen from the Chalk of Grays, in the cabinet of Mr. Morris.

[1] *Ter. albensis* is supposed by Prof. Bronn and M. D'Orbigny to be a variety of *T. obesa*, but the proportions of the foramen in the two forms is so different, as hardly to warrant such a conclusion.

Plate VIII, fig. 13. *Ter. semiglobosa*, from the Lower Chalk of Charing (Kent), in the collection of Mr. Harris.

„ fig. 14. „ a very elongated specimen, probably a malformation, from the Chalk of Gravesend, in the collection of Mr. Bowerbank.

„ fig. 15. „ a specimen from Glyndebourn, near Lewes, (*T. albensis*, Leymerie).

„ fig. 16. „ from the Lower Chalk of Charing, in the collection of Mr. Harris.

„ fig. 17. „ from the Red Chalk of Hunstanton (*T. subundata*, Sow.), collection of Mr. Morris.

„ fig. 18. „ a specimen from the Chalk, with quartz grains, Evershot (Dorsetshire), in the collection of Mr. Morris.

28. TEREBRATULA CARNEA, *Sowerby*. Plate VIII, figs. 1—4.

TEREBRATULA CARNEA, *Sowerby*. Min. Con., vol. i, p. 47, tab. xv, figs. 5, 6, 1812.
— — *Lamarck.* An. sans Vert., vol. vi, p. 248, No. 14, 1819.
— — *Brongniart.* Desc. Géol. des Environs de Paris, pl. iv, fig. 7, 1822.
— — *Parkinson.* An Introd. to the Study of Organic Remains, p. 234, 1822.
— ELONGATA, *Sow.* Min. Con., vol. v, p. 49, tab. 435, figs. 1 & 2, 1825.
— OVATA, *Nilsson.* Petref. Suec., p. 34, pl. iv, fig. 3, 1827 (not *T. ovata*, Sow.).
— LENS, ? *Nilsson.* Ib., p. 35, pl. iv, fig. 6.
— CARNEA, *Defrance.* Dic. d'Hist. Nat., vol. 53, 1828.
— — *Fleming.* A Hist. of British Animals, p. 369, 1828.
— — *Deshayes.* Encycl. Method., iii, p. 1028, No. 20, 1828.
— — and T. ELONGATA, *Woodward.* A Synoptical Table of Organic Remains, p. 21, 1830.
— — *Schlotheim.* Syst. Vers., No. 64, 1832.
— — *Mantell.* Geol. of the South-east of England, p. 127, 1833.
— — *Von Buch.* Classification et Descriptions des Terebratules, Mém. Soc. Géol. de France, vol. iii, p. 203, pl. xix, fig. 2, 1834.
— — *Bronn.* Leth. Geog., pl. xxx, fig. 13, p. 654, 1837.
— — *Pusch.* Polens Pal., p. 18, t. iii, fig. 12, 1837.
— SUBROTUNDA, *D'Orbigny.* Pal. Franç., Ter. Crétacés, vol. iv, 1837 (not Sowerby).
— OVATA, *Hisinger.* Leth. Suec., p.82, pl. xxiv, fig. 3, 1837 (not *T. ovata*, Sow.).
— CARNEA, *Brown.* Illustrations of Fossil Conch. of Great Britain, pl. liv, figs. 30—33, 1838.

TEREBRATULA CARNEA, *Geinitz.* Char. Kreid., p. 16, 1839.

— — *Rœmer.* Die Vers. Nord. Kreid., 1840 (but not all his Synonymes).

— OVATA, *Rœmer.* Ib. (but not *T. ovata*, Sow.).

— CARNEA, *Morris.* Catalogue, 1843.

— — *D'Orbigny.* Russia and Oural, vol. ii, p. 494, pl. xliii, figs. 21—25, 1845.

— — *Reuss.* Bohem. Kreid., p. 50, No. 14, tab. xxvi, figs. 10, 11, 1846 (but not all his Synonymes).

— — *D'Orbigny.* Pal. Franç., Ter. Crétacés, vol. iv, p. 103, pl. 515, figs. 5—8, 1847, and Prodrome, vol. ii, p. 258.

— — *Bronn* (part). Index Pal., vol. ii, p. 1232, 1848 (but not all his Synonymes).

— — *Alth.* Geol. Lemberg (in Haidinger's Abhandl.), p. 258, tab. xiii, fig. 8, 1850.

— — *Quenstedt.* Handb. de Petref., p. 473, tab. xxxviii, figs. 3, 4, 1851.

— — A Catalogue of the Terebratulæ in the British Museum, p. 21, 1853.

Diagnosis. Shell ovate, circular, elongated oval, or obtusely five sided, with somewhat depressed and almost equally convex valves; beak short, more or less incurved, and perforated by a small circular foramen, partly surrounded, and separated from the hinge line by a wide concave triangular deltidium, transversely wrinkled; margin nearly straight all round. Surface smooth, marked only by a few concentric lines of growth. Shell structure minutely punctated. Colour of a light or dull red. In the interior of the smaller or dorsal valve, the cardinal process is more or less produced; the loop short and simple, rarely exceeding a fourth of the length of the socket valve.

Dimensions very variable: length 17, width 15½, depth 10 lines;

„ 17, „ 12 „ 9 „

„ 20, „ 19 „ 11 „ &c.

Obs. This well-known shell was first described and figured by Sowerby, under the name of *T. carnea*, on account of the fleshy red tinge presented by many specimens; and which is no doubt remains of the original colour, which was in all probability similar to that still observable in several recent Terebratulæ, such as *T. lenticularis* so abundantly found in the deep sea of Fauveau Straits, New Zealand. *T. carnea* varies more or less in external shape; to the lengthened example, Sowerby applied the name *T. elongata*.

Several Palæontologists, among whom we may mention M. D'Orbigny, have placed the so-termed *T. subrotunda* of Sowerby ('Min. Con.,' tab. xv, figs. 1 and 2), among the synonymes of *T. carnea*, but it is doubtful whether this determination be correct; the figure in the 'Min. Con.' represents a circular Terebratula bearing external resemblance to some varieties of *T. carnea*, but, as the locality and bed mentioned is "*the hardest chalk about Hornisham, in Wiltshire*," it seems probable that the illustration was not taken from a specimen of *T. carnea*, but from a flattened variety of *T. semiglobosa*; and the author has rendered his species the more problematical, by adding, that his friend Mr. Meade had sent him specimens *from the Cornbrash*, $1\frac{1}{4}$ inch in length! Dr. Mantell refers to *T. subrotunda*, Sow., a shell from Hamsey and Eastbourn (Sussex),[1] but Messrs. Waterhouse and Woodward, who have seen the original, have pronounced it to be simply a depressed young individual of *T. subundata* or *T. semiglobosa*, and agreeing with Sowerby's type. I may add, that I likewise possess specimens of *T. semiglobosa*, from Glyndebourn, near Lewes, quite as circular and depressed as the figure of *T. subrotunda* in the 'Min. Con.'

Some authors[2] have likewise erroneously described *T. ovata* (Sow.) as a synonyme of *T. carnea*, a mistake principally referable to Dr. Mantell,[3] who does not appear to have been acquainted with Sowerby's type, which was stated to occur at Chute, near Heytesbury, in Wiltshire, an Upper Green Sand locality, where no true specimen of *T. carnea* has been discovered.

The great resemblance *T. carnea* bears to some examples of the recent *T. vitrea*, did not escape the observation of the late Baron Von Buch;[4] but I am disposed to coincide with M. D'Orbigny, in the belief, that they are specifically distinct. *T. vitrea* never presents the colour with which we believe *T. carnea* was tinted.

The foramen in some examples is so small as hardly to afford space for the passage of a hair; but in the generality of individuals the aperture, although always small, is far from presenting such minute proportions.

In the neighbourhood of Norwich, a great number of internal siliceous or flint casts of this species have been collected by Mr. Fitch, on which the muscular and other impressions are beautifully represented.

Sowerby mentions that he found *Ter. carnea* in the soft Chalk of Towse, near Norwich, and from that locality many beautiful examples have been procured by Messrs. Fitch, Woodward, Image, and others; it likewise occurs at Trimmingham, Brighton, in Ireland, and in many other Chalk localities. On the Continent, it is very common in similar deposits at Meudon, near Paris; Halden, Westphalia, in Russia, &c.; but seems to be very rare in Lower Chalk beds and localities characterised by *T. semiglobosa*.

Plate VIII, fig. 1. A typical specimen of *T. carnea* from the Chalk of Trimmingham.
      „   fig. 2. Interior of the larger or ventral valve.
      „   fig. 2[a]. Interior of the smaller or dorsal valve, with the loop.

---

[1] Geol. of Sussex, p. 130.

[2] Among these, we may mention M. D'Orbigny (see 'Pal. Franç., Terrains Crétacés,' vol. iv, p. 103, 1847).—Dr. Bronn ('Index Pal.,' vol. ii, p. 1232).—See also Nilsson and Hisinger.

[3] Geology of Sussex, 1822.

[4] Mémoirs de la Soc. Géol. de France, vol. iii, p. 204.

Plate VIII, fig. 3.   A lengthened variety, *T. elongata*, Sow., from the Chalk of Norwich, in the cabinet of Mr. Fitch.

„   fig. 4.   A large circular variety from the same locality, in the collection of the Rev. T. Image.

„   fig. 5.   Another example from the same locality.

29. Terebratula depressa, *Lamarck*.   Plate IX, figs. 9—24.

Terebratula depressa,	*Lamarck.*	Hist. des An. sans Vert., vol. vi, p. 249, 1819,[1] and *Dar. Notes on an examination of Lamarck's species of Fossil Terebratula*, Annals and Mag. of Nat. Hist., vol. v, 2d ser., pl. xiii, fig. 15, 1850.
—	ovalis, *Morris.*	Quart. Journal Geol. Soc., Nov., 1846 (not *T. ovalis*, Lamarck).
—	nerviensis, *D'Archiac.*	Mém. Soc. Géol. de France, vol. ii, 2d ser., p. 313, pl. xvii, figs. 2—10, 1847.
—	viquesneli, *D'Archiac.*	Ib., p. 316, pl. xviii, fig. 1.
—	depressa, *Bronn.* (part).	Index Pal., vol. ii, p. 1234, 1848.
—	— *D'Orbigny.*	Prodrome, vol. ii, p. 172, 1850.
—	— *Sharpe.*	Quart. Journ. of the Geol. Soc., vol. x, p. 191, 1853.

Diagnosis.   Shell depressed, oblong oval, tapering at the beak; valves almost equally deep, externally smooth, and marked by a few concentric lines of growth.   *Dorsal* or socket valve either regularly convex or interrupted by a mesial fold of moderate elevation; *ventral* valve with, or without, a shallow longitudinal depression; beak nearly straight, more or less produced, and truncated by a large circular foramen, partly margined by a wide deltidium, in one piece; beak ridges undefined, lateral margins moderately flexuous, frontal edge of the ventral valve indenting to a greater or lesser extent that of the dorsal one.   Loop short and simple, not exceeding a third of the length of the smaller or dorsal valve.

Dimensions variable: length 22, width 17, depth 14 lines;

„   20,   „ 20,   „ 11 „

Obs.   This Terebratula is one of our largest Cretaceous forms, varying very considerably in shape and comparative dimensions.   When quite young it is much depressed, with the margins straight all round; but with age, the valves either continue to remain regularly convex without any defined mesial fold, or with one of moderate elevation.   The beak is also more or less produced, and at times much elongated, with a large deltidium bearing resemblance to that peculiar to *Ter. longirostris* of Wahlenberg; but the Lamarckian species seems separable by its greater width, and by the almost total absence

---

[1] "*T. testa oblonga transversim dilatata, supra coarctata et oblusa, striis concentricis lævibus; nate producta non incurva : foramine magno.*"

of the longitudinal depression or biplication so strongly marked in adult examples of the Swedish type.

In Belgium, *T. depressa* is perfectly characterised, and abounds in the *Tourtia* of Tournay, Montignies-sur-Roc, and Gussignies, whence Lamarck obtained his specimens;[1] but in England, we are at present only acquainted with the single locality of Farringdon, where the race does not seem to have attained the large dimensions of the Belgian type. Our British examples are commonly shorter, and more stinted in their growth, and the mesial fold is at times more produced than in the generality of Belgian individuals.

T. depressa does not seem to have been yet discovered in France, no mention of it being made in the 'Palæontologie Francaise.'

In 1847, Viscount d'Archiac believed the species new, and named it *Terebratula nerviensis*.[2] In 1846, the same shell was mistaken by Mr. Morris, for *T. ovalis* (Lamarck), which last belongs to the Jurassic epoch, and is specifically distinct.[3] And in 1848 Professor Bronn considered *T. longirostris* to be a synonym of the Lamarckian *T. depressa*; an opinion in which I am unable to concur.

Many varieties might be noticed, but from these passing one into the other they do not in my opinion require distinguishing denominations. *Ter. Viquesneli*, D'Archiac, is one of them, it is found likewise at Farringdon, but appears to be simply a young state of the Lamarckian species.

Plate IX, fig. 9. One of the largest examples hitherto discovered in the Upper Green Sand of Farringdon. The valves are commonly found detached.

" fig. 10. A ventral valve, from the collection of Mr. Sharpe.

" fig. 11. A specimen drawn from two separate valves, in the collection of Mr. Sharpe.

" figs. 12—21. Different examples, ages, and varieties, of *T. depressa*; figs. 14 and 17 from the collection of Mr. Lowe; figs. 18 to 21, from that of Mr. Sharpe.

" figs. 22—23. A young specimen of *T. Viquesneli*, D'Archiac, from Farringdon, in the collection of Mr. Sharpe.

" fig. 24. A copy of the Belgian figure of *T. Viquesneli*, published by Viscount d'Archiac, for the sake of comparison.

[1] The original specimens are now in Baron Delessert's collection, and were figured by myself in the 'Annals' for 1850.

[2] This species was admirably described and figured by the distinguished French author in his "*Rapport sur les Fossiles du Tourtia*," 'Geol. Trans. France,' 1847.

[3] *Ter. Repeliniana*, D'Orb. ('Prodrome,' vol. ii, p. 25, 1850), from the White *Coral Rag*, near Vurey (Isère), &c., bears more resemblance to *Ter. depressa* (Lamarck), than any other Terebratula with which I am acquainted.

30. TEREBRATULA CARTERI, *Dav.* Pl. VII, fig. 3.

Diagnosis. Shell elongated oval, or unequally five sided, and somewhat compressed; *ventral* valve deeper than the dorsal one, with a shallow longitudinal sinus corresponding with a mesial fold of moderate elevation; beak short, slightly incurved, and truncated by a foramen of moderate dimensions; deltidium almost inconspicuous. Surface smooth, marked by a few concentric lines of growth. Lateral margins slightly flexuous, with the ventral frontal edge indenting that of the dorsal valve. Loop unknown, but in all probability short and simple. Length 20, width 15, depth 10 lines.

Obs. I obtained this shell some years ago from the Gray Chalk, near Dover, and have been unable to identify it with any of the other British Cretaceous forms that have come under my observation, it bears much of the outward aspect of some Jurassic Terebratulæ, although perfectly identical with none of those with which I have compared it. I therefore take great pleasure in naming it after Mr. Carter, who has afforded such valuable assistance in the working out of the Cambridge *Upper Green Sand* species.

Plate VII, fig. 3. *Ter. Carteri*, from the Gray Chalk in the vicinity of Dover.

31. TEREBRATULA ROBERTONI, *D'Archiac.* Plate IX, fig. 25.

> TEREBRATULA ROBERTONI, *D'Archiac,* Mém. Soc. Géol. de France, vol. ii, 2d series,
> p. 315, pl. xviii, fig. 2, 1847.
> — — *Sharpe,* Quart. Journ. of the Geol. Soc., vol. x, p. 192,
> 1853.

Diagnosis. Shell of an elongated oval shape; valves regularly convex, the ventral one rather the deepest; surface smooth, marked by concentric lines of growth; beak moderately produced, incurved and truncated by a large circular foramen, partly margined by a short deltidium; *ventral* valve somewhat keeled; beak ridges undefined, lateral margins flexuous; the frontal edge of the ventral valve slightly indenting the opposite one. Loop unknown, probably short and simple. Length 11, width 8, depth 5½ lines.

Obs. Viscount D'Archiac states[1] that *Ter. Robertoni* differs from *T. depressa,* Lamarck (his *T. nerviensis*), by its more regularly rhomboidal form, unequal depth of the valves, almost entire absence of sinuosity in front, and above all by the inflated extremity of its prominent and incurved beak. M. D'Orbigny appears to consider *T. Robertoni* as a simple synonyme of *T. biplicata;*[2] but this appears far from being the case, since it does not present any trace of biplication, a character always more or less visible in Brocchi's species, and it differs likewise in external shape. *T. Robertoni* was found in the Sponge Gravel of the Upper Green Sand of Farringdon, by Mr. Lowe, it agrees exactly both in

[1] 'Rapport sur les Fossiles du Tourtia, p. 315.
[2] 'Prodrome,' vol. ii, p. 172.

shape and dimensions, with the type figured by Viscount D'Archiac, from the Tourtia near Tournay (Belgium). The same French author states to have likewise collected the species in a bed of *Upper Green Sand*, above the *Gault*, near Wissant (Pas-de-Calais), France.

Plate IX, fig. 25. *Ter. Robertoni*, from Farringdon, in the collection of Mr. Lowe.

32. WALDHEIMEA (TEREBRATULA) CELTICA, *Morris*. Plate IX, figs. 32—35.

TEREBRATULA LONGA, *Rœmer*. Verst. Nordd. Ool., p. 50, pl. ii, fig. 11, 1836; Kreid., p. 44, No. 50, 1840. (Not *T. longa*, Zieten, 1832.)

— — *Morris*. Annals Nat. Hist., vol. xx, p. 255, pl. xix, fig. 1, 1847.

— FABA, *D'Orbigny*. Pal. Franç., Ter. Crétacés, vol. iv, p. 77, pl. xiv, figs. 10. (Not *T. faba*, Sow., Trans. Geol. Soc., vol. iv, p. 14, fig. 10, 1836.)

WALDHEIMEA CELTICA, *Morris*. MS. Catalogue of the Terebratulæ in the British Museum, p. 62, 1853.

Diagnosis. Shell oblong, elongated oval, ventricose posteriorly, becoming rather attenuated anteriorly, and subtruncate; valves nearly equally convex, ventral valve some-what keeled; beak slightly produced, and obliquely truncated by a foramen of moderate dimensions, partly surrounded and separated from the hinge line by a small deltidium in two pieces; beak ridges more or less defined; *dorsal valve* most inflated near the umbo; margins even; surface smooth, marked only by a few concentric lines of growth. Loop elongated, reaching to near the frontal margin before becoming reflected. Shell structure punctated. Length 17, breadth 10, depth 8 lines.

Obs. This form was described and figured by Rœmer, in 1836, under the name of *T. longa*;[1] but Zieten had already made use of the same denomination to designate a Jurassic Terebratula from Donsdorf.[2]

In 1847, the species under notice, was discovered by Dr. Fitton and Mr. Morris, in hard ferruginous nodules of the Lower Green Sand at Horseledge and Yellowledge, near Shanklin Bay (Isle of Wight), and published under Rœmer's denomination; but Mr. Morris having subsequently found that the shell differed specifically from the Jurassic type, has proposed for it the name of *Ter. Celtica*.

M. D'Orbigny commits another mistake, while considering *T. longa* (Rœmer) the same as *T. faba* of Sowerby; but the author of the 'Palæontologie Française' was not probably aware that the so-termed *T. faba* (Sow.) is itself only a variety or dwarf example of the well-known *T. biplicata* (Brocchi), from the Upper Green Sand of

---

[1] Rœmer appears to have figured and described in his 'Die Verst. der Nord. Oolithen Gebirges,' 1836, a series of Fossils belonging to the Hils Conglomerate, but which he considered at that epoch, as Jurassic, among these, we find his *T. longa*. The author has subsequently corrected this mistake in his monograph of the Chalk of that country.

[2] 'Die Verst Wurtembergs, pl. xxxix, fig. 7.

Warminster, and therefore not only specifically and stratigraphically distinct from Rœmer's *T. longa*, but belonging to a different section of the great genus *Terebratula*.

T. Celtica appears easily to be distinguished from other Cretaceous species by its peculiar elongated shape.

Rœmer's specimens are said to be from the Hilsthorn of Elligser Brinkes.

Plate IX, figs. 32—34. Examples from the Lower Green Sand, near Shanklin Bay, Isle of Wight, in the collection of Mr. Morris; several fine specimens have also been collected in the same locality by Mr. S. Saxby, of Bonchurch.

33 WALDHEIMEA (TEREBRATULA) TAMARINDUS, *Sowerby*. Plate IX, figs. 26—31.

TEREBRATULA TAMARINDUS, *Sowerby*. Trans. of the Geol. Soc., vol. iv, p. 338, pl. xiv, fig. 8, 1836.

— — *Morris*. Catalogue, 1843.

— SUBTRILOBATA, *Leymerie*. Mém. Soc. Géol. de France, vol. v, p. 12, pl. xv, figs. 7, 9.

— TAMARINDUS, *Bronn*. Index Pal., p. 1253, 1848.

— — *D'Orbigny*. Pal. Franç., Terrains Crétacés, vol. iv, p. 72, pl. 505, figs. 1—10, 1847 : Prodrome, vol. ii, p. 85, 1850.

WALDHEIMEA TAMARINDUS. A Catalogue of the Fossil Terebratulæ in the British Museum, p. 62, 1853.

TEREBRATULA TAMARINDUS, *Sharpe*. Quart. Journ. Geol. Soc., vol. x, p. 191, 1853.

Diagnosis. Shell very variable in shape, nearly orbicular, oval or obtusely five-sided; surface smooth, marked by a few concentric lines of growth. Valves almost equally convex, without either sinus or mesial fold; the *ventral* or perforated valve is generally the deepest; margin very obtuse and slightly flexuous, forming a small convex curve in front; beak moderately incurved, and truncated by a circular foramen, partly surrounded, and slightly separated from the hinge line by a deltidium in two pieces; beak ridges incurved, so as to approach the hinge margin. Loop elongated, reaching to near the frontal margin before becoming reflected. Shell structure largely punctated. Length 7, width 6, depth 4 lines.

Obs. The dimensions of *Ter. tamarindus* do not appear to have ever greatly exceeded seven lines in length. It occurs in the *Lower Green Sand* of the Isle of Wight; *Kentish Rag*, near Sandgate, and in the *Upper Green Sand* of Farringdon. On the Continent, it is mentioned, as occurring in the *Lower Nécomien* of Auxerre (Yonne), Bettancourt-la-Ferrée, at Wassy, Saint-Dizier, &c. It was also discovered by M. De Verneuil, in Spain. The margin is often considerably thickened. It is a rare British Cretaceous Fossil.

Plate IX, fig. 26. A specimen from the Kentish Rag, near Sandgate.

Plate IX, figs. 27 and 28. Two examples from the *Upper Green Sand* of Farringdon, in the collection of Mr. Sharpe; this shell is one of the rarest in the locality.

„   figs. 29 and 30. Two specimens from the *Lower Green Sand* of the Isle of Wight, from the collection of Mr. Morris.

„   fig. 31   A pentagonal specimen, with a very thickened edge, Isle of Wight.

Genus—RHYNCHONELLA, *Fischer*, 1809.

Obs. This genus having been described in p. 93 of the "Introduction," and in Part III, p. 65, it will not be necessary to repeat those details in the present Monograph.

The most active researches among the British Cretaceous *Rhynchonellæ* have not brought to light a single unpublished species, and so numerous are the varieties and passages from one form into another, that it is often almost impossible to draw up a diagnosis embodying the character of every variety. We have admitted the following fourteen species, as well as some few named varieties :—

1. RYNCHONELLA PLICATILIS, *Sow.*	9. RHYNCHONELLA SULCATA, *Parkinson.*	
— var. OCTOPLICATA, *Sow.*	10. — MANTELLIANA, *Sow.*	
— var. WOODWARDII, *Dav.*	11. — CUVIERI, *D'Orb.*	
2.? — LIMBATA, *Schloth.*	12. — MARTINI, *Mantell.*	
3. — COMPRESSA, *Lam.*	13. — GRASIANA, *D'Orb.*	
4. — LATISSIMA (LATA), *Sow.*	14. — LINEATA, *Phillips.*	
5. — GIBBSIANA, *Sow.*		
6. — PARVIROSTRIS, *Sow.*		
7. — DEPRESSA, *Sow.*		
var. A.		
var. B.		
8. — NUCIFORMIS, *Sow.*		

Some Palæontologists may, perhaps, consider *R. limbata*, Schl. (= *sub-plicata*, of Mantell) as only a variety of *R. plicatilis*, and it may be still a question whether *R. Grasiana* is more than the adult condition of *R. Martini*, Mantell. Mr. S. P. Woodward considers *R. octoplicata* as specifically distinct from *R. plicatilis*, and Mr. Sharpe admits *R. triangularis*, Wahl., among our British species.

34. RHYNCHONELLA PLICATILIS, *Sow.*, Sp.   Plate X, figs. 37—42.

> var. OCTOPLICATA, *Sow.*   Plate X, fig. 1—17.
> var. WOODWARDII, *Dav.*   Plate X, figs. 43—46.

> TEREBRATULA PLICATILIS, *Sowerby.*   Min. Con., vol. ii, tab. 118, fig. 1, 1816.
> —   OCTOPLICATA, *Sowerby.*   Ib., fig. 2, 1816.

TEREBRATULA PLICATILIS, *Mantell.* Fossils of the South Downs, p. 210, 1822.

— OCTOPLICATA, *Brongniart.* Desc. Géol. des Environs de Paris, pl. 4, fig. 8, 1822.

— — and PLICATILIS, *Parkinson.* An Introduction to the Study of Organic Remains, p. 234, 1822.

— — *Dalman.* Vet. Acad. Handl., p. 53, 1827.

— PLICATILIS, *Defrance.* Dic. Sc. Nat., vol. liii, p. 159, 1828.

— OCTOPLICATA, *Fleming.* A Hist. of British Animals, vol. i, p. 372, 1828.

— PLICATILIS and T. OCTOPLICATA, *Woodward.* A Syn. Table of Org. Remains, pp. 21 and 22, 1830.

— — *Deshayes.* Coq. Caractéristiques, p. 114, pl. ix, figs. 3, 4, 1831.

— — *Deshayes.* Ency. Meth., iii, p. 1026, No. 11, 1832.

— OCTOPLICATA and T. PLICATILIS, *Schlotheim.* Syst. Vers. Petref., 1832.

— — — *Mantell.* Geol. of the S. E. of England, p. 127, 1833.

— — *V. Buch.* Class. des Térébratules, 1834; and Mém. Soc. Géol. France, vol. iii, p. 147, pl. xv, figs. 18—24,

— PLICATILIS, *V. Buch.* (part). Ib., 1834; and ib., vol. iii, p. 153, tab. xv, fig. 24, 1838.

— — and T. OCTOPLICATA, *Deshayes.* Nouv. Ed. de Lamarck, vol. vii, p. 356–7, 1836.

— OCTOPLICATA, *Hisinger.* Leth. Suec., p. 79, pl. xxii, fig. 12, 1837.

— — *D'Archiac.* Mém. Soc. Géol. de France, vol. iii, p. 295, 1839.

— PLICATILIS, *Geinitz.* Charact. Petref. Kreid., p. 14, 1839.

— OCTOPLICATA, *Bronn.* Leth. Geog., pl. xxx, fig. 9, 1837.

— PLICATILIS and T. OCTOPLICATA, *Rœmer.* Die Vers. Nord. Kreid., p. 38, fig. 9, 1840.

— OCTOPLICATA, *Geinitz.* Charat. Kreid., pl. xvi, fig. 16, 1840.

— PLICATILIS and T. OCTOPLICATA, *De Hagenow.* N. Jahrb. F. Mineral., 1842.

— — *Morris.* Catalogue, 1843.

— — *Geinitz.* Grundriss. die Vers., pl. xxi, fig. 3, 1844.

— — and T. OCTOPLICATA, *Reuss.* Bohem. Kreid., p. xxv, figs. 10—16, 1846.

— OCTOPLICATA, *D'Orbigny.* Russia and the Oural, vol. ii, p. 492, pl. xliii, figs. 15—17, 1845.

RHYNCHONELLA PLICATILIS, *D'Orbigny.* Pal. Franç., Ter. Crétacés, vol. iv, pl. 499, figs. 9—12; and Prodrome, vol. ii, p. 257, 1850.

TEREBRATULA OCTOPLICATA and T. PLICATILIS, *Tennant.* A Strat. List. of Brit. Org. Remains, p. 47, 1847.

— — *Dixon.* Geol. of the Fossils of the Tertiary and Cretaceous Formation, pl. xxvii, fig. 16, 1850.

— PLICATILIS, *Kner.* Natur. Abhand., tab. v, figs. 5, 6, 1850.

— OCTOPLICATA, *Alth.* Ib., p. 255, 1850.

Diagnosis. Shell transversely oval, with its greatest width towards the centre; *ventral*

or *dental* valve less inflated than the opposite one, with a shallow sinus; beak short, acute, and moderately incurved; foramen small, almost contiguous to the umbo, and entirely surrounded by the deltidium and its tubular expansions; beak ridges well defined, leaving a flattened space or false area between them and the hinge line. *Dorsal* or *socket* valve, generally more gibbose than the opposite one, its uniform convexity being interrupted from about the middle of the valve to the front, by a widely slightly produced mesial fold. The hinge line of the ventral valve indents the lateral portions of the umbo; margins flexuous on approaching the front, become sharply bent at almost right angles, indenting to a considerable extent the frontal edge of the dorsal valve. Externally, each valve is ornamented by from fifty to sixty plaits; these commonly, on approaching their terminations, become flattened, and, as if divided by a narrow longitudinal split or depression. In the interior of the dorsal valve, two curved processes exist for the support of the spirally coiled extensile arms. Shell structure impunctate.[1]

Dimensions variable: length 12, width 15, depth 9 lines.

,, 11, ,, 12, ,, 12 ,,

Var. OCTOPLICATA, *Sow.*   Plate X, figs. 1—17.

This variety agrees in general character with *R. plicatilis* (type), but differs more often by its plaits, which, on approximating, the front and lateral margins unite two by two, forming fewer and larger costæ; these last are also commonly acute, and not flattened or split, as is very often the case in the typical specimens of *R. plicatilis*.

Var. WOODWARDII, *Dav.*   Pl. X, figs. 43—46.

TEREBRATULA GALLINA, *Woodward*. An Outline of the Geol. of Norfolk, tab. iv, fig. 12, 1833.

Diagnosis. Shell transversely oval: valves moderately convex, with a shallow sinus in the ventral, and slightly produced mesial fold in the opposite one. Externally each valve is ornamented by from 24 to 44 simple plaits, often split close to the margin; length 9, width 10 to 12, depth 6 to 7 lines.

Obs. It was justly observed by the author of the 'Pal. Française,'[2] that Sowerby's

---

[1] Dr. Carpenter has described and figured the remarkable shell structure observable in this species, in his very valuable memoir "*On the Microscopic Structure of Shells.*" ('British Association' for 1844, pl. xiv.) See also the "Introduction" to this work, pl. v, fig. 6.

[2] 'Terrains Crétacés,' vol. iv, p. 58, 1847; but prior to M. D'Orbigny, Dr. Mantell had stated 1822, that *R. octoplicata* was only a var. of *R. plicatilis*, "the specimens in my possession vary so much in the number of plica, and the convexity of the valves, and the characters of each are so equally blended, in many examples, that I have been obliged to consider them as only a variety of the same species (Fossils of the South Downs). Mr. Morris and Dr. Bronn have likewise arrived at similar conclusions, but we cannot admit all the synonyms mentioned by the learned German author, viz., *T. latissima, parvirostris, Martini,* and *nuciformis.*" Geinitz likewise looks upon *R. octoplicata*, as a var. of *R. plicatilis*, but erroneously adds *T. pisum,* Sow. and *T. Mantelliana,* Sow., two well-distinguished species.

descriptions and figures of *Ter. plicatilis* and *T. octoplicata* are so entirely similar, that no one is able to perceive in them distinguishing features. *R. plicatilis* varies to a considerable extent, as do all Brachiopoda, and it is not unusual to meet perfectly adult individuals of various dimensions as well as convexity, due, no doubt, to more or less favorable conditions of existence. After much examination I entertained a similar opinion to that already expressed by several Palæontologists, viz., that those examples in which, at a certain age, the plaits became united two by two near the front and literal margins, could only constitute a simple variety of *Rh. plicatilis*, especially as a similar tendency is common to individuals of various forms, as for instance, *R. latissima*, Sow., &c. This complex plication does not always take place only in those examples in which the plaits are acute to the very edge. Nor do all specimens of true *Rh. plicatilis* present the split condition of the plicæ above described, although such may be the prevalent character in most examples. The term *octoplicata* is in itself essentially ill-chosen, from the positive fact that nothing is more variable than the number of plaits, and as an illustration of which I have figured in Pl. X a series of examples collected in the same locality by Mr. Fitch, with 3, 5, 6, 7, 8, and 9 plaits on the mesial fold, and specimens with exactly 8 plaits are by no means the most abundant. Although I have not yet obtained in England examples of *R. plicatilis* quite as large as some of its variety *octiplicata*, still in other countries typical specimens of *R. plicatilis* have been found equalling in dimensions any of those of the variety (Pl. X, figs. 1 and 3).

I must, however, here observe that Mr. S. P. Woodward differs from the conclusions we have arrived at, and is of opinion that *R. octoplicata* can be distinguished and should be preserved as a seperate species from *R. plicatilis*. Associated with the shell last mentioned we often find another form, figured in 1833 by Woodward as *Tereb. Gallina* (Brong.),[1] but which, although somewhat similar in external contour to the Sowerby species, appears to possess a facies of its own, and, if not specifically distinct from *R. plicatilis*, would, at any rate, constitute a well-marked variety, being distinguished by fewer plaits, which are proportionally wide, with flattened ridges, and usually split near the margins.

Plate X, figs. 37 to 39. Typical example of *R. plicatilis*, from the Chalk of Brighton.

„   fig. 40.   Front view of a very inflated individual, from the Kentish Chalk, in the collection of Mr. Bowerbank.

„   figs. 41, 42.   An unusually expanded var., likewise from the Kentish Chalk.

„   figs. 1 to 11.   A series of examples of the *var. octoplicata* (of authors), from the Norwich Chalk, in the collection of Mr. Fitch; figs. 1 and 3 are the largest British specimens I have seen.

„   figs. 12, 13.   A young flattened example, from the Chalk at Royston (Cambridgeshire), in the British Museum.

„   fig. 16.   A specimen, exhibiting spots attributed to colour? from the Chalk of Norwich.

[1] 'An Outline of the Geol. of Norfolk, pl. vi, fig. 12.—See likewise our Pl. X, figs. 43 to 46.

Plate X, fig. 17. A fragment of the beak (enlarged) to illustrate the tubular expansion
of the deltidium.

„   fig. 14. Interior of the dorsal valve, from a specimen in the British Museum.

„   fig. 15. Interior of the ventral valve, ib.

„   figs. 43, 44. *Var. Woodwardii*, from the Chalk of Norwich, in the collection of Mr. Fitch.

„   figs. 45, 46. Ib. From the Chalk of Charing, in the cabinet of Mr. Harris.

35. RHYNCHONELLA LIMBATA, *Schlotheim*, Sp.   Pl. XII, figs. 1—5.

TEREBRATULITES LIMBATUS, *Schlotheim*. Leonhard's Tash., vol. vii, p. 113, 1813;
Petrjk. i, 286, reference Faujas, Mont St. Pierre,
pl. xxvi, fig. 4, 1799.

TEREBRATULA SUB-PLICATA, *Mantell*. Fossils of the South Downs, p. 211, tab. xxvi,
fig. 5, 1822.

—   —   *Woodward*. A Synoptical Table of Brit. Org. Remains,
p. 22, 1830.

—   LENTIFORMIS, *Woodward*. Geol. of Norfolk, tab. vi, fig. 11, 1833.

—   SUBPLICATA and LENTIFORMIS, *Morris*. Catalogue, 1843.

RHYNCHONELLA SUBPLICATA, *D'Orbigny*. Pal. Franç., Terrains Crétacés, vol. iv, p. 48,
pl. 499, figs. 13—17 (under the false name of *Rhyn.
dutempleana*), 1847.

TEREBRATULA LIMBATA, *Bronn*. Index Pal., vol. 2, p. 1246, 1848.

Diagnosis.   Shell more or less transversely oval; somewhat trigonal or circular
when young; beak short, narrow, and incurved; foramen minute, close under the
acute extremity of the beak, and entirely surrounded by the deltidium and its tubular
expansions. A flattened space occurs between the beak ridges and hinge line: valves
moderately convex, with a longitudinal sinus in the ventral valve, to which corresponds a
mesial fold in the opposite one: external surface entirely smooth when young, and often
remaining so to an advanced age: from 10 to 20 short rounded plates ornament the
vicinity of the margin; 3 to 5 occupying the mesial fold and sinus.

Dimensions variable: length 9, width 12, depth 6 lines;

„   8,   „   9,   „   5   „

„   5,   „   5,   „ $2\frac{1}{2}$ „   (*T. lentiformis*, Woodward.)

Obs. Faugas St. Fond appears to have been the first author who figured this form,
but without a name.   In 1813, Schlotheim applied to it the denomination of *Terebratulites
limbatus*, referring at the same time to Faugas's figure; this name is therefore the oldest
we are acquainted with, and has a right to priority, as admitted by Prof. Bronn.

In 1822, the same species was described and figured by Dr. Mantell, under the name of
Ter. subplicata, by which denomination it is known to the greater number of British and
Foreign Palæontologists.   Dr. Mantell states it to be well characterised by its smooth

11

surface and elevated plicated front.   *Rh. limbata* is, however, very nearly related to *R. Octoplicata*, Sow., of which it may perhaps only constitute a marked variety, in which the greater portion of the surface is either entirely smooth, or indistinctly plicated, except towards the front and lateral margins.   In 1833, a small race, almost completely circular in shape, and of the dimensions of a flattened pea, was named *Ter. lentiformis* by Woodward.

In the 'Pal. Franc.' vol. iv, p. 46, M. D'Orbigny considers the name *subplicata* to be a synonyme of *Rh. octoplicata;* but in p. 48 of the same work, he admits the species to be distinct, and in both cases refers to Dr. Mantell's name and figure.   *Rh. limbata* abounds in the Upper Chalk of many localities, always associated with *Rh. octoplicata* (Sow.).   It has been collected at Norwich, in Kent, Sussex, in Ireland, at Meudon and Chavot (France), &c.   Ciply, in Belgium, is the locality from which Faugas's figured specimen was obtained, &c.

Plate XII, figs. 1, 2, 3. Specimens from the Norwich Chalk, in the collection of Mr. Fitch; fig. 1, enlarged.

„   figs. 4, 5. Young specimens, or a dwarf race (*Ter. lentiformis*, Woodward).

36. RHYNCHONELLA COMPRESSA, *Lamarck*, Sp.   Pl. XI, figs. 1—5, and Pl. XII, fig. 25.

TEREBRATULA COMPRESSA, *Lamarck.* An. sans Vert., vol. vi, p. 256, No. 54, 1819; and *Davidson,* "*Notes on an Examination of the Lamarckian Species of Fossil Terebratulæ,*" Annals and Mag. of Nat. Hist., June, 1850, pl. xv, fig. 54.

—   DIFFORMIS, *Lamarck.* Ib., vol. vi, No. 48, 1819 (Encycl. Méth., pl. 242, fig. 5, 1789); and *Dav.* Ib., June, 1850, pl. xv, fig. 48.

—   DIMIDIATA, *Sowerby.* Min. Con., tab. 277, fig. 5, 1821.

—   —   *Parkinson.* An Introd. to the Study of Organic Remains, p. 234, 1822.

—   GALLINA, *Brong.* Desc. Géol. des Environs de Paris, p. 84, pl. ix, fig. 2, 1822.

—   COMPRESSA, *Defrance.* Dic. des Sc. Nat., vol. liii, p. 158, 1828.

—   DIMIDIATA, *Fleming.* A Hist. of British Animals, p. 372, 1828.

—   DIFFORMIS, *Defrance.* Dic. Sc. Nat., vol. liii, p. 160, pl. v, fig. 3, 1828.

—   DIMIDIATA, *Woodward.* Synoptical Table of Br. Organic Remains, p. 21, 1830.

—   DIFFORMIS, *Deshayes.* Encycl. Meth., iii, p. 1029, No. 22, 1832.

—   ALATA, *V. Buch.* Class. des Térébratules, Mém. Soc. Géol. de France, vol. iii, p. 150, pl. xv, fig. 21, 1834.

—   COMPRESSA, *Deshayes.* Nouv. Ed. de Lamarck, vol. vii, p. 345, No. 54, 1836.

—   DILATATA, *Sowerby, in Fitton.* Trans. of the Geol. Soc. of London, vol. iv, p. 343, pl. xviii, fig. 2, 1836.

—   DIFFORMIS. *Des.* Nouv. Ed. de Lamarck, vol. vii, p. 343, No. 48, 1836.

—   DIMIDIATA, *Morris.* Catalogue, 1843.

RHYNCHONELLA COMPRESSA, *D'Orbigny.* Pal. Franç., Ter. Crétacés, vol. iv, p. 35, pl. 497, figs. 1—6, 1847.
— DIFFORMIS, *D'Orbigny.* Ib., vol. iv, p. 41, pl. 498, figs, 6—9, 1847.
TEREBRATULA COMPRESSA, *Bronn.* Index Pal., p. 1233, 1848, (but not a Syn. of *T. limbata.*)

Diagnosis. Shell depressed, elongated oval, wider than long, angular at the cardinal, dilated towards the pallial, region, somewhat indented in front; the greatest width and depth lying towards the middle of the shell: valves unequally convex, the dorsal one generally the deepest, with a wide, slightly produced, and flattened mesial fold, occupying about one third of the width of the shell: in the *ventral* valve, a corresponding wide longitudinal sinus: beak acute, moderately produced, and incurved: foramen rather small, and entirely surrounded by the deltidium: beak ridges sharply defined, leaving a flattened space between them and the hinge line: externally each valve is ornamented by from 32 to 48 strong simple plaits, 8 to 11 of which compose the mesial fold and sinus.

Dimensions variable: length 17, width 23, depth 8 lines.
,,   13, ,,   18, ,, 10 ,,

Obs. This fine species was described by Lamarck, in 1819, from specimens derived from the Upper Green Sand of Mans (France): it varies greatly in degree of compression, some examples being considerably flattened, while others are more convex, and this last variety is the one commonly found both at Chute, near Warminster, and Cap-la-Heve, near Havre (France). A similar shell was described at a later period (1836) by Sowerby, under the name of *Ter. dilatata*, and *Ter. Gallina* (Brongniart) seems likewise to belong to the same type. *R. compressa* is not always regularly trilobed, but often unsymmetrical, from the mesial fold becoming totally or partially shifted either to the one or other side; the shell then appears divided, as in *Rh. inconstans*, into two portions, one half occupying a higher level than the other, or with one edge turned up and the other down; a malformation so common among the *Rhynchonellæ* that it cannot be made use of as a character of any specific importance: thus *Terebratula difformis*[1] (Lamarck), and *Ter. dimidiata* (Sowerby), are nothing more than irregularly developed examples of *R. compressa*, of which any one will become convinced who may examine the typical specimens in

[1] M. D'Orbigny seems to consider *Rhyn. difformis* (Lamarck) to be specifically distinct from *R. compressa* of the same author, and states p. 42, vol. iv, of the 'Pal. Franç.,' "Cette espèce (*R. difformis*) se distingue du *T. contorta* par ces côtes plus grosses. Lorsqu'elle est régulière, elle se rapproche du *R. compressa*, mais elle diffère par sa forme plus renflée encore, est plus courte, moins dilatée latéralement, c'est une espèce bien séparée, mais très variable dans sa forme." Lamarck observes, that his specimens of this shell were derived from the Green Sand of Cap-la-Heve (near Havre), and likewise from Mans. And in both localities I have had the opportunity of examining and collecting specimens, uniting these malformations by insensible passages to the regularly developed condition of *R. compressa*; and both in the French and British localities we find unsymmetrical individuals likewise more or less flattened, as is the case with well-shaped examples.

the Lamarckian and Sowerby collections, or from the study of a series of specimens derived from the localities in which *R. compressa* occurs.[1]

In England, *R. compressa* is found in the Upper Green Sand of Chute Farm, near Horningsham, at Halldown, in the Chloritic Marl of Chard, and other localities. On the Continent, it abounds at Havre, Mans, and la Fléche (Jarthe). M. D'Orbigny also states he has found it at Lattes, La Malle, and Escragnolles (Var.), at l'Ile Madame, Ile d'Aix, and at the Pont des Barques (Charente Inférieure).

Plate XI, fig. 1.   A well-shaped example, from the Upper Green Sand near Warminster, in the collection of Mr. Cunnington; it is identical in shape to some found at Havre.

  „   fig. 2.   A very large individual, from the same locality, similar to the one figured by Dr. Fitton as *T. dilatata*, Sow. (Geol. Trans., vol. iv, pl. xiv, fig. 2.)

  „   fig. 3.   Front view of a specimen from Chute Farm, in which the mesial fold is shifted to one side, from the cabinet of Mr. Cunnington.

  „   fig. 4.   A malformation from the Chloritic Marl of Chard, which entirely agrees with Sowerby's type of *Ter. dimidiata*.

  „   fig. 9.   Another similar example, from the Upper Green Sand near Warminster.

Plate XII, fig. 25.   A specimen from the Chloritic Marl of Chardstock, in the collection Mr. Th. Walrond.

37. RHYNCHONELLA LATISSIMA (*lata*), Sow., Sp.   Pl. XI, fig. 6—22, and Pl. XII, fig. 24.

TEREBRATULA LATA, *Sow.* Min. Con., vol. v, p. 165, tab. 502, fig. 1, 1825, changed afterwards (1829) to *Ter. latissima*, by the same author (not *Ter. lata*, Sow., Min. Con. vol. i, pl. 100, fig. 2, 1812).

—   ALATA? *Nilsson.* Petrefacta Suecana, pl. iv, fig. 9, 1827.

—   LATA, *Sow. (in Fitton).* Trans. Geol. Soc., vol. iv, pl. xiv, fig. 11, 1836.

—   CONVEXA, *Sow.* Ib., pl. xiv, fig. 12, 1836.

—   LATISSIMA, *Rœmer.* Die Vers. Nord. Kreid., pl. vii, fig. 4, 1840.

—   —   *Morris.* Catalogue, 1843.

—   —   *D'Archiac.* Mém. Soc. Géol. de France, vol. ii, 2d ser., p. 330, pl. xxi, fig. 7, 1847.

—   SCALDINENSIS. Ib., pl. xx, fig. 11.

---

[1] Some authors have attributed to this species *Rhyn. alata* (Lamarck, sp.); but as observed by MM. D'Orbigny, Deshayes, and myself, *T. alata* is nothing more than a synonyme of *Rhynchonella* (Anomya) *vespertilio* of Brocchi ('Conchologia Fossile,' 1814). Lamarck refers to pl. 245, figs. 2, *a, b,* of the 'Ency. Méthodique.' which figure certainly represents a shell indentical with the one illustrated by the Italian author. *R. vespertilio*, at times, bears some resemblance to *R. compressa*, but is in general more regularly convex and trilobed, with a much deeper sinus, and a more elevated mesial fold.

RHYNCHONELLA LATA, *D'Orbigny.* Pal. Franç., Ter. Crétacés, vol. iv, p. 21, 1847 (but perhaps not all his list of Synonymes).
TEREBRATULA PLICATILIS, *Bronn.* Index Pal., p. 1246, 1848 (but not *T. plicatilis,* Sow., nor the generality of Bronn's other synonymes).
— LATA, *Austen.* Quart. Journ. Geol. Soc., vol. vi, p. 477, 1850.
— LATISSIMA, *Sharpe.* Quart. Journ. of the Geol. Soc., vol. x, p. 192, 1853.

Diagnosis. Shell transversely oval or unequally five sided, with rounded angles: *ventral* or dental valve moderately convex: beak acute, slightly produced and incurved: foramen circular, entirely surrounded by the deltidium and its tubular projections: beak ridges well defined, leaving a flattened space between them and the hinge line: the regular convexity of the valve is interrupted by a sinus of moderate depth, commencing towards the middle and extending to the front. The *dorsal* or socket valve is either convex and regularly arched, or somewhat flattened, with a mesial fold not rising much above the uniform convexity of the shell. Externally, the surface of each valve is ornamented by from 50 to 80 plaits.

Length 12, width 16, depth 7 lines;
„ $10\frac{1}{2}$, „ 12, „ 7 „
„ 9, „ 10, „ 6 „ &c.

Obs. The shells here described may perhaps only constitute a variety of *R. compressa,* Lamarck: but they seem to be distinguished by a less expanded, and in general more regularly transverse oval shape; also by the number and quality of their plaits, which are more numerous in the shells under notice than in the Lamarckian type, which does not appear to present the complex condition at times observable in *R. latissima* (Sow.). So much so, that some examples illustrated in my Pl. XI, figs. 19—22, have been by some authors supposed to belong to another species, viz., *R. antidichotoma* of Buvignier,[1] but after having examined a numerous series of specimens collected by Messrs. Sharpe, Lowe, Waterhouse, Cunnington, myself, and others, I was able to convince myself in a most satisfactory manner, that all the examples illustrated in Pl. XI, figs. 6—22, belonged to the same species. In the extensive series principally derived from Warminster and Farringdon, every possible variation in the plication may be perceived: in almost all, the plaits are few in number in the young, but soon augment at variable distances from the extremities of the beak and umbo by the intercalation of a fresh plait between those already formed. In many examples, the last as well as the original ones proceed uninterruptedly to the margin, while in others, some of the intercalated ribs are lost, or disappear between their immediate neighbours before reaching the front or margin, while in some cases only a few while in other examples almost every two of the plaits unite, and form a belt of larger costæ near the front and margin. All these complex characters are accu-

---

[1] This shell differs from *R. latissima* by its general shape and small foramen, which is widely separated from the hinge line by a largely developed deltidium; also to some extent by the character of its plaits. See Pal. Franc. Terrains Crétacés, vol. iv, p. 500, fig. 1—4.

rately illustrated in figures 6—22, and from them it may be observed, that in most cases only a few of the plaits here and there disappear or become united, while in others from 50 to 60 smaller ones may be counted towards the middle of the shell, and only from 30 to 35 larger ones near the margin. Certain conditions and localities seem to have favoured this kind of malformation; thus, few occur in the Upper Green Sand of Chute Farm, near Warminster, while they are more plentiful at Farringdon : exceptional examples of *R. latissima* bear some resemblance to *R. sulcata* (Parkinson), but this last is in general less transverse, and much more largely plicated, the number of the plicæ being likewise less numerous; but it would not be very difficult, I think, to find extreme and exceptional examples connecting *R. compressa*, Lam., with *R. latissima*, Sow., and the last to *R. sulcata*, Parkinson. However the generality of the individuals of each seems to be sufficiently distinguished by prevailing and peculiar characters to make it desirable, at least for the present, to describe each of them under a separate head. Young shells of *R. parvirostris* might also be confounded with others of the same age of *R. latissima*, although adults of both can be easily separated.

As in the case of *R. compressa*, *R. latissima* is at times more or less unsymmetrical from the mesial fold becoming shifted to the one or other side, this may be seen in Pl. XI, fig. 6, and to a greater extent in fig. 12.

In 1825, Sowerby described and figured the shells we are commenting upon by the name of *Ter. lata*, but in the index to the 'Min. Con.,' published at a subsequent period (1829), he changed his first denomination to that of *T. latissima*, from having observed that he had already made use of his former appellation for another species: the first term might strictly still be retained, because his original *Ter. lata* is only a synonyme of *Ter. ovoides*, while the second *Ter. lata* belongs to another form and even genus (*Rhynchonella*), but to avoid any possible confusion I have adopted Sowerby's later denomination.

In 1836, varieties of the same shell received from Sowerby the names *R. convexa* and *R. elegans*, but which names must be added to the synonymes. A few authors have likewise considered *R. latissima* to be synonymous with *R. alata*, Lamarck, but the last is only itself a synonyme of *R. vespertilio* (Brocchi), as has been already explained under *R. compressa*.

Dr. Bronn goes the length of considering *R. latissima* as simply a variety or synonyme of *R. plicatilis*, Sow., but in this opinion I believe the distinguished German author will find few supporters.

In England, *R. latissima* is essentially an *Upper Green Sand* species, abounding both at Chute Farm, near Warminster, Farringdon, &c.; while in France, according to M. D'Orbigny, it would be a *neocomien* shell; but I fear some mistake either in the identification with our British specimens, or with the age of the bed, has been committed, as I feel certain that Sowerby's type occurs in the Upper Green Sand of Mans, in France, as well as in the Tourtia of Belgium. One of M. D'Orbigny's illustrations ('Pal. Franc.' Pl. 491, fig. 8) completely agrees with some of our British Upper Green Sand examples, and it is possible that the author of the 'Pal. Franc.' may have considered as belonging to the same species, shells named *R. parvirostris* and *R. Gibbsiana*, which appear with us distinct, and peculiar to the Lower Green Sand (Neocomien of the French).

The figures published by Viscomte D'Archiac of his *Rh. Scaldinensis* entirely agree with our typical examples of *R. latissima*, and much more closely so even than the figures he attributes to the Sowerby species. The celebrated author of the 'Histoire des Progrès de la Géologie,' mentions that his *Ter. Scaldinensis* numbers as many as 65 plaits, while Sowerby only mentions 40 to his *R. latissima*, but from what I have said above, it may be seen that the plaits in our British species vary very much, and are often as numerous as in *R. Scaldinensis* of the French author.

Plate XI, figs. 6, 7, 8, and 14. Different specimens and shapes of *R. latissima*, from the Upper Green Sand of Warminster.

„   figs. 9, 10, 11. Young specimens, from same locality.

„   fig. 12. Unsymmetrical and aged example, same locality.

„   fig. 13. A malformation, from Warminster, from the cabinet of Mr. Cunnington.

„   fig. 15. A ventral valve from Farringdon.

„   fig. 16. A convex variety, same locality.

„   fig. 17. Profile and front view of another Farringdon specimen, in the collection of Mr. Lowe. Fig. 17', an enlarged illustration, to show exactly the complex condition of the plaits.

„   fig. 18. A young specimen from Farringdon.

„   figs. 19 and 20. A specimen from the same locality, in the collection of Mr. Lowe, in which the complex plication above described is well exemplified, and especially so in the enlarged figure 19ᶜ.

„   figs. 21 and 22. Two other examples from the same locality, in the collection of Mr. Sharpe. *R. Antidichotoma*, Buv.? according to Mr. Sharpe.

Plate XII, fig. 24. A very remarkable example, from the Chloritic Marl of Chardstock, in the collection of Mr. Wiest.

38. RHYNCHONELLA SULCATA, *Parkinson*, Sp.? Plate X, figs. 18—36.

TEREBRATULA SULCATA, *Parkinson*. Trans. Geol. Soc., vol. i, p. 347, 1811, and vol. v, p. 57, 1821 (but neither figured nor described).

—   — *Morris*. Catalogue, 1843.

RHYNCHONELLA SULCATA, *D'Orbigny*. Pal. Franç., Ter. Crétacés, vol. iv, p. 36, pl. 495, figs. 1—7, 1847.

Diagnosis. Shell transversely oval, wider than long; valves more or less unequally convex, *ventral* or dental valve in general the deepest, with a shallow longitudinal sinus to which a moderately produced mesial fold corresponds in the opposite one. Beak short, acute, entire, and but little incurved; foramen small, surrounded by a deltidium in two pieces; beak ridges well defined, the hinge line not encroaching on that of the dorsal valve; lateral margins slightly flexuous; the frontal edge of the ventral valve indenting more or less that of the dorsal one. External surface ornamented by a number of simple radiating plaits, from 30 to 40 on each valve.

Length 12, width 15, depth 10 lines. (This species at times attains somewhat larger dimensions.)

Obs. In 1811, Parkinson simply mentioned the name *Terebratula sulcata*, without description or figure. And in another paper, read before the Geological Society in 1818, but published only in 1821, we find the same name repeated, as follows:

"FOSSILS IN THE BLUE MARL. *Terebratula sulcata*, found near Dover, Folkstone, and Cambridge," but no figure or description is given, so that this appellation is in reality equivalent to a MS. denomination, and the author may have intended the shell for the one afterwards named *T. Mantelliana* by Sowerby, and which is found in those localities.

In the 'Geology of Sussex,' p. 130, 1822, Mantell likewise describes a Rhychonella by the name of *sulcata*, from the Chalk of Hamsey and Stoneham in Sussex, but also without figure, and to this species the name *T. Mantelliana* was subsequently appended by the author of the 'Min. Con.,' that of *Ter. sulcata* being retained for another shell found abundantly in the Upper Green Sand of Cambridge.

In 1843, Mr. Morris mentioned *Ter. sulcata* as from the Gault of Folkstone and Cambridge: and in 1847, M. D'Orbigny describes the Upper Green Sand Cambridge species as that of Parkinson; considering at the same time *Rh. Gibbsiana* (Sow.) a synonyme; but here the learned author of the 'Pal. Franc.' seems to be evidently mistaken. The *R. Gibbsiana* (Sow.) occurs, it is true, in the vicinity of Folkstone, Sandgate, Hythe, &c., but in another bed, viz., Lower Green Sand (Neocomien), and cannot, I believe, be confounded with the Upper Green Sand species, now so well known to collectors as the true (?) *R. sulcata* of Parkinson. In a catalogue of the *Lower Green Sand* fossils in the museum of the Geological Society,[1] Professor Forbes stated that *R. sulcata* occurs in the *Lower Green Sand* of Hythe, and mentioned as his var. β, *R. parvirostris* of Fitton, a view I can hardly admit. Professor Bronn,[2] while adopting the term *sulcata*, states it to be his opinion that *R. depressa* (Sow.), *inconstans, rostralina, plicatella*, and *multiformis* (Rœmer), as well as *T. parvirostris* and *elegans* (Sow.) belong all to the same type; and although perhaps some of the shells mentioned may bear a resemblance to our Upper Green Sand species, neither *R. elegans, parvirostris*, nor *depressa* can I think, with propriety, be united to the Cambridge *R. sulcata*.

Rh. Gibbsiana is more triangular in its external aspect, its sinus and fold much more

---

[1] Quarterly Journal of the Geol. Soc., vol. i, p. 345, 1845.
[2] Index Pal., vol. ii, p. 1852; ——— 1848.

developed, and the plaits always smaller and more delicate than those observable on the Cambridge shell, which it has been agreed to term *R. sulcata*.

Few species vary more in external shape or detail than the one under consideration, as may be seen from the series of illustrations I have selected from among several hundred individuals assembled from a single locality by Mr. Carter. The mesial fold and sinus does not always occupy the middle of the shell, nor in all cases is it symmetrical, for out of ten examples eight or nine will have their fold and sinus shifted more to the one or the other side, as seen in figs. 23, 25, and 27 of our Plate, while in some examples the one half of the valve is more elevated than the other, being twisted indifferently to the right or to the left, as is so common to *Rh. inconstans*, and to those malformations of *Rh. compressa* to which Sowerby had applied the term *R. dimidiata*. The plaits are generally simple, but in some instances, although rarely, bifurcate here and there.

Rh. sulcata abounds in the *Upper Green Sand* near Cambridge, is less commonly met with in the neighbourhood of Warminster, and was found by Mr. Bean in the Speeton Clay of Yorkshire. Some rare examples have likewise been found in the Gault of Folkstone, and in the corresponding bed at Wissant, on the French coast. M. D'Orbigny mentions the species as abounding in his TERRAIN ALBIEN at Grandpré, and Fleville (Ardennes), Gérodot (Aube), at the Perte du Rhône (Ain), and Clausayes (Drome), &c.

Plate X, figs. 18—20, and 23—36. Illustrate a series of specimens from the Upper Green Sand of Cambridge, in the cabinet of Mr. Carter. Figs. 18, 21, are regular in shape, the others show some of its innumerable malformations. Figs. 15 and 36 are internal casts, on which the muscular and vascular impressions are well preserved.

„   figs. 21—22. From the Speeton Clay, in the collection of Mr. Bean.

39. RHYNCHONELLA MANTELLIANA, *Sowerby*, Sp.   Plate XII, figs. 20—23.

TEREBRATULA MANTELLIANA, *Sowerby*. Min. Con., vol. vi, p. 72, tab. 537, fig. 5, 1825.
—   —   *Fleming*. A Hist. of British Animals, vol. i, p. 374, 1828.
—   —   *V. Buch*. Mém. Soc. Géol. de France, vol. iii, p. 154, pl. xv, fig. 26, 1838.
—   —   ? *Geinitz*. Char. Kreid., p. 15, 1839.
—   —   *Morris*. Catalogue, 1843.
RHYNCHONELLA MANTELLIANA, *D'Orbigny*. Pal. Franç., Ter. Crétacés, vol. iv, p. 40, 1847, (the illustrations given by this author, pl. 498, figs. 1—5, do not recall the common aspect of the Sowerby species.)

Diagnosis. Shell transversely obovate, rather wider than long; valves almost equally convex, with a shallow longitudinal depression or sinus in the dental or ventral valve, which corresponds to a slightly produced mesial fold in the opposite one. Beak short, entire, not much incurved, foramen small, and entirely surrounded by the deltidium : lateral margins almost straight : the frontal edge of the ventral valve encroaches on that of the dorsal one. Externally each valve is ornamented by from 15 to 18 wide simple plaits, 3 or 4 forming the mesial fold. Dimensions very variable : length 7, width 8, depth 4½ lines.

Obs. This species was accurately described and figured by Sowerby, in the ' Mineral Conchology,' under the name of *Ter. Mantelliana;* and it is most probably one of the shells intended by Parkinson as the type of his *Ter. sulcata,* but as the last-named author neither described nor figured his form, Sowerby's denomination must be retained for the well-known species under consideration.

Rh. Mantelliana is commonly a small shell, of about the dimensions above given, but it has been found sometimes, although rarely, of larger dimensions, as proved by the fine specimen (Pl. XII, fig. 23) found in the Lower Chalk near Lewes, by Mantell, and it forms part of his collection in the British Museum; it abounds in the Lower Chalk and Chalk Marl between Dover and Folkstone, at Hamsey, and in many other localities, and has been collected, although much more rarely, in the Upper Green Sand of the neighbourhood of Warminster, and in the Chloritic Marl of Bonchurch (Isle of Wight), by Mr. S. H. Saxby. On the Continent, it occurs in beds of a similar age to those above mentioned, both in France and Belgium. *Rh. Mantelliana* is well distinguished from *Rh. Cuvieri,* by its larger and less numerous plaits, as well as by its greater width.

Plate XII, fig. 20 and 21. Two examples from the Gray Chalk of Folkstone and Hamsey.

,, fig. 22. A specimen from the Upper Green Sand of Chute Farm, near Warminster, in the collection of Mr. Cunnington.

,, fig. 23. A very large specimen, from the Lower Chalk of Lewes, in the Mantellian collection in the British Museum. It measures: length 10, width 11, depth 6½ lines.

40. RHYNCHONELLA CUVIERI, *D'Orbigny.* Plate X, figs. 50—54.

TEREBRATULA PISUM, *Geinitz.* Kreide, pl. xvi, fig. 18, 1840 (but not *Ter. pisum,* Sow.).

RHYNCHONELLA CUVIERI, *D'Orbigny.* Pal. Franç., Ter. Crétacés, vol. iv, p. 39, pl. 497, figs. 12—15, 1847.

Diagnosis. Shell small, transversely or longitudinally oval, length and width often the same : valves regularly convex, and of nearly equal depth, a shallow depression or sinus existing towards the front of the dental or *ventral* valve, to which a similar slight elevation corresponds in the opposite one ; beak small, acute, and entire ; foramen minute, completely surrounded and removed from the contiguity of the hinge line by a deltidium and its largely expanded tubular expansions. The lateral margins are but slightly sinuous, the frontal edge of the ventral valve indenting that of the opposite one. Externally, each valve is ornamented by from 30 to 34 equal and simple plaits: length 6½, width 6½, depth 5 lines.

Obs. M. D'Orbigny mentions, that this species approaches *Rh. Grasiana* by general aspect, but is not quite as wide, usually it possesses fewer plaits, and is more circular in shape: he found it at Fécamps, near Rouen (France), along with *Inoceramus problematicus*, or in other words, in his ETAGE TURONIEN; examples were likewise obtained from Cap-Blanc-Nez (Pas de Calais) and at la Fleche (Sarthe). In England it has been found by Mr. Baber, in the Chloritic Marl, with quartz grains, near Chard and Chardstock; it abounds in the Lower Chalk of Lewes, at Glynde Bourn (Sussex), and two specimens (probably from the same locality) are figured, without name or description, in Dixon's work, 'Geology and Fossils of the Tertiary and Cretaceous Formations of Sussex' (pl. xxvii, figs. 15, 16, 1850). It has likewise been procured by Mr. Carter in Chalk of a similar age near Cambridge, and in that near Norwich by Mr. Fitch.

Some examples are so uniformly convex as hardly to present any trace of sinus or mesial fold.

Plate X, figs. 50—52. Enlarged illustration of an elongated specimen from the Lower Chalk, near Cambridge, in the collection of Mr. Carter.

„ figs. 53, 54. A wider example from a bed of similar age at Glynde Bourn, near Lewes (Sussex.)

41. RHYNCHONELLA DEPRESSA, *Sow.*, Sp. Plate XI, figs. 28—32; and Plate XII, fig. 26.

ANOMITES TRIANGULARIS, *Wahlenberg??* Petrif. Tellures Suecana nova acta Soc. Scientiarum Upsaliensis, vol. viii, tab. iii, figs. 11—13, 1821.

TEREBRATULA DEPRESSA, *Sowerby.* Min. Con., vol. v, p. 165, tab. 502, 1825 (not *D'Orbigny*, Pal. Franç., vol. iv, pl. 491, fig. 17, 1847).

— — *Woodward.* A Synoptical Table of British Organic Remains, pl. xxi, 1830.

— — *V. Buch.* Class. des Térébratules, Mém. de la Soc. Géol. de France, vol. iii, p. 137, pl. xiv, fig. 6, 1838.

— — *Morris.* Catalogue, p. 133, 1843.

— — *Tennant.* A Strat. List of British Fossils, p. 47, 1847.

— — *Bronn* (part). Index Pal., vol. ii, p. 1234, 1848, (not all his Synonymes).

— — *Austen.* Quart. Journ. of the Geol. Soc., vol. vi, p. 477, 1850.

RHYNCHONELLA DEPRESSA, *Sharpe.* Quart. Journ. of the Geol. Soc., vol. x, p. 192, 1853.

— TRIANGULARIS, *Sharpe.* Ib., p. 192, 1853.

Diagnosis. Shell depressed, triangular, wider than long, valves almost equally convex;

the *ventral* or dental one with a wide longitudinal sinus, to which corresponds a slightly raised mesial fold in the opposite or dorsal valve: beak acute, tapering and but slightly incurved, ridges sharply defined, leaving a wide flattened space or false area between them and the hinge line, which last indents the lateral portions of the umbo: foramen comparatively large, entirely surrounded, and more or less removed from the hinge line by the deltidium and its tubular prolongations; lateral margins moderately sinuous; the frontal edge of the ventral valve indents the opposite one to a lesser or greater extent. Externally, 17 to 30 plaits ornament each valve, 6 to 10 forming the mesial fold. Dimensions variable: length 8, width 9, depth 6 lines;

       ,,   7, ,, 7½, ,, 4 ,,

Obs. The shell above described has been distinguished and admitted by British geologists as *Ter. depressa* of Sowerby,[1] although misunderstood by several continental authors. It may, however, remain a question whether *R. depressa* be really distinct from the *Anomites triangularis* of Wahlenberg,[2] a point I have been unable to determine, from the figures published by the Swedish author not conveying a sufficiently satisfactory resemblance to Sowerby's species and specimens, being too circular, and exhibiting no trace of mesial fold or sinus, which is always visible in examples of similar dimensions of *R. depressa.* Nilsson[3] describes and reproduces Wahlenberg's figures, but does not throw further light on the contested question. The beak, foramen, and deltidium are both inaccurately and vaguely represented, for which reasons I did not consider it advisable to remove Sowerby's denomination until more positive evidence can be obtained by the inspection of Swedish specimens.

Mr. Sharpe considers that among the shells found at Farringdon, and referred by myself to *R. depressa*, the two species do occur, and may be distinguished; but after a minute study of all the specimens collected by that distinguished Palæontologist, as well as of those assembled in the locality by myself and others, I felt unable to arrive at a similar conclusion, from finding that all possessed (to my eyes) the same *essential* specific character. According to Mr. Sharpe, the young shell (Pl. XI, fig. 32) would represent *R. triangularis*, while the figs. 29 and 30, represent *R. depressa;* in these, however, we observe the same general shape, the same character of plication, with many of the plicæ augmenting by intercalation, and varying in number; a peculiarity common to specimens of every species

[1] "Triangular, depressed, regularly plaited, front elevated, lateral angles rounded, beaks produced, plaits 20; when so young that the front is hardly elevated, this shell is almost orbicular: in which circumstance it differs from the last (*Ter. lata*), the proportions of which do not vary much by age; the plaits are sharp, about eight of them are raised with the front. Found at Farringdon. ('Min. Con.,' vol. v, p. 165, tab. D ii, fig. 2.)

[2] 'Petrifacta Telluris Suecana, Nova Acta Regiæ Societates Scientiarum Upsaliensis, vol. viii, tab. iii, figs. 11, 12, 13, 1821.

[3] 'Petrifacta Suecana,' p. 36, tab. iv, fig. 10, 1827. "T. testa ovato-triangulari, longitudinaliter sulcata; sulcis et interstriis numerosissimis æqualibus; valva minore convexiore; rostro acutangula subrecto; margine superiore; bex sinuato. Locality—Balsberg," where it is stated to be rare.

of Rhynchonella, and, had space permitted, illustrations could have been introduced exhibiting every passage uniting such shells as figs. 28 and 30. In all, the umbo of the dorsal valve is much incurved, its extremity being to a lesser or greater extent concealed under the development or encroachment of the deltidium. On well preserved examples the concentric lines of growth are numerous and in close approximation, giving to the upper ridge of each plait a somewhat granulated appearance, but which is more deceptive than real, since these projections form part of an uninterrupted and continuous concentric line or ridge. In young individuals, no trace of sinus or mesial fold can be perceived, the frontal line being straight, but with age both the sinus and fold gradually appear, and always exist to a greater or lesser degree in adult individuals. Geinitz published two figures representing the exterior of the ventral valve of a shell he terms *Ter. Triangularis*,[1] and which in external contour appears to somewhat resemble our British examples, but the profile view would almost indicate a different species. M. D'Orbigny's figures[2] of his so-termed *Rh. depressa* do not appear to resemble Sowerby's shells, and belong (I have little doubt) to a distinct species, although the description published in the 'Pal. Française,' would denote a shell different from that figured in his plate.

R. depressa abounds in the Upper Green Sand of Farringdon, along with *R. nuciformis*.

Plate XI, fig. 28. A specimen of *R. depressa*, Sow., from the Upper Green Sand of Farringdon, in the cabinet of Mr. Lowe, it presents 11 plaits on the mesial fold; 28ᵇ ᶜ are enlarged representations.

„ fig. 29. Another example from the same locality, in which the central plaits are narrower than the lateral ones.

„ fig. 30. A specimen with an unusually small number of plaits, from the collection of Mr. Sharpe; fig. 3ᵈ, enlarged.

„ fig. 31. A young individual, from the same locality.

„ fig. 32. A young and somewhat elongated example, believed by Mr. Sharpe to represent *R. triangularis*; fig. 32ᵇ a magnified illustration, from the collection of Mr. Sharpe.

Plate XII, fig. 26. A large transverse specimen, in which the beak is not so much produced as in those figured in Pl. XI, locality Farringdon.

In the Upper Green Sand of Warminster, in that of the Isle of Wight, and in equivalent beds at Chardstock, are found numerous examples of two forms represented in Pl. XII, figs. 28 and 30. They appear to constitute (if not separate species) well-marked varieties of *Rh. depressa* of Farringdon. I will therefore briefly mention them under the head of *varieties* A and B.

[1] 'Charact. der Schichten und Petref.,' pl. xix, figs. 1—3, 1842.
[2] 'Pal. Franç., Terrains Crétacés,' vol. iv, p. 18, pl. 491, figs, 1—7. M. D'Orbigny states that his specimens were obtained in the Terrain Néocomien of France.

RHYNCHONELLA DEPRESSA, *Var.* A. Pl. XII, fig. 30.

This shell presents much of the general contour of the type species, and especially of those examples in which the beak is short, with the foramen close to the hinge line (Pl. XII, fig. 26). Each valve is ornamented by from 20 to 22 large plaits, a few of which are due to intercalation: the surface is likewise covered by numerous concentric lines or ridges of growth, similar in character to those visible on well preserved Farringdon specimens. This variety has been found in the Green Chloritic Beds of Chardstock by Mr. Wiest, and in the Upper Green Sand of the neighbourhood of Warminster by Mr. Sharpe: length 5, width 6, depth $3\frac{1}{2}$ lines.

Plate XII, fig. 30. A specimen from Chardstock; $30^{a\,b}$, enlarged illustrations of the same.

RHYNCHONELLA DEPRESSA, *Var.* B. Pl. XII, fig. 28.

This variety is distinguished from the preceding one, as well as from typical examples of *R. depressa*, by a greater number of plaits, 45 to 48 ornamenting the surface of each valve; these are likewise intersected by small approximate concentric ridges of growth, visible only on well-preserved specimens. When young, the shape is triangular and identical to that of individuals of a similar age at Farringdon (Pl. XI, fig. 31). This shell has been collected by Messrs. Sharpe and Wiest in the same beds and localities along with the preceding variety (A); it appears distinguished from *R. nuciformis* by its more dilated and compressed appearance, and its plaits do not exhibit towards their extremities that split condition observable in so many examples of the last-named species. Dimensions variable.

Plate XII, fig. 28. From Chardstock, Upper Green Sand

„   fig. $28^{b\,c}$. Enlarged illustration of the same.

„   fig. 28^{d}. A young triangular individual.

I am still uncertain whether specimens similar to the one Pl. XII, fig. 29, should constitute a variety of the present species. It was found in the Upper Green Sand of Shaftesbury by Mr. S. P. Woodward, and I have picked up similar specimens in equivalent beds in Normandy (France). Some extreme examples of *R. depressa* and *R. nuciformis* can hardly be distinguished.

42. RHYNCHONELLA NUCIFORMIS (*Sowerby*, Sp.).   Plate XI, figs. 23—27, and Plate XII, fig. 27.

TEREBRATULA NUCIFORMIS, *Sow.*   Min. Con., vol. v, p. 166, tab. 502, fig. 3, 1825.
—   —   *Woodward.* A Synopt. Table of British Org. Remains, p. 21, 1830.
—   —   *Morris.* Catalogue, 1843.
—   —   *Tennant.* A Strat. List of British Fossils, p. 47, 1847.
—   —   *Austen.* Quart. Journ. of the Geol. Soc., vol. vi, p. 477, 1850.
—   —   *Sharpe.* Quart. Journ. of the Geol. Soc., vol. x, p. 192, 1853.

Diagnosis.   Shell more or less transversely oval and inflated: valves unequally convex, the dorsal one more often the deepest: beak acute, moderately produced and incurved: foramen almost contiguous to the umbo, of moderate dimensions, and entirely surrounded by the tubular prolongations of the deltidium: between the beak ridges and hinge line exists a flattened space, which slightly indents the lateral portions of the umbo.

The ventral or dental valve presents a longitudinal depression or shallow sinus, to which, in the opposite valve, a mesial fold corresponds of variable elevation: externally each valve is ornamented with from 30 to 40 plaits, 7 to 12 occupying the mesial fold or sinus, the ridges of the plaits are more or less acute, but, on approaching the front and lateral margins, often become flattened, with a longitudinal indented line along their centre.   Dimensions and relative proportions very variable: length, 6½, width, 7, depth, 7 lines.
„ 7   „ 9   „ 6   „
„ 8   „ 8½ „ 6   „

Obs.   *Rh. nuciformis* was stated by Sowerby to be *a globose shell, smaller than a hazelnut, the edges of the plaits being rounded, and near the front often with a sunk line upon them* (loc. Farringdon); and although distinguished in England from other *Rhynchonellæ*, has, on the Continent, been very generally confounded with other forms.   M. D'Orbigny places it as a synomyn of *R. depressa* (Sow. sp.),[1] but from which it appears to differ by its general shape, which is transversely or oblongly oval, and at times almost circular, with its plaits often split near the front and margins, as is so well exemplified in the Palæozoic *Rh. Wilsoni* and other similar forms.   While *R. depressa* (Sow.), as its name implies, is a depressed shell with imbricated plaits, this last character not having been observed in true *R. nuciformis*.   Prof. Bronn commits another mistake, by considering the shell we are describing to be the same as *R. plicatilis*,[2] from which it appears removed by more than

[1] The author of the 'Pal. Franç.,' does not appear to have been acquainted with Sowerby's *R. nuciformis* and *depressa*, for his figure of this last ('Pal. Franç.,' vol. iv, pl. 491, figs. 1, 7), does not agree with any of the examples found in England.
[2] 'Index Pal.,' vol. ii, p. 1246.

one character, as may be easily perceived by comparing the figures or examples of the two species.

R. *nuciformis* is not very rare in the Upper Green Sand of Farringdon; it also occurs in the Chloritic Beds of Chardstock, where it has been collected by Mr. Wiest, &c.

Plate XI, figs. 23 and 24.   Two typical examples from the Upper Green Sand of Farringdon (collection of Mr. Sharpe); fig. 23ᵃ ᵇ ᶜ, are enlarged representations to show the character of the plaits.

„   fig. 25.   A transverse and less gibbose specimen, from the same locality and collection.

„   figs. 26 and 27.   Two other examples from Farringdon, in the cabinet of Mr. Lowe.

Plate XII, fig. 27.   From the Upper Green Sand of Niton, Isle of Wight, in the collection of Mr. S. H. Saxby.

43. RHYNCHONELLA MARTINI, *Mantell*, Sp.   Plate XII, figs. 15, 16.

TEREBRATULA MARTINI, *Mantell*.   Geol. of Sussex, p. 131, 1822.
—   PISUM, *Sowerby*.   Min. Con., vol. vi, p. 70, tab. 536, figs. 6, 7, 1826.
—   —   *Fleming*.   A Hist. of British Animals, vol. i, p. 374, 1828.
—   —   *Woodward*.   A Synop. Table of British Org. Remains, p. 21, 1830.
—   —   *V. Buch*.   Class. des Térébratules, Mém. de la Soc. Géol. de France, vol. iii, p. 148, pl. xv, fig. 18 *bis*, 1838.
—   BREVIROSTRIS, *Rœmer*.   Die Vers. Nord. Kreid., pl. vii, fig. 7, 1840.
—   MARTINI, *Morris*.   Catalogue, 1843.
RHYNCHONELLA PISUM, *D'Orb*.   Prodrome, vol. ii, p. 171, 1850.

Diagnosis.   Shell sub-orbicular, longer than wide, nearly square in front: valves almost equally convex, with the greatest depth at a short distance from the beaks, a slight longitudinal depression existing towards the front of either valve: no regular sinus nor mesial fold: margin nearly straight all round or slightly raised in front: beak short, acute, and moderately incurved, with a flattened space between the beak ridges and hinge line: foramen small, contiguous to the umbo, and entirely surrounded by a deltidium and its tubular prolongations: externally each valve is ornamented by from 30 to 40 delicate plaits, intersected by numerous concentric lines of growth.   Length $4\frac{1}{2}$, width 4, depth $2\frac{1}{2}$ lines.

Obs.   R. *Martini* is a small shell, never greatly exceeding the dimensions above stated, and more often not as large.   The plaits are narrow, delicate, and augment here and there by intercalation: the numerous, closely packed, and slightly raised concentric lines of growth gives to the ridges of the plaits a granulated aspect, which is more deceptive than real.

This species was, for the first time, named and described by the celebrated author of the 'Geology of Sussex,'[1] and although unfigured at that period, may be easily identified, being a shell well known to British geologists. In 1826 Sowerby figured and described the same species under the new appellation of *Terebratula pisum*, the same as had been given some years before by Mantell to a similar shell, mentioning Hamsey as his locality; and it seems singular that the greater number of subsequent authors preferred the Sowerby denomination, and it was only in 1843 that Mr. Morris, in his 'Catalogue,' reestablished Mantell's claims, by placing *T. pisum* as a synonym. *Ter. brevirostris* (Rœmer, 1840) has no better claims, being identical, both in shape and character with the Mantellian type. V. Buch adopts Sowerby's name, stating that the species does not appear to differ essentially from *R. octoplicata* of the same author! but this will require confirmation before being admitted, as the species seems to be little known to continental authors, he mentions several localities.

R. Martini abounds in the Chalk Marl and Grey Chalk of Hamsey and Folkstone, it has likewise been obtained from the "Chalk detritus" of Charing (Kent) by Mr. Harris; and some rare individuals have also been discovered in the Upper Green Sand of Horningsham, near Warminster, associated with another small species, which has since been termed *Rh. Grasiana* by M. d'Orbigny; the last-named shell seems to differ from the true *R. Martini* by its greater breadth and gibbosity, as well as by the frontal margin of the ventral valve greatly indenting that of the dorsal one. Mr. S. P. Woodward seems inclined to consider *R. Grasiana* as the adult state of *R. Martini* an opinion which may perhaps prove to be correct, but which I do not yet consider sufficiently demonstrated, from never observing among the numerous examples of *R. Martini*, found at Hamsey and Folkstone, specimens presenting the characters assigned to *R. Grasiana:* it may, therefore, for the present, be desirable to describe both separately; but if future observers should decide on the two being considered as one, then M. d'Orbigny's name will require to be placed as the synonym, on account of Mantell's priority, and it is but just to observe that, while proposing his name, *R. Grasiana*, the distinguished French Palæontologist did not omit to remark that, "*perhaps his species is the* T. pisum *of Sow., but which he was unable to affirm, on account of the differences which he remarks between his specimens and those figured by Sowerby*" ('Pal. Franc.' Ter. Cret., vol. iv., p. 38); but, although fully admitting the difficulty, still specimens of the true *R. Martini*, perfectly agreeing with the figures published by Sowerby of *T. pisum*, occur in France, and have been collected more than once, both by M. Bouchard and myself, at Cap Blanc Nez, near Calais.

[1] Page 131, "*Ter. Martini*, subscrotiform, longitudinally striated, margin finely serrated; both valves slightly depressed in front, beaks very small. This is a minute and delicate species, scarcely 0·3 inch either in length or width; each valve is marked with upwards of 30 longitudinal striæ, and both equally convex. The margin is finely serrated by the terminations of the striæ, and is nearly straight in front, the sides are not waved, as in the last species (*T. sulcata*), named after W. Martin. Locality—Hamsey."

Plate XII, fig. 15.   A typical example (fig. 15 *a, b,* enlarged) from the Grey Chalk in the vicinity of Folkstone, I avail myself of this occasion to thank Mr. Mackie for the opportunity he has kindly afforded me in the examination of an extensive series of this and other species from his locality.

„   fig. 16.   A specimen from the Chalk Detritus of Charing (Kent), in the collection of Mr. Harris:—fig. 16ᵃ is an enlarged illustration, to show how the plaits augment at times by intercalation; this is, however, an extreme case, as in the generality of specimens the plaits appear more regular.

„   fig. 16ᵈ.   A specimen from the Upper Green Sand of Chute, near Warminster; from the cabinet of Mr. Cunnington.

44. Rhynchonella Grasiana, *D'Orbigny.*   Plate XII, figs. 17, 19.

Rhynchonella Grasiana, *D'Orbigny.* Pal. Franç., Ter. Crétacés, vol. iv, p. 497, figs. 7—11, 1847.

Diagnosis.—Shell transversely oval, somewhat obtusely five-sided; slightly indented in front. Valves unequally convex, the *dorsal* or socket one commonly the deepest, without a produced prominent mesial fold. In the *ventral valve* a wide longitudinal sinus of moderate depth, extends from near the centre of the valve to the front, where the margin indents considerably that of the opposite valve; beak short, acute; foramen small, and entirely surrounded by the large tubular expansions of the deltidium; a flattened space exists between the beak ridges and hinge line, which last slightly indents the lateral margins of the umbo. Externally each valve is ornamented by from 46 to 56 small plaits, at times augmented by the intercalation of smaller ones at various distances from the beak and umbo. Length 5, width 5, depth 3 lines.

Obs. We need not repeat the observations relative to this form noticed under *R. Martini* (Mantell); but it would appear that some French examples have attained larger dimensions than any hitherto obtained from our British localities; thus a well formed specimen from the Basse Alpes (for which I am indebted to M. d'Orbigny), measures, length 7, width 7, depth 5 lines; and I have seen a few even exceeding those measurements. The author of the 'Pal. Franc.' justly observes, that certain examples of this species somewhat resemble *R. Cuvieri* (D'Orb.) in their external contour, but may be distinguished by the greater number of plaits, position of the foramen, and less sinuous margins.

R. Grasiana abounds in the Upper Green Sand of Chute, near Warminster, in Ferruginous Beds of the same age near Clifton Hampden (Mr. Sharpe's collection), and in the Chloritic beds of Chardstock. In France, it is found in the Upper Green Sand near Havre (Seine Inf.), and in the neighbourhood of Grasse.

Plate XII, fig. 17. A specimen, natural size, from the Upper Green Sand of Chute Farm, near Warminster.

„   fig. 18. Another example, greatly enlarged.

„   fig. 19. A specimen from the same locality, in the British Museum.

„   fig. 19^d. Magnified illustration, to show the large development of the tubular expansions of the deltidium.

„   fig. 19^e A very transverse and fine example from the Upper Green Sand, near Warminster, in the collection of Mr. Sharpe.

45. RHYNCHONELLA PARVIROSTRIS, *Sow.* Sp.   Plate XII, figs. 13—14.

TEREBRATULA PARVIROSTRIS, *Sowerby, in Fitton.* Trans. Geol. Soc., vol. iv, 2d ser., pl. xiv, fig. 13, 1836.

—   —   *Morris.* Catalogue, 1843.

—   —   *Forbes.* Cat. of Lower Green Sand Fossils, Quart. Journ. Geol. Soc., vol. i, p. 345, 1845.

Diagnosis. Shell imperfectly tetrahedral, wider than long; valves unequally convex; beak narrow, slightly incurved, tapering rapidly to an acute point; foramen rather small, contiguous to the hinge line, but entirely surrounded by a deltidium; beak ridges well defined, a flattened space existing between them and the hinge line. Near to the extremity of the beak commences a deep longitudinal sinus, which extends to the front; the middle portion of the valve presents a somewhat concave curve, from the plaits being bent upwards near their extremities. The *dorsal* or socket valve is more inflated than the opposite one, with a produced mesial fold, which rises rapidly to within a short distance of the front, where it bends downwards to meet the edge of the sinus of the ventral valve. Externally, from 35 to 40 plaits ornament each valve, 8 to 10 of these forming the mesial fold and sinus. Dimensions variable: length 8, width 10, depth 6½ lines.

Obs. Adult individuals of this species seem well distinguished from other Cretaceous *Rhynchonellæ* by the peculiar bend of their valves and plaits; the hinge line is likewise less oblique, and more obtuse than what is seen in the generality of species; young examples are, however, much depressed, and not always readily to be distinguished from specimens of *R. latissima*, Sow., of a similar age.

R. parvirostris is found in the Lower Green Sand of Shanklin (Isle of Wight), where it has been obtained by many Palæontologists.

In 1845, Professor Forbes considered *R. parvirostris* to be a variety of *R. sulcata* (Parkinson); but I could not trace a sufficient resemblance to admit the conclusion arrived at by that distinguished Palæontologist.

Plate XII, fig. 13. An adult individual, from the Lower Green Sand of the Isle of Wight.

„   fig. 14. A younger shell, from the same bed and locality.

46. RHYNCHONELLA GIBBSIANA, *Sow*. Sp.   Plate XII, figs. 11, 12.

TEREBRATULA GIBBSIANA, *Sowerby*.   Min. Con., vol. vi, p. 72, tab. 537, fig. 4, 1829.
— GIBBSII, *Woodward*.   A Synoptical Table of British Organic Remains, p. 21, 1830.
— GIBBSIANA, *Morris*.   Catalogue, 1843.

Diagnosis. Shell somewhat obtusely triangular, generally wider than long, with a moderately developed mesial fold in the *dorsal* valve, to which corresponds a sinus in the *ventral* one; beak acute, not much produced or incurved; foramen rather small, and surrounded by a deltidium; beak ridges sharply defined, with a flattened space between them and the hinge line, margins sinuous, the frontal edge of the ventral valve greatly indenting that of the opposite one. Each valve is exteriorly ornamented by from 45 to 50 small delicate plaits, 10 or 12 of these occupying the mesial fold and sinus. Dimensions variable, the greatest depth near the umbo; length $7\frac{1}{2}$, width 9, depth 5 lines;

„ 8 „ 8, „ $4\frac{1}{2}$ „

Obs. M. d'Orbigny considers this shell to be a variety of *R. sulcata*, but I do not feel prepared to admit that conclusion; *R. Gibbsiana* appears to me to be a much more triangular shell, with a deeper sinus, and externally ornamented by smaller or more delicate plaits; it seems peculiar to the *Lower Green Sand*, of Sandgate, Hythe, Pluckley, Peasmarsh, as well as at Sandown and Atherfield, Isle of Wight, (collection of Mr. S. Saxby). Some exceptional specimens found at Hythe bear resemblance to *R. latissima*, Sow.

The author of the 'Pal. Française,' moreover states, that the reference to this species published by Professor Forbes in his 'Catalogue of the Lower Green Sand Fossils' is incorrect; but on what grounds this assertion is made, I am at a loss to understand; it would, on the contrary, appear to me, that the French author had himself erroneously identified our British type, which he refers to his TERRAIN ALBIAN, which is not the age of our fossil in the vicinity of Folkstone, or other British localities.

Professor Bronn seems to be still further from the mark, while stating, in his 'Index Palæontologicus,' that *R. Gibbsiana* is nothing more than a variety or synonyme of *R. plicatilis*, Sow.!

Plate XII, fig. 11.   A specimen from the *Lower Green Sand* of Sandown, Isle of Wight.
„   fig. 12.   A rather enlarged example, from the Lower Green Sand, at Pluckley, in the collection of Mr. Harris.
„   fig. 12ᶜ   Another specimen, nat. size, from Hythe.

47. RHYNCHONELLA LINEOLATA, *Phillips*, Sp.   Plate XII, figs. 6—10.

TEREBRATULA LINEOLATA, *Phillips*.   Geol. of Yorksh., vol. i, p. 178, pl. ii, fig. 27, 1835.
— LINEOLATA, *Morris*.   Catalogue, 1843.

TEREBRATULA SUBLINEARIS, *Münster?* MS. (Cambridge Museum.) [1]
— LINEOLATA, *Bronn.* Index Pal., p. 1240, 1848.
— — *D'Orbigny.* Prodrome, vol. ii, p. 120, 1850.

Diagnosis. Shell ovate, more or less elongated; unequally convex and flattened; *ventral* valve commonly the deepest, with a slight and shallow longitudinal sinus; beak acute, moderately incurved; foramen small, circular, and surrounded by a deltidium. The *dorsal* valve is regularly convex to within a short distance of the front, where the surface exhibits either two or more plaits, with a mesial depression along the middle; externally, each valve is ornamented by numerous minute longitudinal striæ, which sometimes dichotomise near their extremities, or unite towards the front and lateral margins, forming a series of larger ribs; the lateral margins are but slightly flexuous, the frontal edge of the ventral valve indenting more or less that of the dorsal one.

Dimensions variable: length 8, width 7½, depth 6 lines;
„ 4, „ 4, „ 3 „

Obs. This remarkable species was insufficiently figured and named (but not described) by Professor Phillips, from an unusually large example (Pl. XII, fig. 6,) obtained by Mr. Bean, in the Specton Clay of Knapton, Yorkshire. The original type, still in the possession of Mr. Bean, was kindly forwarded for my examination, and I at once perceived that it belonged to the genus *Rhynchonella;* the external surface is entirely covered by delicate longitudinal striæ, 7 or 8 occupying the breadth of a line.

[1] Both Münster and Goldfuss have given many MS. names to species of Brachiopoda in the shape of catalogues, as well as to specimens in the Museum of Bonn; but these can claim no right to priority, never having been described nor figured, but in some cases having been adopted and referred to by Palæontologists. It may not prove devoid of interest to add some particulars kindly communicated by Dr. F. Rœmer, of Bonn.

In a catalogue intitled *Verzeichniss der Versteinerungen welche in der Kreis-Naturalien-Sammlung zu Bayreuth Vorhanden sind Bayreuth im September,* 1833, 8vo, pp. 115, Count Münster names the following species, the majority of which are only MS. denominations:

P. 44. *Ter. septemplicata,* from the Jurassic Limestone of Ebermanstadt.
P. 45. *Ter. striato-plicata,* from the Jurassic Limestone of Streitberg.
P. 46. *Ter. alaria,* from the ferruginous Oolite of Rabenstein and Thurnau.
P. 47. *Ter. canaliculata,* from the Jurassic Limestone of Würgan.
— *Ter. pentaëdra,* from the Jurassic Limestone of Oberfellendorf.
P. 48. *Ter. nana,* from the Jurassic Limestone of Streitberg.
— *Orbicula? dubia,* and *O.? semilunaris,* from Streitberg and Obermönchau.
— *Crania? paradoxa,* ibidem; *C.? pileus,* and *C. obscura,* Streitberg.
P. 73. *Ter. pentagona, subovoides, angularis,*
P. 74. *Ter. striato-plicata, semiplicata, subdecussata, quadrifida, Delthyris acuticosta,* } from the Lias of
Del. speciosa, } Franconia.
P. 101. *Ter. Schlotheimii, sublata, subelongata,* from the Productus Limestone of Regnitzlosau.
— *Ter. coiculam,* from the Dev. Limestone of Elbersreuth.
— *Ter. gracilis, reflecta,* ibidem.
P. 102. *Ter. subcrumena,* from the Dev. Rocks of Geroldsgrün.
— *Atrypa dubia, subcurvata,* from the Dev. Rocks near Hoff, in Franconia.
— *Atrypa glabra, A. rugosa,* from the Productus Limestone of Regnitzlosau.
— *Gypidia pelargonata,* ibidem.
P. 103. *Delthyris alata, Lept. polymorpha,* from the Productus Limestone near Hoff.
P. 104. *Lept. concentrica, aculeata, linearis, setosa, subrucosa, speluncaris,* from the Productus Limestone of Regnitzlosau.

GOLDFUSS added a long list of names of Brachiopoda (many of which can be only made out by com-

In the *Upper Green Sand* of Cambridge, a dwarf race of the same species has been plentifully collected by Mr. Carter, and appears to agree in all essential characters with the Speeton Clay specimen, except in dimensions. Mr. Carter considers the shell denominated *Ter. sublinearis*, by Count Münster, in the Cambridge Museum, to be specifically distinct from the one found in the Cambridge Upper Green Sand.

I collected the same shell in the *Tourtia*, near Tournay, in Belgium, it is exactly similar, but a little larger than those commonly met with at Cambridge; the last are also comparatively much more coarsely striated than the single example I have seen from Knapton, where the species would appear to be very rare.

Plate XII, fig. 6. The original type specimen from the Speeton Clay of Knapton.

„   fig. 6^{cde}. Are enlarged illustrations.

„   figs. 7, 8, 9. Specimens of a dwarf race from the *Upper Green Sand* of Cambridge, in the collection of Mr. Carter. These illustrations are enlarged, the vertical line indicates the natural size.

ARGIOPE.

In p. 16 of the present Monograph, will be found described as *Argiope decemcostata*, Rœmer, Sp., a shell common to the Chalk of England, France, Belgium, Prussia, &c.; but after the publication of the first portion of this monograph, I was informed by M. de Hagenow, that he still doubted the shells I had figured (Pl. III, figs. 1, 13), as

paring the originals in the Museum of Bonn), in the *Handbuch der Geognosie*, von H. T. de la Beche: *nach der zweiten Auslage des Englischen originals bearbeitet*, von H. von Dechen, Berlin, 1832. These are:

P. 382. *Ter. impressa*, from the white Jurassic Marls at Hohenzollern, Stufrnberg Urach.

P. 523. *Lept. convoluta, furcata, capillata, minuta*, from the Dev. Limestone of the Eifel.

— *Lept. striata, pectinata*, from the Dev. Greywacke of Coblentz.

— *Lept. loevis*, from the Carb. Limestone of Visé.

P. 524. *Lept. corrugata, considea*, ibidem.

P. 525. *Orthis radiata*, Eifel; *O. costata*, Kentucky; *O. granulosa*, Catskill Mountains; *O. fasciculata*, Eifel; *O. nodosa*, Eifel; *O. undulata*, Albany.

— *Delthyris microptera*, Eifel, Glocestershire, Herefordshire; *D. compressa (D. triangularis*, Sow.), Bansberg, Derbyshire; *D. heteroclyta (Calceola heteroclyta*, Def.), Eifel; *D. macroptera*, Eifel, Catskyll Mountains.

P. 526. *Delt. ceptoptera*, Eifel; *D. pachyoptera*, N. York; *D. dorsata*, Straberg; *D. bisulcata*, Vise; *D. canalifera (Ter. aperturata*, Schl.; *T. canalifera*, Sow.), Bensberg; *D. canaliculata*, Bensberg; *D. polymorpha*, Ratingen; *D. incisa*, Ratingen, Moskau; *D. symmetrica*, England; *D. curvata (Ter. curvatus*, Schl.), Eifel; *D. biplicata, D. radiata, D. thecaria*, Ratingen; *D. striatula (Ter. striatulus*, Schl.), Eifel.

P. 527. *D. vestita (Ter. vestitas*, Schl.), Ratingen; *D. concentrica*, ibid.; *D. imbricata (T. imbricata*, Sow.) ibid.; *Gypidia gryphoides (Uncites Gryphus*, Def.), Paffrath; *G. loevis*, ibid.; *Strigocephalus striatus*, Eifel; *Atrypa nitida*, Lake Simcoe.

P. 528. *Ter. triloba, T. canaliculata, T. quinquelatera, T. dichotoma, T. pentagona, T. Wahlembergii, T. subglobosa, T. bifida, T. clavata*, from the Eifel; *Ter. Dalmanni*, Carb. Limestone of Ratingen; *T. Anygdale*, Eifel; *T. complicata*, Dev. Rocks of Irelohn.

The only Brachiopoda described and figured by Goldfuss are those published in his great work, 'Petrefacta Germaniæ.'

belonging to the Essen species; that they were identical with those he had published in 1842, under the names of *Orthis (Argiope) Bronnii* and *O. (Argiope) Buchii*. This view may probably prove correct, and I may have been led into error from having placed too much value on the fact, that many individuals of the Chalk species possess ten costæ, and thus so far agreeing with Rœmer's Green Sand species. In 1852, I was acquainted with no other British species or specimens of *Argiope* lower down than the Soft or Upper Chalk, but since that period, I have been able to examine several others from the *Lower Chalk* of Kent and *Upper Green Sands* of Warminster and Cambridge. It will therefore be desirable to reconsider our British Cretaceous Argiopes.

48. ARGIOPE MEGATREMA, *Sowerby*, Sp.   Plate XII, figs. 31—32, and 34—36.

TEREBRATULA MEGATREMA, *Sowerby, in Fitton.* Trans. Geol. Soc., vol. iv, 2d series,
                       p. 343, pl. xviii, fig. 3 (read in 1827, printed in 1836.)
   —   —   *Morris.* Catalogue, 1836.
   —   DECEMCOSTATA, *Rœmer.* Vers. Nord. Kreid, p. 41, tab. 13, 1840.
   —   MEGATREMA, *Bronn.* Index Pal., p. 1241, 1848.
   —   —   *D'Orbigny.* Prodrome, vol. ii, p. 172, 1850.
   —   —   A Cat. of the Terebratulæ in the British Museum, p. 56, 1853.
   —   DECEMCOSTATA, *Dr. Fr. Rœmer.* Die Kreid, Westphalans, p. 71, 1854.

Diagnosis. Shell transversely obovate, or obtusely pentagonal; valves almost equally and moderately convex, deepest near the umbo; beak produced, nearly straight, truncated by an oblique and large foramen; beak ridges defined, leaving between them and the hinge line a more or less developed triangular area; hinge line as long or shorter than the greatest width of the shell. The external surface of each valve is ornamented by from ten to twelve ribs, with flattened interspaces between. The costæ correspond in each valve, a few smaller ones being intercalated between the larger ones near the frontal margin. In the interior of the *dorsal valve*, the loop first fixed to the base of the dental sockets is folded into two lobes, and attached to a single mesial plate.[1] Length $2\frac{1}{2}$, width 3, depth $1\frac{1}{2}$ lines.

Obs. So rare is this little shell, that for many years I was unable to trace a single example; and it was only during a recent visit to the Bristol Institution Museum, that I had the good fortune to discover in that collection one example, which proved, on examination, to agree with the shell described and figured by Sowerby, in the 'Geological Transactions:' it is a true *Argiope*, not a *Terebratula*, as generally supposed. Rœmer's figure of *A. decemcostata*, Pl. XII., fig. 35, and another, fig. 36, furnished by M. de Hagenow, might at first sight indicate a distinct species on account of the length of the hinge line and area, but having recently received from Dr. Fr. Rœmer a specimen collected

---

[1] The interior of this species has been admirably described and figured by M. Suess, in his excellent memoir entitled 'Ueber die Brachial-Vorrichtung bei den Thecideen,' pl. iii, fig. 1 (Aus dem Decemberhefte des Jahrganges, 1853) der Sitzungsberichte der mathem-naturw. Classe der kais Akademie der Wissenschaften (xi Bd., s. 991), besonders abgedruckt.

by himself at Essen, I became convinced that this long hinge line was exceptional, and that the true characters of the species are similar to those assumed by *A. megatrema*, Sow., or intermediate between figs. 32 and 34, five or six examples of the last having been discovered by Mr. Carter in the Upper Green Sand near Cambridge.

Plate XII, fig. 31.   The original figure of *Terebratula Megatrema*, Sowerby, from the 'Geological Trans.,'[1] vol. x.—Fig. 31ᵃ. An enlarged illustration.

„   fig. 32.   A specimen from the Upper Green Sand of Warminster, in the Bristol Institution Museum.—Fig. 32ᵃᵇ. Enlarged figures.

„   fig. 34.   A British example, from the Upper Green Sand of Cambridge, in the collection of Mr. Carter.—34ᵃᵇᶜ, enlarged. This specimen shows the intercalated ribs.

„   fig. 35.   Rœmer's published figure of *Ter. decemcostata* considerably enlarged, from the Green Sand of Essex, and introduced here to facilitate comparison.

„   fig. 36.   Another example from the same locality, drawn by M. de Hagenow.—36ᵃ, enlarged illustration.

ARGIOPE BRONNII, *De Hagenow*, Sp.   Plate III, figs. 1—13, and Plate XII, figs. 37, 38.

(Described as ARGIOPE DECEMCOSTATA, *Rœmer*, in p. 16 of the present Monograph.)

ORTHIS BRONNII, *V. Hag.*   Neuv. Jahrb. F. Mineral, pl. ix, fig. 7, 1842.
—   BUCHII, *V. Hag.*   Ib., pl. ix, fig. 8, 1842.
TEREBRATULA DUVALII, *Dav.*   London Geol. Journal, p. 113, pl. xviii, figs. 15—18, 1847.
MÉGATHYRIS CUNEIFORMIS, *D'Orb.*   Pal. Franç., Ter. Crétacés, vol. iv, p. 147, pl. 521, figs. 1—11, 1847.

Having already fully described this species, I will simply remark that it is found in the Upper Chalk of Northfleet, Gravesend, Meudon (France), and in Prussia.   It possesses in general fewer ribs than the Upper Green Sand species, and its dorsal valve is likewise less convex.   M. de Hagenow having kindly presented me with specimens of both his *A. Bronnii* and *A. Buchii*, I was able to convince myself that they belong to a single species, identical in shape and character with those found in England and France; and to facilitate comparisons, I have reproduced, in Plate XII, the original published figures of M. de Hagenow's two species.

Plate XII, fig. 33.   Illustrates the only specimen of *Argiope* hitherto discovered in the *Lower Chalk* of Kent (British Museum).

„   fig. 33ᵃ.   A magnified illustration of the same.

[1] Sowerby describes his species, as follows:—*Ter. megatrema*: moderately convex, transversely obovate, with a few distinct ribs.   The beak is large and produced with a very large perforation, whence the name.

49. CRANIA CENOMANENSIS, *D'Orbigny*. Plate XII, figs. 40, 41.

CRANIA CENOMANENSIS, *D'Orbigny*. Pal. Franç., Ter. Crétacés, vol. iv, p. 138, pl. 524,
figs. 1—4, 1847.

— — *Sharpe*. Quart. Journ. of the Geol. Soc., vol. x, p. 193, 1853.

Diagnosis. Shell unsymmetrical, transversely oval: *lower* or *ventral* valve thick, and almost flat, attached to marine objects by a large portion of its external surface; interiorly a raised margin surrounds the shell, four principal muscular impressions occupy the posterior half of the inner disk, the posterior adductor pair are large and oval, slightly produced, and placed obliquely close to each other and to the cardinal edge. The anterior adductor impressions are almost approximate at their base, and situated close to the centre of the shell, with their outer extremities directed upwards, so that a lozenge-shaped depression remains between the four large impressions above described; towards the centre of the shell there likewise exists a small elongated projection, the remaining portion of the inner disk exhibiting distinct imprints of the vascular system. The *dorsal* or upper valve is thin, conical, or patelliform; the vertex sub-central, rough externally. The interior is deep, with a thin concave border, which fits upon and over the raised margin of the opposite valve, a small inner ridge surrounds the shell at a short distance from the edge. The posterior adductor scars are oval and widely separate; the anterior pair are placed near the centre of the valve, and in contact at their base, with their outer extremities directed upwards and towards the cardinal angles of the valve.

Dimensions variable: length 4, width 5;

,, 4½, ,, 7, height 2 lines.

Obs. The only British specimens of this species I have been able to examine were discovered by Mr. Sharpe in the Upper Green Sand or gravel, of Farringdon, and belong to one upper and lower valve of two different individuals. The ventral valve (Pl. XII, fig. 40) is very flat, with a serpula covering a portion of its outer surface, and this is likewise the first example of the attached valve hitherto discovered, M. d'Orbigny being only acquainted with the dorsal one (Pl. XII, fig. 41), with which Mr. Sharpe's specimen perfectly agrees; I must, therefore, dissent with the last-named gentleman, who considers his upper valve to belong perhaps to the large chalk *Crania*, found at Ciply, in Belgium, termed *C. Parisiensis* by Mr. Sharpe,[1] but which appears to me specifically distinct from the species which bears that name.

Plate XII, fig. 40. *Lower valve* from Farringdon, in the collection of Mr. Sharpe.

,, fig. 40ª. Enlarged illustration.

,, fig. 41. *Upper valve* from the same locality and collection; 41ª, enlarged.

[1] Quart. Journal Geol. Soc., vol. x, p. 192, 30th Nov., 1853.

TABLE ILLUSTRATING THE GEOLOGICAL DISTRIBUTION OF THE BRITISH CRETACEOUS BRACHIOPODA.

Genus or Sub-genus. Species.	Author.	Date.	Reference to the Plates and Figures in this Monograph. (Part II.)	Lower Green Sand.	Speeton Clay.	Gault.	Red Chalk.	Upper Green Sand.	Farringdon Sponge Gravel.	Chlor. Marl or Chalk with Silic. grains.	Lower Chalk or Chalk Marl.	Chalk.
Lingula truncata	Sowerby	1827	pl. i, figs. 27, 28, and 31	*								
„ sub-ovalis[1]	Davidson	1852	pl. i, figs. 29, 30					*				
Crania Parisiensis	Defrance	1819	pl. i, figs. 1—7									*
„ Ignabergensis[2]	Retzius	1781	pl. i, figs. 8—14								*	*
„ cenomanensis	D'Orbigny	1847	pl. xii, figs. 40—41							*		
Thecidium Wetherellii[3]	Morris	1851	pl. i, figs. 15—26; & pl. xii, fig. 39							*	?	*
Argiope megatrema	Sowerby	1836	pl. xii, figs. 31—36						*		?	
„ Bronnii[4]	De Hagenow	1842	pl. iii, figs. 1—13; & pl. xii, figs. 37, 38								?	
Magas pumila	Sowerby	1816	pl. ii, figs. 1—10 and 23?							*		*
Terebratella Menardi[5]	Lamarck	1819	pl. iii, figs. 34—42							*		
„ pectita	Sowerby	1818	pl. iii, figs. 29—33							*		
Trigonosemus elegans	Kœnig	1825	pl. iv, figs. 1—4							*		
„ incertum[6]	Davidson	1852	pl. iv, fig. 5							*		
Terebrirostra lyra	Sowerby	1818	pl. iii, figs. 17—28							*		
Megerlia lima[7]	Defrance	1828	pl. iv, figs. 15—28; and pl. v, figs. 1—4						*		*	*
Terebratulina striata	Wahlenberg	1821	pl. ii, figs. 18—28			*?			*		*	*
„ gracilis, and	Schlotheim	1813	pl. ii, figs. 13—16			?	*				*	*
var. rigida	Sowerby	1829	pl. ii, fig. 17				*					
Terebratula? capillata?[8]	D'Archiac	1847	pl. v, fig. 12						*			

[1] This species has been found by Mr. S. H. Saxby, in the Blue Rag of the Upper Green Sand series, at Bonchurch (Isle of Wight).

[2] Erroneously spelt Egnabergensis, in p. 11. This is *Numulus minor* of Stobœus, 1732.

[3] Several ventral or attached valves of a *Thecidium*, considered to belong to *T. Wetherelli* (Pl. XII, fig. 39), have been collected by Messrs. Sharpe, Wright, Morris, and myself, in the Sponge Gravel of Farringdon. They have been found adhering to valves of *Ter. depressa* (Lamarck), to the *Actinopora papyracea, Manon Farringdonensis*, &c. Specimens with both valves are found in the Chalk of Brighton.

[4] This shell has (according to M. de Hagenow) been erroneously described and figured in this Monograph, under the name of *Arg. decemcostata* (p. 16, and Pl. III, figs. 1—13), but corrected again in p. 102. Although I do not yet consider the question finally settled.

[5] Several examples of this shell have been found by Mr. Wiest, in the Chloritic bed of Chardstock, along with *T. pectita;* the last-named species has likewise been found by Mr. S. H. Saxby, in the *Chloritic Marl* of the Isle of Wight.

[6] In 1852, only one example was known, but since that period several others have been discovered in the neighbourhood of Chardstock, by Mr. Wiest.

[7] In p. 40, I proposed to established a *Sub-genus* KINGINA for the reception of this and other similarly organised forms, based upon certain modifications, in the shape and attachment of the apophysary skeleton or loop; but from its having been subsequently observed that these differences were not *essentially* distinct from what we find in Professor King's sub-genus *Megerlia*, I am ready to abandon my former view, as it is *always incumbent to simplify the nomenclature*, when it can be achieved without serious effects.

[8] M. de Koninck has informed me, that this species was named *Spondylus undulatus*, by Geinitz, in 1839, and figured in the 'Die Vers., von Kieslingswalda,' pl. vi, fig. 8, 1843. I have seen the original example, now at Liege, in M. de Koninck's collection, and it is certainly the same shell as subsequently described by the name of *T. capillata*, by Viscount d'Archiac.

Genus or Sub-genus.	Species.	Author.	Date.	Reference to the Plates and Figures in this Monograph. (Part II.)	Lower Green Sand.	Speeton Clay.	Gault.	Red Chalk.	Upper Green Sand.	Farringdon Sponge Gravel.	Chlor. Marl or Chalk with Silic. grains.	Lower Chalk or Chalk Marl.	Chalk.
Terebratula? ovata		Sowerby	1812	pl. iv, figs. 6—13 .	·	·				*	*		
,,	? rugulosa	Morris	1847	pl. iv, fig. 14 .	·	·					*		
,,	? squamosa	Mantell	1822	pl. v, figs. 5—11	·	·				*	*	*	
,,	? oblonga	Sowerby	1826	pl. ii, figs. 29—32 .	*				*	*	*	*	
,,	obesa	Sowerby	1823	pl. v, figs. 13—16						*		*	*
,,	biplicata	Brocchi	1814	pl. vi, figs. 1—49; and pl. ix, figs. 37 and 40?	?	*	*	*	*	*		?	
,,	prælonga	Sowerby	1836	pl. vii, figs. 1, 2 .	*								
,,	sella	Sowerby	1823	pl. vii, figs. 4—10 .	*	·	*						
,,	Tornacensis? var. Roemeri	D'Archiac	1847	pl. vii, figs. 11—16; and pl. ix, figs. 1—8 .	·	·				*			
,,	sulcifera	Morris	1847	pl. vii, figs. 17—20	·	·				·		*	
,,	semiglobosa and varieties	Sowerby	1812	pl. viii, figs. 6—18 .	·	·			*		*	*	*
,,	carnea	Sowerby	1812	pl. viii. figs. 1—5 .	·	·				·	·		*
,,	Robertoni	D'Archiac	1847	pl. ix, fig. 25 .	·	·				*			
,,	depressa	Lamarck	1819	pl. ix, figs. 9—24 .	·	·				*			
,,	Carteri	Davidson	1854	pl. vii, fig. 3 .	·	·				*			
Waldheimea celtica		Morris	1853	pl. ix, figs. 32—35 .	*	·							
,,	tamarindus	Sowerby	1836	pl. ix, figs. 26—31 .	*	·							
Rhynchonella plicatilis		Sowerby	1818	pl. x, figs. 37—42 .	·	·						*	*
,,	var. octoplicata	Sowerby	1818	pl. x, figs. 1—17	·	·						*	*
,,	var. Woodwardii	Davidson	1854	pl. x, figs. 43—46 .	·	·							*
,,	limbata[1]	Schlotheim	1813	pl. xii, figs. 1—5	·	·							*
,,	compressa	Lamarck	1819	pl. xi, figs. 1—5; and pl. xii, fig. 25	·	·				*	*		
,,	latissima	Sowerby	1825	pl. xi, figs. 6—22; and pl. xii, fig. 24	·	·				*	*		
,,	depressa	Sowerby	1825	pl. xi, figs. 28—32; and pl. xii, fig. 26	·	·			*	*			
,,	var. A	Davidson	—	pl. xii, fig. 30	·	·				*	*		
,,	var. B	Davidson	—	pl. xii, fig. 28	·	·				*	*		
,,	sulcata	Parkinson	1811	pl. x, figs. 18—36 .	·	·	*	*	*				
,,	Mantelliana	Sowerby	1826	pl. xii, figs. 20—23	·	·			·			*	
,,	Cuvieri	D'Orbigny	1847	pl. x, figs. 50—54 .	·	·					*		
,,	nuciformis	Sowerby	1825	pl. xi, figs. 23—27 .	·	·			*	*			
,,	Martini	Mantell	1822	pl. xii, figs. 14—16 .	·	·				?	*	·	
,,	Grasiana	D'Orbigny	1847	pl. xii, figs. 17—19 .	·	·				*	·	*	
,,	parvirostris	Sowerby	1836	pl. xii, figs. 13, 14 .	*	·				*	·		
,,	Gibbsiana	Sowerby	1826	pl. xii, figs. 11, 12 .	*	·							
,,	lineolata	Phillips	1836	pl. xii, fig. 6 .	·	*							
,,	var.	—	—	pl. xii, figs. 7—10	·	·			·	*			

[1] This is the *sub-plicata* of Mantell, and it may still remain a question whether it should not merely constitute a var. of *R. plicatilis* or *octoplicata*, Sowerby.

SUPPLEMENTARY OBSERVATIONS ON THE STRATIGRAPHICAL DISTRIBUTION OF THE SPECIES.

The general results in the present Monograph differ from those published by several distinguished contemporaneous authors; it will therefore be necessary to explain the cause of this apparent difference of opinion.

The most recent Catalogue of *British Cretaceous Brachiopoda*, is that published in 1847, by Mr. Tennant,[1] in which forty-nine specific names have been enumerated; but a critical examination has led me to place about twenty[2] of these, either among the synonyms of species already mentioned, or as hitherto undiscovered in Great Britain; so that the entire list published by the above-mentioned author would not exceed some *thirty species*.

It has been stated in the fifth volume of the 'Histoire des Progrès de la Geologie' (p. 109, 1851), that fifty-two species of Brachiopoda have been recorded as existing in *British Cretaceous Strata*: the learned author mentioning at the same time that his results and identifications are chiefly based on those already published in Mr. Morris's 'Catalogue of British Fossils' (1843).

[1] 'A Stratigraphical List of British Fossils,' p. 47, 1847. In January, 1854 (the period at which my Table had been completed, see 'Bull. Soc. Geol. de France,' vol. xi), the new edition of Mr. Morris's 'Catalogue' had not appeared.

[2] These are:

1. *Crania ovalis,* Woodward = *C. Ignabergensis,* Retzius.
2.   ,,   *spinulosa,* Neilsson (not hitherto found in England).
3.   ,,   *striata,* Sowerby = *C. Ignabergensis.*
4. *Lingula ovalis* (a Jurassic shell).
5. *Orbicula lævigata,* Deshayes (not a Brachiopod).
6. *Terebratula brevirostris,* Roemer = *Rhynchonella Martini,* Mantell.
7. *Magas truncata,* Sowerby = var. of *M. pumila,* Sowerby.
8. *Terebratula crysalis,* Schlotheim — young of *T. striata,* Wahlenberg.
9.   ,,   *convexa,* Sowerby = *R.* (*lata* or) *latissima,* Sowerby.
10.   ,,   *dilatata,* Sowerby = *R. compressa,* (var.) Lamarck.
11.   ,,   *dimidiata,* Sowerby = *R. compressa,* (var.) Lamarck.
12.   ,,   *elegans,* Sowerby = *R. latissima,* Sowerby.
13. *Terebratula elongata,* Sowerby = *T. carnea,* Sow.
14.   ,,   *faba,* Sowerby = (dwarf) *T. biplicata,* Brocchi.
15.   ,,   *obliqua,* Sowerby (not Cretaceous).
16.   ,,   *octoplicata,* Sowerby = var. of *R. plicatilis,* Sowerby.
17.   ,,   *pentagonalis,* Phillips = var. of *T. striata,* Wahl.?
18.   ,,   *quadrata,* Sowerby = *T. oblonga,* Sowerby.
19.   ,,   *rigida,* Sowerby = var. of *T. gracilis,* Schlotheim.
20.   ,,   *rostrata,* Sow. (a Jurassic shell).
21.   ,,   *striatula,* Mantell = *Ter. striata,* Wahlenberg.
22.   ,,   *subrotunda,* Sowerby = var. of *T. semiglobosa,* Sow.
23.   ,,   *subundata,* Sowerby = var. of *T. semiglobosa,* Sow.

Viscount d'Archiac has accordingly presented us with the following Table:

Distribution of the Class according to Viscount d'Archiac.	Number of the Genera.	Number of the Species.	Upper Chalk (*Craie Blanche*).	Lower Chalk and Chalk Marl (*Craie Tufeau*).	Upper Green Sand.	Gault.	Lower Green Sand.	Green Sand of Devonshire.	Species common to the *Upper Chalk* and to the *Chalk Marl*.	Species common to the *Upper Chalk* and to the *Upper Green Sand*.	Species common to the *Chalk Marl* and to the *Upper Green Sand*.	Species common to the *Upper Green Sand* and to the *Gault*.	Species common to the *Gault* and *Lower Green Sand*.	Species common to the *Gault*, and to the *Lower Green Sand*.	Species common to the *Upper Green Sand* and to the *Lower Green Sand*.
BRACHIOPODA . .	5	52	18	11	10	5	22	5	5	—	2	—	1	2	3

And I now place into corresponding Columns the results arrived at during the publication of the present Monograph.

According to Mr. Davidson. BRACHIOPODA . .	Genera or Sub-Gen.		Upper Chalk	Lower Chalk	Upper Green Sand	Gault	Lower Green Sand								
	13	about 49	14 or 15	15	28 or 30	7 or 8	8	—	7	6	8	6	1	2	2 or 3

Although fifty-two species are recorded in the French author's table, and only forty-nine in our own, still in reality the last number greatly exceeds that presented by the Viscount, because, at least twenty-two to twenty-four of his names are synonyms, while my list contains a number of species new to England, and mentioned in no other publication. But I must at the same time hasten to announce, that notwithstanding all the care, researches, and consultations undertaken in the identifications of the species, I have not always arrived at results which can be considered finite, and possibly forty-five good species may comprise all that have been hitherto discovered in *British Cretaceous Strata*.

Much additional investigation will likewise be required before the exact stratigraphical repartition of certain forms can be definitely established; and to arrive at this most important geological desideratum, it will be necessary to settle in a definite manner the *comparative age* of certain beds above the Gault in the Isle of Wight, at Cambridge, Farringdon, Chardstock, and in a few other localities.

Thus, according to the generality of British geologists, the *Farringdon Sponge Gravel* would belong to the age of the *Lower Green Sand;* by myself, to that of the *Upper Green Sand* or *Tourtia:*[1] and by Mr. Sharpe, more modern than the Chalk, or in other words, to the *Upper Maestricht* (*Cretaceous*) *beds* and Pisolitic Limestone of Laversine (France).

[1] See 'Bull. de la Soc. Geol. de France,' vol. xi, Feb. 1854. I believe this view is likewise sanctioned by Mr. S. P. Woodward.

And as the stratum in question contains some twelve or more species of Brachiopoda, by casting these into one or other deposit, the number of forms peculiar to each is necessarily materially modified.   This will in great measure explain why in my table only eight species are recorded from the *Lower Green Sand*, while there would exist twenty-two, according to Viscount d'Archiac and others.

I mentioned in 1852 (page 2 of this Monograph), "*that the age of the Farringdon beds may yet afford a subject of discussion, although several geologists state them to be Lower Green Sand.*"   Since then, Mr. Sharpe has renounced his share in the views entertained by Messrs. Austen, Forbes, and others,[1] and has lately published a very interesting memoir,[2] wherein he exposes his present opinion, which is chiefly founded upon the examination of 111 species he had been able to assemble from that celebrated locality. But as these results are not in accordance with those of other geologists, and differ likewise with my own, I will endeavour in a few short observations to explain wherein we disagree.

Mr. Sharpe records his palæontological inferences in the following table:

	Sponges.	Bryozoa.	Brachiopoda.	Lamelli-branchiata.	Echinoder-mata.	Sundries.	Total.
Species peculiar to the deposit . . .	4	1	1?	3	4	1	14
Maestricht Sands	2	3	4	3	—	—	12
Upper Chalk	1	20	3	5	—	3	32
Lower Chalk	1?	2	3	2	—	4	12
Upper Green Sand, including Tourtia . .	11	17	11	—	—	3	50
Gault	—	1	1?	2	—	1	5
Lower Green Sand[3]	—	4	2	3	3	1	13
Total number of Species examined . . .	16	44	19	18	7	7	111

[1] Quarterly Journal of the Geol. Soc., vol. vi, p. 454, 1850.

[2] Ib., vol. x, p. 176, Nov., 1853.

[3] Mr. Sharpe mentions the following Lower Green Sand species as occurring in the Farringdon Gravels:

"*Reptomulticava micropora*.

        ,,   *collis*, also found in the White Chalk.

"*Heteropora cryptopora*, also found in the Maestricht Sand.

"*Proloscina marginata*.

"*Terebratula tamarindus*.

        ,,   *oblonga*, also found in the Upper Green Sand.

"*Ostrea macroptera*, also in the Gault and Upper Green Sand, and perhaps in the Chalk.

"*Pecten Dutemplii*, also in the Upper Green Sand.

"*Pecten interstriatus*, perhaps identical with *P. Dutemplii*.

"*Nerpula quinque-angulata*.

And further observes, "that the examination of the whole list of species, or that of the families separately, gives the results, that the Farringdon Gravels contain species hitherto thought characteristic of every bed, from the Lower Green Sand to the Maestricht Sand inclusive; but that those referable to species found elsewhere, above the Gault, predominate nearly ten to one in the number of species, and still more so in that of individuals; so that we need only consider to what part of the Cretaceous series above the Gault this deposit belongs.

"This conclusion limits our choice to the *Upper Green Sand*, or to a place altogether *above the Chalk*; for no one could seriously propose to place it on a level with the Chalk. . . . . . . That nowhere is there any trace of gravel nor any ferruginous bed in the Upper Green Sand of this part of England. Very few of the organic remains of Farringdon which are referable to the Upper Green Sand are found in that deposit in this neighbourhood; Warminster being the nearest spot which affords any large numbers of these species, and then only in the uppermost bed of the formation; but for the counterparts of the greater number we must travel to the Tourtia of Belgium, or to Essen in Westphalia.[1] That it might lead to erroneous results if we drew our conclusions from the Bryozoa, ... their geological range in this country being little known, but the remaining classes furnish safe grounds of comparison: of these, the Farringdon Gravel contains thirty-three species, found either in the *Upper Green Sand*, the *Tourtia*, or in the *Craie Chloritée* of France; but of these thirty-three species, only thirteen are known in the Upper Green Sand of England, and most of them range upwards into higher strata." The author concludes "that we are driven step by step, by the exhaustion of all other alternatives, to class the Farringdon Sponge Gravel as more modern than the Chalk (but not in the Tertiaries); other relics of the Upper Cretaceous deposits being found in the Limestones of Faxoe, the Calcareous Sands and Sandstones of Maestricht, Ciply, and the *Pisolitic Limestones* of Laversine and Vigny."

The attentive examination of Mr. Sharpe's table of species will, in my humble opinion,

---

"*Salenia punctata*, of Atherfield.

"*Goniopygus peltatus*, of Switzerland.

"*Diadem dubium*,   ,,

Besides these, I picked up in the locality, an Urchin, which Dr. Wright states to be unquestionably the *Nucleolites Neocomiensis*, Ag. It is a little larger than the type specimens, and much larger than the French ones; but its characters are so well marked, that it cannot be mistaken.

Several Oolitic Fossils have been long since detected in the Farringdon Gravel Beds, but which Mr. Sharpe justly regards as strangers brought in the fossil state to the locality. There are Coral Rag and Kimmeridge Clay Fossils, over which the Farringdon Gravel appears to extend.

[1] In the fourth volume of the new edition of Bronn's 'Lethea. Geog.,' p. 25, &c., the Green Sand of Essen is considered to be parallel to the Tourtia and Chalk Marl? of Sussex. See also Dr. Fr. Roemer, Memoir, *die Kreid Westphalens*, 1854. Wherein the following Brachiopoda are mentioned, several of which are common to our English Upper Green Sand under other names: *Thecidea digitata*, Bronn.; *Th. hippocrepis*, Goldf.; *Th. hieroglyphica*, Goldf.; *Ter. gallina*, Brong.; *T. latissima*, Sow.; *T. paucicosta*, Roemer; *T. nuciformis*, Sow.; *T. Beaumonti*, d'Arch; *T. auriculata*, R.; *T. radians*, R.; *T. nerviensis*, d'Archiac; *T. Tornacensis*, d'Arch.; *T. pectoralis*, R.; *T. arcuata*, R.; *T. canaliculata*, R.; *T. decemcostata*, R.

go far to bias many in favour of the view here taken, viz., that the *Farringdon Sponge Sand* and *Gravel* has greater claims to the age of the *Upper Green Sand* than to any other hitherto mentioned, for out of the 111 species, fifty are allowed to be forms of that period, besides fourteen stated to be peculiar to the locality. Any one who visits the quarries of Little Coxwell, will arrive at the conclusion, that the *Sponges* and the *Brachiopoda* (*almost all Upper Green Sand* or *Tourtia* forms) are the *truly abundant and characteristic fossils of the locality*, and that the sprinkling from other classes are *the rarities*, and should not therefore supersede the higher claims of the first. Among the Brachiopoda, two only, *Ter. tamarindus* and *T. oblonga*, are, properly speaking, *Lower Green Sand* fossils, but the last has also been found *as an exception* in the *Upper Green Sand*, near Warminster; and so very rare are these two species at Farringdon, that during two long days' search, I was unable to obtain a *single fragment*, nor do I believe that in all the collections, half-a-dozen examples could be assembled, while specimens of most of the other species may be collected by thousands.[1]

Mr. Sharpe publishes the following list of species:

1. *Terebratula depressa*, Lamarck.	11. *Terebratella Menardi*, Lamarck.
2. „ *nerviensis* (var. F), D'Archiac.	12. *Rhynchonella latissima*, Sowerby.
3. „ *Boubei*, D'Archiac.	13. „ *depressa*, „
4. „ *Roemeri*, „	14. „ *nuciformis*, „
5. „ *Keyserlingi*, „	15. „ *triangularis*, Wahlenberg.
6. „ *biplicata*, Sowerby.	16. „ *antidichotoma*, ? Buv.
7. „ *tamarindus*, „	17. *Crania cenomanensis*, D'Orbigny.
8. „ *oblonga*, „	18. „ *Parisiensis*, ? Defrance.
9. „ *revoluta*, D'Archiac.	19. *Thecidia Wetherellii*, Morris.
10. „ *Robertoni*, „	

Having also had the opportunity of examining several hundred specimens from the locality, in addition to all those assembled by Mr. Sharpe, and kindly placed by that gentleman at my disposal for publication, I am tempted to suggest a few alterations to Mr. Sharpe's list, having considered it a duty throughout this work to frankly express the results of my own investigations, which are also open to correction and criticism. Thus, according to my impression, the Brachiopoda hitherto obtained from Farringdon would belong to the following species:

1. *Crania cenomanensis*, D'Orbigny. *One upper valve* (*C. Parisiensis* of Mr. Sharpe's list), which perfectly agrees with the type specimens from the *Upper Green Sand* of Mans (France), described by M. D'Orbigny; one lower valve (referred by Mr. Sharpe to *C. cenomanensis*), but as it is the only *lower valve* hitherto discovered, I cannot so positively affirm that it belongs to D'Orbigny's species, although it probably is so.

---

[1] Generally in single valves; bivalve examples are less abundant. It was evidently a littoral deposit accumulated in water much agitated, the dislocation of the valves of the Brachiopoda, and fractured condition of the test of the cidaris, &c., as well as the rolled state of the gravel, attest sufficiently its formation.

2. *Thecidium Wetherellii*, Morris? (not abundant).   This identification is founded on the examination of several *ventral* or *dental valves* found adhering to Tereb. depressa (Lamarck), Briozoa, &c.   The characteristic or *dorsal valve* not having hitherto been discovered in the locality, the identification may still be considered incomplete, although there is every probability of its being the same as that described by Mr. Morris from the Chalk of Kent.

3. *Terebratella Menardi*, Lamarck, sp. (common).   This *Upper Green Sand* species has also been discovered (by Mr. Wiest) in the *Scaphites bed* at Chardstock, in company with other well-known *Upper Green Sand* Brachiopoda; but I differ with Mr. Sharpe in his assertion that Leymerie's var. *oblongata* ('Mém. Soc. Geol. France,' vol. v, pl. 15, fig. 12) is found at Farringdon. *T. Menardi* (*T. truncata*, Sow.), from Little Coxwell, is a less developed race, but perfectly agrees in character with the true Lamarckian type found at Mans (France).

4. *Terebratula biplicata* (Brocchi).   Not abundant, but exactly similar to those so common in the *Upper Green Sand* of Warminster, and the *Chloritic Marl* of the Isle of Wight.

5.   „   *Tornacensis*, var. *Rœmeri?* D'Archiac (very common).   Mr. Sharpe's views on this point differ somewhat from my own.   That gentleman considers Viscount d'Archiac's *T. Rœmeri* specifically distinct from the same author's *T. Tornacensis;* but I agree with M. d'Orbigny and others while stating that *T. Rœmeri* is only a difference of age or variety of *T. Tornacensis*, and am of opinion that the shells described by Mr. Sharpe as *T. Rœmeri*, from Farringdon, are nothing more than a dwarf race or variety of the full-grown *T. Tornacensis;* this view may perhaps prove erroneous, as well as that of my placing *T. Boubei, Keyserlingi*, and *revoluta* of Mr. Sharpe's list, among the varieties of *T. Tornacensis*.

6.   „   *depressa*, Lamarck (common; a well-known Tourtia species).   *T. Nerviensis*, var. *E* of Mr. Sharpe's list, appears to me to be nothing more than a variation due to age.

7.   „   *Robertoni*, D'Archiac (rare).   A Tourtia shell.

8.   „   *tamarindus*, Sow. *Very rare*, only hitherto known in Lower Green Sand.

9.   „   *oblonga*, Sow. Equally rare, and, properly speaking, a *Lower Green Sand* fossil.

10. *Rhynchonella latissima*, Sow. (abundant). A well-known *Upper Green Sand* shell, plentiful at Warminster and at Chardstock, also in the Tourtia, &c. *R. antidichotoma*, Buv., of Mr. Sharpe's list, is nothing more than a variety or accident in the plication of Sowerby's shell.

11.   „   *depressa*, Sow. This species has been likewise obtained from the *Upper Green Sand* near Warminster, Chardstock, the Isle of Wight, &c., and varies greatly in shape and number of the plaits.

It may, perhaps, be *Rh. triangularis* of Wahlenberg, but I feel unable to concur in the opinion expressed by Mr. Sharpe that both Wahlenberg's and Sowerby's species occur at Farringdon, if they be distinct.

12. *Rhynchonella nuciformis*, Sow. (not quite so common as *R. depressa*), is a well-known *Upper Green Sand* shell, and is found both at Chardstock and in the Isle of Wight, &c.

I have not, therefore, been able to recognise positively more than twelve out of Mr. Sharpe's nineteen species, but admit, at the same time, that there did exist among the specimens obtained some few valves which might admit of doubt; but the material in our possession was not sufficiently perfect to warrant a positive identification. No locality seems (palæontologically speaking) more anomalous than Farringdon, and from the problematic condition of its sands and gravel, is worthy of still further investigation.

The only well authenticated British *Lower Green Sand* species I have been able to examine, are :

1. *Lingula truncata*, Sowerby (p. 6).	5. *Terebratula celtica*, Morris (p. 73).
2. *Terebratula oblonga*, Sowerby (p. 51).	6. „ *sella*, Sowerby (p. 59).
3. „ *prælonga*, Sowerby (p. 58).	7. *Rhynchonella Gibbsiana*, Sowerby (p. 98).
4. „ *tamarindus*, Sowerby (p. 74).	8. „ *parvirostris*, Sowerby (p. 97).

But, besides these eight, imperfect examples of one or two doubtful shells were placed into my hands, which I was unable to identify.

In the Catalogue of *Lower Green Sand* fossils, published by Professor Forbes,[1] mention is not made of other species, if we except those quoted from Farringdon, and which have been found in no authenticated *Lower Green Sand* locality with which I am acquainted. Mr. S. H. Saxby, who has minutely explored the Cretaceous strata of the Isle of Wight, informs me (while placing his whole collection at my disposal for examination) that "Brachiopoda are by no means prominent in *Lower Green Sand* catalogues; indeed there are only two species that can be termed common, viz., *T. sella* and *R. parvirostris*, and these, especially the former, in such profusion as only Brachiopoda know how to lavish. It is, moreover, not a little remarkable that the *Cracker bed*, which furnishes the most

---

[1] 'Quarterly Journal of the Geol. Soc.,' vol. i, p. 345, 1845.

characteristic and beautiful fossils of the series, is totally barren of them; I never knew of the occurrence of the least trace of a Brachiopod for the first hundred and fifty feet as we ascend, except in the five or six feet of *Perna bed* at the base. In the upper beds, to the depth of more than 200 feet, nearly 300 perhaps, there is a great dearth of fossils of any description; I have, however, obtained a specimen of *Rh. Gibbsiana* from the uppermost portion. The nodular beds of Horseledge and Black Gang Chine present a very curious analogy with the Cracker nodules, in the repetition of ferruginous casts of those fossils which we saw in the latter as delicate shells, and which in very many instances do not seem to appear between these points in the series, a distance of perhaps 500 feet. Now, the *Horseledge* or *Yellow Ledge* beds contain *Lingula truncata, Ter. celtica, T. tamarindus, T. sella,* and *Rh. parvirostris,* while the Crackers are apparently deficient of these species."

It is, therefore, evident that the *Lower Green Sand* in England contains the *minimum,* and not the *maximum* of species.

THE GAULT and its dependencies dispose but of few species of Brachiopoda, and these no where numerically abundant, and, with the exception of *T. sella,* appear specifically different from those common to the Lower Green Sand.

The following is a list of the forms I have hitherto been able to examine:

1. *Terebratulina striata,* Wahl., var. *pentagonalis,* Phillips. One example from the Speeton Clay.
2.   ,,   *gracilis,* Schl., var. Common, according to Mr. C. B. Rose, in the *Blue Gault* of West Norfolk.
3. *Terebratula capillata?* D'Archiac (rare). Three or four examples from the *Red Chalk* of Hunstanton Cliffs (Norfolk).
4.   ,,   *biplicata,* var. Sow. Occasionally met with, but nowhere abundantly, in the *Speeton Clay, Gault,* and *Red Chalk.*
5.   ,,   *sella,* Sow. (very rare). In the *Gault* near Maidstone.
6.   ,,   *semiglobosa,* Sow., var. *subundata* (rare). In the *Red Chalk;*[1] and this appears to be the first appearance of the species which continues to be represented in each successive deposit up to the Chalk.
7. *Rhynchonella sulcata,* Parkinson. A few rare examples from the *Speeton Clay* and *Gault.*
8.   ,,   *lineolata,* Phillips, sp. A single specimen from the Speeton Clay of Yorkshire.

Over the GAULT, in natural succession, we arrive at a very variable series of sandy and marly beds, with occasional bands of limestone, known under the appellations of *Upper Green Sand, Chloritic Marl, Tourtia* and *Craie et Sables Chlorités* by the French; and it is in the direct succession or equivalents in different localities of the layers forming this series of beds, between the *Gault* and the *Chalk Marl,* that some further investigation

---

[1] Mr. C. B. Rose states that the *red bed,* (improperly called *Chalk*), occupies the place of the *Gault* lying immediately upon the Lower Green Sand; the fossils are similar to those met with in other *Gault* districts (see 'Ed. Phil. Journal,' Nov. and Dec. 1835, and Jan. 1836): at Hunstanton Cliff the *red bed* measures about 3 feet 10 inches.

is required: but viewing all the beds as one whole, we recognise therein from twenty-eight to thirty species, or the *maximum* of specific and numerical development of the Brachiopoda in the Cretaceous system in Great Britain.

In the Isle of Wight two well-separated beds may be traced above the Gault. First, the *Upper Green Sand*, from which Mr. S. H. Saxby procured *Lingula sub-ovalis* (Dav.), a Terebratula, probably *T. squamosa* (Mantell), *Ter. biplicata*, var. (Sow.), *Rh. nuciformis* (Sow.), and *R. depressa* (Sow.). The second bed is termed the *Chloritic Marl*, and from which the same gentleman obtained *Ter. biplicata* (Brocchi), *T. pectita* (Sow.), *Rh. Grasiana* (D'Orb.), *R. compressa* (Lamarck), and *R. Mantelliana* (Sow.). The fossils from both being identically similar to those so abundantly found in the neighbourhood of Warminster.[1]

One of the most interesting localities in England for the study of the beds under notice is the neighbourhood of Chard and Chardstock, and I am particularly indebted to Mr. J. Wiest for the following details.

"The formations about to be described are situated chiefly, if not exclusively, on each side of the valley of the Kit, and consist of the following beds taken in the descending order:"

I. LOWER CHALK without flints, from two to thirty feet in thickness at different places; its fossils are those common to that deposit and age.

II. CHALK MARL or *Discoidean stratum*, from two to three feet in thickness, representing a homogeneous mass, with fine siliceous and chloritic grains; it spreads out more equally than No. I, but does not appear to be present where the last attains a considerable thickness. *Am. Mantelli*, *Discoidea cylindrica*, and *Ananchites subglobosa*, are amongst its fossils.

III. The SCAPHITES BED, harder than I and II., and, before becoming exposed to air and damp, a compact accumulation of fossils, from three to nine inches in thickness; siliceous grains, more numerous and rough than in II, are interspersed in the mass. *Scaphytes*, *Nautilus triangularis*, Montfort,══*Fleuriausianus*, *N. lævigatus*, *Am. varians*, *A. obtectus*, and *Galerites*, are amongst its fossils.

IV. GREEN BED, near Chardstock, distinctly separated from III, forms a hard compact mass of rocks, with abundant siliceous and chloritic grains; from six inches to three feet in thickness, and containing the greatest variety of fine fossils. *Nucleolites Morrisii*, *N. lacunosus*, *Ter. lyra*, and other Upper Green Sand fossils characterise that deposit.

V. CRUSTACEAN STRATUM, less cemented than IV, with siliceous grains predominant; one to three feet in thickness. It contains a few *Terebratulæ*, *Pectens*, but chiefly remains of Crabs.

---

[1] Consult also the valuable 'Memoirs on the Geology of the Isle of Wight,' &c., by Dr. Fitton, Professor Forbes, and others, published in the 'Transactions and Quarterly Journal of the Geol. Soc. of London.'

VI. NAUTILUS LÆVIGATUS LAYER, a loose sand, nearly one foot in thickness, and containing but few fossils. *N. lævigatus* is also found in the Farringdon Sand and Gravel.

These details were accompanied by a numerous collection of all the Brachiopoda peculiar to the locality, and after minute examination, I have referred the individuals to the following species. The stratigraphical distribution is given on Mr. Wiest's authority.

	I.	II.	III.	IV.	V.
Terebratula semiglobosa, *Sowerby*	..	*	*	* ?	
" squamosa, *Mantell*	*	*	*		
" rugulosa, *Morris*	..	*	*		
" ovata, *Sowerby*	..	..	..	*	
Megerlia lima, *Defrance*	..	..	..	*	
Terebratella Menardi, *Lamarck*	..	..	*		
" pectita, *Sowerby*	..	..	..	*	*
Terebrirostra lyra, *Sowerby*	..	..	..	*	*
Trigonosemus incerta, *Davidson*	..	*	* ?		
Rhynchonella latissima, *Sowerby*	..	*	*	*	
" compressa, *Lamarck*	..	*	*	*	
" depressa, *Sowerby*, var. *A.* and *B.*	..	..	..	*	
" nuciformis, *Sowerby*	..	..	..	*	*
" Grasiana	..	..	..	*	*
" Cuvieri	..	*	*		

Chardstock is, therefore, a particularly interesting locality; it exhibits the British equivalents of the *Craie chloritée* of France and the *Tourtia* of the Belgians. II and III entirely agrees in mineral composition, external aspect, and palæontological contents, with the *Craie Chloritée* of the Mont St. Catherine, near Rouen (France), and in both localities we find *Ter. rugulosa*, *T. squamosa*, as well as the generality of fossils peculiar to that age. III constitutes by its fossils a natural passage into IV, which last contains the greatest number of species common to the *Upper Green Sand* of Warminster; Havre (France), and the Tourtia of Belgium; among the Brachiopoda we may mention *Ter. lyra*, *T. ovata*,[1] *T. pectita*, *Meg. lima*, *Rh. latissima*, *R. compressa*, *R. Grasiana*, *R. depressa*, and *R. nuciformis*.

In England, the neighbourhood of Warminster has yielded by far the greatest number of Upper Green Sand Brachiopoda, it contains—

[1] I have found *Ter. ovata*, *M. lima*, and *Rhyn. lineolata*, in the *Tourtia* near Tournay, species not recorded in Viscount D'Archiac's memoir.

1. *Lingula sub-ovalis* (rare).
2. *Argiope megatrema* (very rare).
3. *Terebratella pectita* (common).
4. *Terebrirostra lyra* (not abundant).
5. *Megerlia lima*.
6. *Terebratulina striata* (var.).
7. *Terebratula squamosa*.
8.   „   *ovata* (common).
9.   „   *obesa*.
10. *Terebratula biplicata* (very common).
11.   „   *oblonga* (very rare).
12. *Rhynchonella compressa*.
13.   „   *latissima* (common).
14.   „   *depressa*, var. *A* and *B*.
15.   „   *sulcata*.
16.   „   *Grasiana* (common).
17.   „   *Mantelliana*.

At Cambridge, immediately above the Gault, we find a bed of *Upper Green Sand*, from which Mr. Carter has obtained—

1. *Argiope megatrema*, Sow., = (*decemcostata*, Roemer) (rare).
2. *Megerlia lima* (not very common).
3. *Terebratulina gracilis*, var. *rigida*.
4. *Terebratula biplicata*, var. *Dutempleana* and var. *obtusa* (common).
5. *Rhynchonella sulcata* (common).
6.   „   *lineolata*.

Other similar British localities have yielded species, but no where different from those already enumerated. Nor have I yet been able to obtain any specimen of Brachiopoda from the celebrated Blackdown beds, although several species have been more than once erroneously mentioned as from that locality.[1]

From the CHALK MARL and LOWER CHALK we have obtained—

1. *Crania Ignabergensis* (rare).
2. *Magas pumila* (rare).
3. *Terebratula squamosa*.
4.   „   *obesa*.
5.   „   *biplicata* (rare).
6.   „   *sulcifera*.
7.   „   *semiglobosa* (common).
8.   „   *Carteri* (rare).
9. *Terebratulina striata*.
10.   „   *gracilis*.
11. *Argiope Bronnii* (1 example).
12. *Rhynchonella plicatilis*.
13.   „   *Mantelliana*.
14.   „   *Cuvieri*.
15.   „   *Martini*.

And from the CHALK WITH FLINTS—

1. *Crania Parisiensis*.
2.   „   *Ignabergensis*
3. *Thecidia Wetherellii*.
4. *Argiope Bronnii*.
5. *Magas pumila*.
6. *Trigonosemus elegans* (rare).
7. *Megerlia lima*.
8. *Terebratulina striata*.
9. *Terebratulina gracilis*.
10. *Terebratula obesa*.
11.   „   *carnea*.
12.   „   *semiglobosa*.
13. *Rhynchonella plicatilis*.
       var. *octoplicata*.
       var. *Woodwardii*.
14.   „   *limbata* (sub-plicata).

And perhaps one or two others which I was unable to determine, so that eight or nine species pass from *Lower* and *middle Chalk* into the *Chalk with flints*.

[1] Mr. Sharpe is of opinion that Blackdown Sands are older than the Upper Green Sand ('Quart. Journ. of the Geol. Soc.,' vol. x, p. 187, 1853).

For collecting the last-mentioned species, the following localities may be named. Norwich, Swaffham, Gravesend, Northfleet, Cambridge, Brighton, Lewes, Folkstone, &c.

France is infinitely richer than Great Britain in Cretaceous Brachiopoda, and this will be easily accounted for when we remember that the ETAGE NÉOCOMIEN (of which our *Lower Green Sand* only constitutes a small part) occupies a considerable portion of France, where it acquires vast thickness and importance, contains there likewise the maximum of Cretaceous species, and in it are found all those beautiful forms which materially help to make up the eighty-nine species or varieties which have been figured and described by M. d'Orbigny in the fourth volume of the 'Paléontologie Francaise, *Terrain Crétacées;*' but it is much to be regretted that the distinguished author had not been better acquainted with some of our British types, as a few of them have therein either received new names or been misunderstood. We possess, however, several forms which have not been as yet recorded in French catalogues, such as *Lingula sub-ovalis, Thecidium Wetherellii, Argiope megatrema, Trigonosemus incertum, Ter. capillata, T. Robertoni, T. depressa, T. Carteri,* and *Rh. lineolata.*

The following is the distribution of the French Cretaceous Brachiopoda according to M. d'Orbigny:

I. Etage Néocomien {	Inferieur ou Néocomien	22
	Superieur ou Ungonien . . .	7
II. Etage Aptien	. . . . . .	5
III. Etage Albien	. . . . . .	11
IV. Etage Cénomenien	. . . . . .	16
V. Etage Turonien	. . . . . .	7
VI. Etage Senonien	. . . . . .	23

But as three are repeated, the total amount of Cretaceous species or varieties known to M. d'Orbigny would be about . . . } 88

PLATE I.

CRETACEOUS SPECIES.

Fig.

1, 2^a.   Crania Parisiensis.   Interior of attached valve, enlarged.

2^b.   ,,   ,,   ,,   of upper valve   ,,

3.   ,,   ,,   A portion of the Interior of attached valve, much enlarged.

4, 5, 6.   ,,   ,,   Exterior of both valves.

7.   ,,   ,,   A group of Cranias (from Meudon).

8.   Crania Egnabergensis, natural size : exterior of upper valve.

8^a.   ,,   ,,   Interior of attached valve, enlarged.

8^b.   ,,   ,,   Interior of upper valve   ,,

8^c.   ,,   ,,   Exterior of attached valve : this specimen was only fixed by a small portion of the vertex; enlarged.

8^d.   ,,   ,,   A profile view of both valves, united and attached to a slender coral.

9.   ,,   ,,   Exterior of attached valve, enlarged, showing the coral to which it was fixed.

10, 11.   ,,   ,,   Young specimens, with few costæ.

12.   ,,   ,,   (*Crania ovalis* of Woodward), a malformation.

13, 13^{ab}.   ,,   ,,   A specimen, fixed by a considerable portion of its lower valve to a shell.

14.   ,,   ,,   Shows the valve not fixed by the whole surface, as in *C. Parisiensis*, figs. 5, 6, 7.

15.   Thecidea Wetherellii, nat. size, fixed by nearly all the surface of the attached valve *to an echinus.*

16.   ,,   ,,   Interior of the smaller valve, enlarged.

17.   ,,   ,,   ,,   larger valve   ,,

18.   ,,   ,,   Exterior, enlarged illustration.

19.   ,,   ,,   Profile view, enlarged.

20 to 26.   ,,   ,,   From the Chalk of Pewsey (Wilts). These different enlarged illustrations show the variable manner in which this species was fixed.

27.   Lingula truncata, nat. size.

28.   ,,   ,,   Profile view.

31.   ,,   ,,   As figured by Dr. Fitton, from Hythe.

29, 30.   ,,   subovalis, nat. size.

29^a.   ,,   ,,   Interior enlarged.

Plate I.

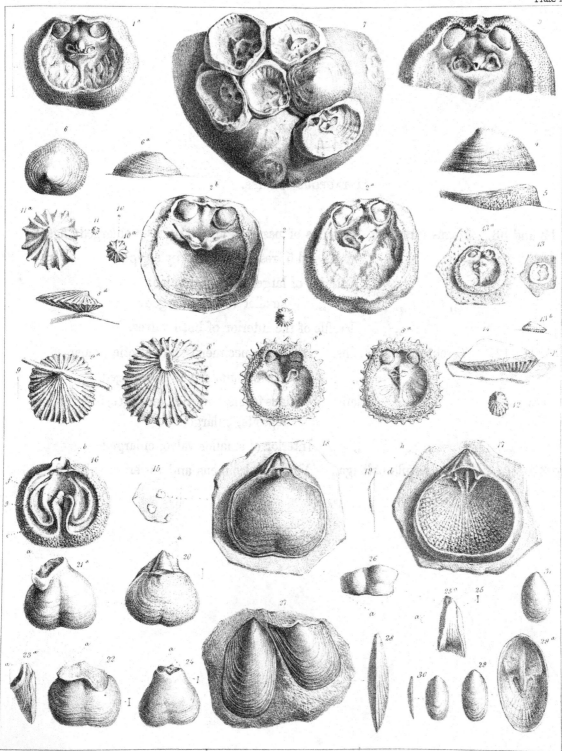

T.Davidson del et lith.

Printed by Hullmandel & Walton

PLATE II.

CRETACEOUS SPECIES.

Fig.

1 to 12 and 33.	Magas pumilus.	A series of passages and varieties—enlarged: figs. 2, 4, and 5, resemble Sowerby's type.
12^a.	,, ,,	Interior of larger or dental valve.
12^b.	,, ,,	,, smaller valve, enlarged.
12^c.	,, ,,	Profile of the interior of both valves.
13, 14, 16, 17.	Terebratulina gracilis.	Different specimens and varieties, enlarged.
15.	,, ,,	Interior of smaller valve, enlarged.
18 to 28.	,, striata.	Various forms, varieties, and ages, from different deposits, enlarged.
19^a.	,, ,,	Interior of smaller valve, enlarged.
29 to 32.	Terebratula oblonga.	Various specimens and varieties, enlarged.

Plate II.

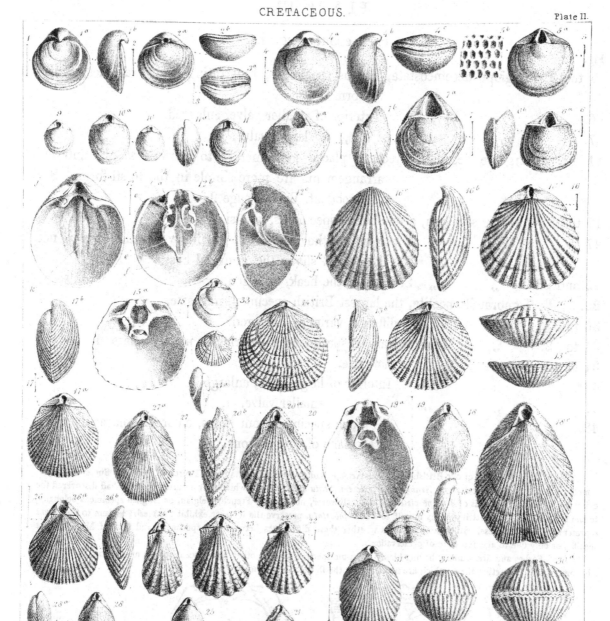

T. Davidson del et lith

Printed by Hullmandel & Walton

PLATE III.

CRETACEOUS SPECIES.

Fig.

1 to 13. Argiope decemcostata, various specimens, illustrating a few varieties of this form.[1]

6.   ,,   ,,   Interior of smaller valve, enlarged.

7.   ,,   ,,   ,,   larger valve,   ,,

14, 15. Argiope decollata, (recent,) placed here merely to show that the internal arrangements are more simple in the fossil form. See the woodcut, Part I, page 9.[2]

17 to 28. Terebrirostra lyra, a number of specimens, nat. size.

27.   ,,   ,, a portion of the interior of smaller valve, the loop is not known; *h* shows the produced boss.

25 and 28.   ,,   ,, Sections of the beak, enlarged.

29. Terebratella pectita, the largest British specimen, nat. size.

30 to 33.   ,,   ,, different varieties, enlarged.

34 to 42.   ,, Menardi, various specimens and varieties, mostly enlarged.

34 to 47.   ,,   ,, nat. size.

40ª.   ,,   ,, Interior of large valve, enlarged.

41.   ,,   ,,   ,, smaller valve, ,,

42.   ,,   ,, A French specimen from Mans, enlarged, placed here for the sake of comparisons.

[1] Since the description of our British Cretaceous Argiope had gone through the press, I received from Mr. Suess, of Vienna, a most interesting communication on *Argiope (Ter.) decemcostata* of Rœmer, in which that accurate observer had discovered the existence of a loop like that of *Argiope cistellula*, &c. However, none of my French, Belgian, or English specimens belonging to the form I figure, and attributed by myself to *A. decemcostata*, preserve the loop, or exhibit the mesial septum to the extent represented by Mr. Suess. It may be doubted whether these differences between my specimens and that of Mr. Suess are specific, or only due to age and state of preservation.

[2] Prof. Forbes and Mr. Cuming having kindly given me specimens of the other three recent species of *Argiope*, I am able to add some details respecting them, which are necessary for the illustration of the genus.

Argiope decollata.   *Argiope Neapolitana.*

I. *Argiope cuneata* (Risso), erroneously mentioned by me as a synonym with *A. decollata*, has only a single median septum; the lobes of the loop are free for one half their extent in the specimen examined, and blend with the shell, as we have noticed in some examples of *A. decollata.*

II. *Argiope Neapolitana* (Scacchi), *T. seminulum*, Phil. In this form the same longitudinal septum exists, but the loop was imperfect in the specimens at my disposal.

III. *Argiope cistellula* (J. Wood). In this species the same mesial septum coexists with a complete two-lobed loop, as represented in *A. Neapolitana*. The animal is preserved, and differs only from the figure given at Part I, in having two lobes instead of four. In justice to M. Philippi, I must state that his figure of the animal of *A. decollata* is essentially correct; but it is so small, that it has been overlooked or misunderstood by all succeeding writers. M. Philippi himself failed to discover the loop, and perceive the nature of the fringed arms which he described and figured.

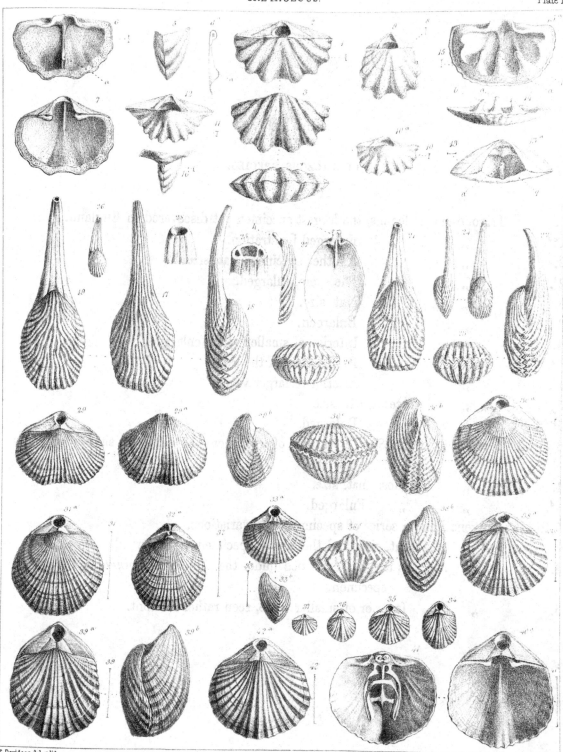

T. Davidson del. et lith.

Printed by Hullmandel & Walton.

PLATE IV.

CRETACEOUS SPECIES.

Fig.

1.   Trigonosemus elegans, the largest specimen yet discovered in England.

1^{cd}.   ,,   ,,   Enlarged illustration.

2.   ,,   ,,   Another specimen, nat. size.

2^{b}.   ,,   ,,   The same, enlarged.

3.   ,,   ,,   Nat. size.

3^{abc}.   ,,   ,,   Enlarged.

4.   ,,   ,,   Interior of smaller valve, enlarged.

4^{a}.   ,,   ,,   Profile view of the same.

4^{b}.   ,,   ,,   Interior of larger valve.

5.   ,,   incerta, nat. size.

5^{abc}.   ,,   ,,   Enlarged.

6 to 13. Terebratula ovata, *Sow.*, a variety of specimens of different shapes and ages, nat. size.

14.   ,,   rugulosa, nat. size.

14^{d}.   ,,   ,,   Enlarged.

15 to 28. Kingena lima, a series of specimens and variations.

22.   ,,   ,,   nat. size, and the largest specimen yet observed.

24.   ,,   ,,   is the var. described under the name of *T. spinulosa*, the original specimen.

25.   ,,   ,,   Interior of smaller valve, seen rather in front.

Plate IV.

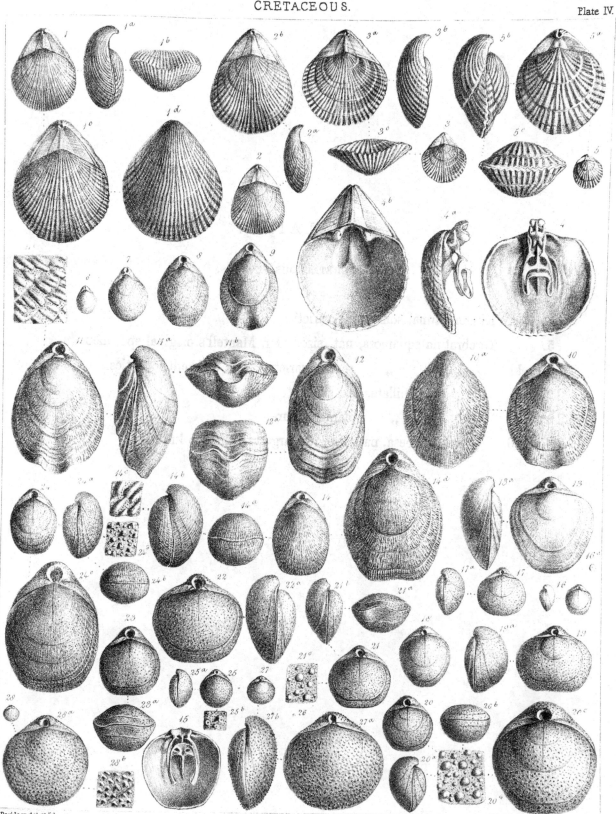

T.Davidson del et lith.

Printed by Hulmandel & Walton.

PLATE V.

CRETACEOUS SPECIES.

Fig.

1 to 4.   Kingena lima, some more varieties.

5.   Terebratula squamosa, nat. size.   Dr. Mantell's original specimen.

6 to 11.   ,,   ,,   several varieties from different localities.

12.   ,,   capillata, nat. size.

12ᵃ.   ,,   ,,   enlarged figure.

13 to 16.   ,,   obesa, nat. size, the largest specimens known.

Plate V.

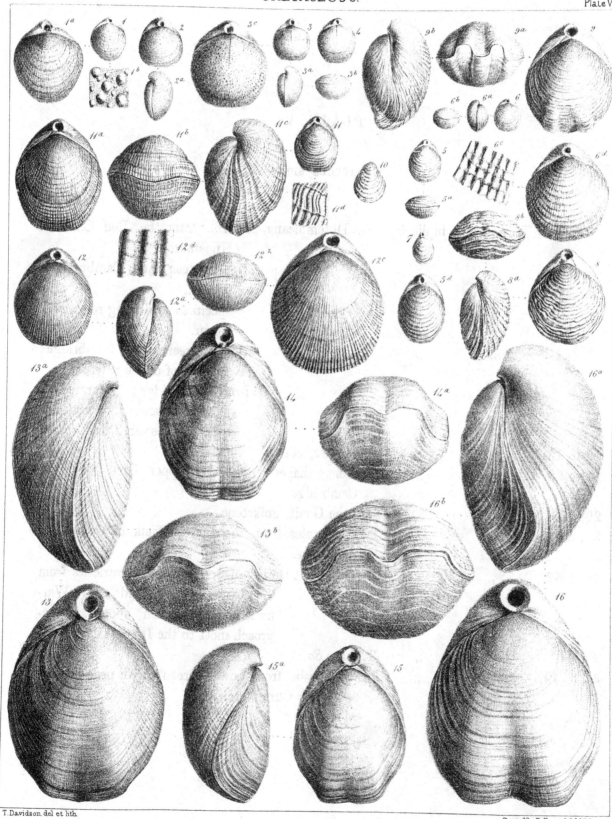

T. Davidson. del et lith.

Printed by Hullmandel & Walton.

PLATE VI.

CRETACEOUS SPECIES.

Fig.

1 to 2.   Terebratula biplicata, *var.* Dutempleana, *D'Orb.*, from the Red Chalk of Hunstanton.

3 to 5.   „   „   „   Upper Green Sand, Cambridge.

6.   „   „   „   „   „   , a specimen exhibiting remains of colour.

7.   „   „   „   Upper Green Sand, Cambridge; interior of dorsal valve.

8, 9.   „   „   „   from Cambridge; intermediate shape between *T. biplicata* (type) and the var. *Dutempleana*.

10 to 13.   „   „   *var.* obtusa, *Sow.*

14 to 28.   „   „   Different shapes from the Upper Green Sand of Cambridge.

29, 30.   „   „   from the Gault, Folkstone.

31, 32.   „   „   Two examples, from the Lower Chalk of Cambridge and Lewes.

33 to 42.   „   „   (*Brocchi*). A series of specimens and varieties from the Upper Green Sand, near Warminster, and which appear to approach most to the Italian figure.

43, 44.   „   „   (T. faba, *Sow.*)

45 to 49.   „   „   ? Stated to be from Lower Green Sand, near Devizes by Mr. Cunnington.

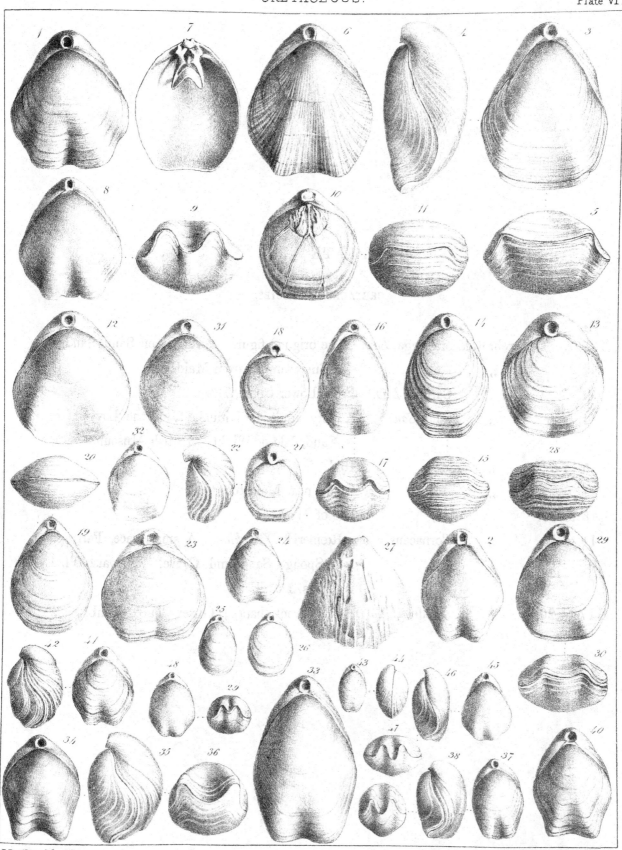

T. Davidson, del. et lith.

Printed by Hullmandel & Walton

PLATE VII.

CRETACEOUS SPECIES.

Fig.

1.   Terebratula prælonga, *Sow*.   The original figure, Lower Green Sand, Sandgate.

2.   „   „   A fine example, from Maidstone.

3.   „   Carteri, *Dav*.   From Lower Chalk, Dover.

4.   „   sella, *Sow*.   A very adult individual, from the Lower Green Sand, Isle of Wight.   British Museum.

5 to 6   „   „   Ibid.

7.   „   „   Gault, near Maidstone

8 to 10.   „   „   Isle of Wight.

11 to 16.   „   Tornacensis, *var*. Rœmeri ? *D'Archiac*.   A small race, Farringdon Sponge Sand and Gravel.   See also Pl. IX, figs. 1 to 8.

17 to 20.   „   sulcifera, *Morris*.   Different shapes; Lower Chalk, Cambridge.

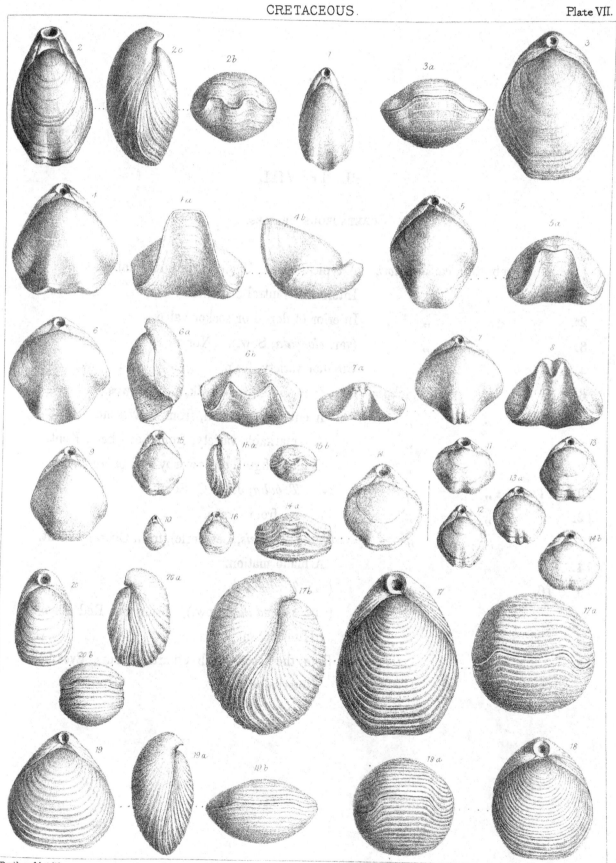

T.Davidson del et lith.

Ford & West Imp. Hatton Garden

PLATE VIII.

CRETACEOUS SPECIES.

Fig.

1. Terebratula carnea, *Sow.* A typical shape, Chalk, Trimmingham.

2.   ,,   ,, Interior of ventral or dental valve.

2ᵃ.   ,,   ,, Interior of dorsal or socket valve.

3.   ,,   ,, (var. *elongata*, Sow.). Norwich.

4 to 5.   ,,   ,, Circular variety, from   ,,

6.   ,, semi-globosa, *Sow.*, from the Lower Chalk, near Lewes.

7.   ,,   ,, A circular specimen, from Gravesend.

8.   ,,   ,, A remarkable variety, from Lewisham, Kent.

9.   ,,   ,, Original figure of Sowerby's *T. subrotunda.*

10 to 11.   ,,   ,, (var. *T. bulla*, J. de C. Sow.).

12.   ,,   ,, A var., from Grays.

13.   ,,   ,, (var. *T. albensis*, Leymerie), from Charing, Kent.

14.   ,,   ,, A malformation.

15 to 16.   ,,   ,, (var. *T. albensis*).

17.   ,,   ,, (var. *subundata*, Sow.), from the Red Chalk, Hunstanton.

18.   ,,   ,, from the Chalk with Quartz Grains, Evershot, Dorsetshire.

PLATE IX.

CRETACEOUS SPECIES.

Fig.

1 to 8. Terebratula Tornacensis, *var.* Rœmeri, *D'Archiac.* A dwarf race, from the Farringdon Sponge Sand and Gravel. See also Pl. VII, figs. 11—16.

9 to 24.   „   depressa, *Lamarck.* A series of shapes, age, and varieties, from the Farringdon Sponge Gravel. 15. Interior of the dorsal or socket valve.

25.   „   Robertoni, *D'Archiac.* Ibid.

26.   „   tamarindus, *Sow.* Lower Green Sand, near Sandgate.

27, 28.   „   „   Farringdon Sponge Sand and Gravel (rare):

29 to 31,   „   „   Different shapes, from the Isle of Wight.

32 to 34. Waldheimia Celtica, *Morris.* Different ages, from the Lower Green Sand, Isle of Wight.

35.   „   „   Interior of the dorsal or socket valve.

36. Terebratula Tornacensis? Farringdon.

37.   „   biplicata, *Brocchi.* Farringdon.

38.   „   ?   Farringdon; only two or three dorsal valves have been found.

39.   „   ?   perhaps, a var. of *T. sella*, from the Lower Green Sand, Maidstone.

40.   „   biplicata: from the highest bed of Lower Green Sand, Folkstone.

Note.—Figs. 36, 38, 39, and 40, are doubtful forms; the want of sufficient material has prevented more positive identification.

PLATE X.

Fig.

1 to 17.   Rhynchonella plicatilis, *var.* octoplicata, *Sow.* A series of specimens and varieties from the Chalk of Norwich.

18 to 20.   „   sulcata, *Parkinson.*   Upper Green Sand of Cambridge.

21, 22.   „   „   Speeton Clay, Yorkshire.

23 to 36.   „   „   Varieties and malformations, from Cambridge.

37 to 40.   „   plicatilis, *Sow.*   Chalk, Brighton.

41, 42.   „   „   Var. (Kent).

43 to 46.   „   „   (var. *Woodwardi*, Dav.).

47 to 49.   „   ?   Chalk, Charing.

50 to 54.   „   Cuvieri, *D'Orb.*   Lower Chalk.

Plate X.

T. Davidson, del. et lith.

Printed by Hullmandel & Walton.

PLATE XI.

Fig.

1.   Rhynchonella compressa, *Lamarck.* Upper Green Sand, near Warminster. (See also Pl. XII, fig. 25.)

2.   „   „   (*Ter. dilatata*, Sow.).

3.   „   „   A malformation viewed from the front.

4.   „   „   „   (*Ter. dimidiata*, Sow.), from Chard.

5.   „   „   „   from Warminster.

6 to 14.   „   latissima (lata) *Sow.* Different examples from the Upper Green Sand of Warminster. (See Plate XII, fig. 24.)

15 to 22.   „   „   Several varieties from Farringdon (*R. antidichotoma* ? Sharpe).

23 to 27.   „   nuciformis, *Sow.* Several examples from the Farringdon Sponge Gravel.

23[a b c].   „   „   Enlarged.

28 to 32.   „   depressa, *Sow.* Farringdon (fig. 32, *R. triangularis*, Sharpe). (See also Pl. XII, fig. 26.)

Pl. XI

T. Davidson del et lith.

Ford & West Imp

PLATE XII.

Fig.

1 to 5.   Rhynchonella limbata, *Schlotheim*, 1813 (*Ter. sub-plicata*, Mantell.   Different varieties, from the Chalk Norwich.   Figs. 4—5. Var. *lentiformis*, Woodward.

6^{ab}.   „   lincolata, *Phillips*.   Drawn from the orginal example, from the Specton Clay, Yorkshire.

6^{cd}.   „   „   Enlarged.

7 to 10.   „   „   Small variety or race.   Upper Green Sand, Cambridge.

11.   „   Gibbsiana, *Sow.*   Lower Green Sand, Isle of Wight

12^{a}.   „   „   Slightly enlarged, from Pluckley, Kent.

12^{c}.   „   „   Another example; nat. size.

13, 14.   „   parvirostris, *Sow.*   Figs. 13, adult; 14, young; from the Lower Green Sand, Isle of Wight.

15.   „   Martini, *Mantell* (= *Ter. pisum*, Sow.).   A typical shape, from the Gray Chalk, Folkstone.

15^{ab}.   „   „   Enlarged.

16^{ab}.   „   „   Chalk detritus, Charing.

16^{c}.   „   „   Enlarged.

16^{d}.   „   „   Upper Green Sand, Warminster.

17, 18.   „   Grasiana, *D'Orb.*   Upper Green Sand, Warminster; fig. 18 enlarged.

19^{abc}.   „   „   Warminster.

19^{d}.   „   „   Enlarged.

20, 21.   „   Mantelliana, *Sow.*   Lower Chalk, Folkstone.

22.   „   „   Upper Green Sand, near Warminster.

23.   „   „   ? Lower Chalk, near Lewes.

24.   „   latissima, *Sow.*   A very remarkable specimen, from the Chloritic Marls of Chardstock.

25.   „   compressa, *Lamarck.*   Ibid.

Th. Davidson del. et lith.
Ford. & West. Imp.

A MONOGRAPH

OF THE

BRITISH FOSSIL BRACHIOPODA.

PART III.
THE OOLITIC AND LIASIC BRACHIOPODA.

BY

THOMAS DAVIDSON,
MEMBER OF THE GEOLOGICAL SOCIETY OF FRANCE.

LONDON:
PRINTED FOR THE PALÆONTOGRAPHICAL SOCIETY.
1851—1852.

A MONOGRAPH

OF

BRITISH OOLITIC AND LIASIC BRACHIOPODA.

PRELIMINARY REMARKS.

UNDER the lowest member of the cretaceous system, we discover an extensive series of strata, consisting of different kinds of sands, sandstones, limestones, and clays, divided into a number of groups, known to geologists under the name of Jurassic, or Oolite and Lias Rocks. These form together a system of great thickness, traversing our island from the Yorkshire Coast to that of Dorsetshire in the shape of a band, varying in breadth from a few miles to between fifty and sixty in its midland course.

Great Britain is considered the typical country whence continental geologists obtained the first clue to those divisions which, with differences only in the nature of their composition, have been traced over many parts of our globe.

The object of the present Monograph is not to enter into the Geology of this system, which has been so ably illustrated by many celebrated British and Continental Geologists, but an endeavour to trace the forms or species of *Brachiopoda* which lived at different periods, while this vast amount of sediment was being deposited in the depths of the sea.

It may, however, be useful for the sake of reference, to give a tabular view of the different members of the Oolite and Lias series. We cannot do better than extract it from Mr. Tennant's 'Stratigraphical List of British Fossils:'

Potland Stone and Sand	Coarse oolitic shelly limestone; sometimes fine-grained or compact, thick-bedded, and with layers of chert, and with subordinate beds of sand.	Isle of Portland; Brill, &c., Aylesbury, Bucks; Thame, Oxon; Tisbury, Wilts.
Kimmeridge Clay . .	Dark blue and grayish laminated clay, with gypsum and bituminous shale.	Kimmeridge, Dorsetshire; near Oxford; Stone and Hartwell, Bucks; near Swindon.

B

Upper Calcareous Grit, Coral Rag, Lower Calcareous Grit .	Coarse shelly limestones, more or less thick-bedded; coarse oolitic limestones abounding in corals, calcareo-siliceous grit.	Headington, Oxon; West-brook, Calne, and Steeple Ashton, Wilts; Malton and Scarboro', Yorkshire.
Oxford Clay, Kelloway Rock . .	Dark blue clay, with *Septaria;* sometimes slaty and bituminous, with a subordinate band of ferruginous sandy limestone (Kel. rock).	Chippenham and Wooton Basset, Wilts; Oxford; Yorkshire, &c.
Cornbrash	Coarse rubbly limestone, thinly laminated with layers of clay.	Stanton, Malmsbury, Ash-ford, Wilts.
Forest Marble	Thinly-laminated shelly limestone, sand, and gritstone, with layers of clay.	Corsham, Box, &c., Wilts; Sapperton, Bradford, Ci-rencester.
Bradford Clay . . .	Layers of clay; sometimes alternating with bands of limestone.	Bradford, Burfield, Pickwick, Tetbury.
Great Oolite	Oolitic shelly limestone, more or less compact and sandy, sometimes thick bedded.	Bath, Bradford, Minchin-hampton Common (very fossiliferous).
On the Yorkshire Coast, the Great or Bath Oolite (*b*), (a hard, blue lime-stone; fine-grained Oolite; hard blueish clay,) is contained between two thick beds (*a, c*) of gritty laminated sandstones and shales, containing an abundance of terrestrial plants.		(*a*) Cayton and Gristhorpe Bays; (*b*) Cloughton and White Nab; (*c*) between Cloughton Wyke and Blue Wick.
Stonesfield Slate . . .	Oolite, shelly, and gritty limestone; slaty.	Stonesfield, Oxon; Seven-hampton Common, &c.
Fullers-earth Clay . .	. . . .	Bath, Box, near Stroud, and Hampton Common.
Inferior Oolite . . .	Two layers of coarse shelly ragstone, with inter-vening bands of marl and soft freestone. Fine-grained sandstone and ironstone.	Dundry, Painswick, Brins-combe, The Cotteswolds, Blue Wick, Yorkshire.
Lias	Alum shale; rubbly and sandy shales, &c. Lower Lias limestone and shales.	Whitby, Redcar, Yorkshire; Gloucestershire; Somerset-shire; Lyme Regis, &c.

These names and divisions are in general use, though each continental country has, in some cases, adopted synonymous denominations, which they always endeavour to attach to the British original type.[1]

The latest innovation is published by M. D'Orbigny, in a small elementary work, where he proposes the following classification :[2]

---

[1] For ample information, consult the works of W. Smith, Conybeare and Phillips, Sir C. Lyell, Sir H. De la Beche, Sir R. I. Murchison, Professor Phillips, Morris, Webster, Buckland, Lonsdale, Buckman, Young, and Bird, &c.; and the works of Elie de Beaumont, Dufresnoy, Baron Von Buch, Alcide D'Orbigny, Bronn, Pusch, Rœmer, D'Archiac, &c.

[2] 'Cours élémentaire de Paléontologie et de Géologie Stratigraphique,' par M. D'Orbigny. 1850.

Terrain Jurassique .
- étages Portlandien.
- „ Kimméridgien.
- „ Corrallien.
- „ Oxfordien.
- „ Callovien.
- „ Bathonien.
- „ Toarcien.
- „ Liasien.
- „ Senemurien.

We are not struck with the superiority of this newly-proposed arrangement over the one in common use, which we have adopted in the following pages, as it is never advisable to burden science with fresh names when old ones serve the same purpose. To the late Mr. Sowerby and other authors we are indebted for numerous correct descriptions of many species from the Oolitic and Liasic series; but the recent impulse given to the study of Palæontology has brought to light so vast an assemblage of new forms and better representatives of several known ones, that it has been deemed advisable to reconsider and publish, under the form of a series of Monographs, all the old species revised, annexing the new forms due to the exertions of local geologists, who have done so much to advance our knowledge of this and other classes, without whose assistance we never could hope to achieve the laudable ends proposed by the Palæontographical Society and their liberal and learned Secretary.[1] Within the last few years, attempts have been made with more or less success to throw light on the class of shells occupying us at present, through an endeavour to classify them by their general affinities. Our object is simply to treat of the Oolitic and Liasic species, comprising only a few genera, because we confine within more general limits as many characters, said to be generic, which we consider to be only specific or sectional differences, and which future discoveries will tend to generalise. How frequently have gaps existing in the zoological chain of affinities[2] been filled up of late years by the discovery of forms unknown to early writers, proving how boundless the field of research is in the kingdom of Nature, and how cautious we should be in assuming the uniform preva-

---

[1] It is therefore pleasing to me to express my grateful feelings and sincere thanks for the kind and liberal assistance I have received in the preparation of this work from MM. Bouchard, Deshayes, Deslongchamps, Buvignier, Milne Edwards, Valenciennes, De Verneuil, Delessert, De Koninck, Morriere, Chenu, &c.; from Messrs. Koïnig, Waterhouse, Woodward, of the British Museum; Sir H. De la Beche; Professor Forbes and Mr. Salter, of the Geological Survey; the Geological Society's Museum; that of Bristol, &c., where I was allowed to make use of the valuable collections deposited there; from Messrs. Bowerbank, Walton, Moore, Morris, J. de C. Sowerby, Hugh Miller, Robertson of Elgin, M'Coy, Rose of Swaffam, Riply of Whitby, Marder, Bean and Leckenby of Scarboro', Tennant, Faulkner of Deddington, Buckman, King, Dr. Wright of Cheltenham, Mr. Cunnington of Devizes, Messrs. Lycett, Lowe, Griesbach, and others, who have communicated to me the Palæontological treasures belonging to their respective collections.

[2] For various reasons we have commenced with the species of the Oolite and Lias; but it is of little importance, as this work is to be composed of a number of distinct monographs. Next year we hope to be able to give the general introduction, in which we shall state our views on the classification and genera.

lence of characters by which we isolate Families or Groups from the remainder of their class.

However much we may feel inclined to dispute the existence in Nature of genera and species, as an abstract proposition in the Philosophy of Zoology, the admission of these terms, and of the ideas we are accustomed to associate with them, are essential to the progress of science. This being the case, we ought at least to found genera in a uniform manner, and to equalize as much as possible the value of the characters upon which they are based. If, among the conchiferous molluscs, some trivial modifications in the position of the processes which protect their valves from dislocation be admitted, as of generic importance, we surely ought not in the Brachiopoda to place together shells of such different structure, as we remark in *T. Carnea* and *T. Concentrica*. M. Deshayes formerly admitted very few genera in the Brachiopoda; but from communications made by him to me, I believe he now considers that a larger number are required from our present more extended knowledge of the differences presented by the internal characters of the shell. The value of these aids in the discrimination of genera is strongly insisted upon by M. Deshayes in his great work upon the Paris Basin, where he states M. Blainville was one of the first to remark, that in the recent Terebratulæ the *Apophysary lamella* which support the soft parts, present peculiar forms in each species, and adds, "Ce serait donc par ce moyen que l'on pourrait déterminer rigoureusement les nombreuses espèces du genre."

For many years our researches have been bent upon the discovery of those internal characters; as we feel convinced that by them alone a rational and permanent classification will in time be arrived at. Through our exertions, and the help of many kind and zealous friends, we have been able to examine the internal structure of most of our British species which disposes us to admit the following genera among them:—

1. Lingula.	5. Leptæna.
2. Orbicula.	6. Spirifer.
3. Crania.	7. Terebratula.
4. Thecidea.	8. Terebratella.
	9. Rhynchonella.

Many more have of late been proposed among our Oolitic species, but which we hope to discuss in the general introduction, and shall merely here observe, that the greater or less length of a simply attached loop in Terebratula cannot be made use of as a generic character, especially when there exists no other distinctive points. The length of the loop may be used as sectional, round which we can group certain species; but who, with any degree of confidence, would place in distinct genera such shells as *Ter. Cornuta, quadrafida, numismalis, obovata, digona, ornithocephala,* &c., which have a simply-attached loop extending to near the frontal margin of the valve, and those such as *Ter. punctata, perovalis, Maxillata, intermedia sphæroidalis, Coarctata,* &c., the loop of which, simply attached to the crura, only extends to less than half the length of the valve, while in others, such as *Ter. Carnea, Dyphia,* &c.; the same process attached to the crura does not extend to more than a fifth of the length of the shell. We have in vain attempted to find out some external character by which we might discover if the species had a long or short loop; it had been imagined that those with a long loop were flatter, depressed in the smaller valve, and with lateral ridges of the beak strongly defined; but this character falls to the ground when we place in comparison such shells as *Ter. perovalis* and *punctata*, that have a short loop, with *Ter. ornithocephala, digona*, and others with a long one, all possessing a similar convexity of valves and other external characters. I have lately observed that the crura differs slightly in those specimens with a short or long process, as may be seen from the figures illustrating the interiors of *Ter. Cornuta, resupinata, ornithocephala*, &c., on one side, and those of *Ter. Maxillata, intermedia, sphæroidalis*, &c., on the other; and it will be perceived that in the elongated looped shells the crural base is much larger, and forms a flat surface, which does not exist in a similar manner in the others.

The punctures visible in Terebratulæ and other genera cannot be used as a general character to distinguish species, as the results of microscopical examination show us that their size and form varies on the same shell, according to the portion placed under examination; thus on one part we find the punctures widely separated becoming closer and larger towards the edge until they assume a completely different appearance; and although some species have a *marked* difference in this respect, others present no distinctions. Professor King, in his valuable 'Monograph on Permian Fossils,' goes too far, in our opinion, while stating that he believes few Brachiopoda unpunctuated, as very many shells in this class show no traces of punctures; but at the same time we believe that many species considered unpunctuated are really so. It had been supposed, till very lately, that the *Liasic* species of spirifers were characterised and distinguished by being punctuated, while other spirifers were not so; this opinion, must however, be abandoned, from the fact observed by Professor King, that many carboniferous spirifers, very different in shape and character, were likewise so, which has been also confirmed by M. De Koninck and others. Before entering into the subject of this Monograph, we think it necessary to state, that the confusion and contradictions we have found in authors were so great that it has in many cases been no easy matter to find out the original types. Many of Sowerby's and Lamarck's species, as well as those of other authors, have been singularly misrepresented, and in fact have been but little understood by the generality of Palæontologists and Geologists, who usually do not attach much importance to these determinations.

While reviewing the Lamarckian species of Terebratulæ, I was surprised to find in that celebrated collection many shells quite differing in species from those intended by the author of the 'Min. Con.;' and these errors seem so current abroad, that even the most distinguished Palæontologists, such as Von Buch, Bronn, D'Orbigny, and others, have placed in their catalogues, as Sowerby's species, shells, differing completely, as will be alluded to during the course of this work. Many of these mistakes are excusable, and might have been expected, from the unfortunate foreshortened position in which that author represented some of his shells. Through the kindness of Mr. J. de C. Sowerby, who in the most liberal manner placed the original collection of the M. C. at my disposal; and with the assistance of Mr. Waterhouse, we have endeavoured to clear up the original types, some of which were merely synonyms, such as *Ter. Triquetra, orbicularis, furcata, lata,* &c. &c.

The difficulty I have encountered, in my endeavours to place the different shells hitherto discovered in their proper places, is such, that I feel convinced numerous errors will have been committed, which I hope other observers and future discoveries may correct. No class of shells is more variable in their form, and the instances in which the former pass into each other are so numerous, that by some one or other character we might readily attach most species together. We often forget, that among the species of Brachiopoda there must have existed different races, as well as in those of other classes, arising from local conditions, and by the little attention paid to these circumstances species are made out of more varieties. It is certain that many species were ornamented by colour, such as we perceive among the recent shells; this colour, which would have been a great help, we are deprived of; its existence, however, is proved by accidental specimens found occasionally possessing traces of it,—we know that some Devonian species were spotted with red. *Ter. Hastata, Communis,* and others irregularly and longitudinally striped, probably also with red, as we see in *Bouchardia rosea;* and among the oolite species it is more than probable, from traces discovered by M. Deslongchamps, that *Rhynchonella spinosa* was of a red colour, as well as some of the cretaceous species. From the limited resources often left us by fossilization we are not always able to trace correct boundaries to species; but we should endeavour to do so where characters are sufficiently marked to be constant, and which may serve as conventional points of comparison and reference.

After many years of researches, M. Deslongchamps[1] published his views on the classification of the Brachiopoda found in Normandy—a paper I have long appreciated, but which, to my surprise, is little noticed by those who have written on a similar subject. In it that distinguished and conscientious author states, that after many different attempts to classify the Brachiopoda, he found the best outward characters to consist in the shape and position of the foramen; since which period Messrs. Morris, D'Orbigny, and others, have thrown much additional light on this point by the discovery that the form of the internal Apophysary system also bore intimate relations to the form and position of the beak, foramen, and deltidium. This, moreover, appears to be the best mode of proceeding where the internal characters are hidden from our view; but, while this character is of generic importance, it does not always help us to separate species. M. Deslongchamps, in his table, proposes to select a type and annex to it a number of allied forms, under the name of varieties: thus of *Ter. resupinata* he mentions twelve varieties, of *emarginata* fifteen, of *perovalis* eleven, of *Concinna* twenty, &c.; and he could have added many more had he operated upon a larger collection; and although I completely agree with M. Deslongchamps as to the reality of his varieties, still this system would, it is to be feared, lead to great confusion,

---

[1] 'Seances publiques de la Soc. Lin. de Normandie,' 1837, p. 30.

from the interminable list of perplexing varieties, impossible to be remembered. Several of these M. Deslongchamps admits, such as *Ter. Triquetra* and *Ornithocephala*, themselves only varieties of age and shape of the same species. We therefore think it preferable, for the sake of convenience, to admit artificially more species and less variety; and, in order to incorporate certain smaller variations, the description of a species should be so framed as to embrace the general idea of the thus limited form rather than the account of a single specimen.

Local causes and malformation, caused by pressure or fracture during the different stages of growth, are not to be wondered at in a class of shells destined in general to sedentary life, living and dying on the same spot attached to rocks, or in circumstances where there existed a want of room for their complete development. These malformations, alluded to by M. Deslongchamps, are much more common than generally imagined, and have often been made into distinct species by different authors, from the inspection of a single specimen. It is likewise certain, notwithstanding what may have been said to the contrary by some more learned Palæontologists, that the limits in vertical range of some species have extended to more than one group of strata, although species commonly are characteristic, and restricted to narrow bounds. We must also allude to some technical denominations made use of to denote the different parts of a shell; thus, for a considerable period, and in many important works, the term *dorsal* has been applied to the larger or perforated valve, and that of *ventral* to the smaller one. I do not wish here to contest the observations which induced the celebrated anatomist, Professor Owen, to reverse the denomination of the valves from the relative disposition of the animal to the valves, which in some genera, such as Orbicula, would be the reverse of that in Terebratula. Because of the immense confusion such a change would unavoidably create in works already published, we have determined to banish completely the terms *dorsal* and *ventral* from our descriptions, and to adopt other terms, also in use, to distinguish the same parts: thus we will use indiscriminately the words *perforated, rostral*, or *large valve*, for the one considered by De Buch and others a *dorsal*, the *ventral* of Professor Owen, and that of *imperforated, upper* or *smaller*, for the lesser valve.

So variable are Brachiopodes in shape, size, gibbosity, &c., that we cannot employ angular measurements proposed by Von Buch, as those characters vary in every specimen; nor do we attach much importance to the dimensions we give to each species. In general we have taken those of the best developed specimen which has come under our notice. Nor are the number of plaits more constant. We find the same species, especially among the *Rhynchonellas*, have one, two, three, four, or more plaits on the mesial fold; thus most species, smooth in the young, are plaited or otherwise ornamented at a more advanced period of growth; it is therefore, in most cases, impossible to determine species from young shells.

In concluding these few preliminary remarks, it is necessary to state, that merely the synonyms and references likely to prove useful are inserted, as it would have been impossible to mention all the authors who have alluded to certain known species without figuring or describing them. Many errors have been committed, and much *confusion* produced, by the rapid manner in which some authors have determined their species; we will not, therefore, refer to those which are published simply under the form of lists.[1]

Genus—LINGULA, *Bruguière.* 1789.

Shell inequivalved, one valve more convex than the other, more or less oval, elongated, tapering and pointed at the beaks, widened at its palleal region, without hinge, valves held together by the adductor muscles; attached to submarine bodies by a long muscular peduncle issuing from between the beaks, a groove existing for its passage in that of larger valves; arms fleshy, without any shelly support; structure horny, covered by an epidermis; two muscular impressions on the one, four on the other valve.[2]

Obs. We are only acquainted with one authentic species of British Oolitic Lingula, *L. Beanii.* Sowerby mentions another, *L. ovalis,* as from Kimmeridge clay, but which appears to belong to the lower green sand. It is worthy of remark, that the genus Lingula, one of the oldest created forms, has persisted, with very little variation in shape, up to the present day, a circumstance very unusual among the Brachiopoda.

1. LINGULA BEANII, *Phillips.* Plate I, figs. 1, 1*a*, 1*b*, 1*c*, 1*d*.

LINGULA BEANII, *Phillips.* 1829. Geol. of Yorksh., Part i, pl. 2, fig. 26.
— — *Morris.* 1843. Catalogue, p. 122.
— — *Dav.* 1847. Lond. Geol. Journal, vol. i, pl. 18, figs. 26—30.
— — *Bronn.* 1849. Index Palæontologicus, p. 655.
— — *D Orb.* 1849. Prodrome, vol. i, p. 286.

Diagnosis. Shell irregularly oblong, oval, rounded in front; valves thin, convex, with numerous concentric lines of growth; internal muscular impressions strongly marked. Dimensions variable; average size 10 lines long by 6 broad.

Obs. Professor Phillips was the first to notice this species in his work on the 'Geology of Yorkshire,' but gives no further description than that it approaches *Ling. mytiloides* of Sowerby. From the great resemblance various species of *Lingula* bear to each other, it is

---

[1] From the necessity of bringing out the work in parts, and from the strong aversion we have to the objectionable practice of giving lists of new species before they are described and figured, we cannot here give the tables and general conclusions as to the distribution of the various species till the conclusion of the series of Monographs on these subjects.

[2] For more ample details, see general Introduction.

often difficult to distinguish and describe them. Professor Phillips states this species to occur in the inferior Oolite of Yorkshire; it has also been found in the marlstone near Bathford, during the cuttings for the Great Western Railway, by Mr. Walton, whence the fine internal specimens figured in Plate I, figs. 1, *a, b, c, d,* were obtained; these figures are enlarged.

I do not know any other English Oolitic species; fragments of a Lingula have been found by Mr. Moore in the upper lias of Ilminster, but not sufficiently perfect to be described or identified.

Figs. 1*a*, 1*b*, represent the larger valve enlarged, with the groove for the passage of the muscular peduncle.

Figs. 2*c*, 2*d*, smaller valve enlarged. It will be seen by these figures the muscular impressions vary slightly in different specimens.

Genus—ORBICULA, *Cuvier.* 1798.

Shell inequivalved, more or less orbicular, upper valve conical, with apex nearer the posterior margin, smaller valve depressed, flat, or slightly convex, affixed to submarine bodies by a tendinous pedicle issuing through a fissure, varying in length and size, extending from its centre to near the margin; no hinge, or calcareous shelly supports; structure horny.

Obs. We are only acquainted with three British Oolitic Orbiculæ, viz., *O. Townshendi, reflexa,* and *Humphresiana; O. Townshendi* is the largest orbicula known. Two other shells have been placed among the orbicula: *O. granulata,* Sow., M. C., Tab. 506, fig. 34, stated to be from the great Oolite of Ancliff, and *O. radiata,* Phillips, Geol. of York, Tab. 4, fig. 12, as from cor. Oolite of Malton; these, however, do not seem to belong to the class of Brachiopoda, but to that of Gasteropodes.

2. ORBICULA TOWNSHENDI, *Forbes,* MS. Plate I, figs. 2, 2*a*, 2*b*.

Diagnosis. Shell bivalve, thin, almost circular, upper valve very convex, regularly rounded; apex near the posterior margin, greatest elevation of the valve towards the central part, the apex lying considerably lower; surface smooth, horny, with irregular circular lines of growth; inferior or attached valve, slightly concave, with deep depression, extending and widening from the centre to within a short distance from the posterior margin, leading to a long, wide, ovular fissure, from which the peduncular fibres issued, and which must, in this species, from the depth of the valve, have been of some length, the fissure measuring 5 lines in length and 3 in breadth, and extending to within 4 lines of the margin of the shell. This valve is ornamented by numerous and regular slightly-elevated concentric striæ, not all forming the complete circle, sometimes extending only to a certain distance, while at other times they dichotomize or unite into strong, wide striæ, especially towards and between the fissure and anterior margin. Interior unknown; length and breadth 19, depth 7 lines.

Obs. This is the largest and finest species of orbicula with which I am acquainted, and only approached by a specimen shown me by M. D'Orbigny, from the lias of France, believed to be distinct by that author, both species forming part of a sub-genus among the orbiculæ proposed by him, under the name of *Orbiculoidea*.

I am indebted to the liberality of the Geological Survey for the loan of this magnificent specimen, forming part of their collection, and named by Professor Forbes after the late Mr. Townshend, who found and bequeathed it to that establishment. Unfortunately the ticket has been lost, but I am assured that it is from the Oxford clay beds of the southern districts of England.

3. ORBICULA REFLEXA, *Sow.*   Plate X, fig. 8.

ORBICULA REFLEXA, *Sow.* 1829.   M. C., vol. vi, p. 4, pl. 506, fig. 1.
—   —   Zool. Journal, vol. ii, p. 321.
—   —   *Morris.* Catalogue, 1843.
—   —   *Bronn.* 1849. Index Palæont., p. 848.
ORBICULOIDEA REFLEXA, *D'Orb.* 1849. Prodrome, vol. i, p. 258.

Diagnosis. Shell bivalve, subelliptical, thin, upper or unattached valve convex, with apex directed towards and near the posterior margin; surface shining and smooth, with the exception of numerous fine concentric lines; structure horny; lower or attached valve nearly flat; aperture for the muscular byssus large and elongated. Length 7, breadth 6, depth 3 lines.

Obs. Two fine specimens of this orbicula are to be seen in the collection of the British Museum, attached to an arca, and said to be from the lias of Northampton. We find Peak Whitby mentioned by Mr. Morris, but never having found the shell in situ can add no other details. It has sometimes been mistaken for a similar but more circular orbicula, found in the coal measures of Coalbrook dale, and it seems even probable that Sowerby's figures were drawn from some specimens of that species.

4. ORBICULA HUMPHRESIANA, *Sow.*   Plate I, figs. 3, 3*a*, 3*b*.

ORBICULA HUMPHRESIANA, *Sow.* 1829.   Vol. vi, p. 5, pl. 506, fig. 2.
—   —   *Morris.* 1843. Catalogue.
—   —   *Bronn.* 1849. Index Palæont., p. 847.

Diagnosis. Shell bivalve, more or less circular, conical, apex elevated at some distance from the posterior margin; surface of upper valve ornamented by numerous longitudinal diverging striæ, from the apex towards the marginal line; lower valve unknown, attached to ostrea deltoidea. Length 6, breadth 5 lines.

Obs. The only specimens I have seen and figured in my Plate I. are from the collection of Mr. Sowerby, and would appear to have been found in the Kim clay, Shotover, Oxon. Fig. 3*a* is an enlarged illustration. Among the recent species we find more than one longitudinally striated. *O. Cummigii*, &c.

Genus—CRANIA, *Retzius*. 1781.

Shell unequivalve, circular or subquadrate, more or less irregular, entirely or partially attached by the substance of smaller valve to rocks, corals, and other submarine bodies; upper valve conical, with lateral or subcentral vertex, without hinge or ligament; lower or attached valve thickest, often irregular, due to the nature of the object to which it is fixed; surface strongly punctured or spongy; four circular depressed, or produced muscular impressions in each valve, the first two, formed by the adductor muscles, are situated near the cardinal edge, the other pair are approximated and placed near the centre, behind which a central prominence is sometimes seen; the space between these and the wide, thicker, granular margin surrounding the shell, is divided by the digitated genito-vascular impressions; arms fleshy, free only at their extremities; no calcareous supports.

Obs. We are only acquainted with two British Oolitic cranias, *C. Antiquior* and *C. Moorei*, and, unfortunately, of these only one of the valves has been hitherto discovered.

5. CRANIA ANTIQUIOR, *Jelly*. Plate I, figs. 4—8.

CRANIA ANTIQUIOR, *Morris.* Catalogue, 1843.
— — *Dav.* London Geol. Journal, vol. i, pl. 18, figs. 21—25. 1847.
— — *Bronn.* Index Palæont., p. 342. 1849.
— — *D'Orb.* Prodrome, vol. i, p. 316. 1849.

Diagnosis. Shell suborbicular, irregular, the lower valve only known; it varies in form, some specimens being almost flat, others more or less concave, and even occasionally patelliform. The muscular impressions are four, more or less; strongly marked in different examples; the two posterior are generally larger and more widely separated than the two anterior ones, which latter usually touch, and are also less circular in form than the upper, and depressed in the centre. In most examples a strongly-marked ridge is seen extending in the mesial line, from the junction of the lower muscular impressions to the margin of the shell, as displayed in fig. 8. The very peculiar spongeous structure, characteristic of the genus Crania, is well marked. In form and general appearance this species somewhat approaches *Crania abnormis* of the tertiary period. Length 7, breadth $6\frac{1}{2}$ lines; some specimens are almost square. From the appearance of the exterior this Crania would not seem to have been much attached.

Obs. A very extensive series of specimens of this species, forwarded to me by

Mr. Walton and Mr. Pearce, of Bath, as well as numerous specimens collected by myself, have enabled me to note the extreme variations exhibited in its general form and shape of the muscular impressions. Sometimes these impressions project in the prominent manner shown by figs. 4, 6, 8, while at other times they are barely distinguishable; in one specimen, belonging to Mr. Pearce, the whole four, where actually depressed, producing a remarkable concavity in the valve. Mr. Pearce was at first disposed to consider these specimens as examples of the upper valve, an opinion, however, he afterwards relinquished; and, curiously enough, although many hundreds have been collected by Messrs. Walton and Pearce, they have not been able to obtain the upper valve in any one instance.

This species was originally discovered by the Rev. H. Jelly, who applied to it the specific appellation "Antiquior," but no figure or description appeared before May, 1847, when I described and figured it in the London Geological Journal, the only published record respecting it being the insertion of this name in Mr. Morris's valuable Catalogue of British Fossils, in 1843.

The Crania Antiquior is found in the great Oolite of Hampton Cliff, near Bath; the specimens figured are from the collection of Mr. Walton.

All the figures are of natural size, except fig. 8, which is enlarged.

6. CRANIA MOOREI, *Dav.* Plate I, fig. 9.

Diagnosis. Shell irregular, transversely oval, suborbicular, truncated posteriorly; upper valve convex and slightly conical, with the vertex near the centre; surface smooth and punctuated; interior presenting four muscular impressions, the posterior ones slightly marked, as well as the anterior two, which are arranged in the form of a V. The digitated genito-vascular impressions hardly visible; interior closely punctuated. Length and breadth about 1 line. The upper valve only is known.

Obs. This small species was found by Mr. Moore, in the upper lias, near Ilminster; its position is higher than that of those beds containing the Leptœnas in the same locality. The figure is drawn from a specimen kindly given to me by Mr. Moore, and I take great pleasure in naming the species after him.

Plate I, fig. 9, natural size. Fig. 9, *a*, *b*, enlarged.

*Genus—*THECIDEA, *Defrance.* 1828.

Shell unequivalved, thickened, more or less irregular; largest valve partially or entirely attached by its own substance, or, when young, in some species by a peduncle issuing from the extremity of the beak to submarine bodies; form longitudinally or transversely oval, sometimes subquadrate; upper valve small, more or less convex, smooth, or otherwise ornamented, granulated, structure punctuated; hinge line more or less straight, with two strong teeth in the attached valve, adapting themselves into corresponding sockets in the smaller valve; beak more or less produced, with long or wide well-defined area and deltideum. Interior of valves variable; in larger valves a longitudinal, central, and two lateral ridges are generally more or less visible, under which two deep muscular impressions are seen; upper valve complicated, more or less deeply and regularly sinuated by an apophysary testaceous ridge, united all round, and leaving a small cavity in the upper portion of the valve free for the body of the animal, these sinuated ridges varying in number, position, and extent in different species; two strong lateral adductor muscles situated under the hinge, no arms, animal small.

Obs. We are acquainted with five British Oolitic species of Thecidea, four from the Lias, a fact hitherto unrecorded, and two in the Inferior Oolite.

7. THECIDEA MOOREI, *Dav.* Plate I, fig. 10.

Diagnosis. Shell irregular, inequivalved, attached by the greatest portion of its inferior valve, almost square; attached or lower valve, modelling itself to the object on which it is fixed, with elevated sides and front rising perpendicularly from the attached part; area well defined, triangular, and receding from its junction with the upper valve; deltidium large, elevated with slight central depression, and marked, as well as the area, with numerous lines of growth. Upper valve almost flat, of a transversely oblong square, with slight depression in the centre, and the sides, except at the hinge, turning sharply over, and forming elevated sides, till they meet the edge of the lower valve, so that the front is very much elevated. Surface strongly punctuated; interior of attached valve only known; hinge line straight, with two strong teeth and elevated mesial ridge. Length and breadth about 2 lines; frontal elevation 1 line.

Obs. Twelve specimens of this remarkable little Thecidea were found attached to a specimen of Rhynchonella serrata, from the marlstone or middle lias in the neighbourhood of Ilminster, along with Thecidea *Bouchardii* and *triangularis*, by Mr. Moore, to whom I feel much pleasure in dedicating the species. These and another, *T. rustica*, are the first specimens of this genus hitherto noticed, as far down as the Liasic period; no mention is made of this genus in M. D'Orbigny's Prodrome; they would appear to be the oldest Thecideas at present known. Mr. Moore also has one specimen of this species, found in the upper lias, from the neighbourhood of Ilminster. *Th. Moorei* is easily distinguished from all other known Thecideas, by its square shape and elevated front.

Plate I, fig. 10, illustrates specimens of natural size, from the collection of Mr. Moore; fig. 10, *a—e*, are enlarged.

8. THECIDEA BOUCHARDII, *Dav.* Plate I, figs. 15, 16.

Diagnosis. Shell irregular, inequivalved, of an elongated transversal form, attached by the greatest part of its lower valve; area in larger valve long and straight, receding from its junction with the upper valves, and at almost right angles to it; deltidium well defined, elevated, and marked by lines of growth, which extend also over the area; upper valve slightly convex, smooth, and strongly punctuated; greatest height at hinge line receding thence to the frontal margin. Interior of attached valve only known; hinge line straight, with two strong teeth and internal elevated mesial ridge, and wide, strongly-granulated margin, leaving two deep depressions on each side of the central elevated ridge. Length 1¼, breadth 2 lines.

Obs. Three specimens of this species were found attached to the same specimen of *Rh. serrata*, or along with *Th. Moorei*, and consequently from the middle lias, and I feel much pleasure in dedicating it to M. Bouchard; its locality is the neighbourhood of Ilminster. M. Tesson, of Caen, showed me a specimen of Thecidea, which approaches this very much in form, and is found in the liasic beds of Fontaine Etoupe Tour, near Caen, in Normandy, and I should not have hesitated in saying it was the same species, had the dimensions of the French specimen not exceeded three times those of our English shell. *Th. Bouchardii* is easily distinguished from the other forms of this genus by its great breadth, and appears a much more delicate species than either T. Moorei or T. triangularis.

Plate I, fig. 15, natural size of a specimen in Mr. Moore's collection; figs. 15a, 16, and 17, enlarged views.

9. THECIDEA DICKINSONII, *Moore*, MS. Plate XIII, fig. 19.

Diagnosis. Shell of an elongated transversal form, unequivalved, and attached by the greatest part of its lower valve; area and cardinal edge straight, and not quite as long as the greatest width of the shell; upper valve slightly convex, smooth, and punctuated. Length 1½, width 2½ lines.

Obs. This is the largest Inferior Oolite *Thecidea*, I believe, as yet known, and is attached to a specimen of *Ter. perovalis*, from Dinnington, belonging to Mr. Moore, who named it after his friend Mr. Dickinson. We find also, in the same locality, *Thecidea triangularis*, which is easily distinguished by its shape.

10. THECIDEA TRIANGULARIS, *D'Orb.* Plate I, figs. 11, 12.

THECIDEA TRIANGULARIS, *D'Orb.* (?) 1849. Prodrome, vol. i, p. 316.

Diagnosis. Shell irregular, inequivalved, attached by the greatest portion of its lower valve, more or less triangular, gibbose, produced behind, and somewhat bilobate in front;

area of attached valve triangular, more or less lengthened, with distinct deltidium; small valve operculiform, convex, with slight depression in centre, punctuated; the interior of attached valve only known. Length 1 line, breadth the same.

Obs. I believe this to be the same species as M. D'Orbigny mentions, but which he neither describes nor figures in his Prodrome, under the name of *Th. triangularis*, as occurring in the Oolitic beds of Ranville, in Normandy, and which name I will here adopt, as, on comparing it with our English specimens, I could find no difference.

In England this species appears to have first appeared in the middle lias, as Mr. Moore found it attached to *Rh. serrata*, and it was afterwards found by Dr. Wright, Mr. Morris, and myself, in the Inferior Oolite of the Cotswold and Leckhampton hills, attached to *Ter. plicata, Rh. Wrightii*, to *corals*, and to probably any other shell in that bed. In Normandy it is found a little higher up.

This species also strongly resembles (except in size) the recent *Thecidea Mediterraneum*, as may be easily perceived on looking at the figure of that species, (Plate I, fig. 13,) which I have purposely placed by the side of *Thecidea triangularis*. Our fossil species does not appear, however, to have ever attained the dimensions of the recent species.

Plate I, fig. 11, natural size of Mr. Moore's specimen from the marlstone; fig. 11, *a, b*, enlarged view of the same. Plate I, fig. 12, natural size of one of the Inferior Oolite specimens; 12*a*, enlarged. Plate I, fig. 13, recent *Thecidea Mediterraneum*.

11. Thecidea Rustica, *Moore*, MS. Plate I, fig. 4.

Diagnosis. Of this small Thecidea we are acquainted with only the upper or unattached valve, and therefore the description must consequently be very incomplete. Unattached valve slightly convex, of a squarish circular form, as long as wide, smooth and punctuated, interior presenting two sockets, in which the teeth of the unattached valve articulate from under an elevated crest or lamella, surrounding the shell at some distance from the edge, and, on reaching the frontal margin, it takes a curve towards the middle of the valve; returning again to the margin, it terminates under the other socket, the position between this ridge and the edge of the shell being strongly granulated, and presenting another smaller ridge, which also joins the sockets, after having gone round the shell. The sinuses observable in *Thecideas*, and which form so many elevations in its centre, are hardly perceptible in this species; there is a slight elevation, strongly granulated, on each side of the first described ridge. Length 1 line, breadth the same.

Obs. The internal organisation of this species is more simple than that generally seen in Thecidea, but it must also be observed that the number of the lamellæ, or ridges, forming the sinuses in this genus, and which represent the calcareous supports in Terebratula, vary very much, as may easily be perceived, in casting a glance over *Thecidea hippocrepis, tetra-*

gona, papillata, hieroglyphica, recurvirostris, digitata, antiqua, &c. This last-named species approaching more than any I know to our *Th. rustica,* and in all good genera we ought to find a graduated scale from the simple to the compound. Thecidea rustica was discovered by Mr. Moore in the upper lias, in the same bed containing *Lept. Moorei, Bouchardii, Liasiana, Pearcei, Sp. Ilminsteriensis,* &c., and therefore higher up than *Th. Moorei* and *Bouchardii.* Mr. Moore has given it the name of *Th. rustica,* which I readily adopt; it is much to be regretted that hitherto only the smaller valve has been found; it does not appear very rare in its bed, if one inclines to take the trouble of seeking it, which can be done only by washing the sandy clay bed where it is found, and then carefully picking them out, after having extended small portions of the washed bed on a plate. In this manner, also, are obtained the *Leptænas,* and numerous Foraminifera with which these beds abound; its locality is the neighbourhood of Ilminster.

Plate I, fig. 14, natural size, from specimens belonging to Mr. Moore; fig. 14, *a, b,* enlarged exterior and interior.

Genus—LEPTÆNA, *Dalman.* 1827.

Shell inequivalved, equilateral, generally transverse, sometimes oval, always compressed; smooth, striated, or exteriorly costated : *larger valve,* more or less convex or concave, sometimes bent or geniculated; beak more or less produced, straight, sometimes recurved, and perforated at its extremity by a very small circular opening; separated from the cardinal edge by a more or less elevated triangular or canaliculated area : *smaller valve,* concave or convex, following the large valve in its different curves, beak of smaller valve not much produced, with or without a linear area : deltideum complete, triangular, with angles more or less open, without reference to the development of the area; often notched at its base for the passage of the tendinous fibres of attachment; hinge transverse, straight, linear; teeth differently disposed, but always provided with two principal diverging teeth on the larger valve, which are received by sockets placed on each side of the central bifid or trifid tooth of the smaller valve : no internal calcareous supports.

Obs. We will not at present discuss the different opinions lately brought forward on this genus, or its subdivisions, each author having his way of thinking on this subject, which has involved us in great confusion; the same shell is thus for some a *Strophomena,* for others a *Leptæna, Leptagonia Chonetes, Productus,* &c. All we shall here state is that the genus was not known to occur above the Palæozoic series before 1847, at which time M. Bouchard and myself described several species from the lower oolitic or liasic deposits; and we are now acquainted with the five following species :—*Leptæna Moorei, L. Piercei, L. granulosa, L. liasiana* and *L. Bouchardii,* all of which are found in England. *Lept. liasiana* (Bouch.) alone having as yet been discovered on the Continent.

12. LEPTÆNA MOOREI, *Dav.*   Plate I, fig. 18.

LEPTÆNA MOOREI, *Dav.* 1847. Annals and Mag. of Nat. Hist., pl. xviii, fig. 1 *a*.
— — *Dav.* Bull. Soc. Géol. France, vol. vi, 2d Series, p. 270.
— — *D'Orb.* Prodome, vol. i, p. 220, 1849.

Diagnosis. Shell small, depressed, wider than long, ornamented by numerous fine costæ, scarcely visible without a lens; larger valve slightly convex; area double, as wide, or rather wider, than the shell; deltidium small, chiefly filled up by the median tooth of smaller valve, which tooth is grooved by four furrows, offering a passage for the muscular fibres of attachment passed outwards.   Length 1½ lines; width 2 lines.

The muscular impressions in the interior of both valves are strongly developed in this species, and indicate that it did not attain larger dimensions than those above assigned to it.

This elegant little species was first discovered in the beds of the upper lias above the marlstone, near Ilminster, by Mr. Moore, to whom it is dedicated.   The following section, forwarded by Mr. Moore, shows the position of the bed containing the *Leptænas*:—

1. Rubbly beds, 6 to 10 feet, with numerous *Ammonites*.
2. Clay, 8 inches.
3. Yellow limestone, 3 to 4 inches.
4. Layers of clay, 18 inches, *Leptæna Pearcei*.
5. Leptæna bed, 1 inch: *L. Moorei*, *L. Bouchardii*, and *L. liasiana*.
6. Marlstone, 2½ inches.
7. Greenish sand, 4 inches, containing numerous *Belemnites*.
8. Marlstone.

I do not understand the reason which has induced M. D'Orbigny to place these *Leptænas*, and other species found in the same beds, in the lower Lias (*Senemurien*, Prodrome, p. 220), where none, to my knowledge, have been discovered according with my printed description and stratigraphical position of the species, which has not been noticed out of England as yet.

Plate I, fig. 18, shows the exact size of an adult specimen of this species.
„   fig. 18 *a—e* are enlarged.

It is not a rare species in its bed, Mr. Moore having found more than one hundred specimens; but from its minuteness it is difficult and tedious to collect.

13. LEPTÆNA PEARCEI, *Dav.*   Plate I, fig. 19.

LEPTÆNA PEARCEI, *Dav.* 1847. Annals and Mag. of Nat. Hist., pl. 18, fig. 4.
— — *Dav.* Bull. Soc. Géol. de France, vol. vi, 2de Série, p. 270, 1850.
— — *D'Orb.* Prodrome, 1849, vol. i, p. 220.

D

The larger valve only of this small species is known; from which it appears, that it differed materially from *L. Moorei,* being much more convex, and the striæ ornamenting its surface having two or three smaller intermediate ones (Plate I, fig. 19*b*); while in *L. Moorei* the costæ appear of the same size; the shape of this shell is also more rounded, and considerably larger than *L. Moorei.*   Length $3\frac{1}{2}$; breadth $4\frac{1}{2}$ lines.

This species appears to be rare; it occurs in a clay stratum above the Leptæna bed: it was discovered by Mr. Moore.

Plate I, fig. 19, nat. size; fig. 19*a, b*, enlarged.

14. Leptæna granulosa, *Dav.*   Plate I, fig. 20.

Leptæna granulosa, *Dav.*   Bull. Soc. Géol. de France, vol. vi, 2de Serie, p. 270, fig. 6, 1850.

Diagnosis. Valve slightly convex, rounded and ornamented by granulous striæ, between which smaller ones are perceived (Plate I, fig. 20*b*); the area appears narrow, and has much resemblance to *L. Moorei.*   Length $1\frac{1}{2}$; breadth 2 lines.

It is a rare species; easily distinguished from *L. Pearcei* and *L. Moorei* by the granulations which ornament its valves: it was found by Mr. Moore in the upper Lias, in the same bed which contains *L. Pearcei.*   Only the larger valve is known, and therefore it can be but imperfectly described.

Plate I, fig. 20, natural size.

  „   fig. 20*a, b*, enlarged.

15. Leptæna liasiana, *Bouchard.*   Plate I, fig. 21.

Leptæna liasiana, *Bouchard.* 1847. Annals and Mag. of Nat. Hist., t. 18, fig. 2, *a—d.*
—   —   *Dav.* 1850. Bull. Soc. Géol. de France, vol, vi, 2de Série, p. 270.
—   —   *D'Orb.* Prodrome, vol. i, p. 220.

Diagnosis. "Shell rounded, inequivalved, equilateral, smooth; larger valve gibbose posteriorly, becoming flatter anteriorly, with a slight longitudinal groove ending in a notch on the front margin of the shell.   Beak small, slightly incurved, truncated at the apex by a minute circular foramen, similar to that which occurs in many other *Leptænas;* for instance, *L. alternata,* of Indiana, North America.   This truncation may also be observed in some species of *Orthis,* from Russia.   Area double, interrupted on the dorsal valve by a large and slightly-convex deltidium, which arises at the apical opening, and gradually enlarges towards the base, occupying one third of the width of the area.   The deltidium is slightly notched, the notch being partly closed by the large median tooth of the smaller valve; the exterior face of which is grooved by four furrows, which afforded a passage for the muscular fibres of attachment, arranged in four bundles.   The smaller valve is deeply concave, following the contour of the larger valve, so that little space remained between them for the body of the animal. Cardinal margin about half the width of the shell. Length 2 lines; breadth the same.

"The general form of this *Leptæna* approaches that of *Productus*. It closely resembles *L. oblonga.* (*Pander*.) It has the same convexity and smoothness, and the beak is similarly truncated by an apicial opening: the area and perpendicular opening has also some analogy to the Russian species, but differs in the contour of the larger valve, and the notch in the front margin." (*Bouchard.*)

Obs. The above description was published by M. Bouchard, in the 'Annals of Natural History,' Oct. 1847, and I have thought it desirable to reproduce it here. At that period this species had not been discovered in England; Mr. Moore was so fortunate as to find in my presence, a short time since, two fine specimens in the Leptæna beds of the upper Lias, in the neighbourhood of Ilminster: its foreign locality being Pic de Saint Loup, near Montpellier (Herault). In the 'Prodrome,' page 220, M. D'Orbigny, besides placing this species and *L. Bouchardii* in the lower Lias, which is a stratigraphical mistake, states it not to be a *Leptæna*. It is to be regretted that, with both perfect figures of the exterior and interior, M. D'Orbigny did not assign its generic position; but both M. Bouchard and myself consider its place for the present to be best among the *Leptænas*, with which they have many similar characters.

Plate I, fig. 21, natural size.

    „   fig. 21*a*, enlarged.

XVI. Leptæna Bouchardii, *Dav.* Plate I, fig. 22.

Leptæna Bouchardii, *Dav.* 1847. Annals and Mag. of Nat. Hist., pl. 18, fig. 3.
   —   —   *Dav.* 1850. Bull. Soc. Géol. de France, vol. vi, 2de Série, p. 270. 1850.
   —   —   ? *D'Orb.* Prodrome, 1849, vol. i, p. 220.

Diagnosis. Shell very small, almost ovular; surface smooth; large valve very convex, smaller valve concave, leaving only a little space for the animal; beak small, not much recurved; area double, and cardinal margin smaller than the greatest width of the shell; deltidium very large. Length 1¼; width about half a line.

Obs. This species is readily distinguished from *Leptæna liasiana* by its more elegant form; the small valve is more regularly concave, and the larger one more convex. When describing this species in the 'Annals,' I had not perceived that the beak was truncated at its extremity by a minute circular foramen, as in *L. liasiana;* but which numerous specimens, sent to me by Mr. Moore, have amply confirmed. This species appears never to have attained larger dimensions; as the internal characters presented by the smaller valve are those belonging to a full-grown shell. *L. Bouchardii* was found, by Mr. Moore, associated with *Leptæna Moorei*, in the Leptæna bed, previously described under that species.

Plate I, fig. 22, natural size.

    „   fig. 22*a*, *b*, *c*, enlarged.

Genus—SPIRIFER, *Sowerby*. 1818.

Shell unequivalved, equilateral, generally transverse, more or less trigonal, and convex. Exterior rarely smooth, more often striated or costated; larger valve always convex, often gibbous, divided by a medio-longitudinal sinus, of more or less depth and width, corresponding with the mesial fold in smaller valve. Beak generally acute and straight, sometimes recurved and obtuse, never truncated, area always triangular, more or less elevated; thrown backwards, flat or concave, and divided by a mesial deltideal fissure, *always covered by a deltideum*, notched at its base, for the passage of the peduncular fibres. Smaller valve always convex, but less so than in larger valve; longitudinally divided by a mesial fold, elevated and proportioned to the sinus in larger valve, to which it corresponds. No area, summit not much developed, extending a little beyond the rectilineal cardinal edge; hinge straight, transverse, formed of two diverging teeth, limiting the base of the deltoid fissure of larger valve, and placed in the sockets existing on each side of the beak of the smaller valve; internal calcareous supports formed by two lamellæ, arising from under the beak of smaller valve, and forming a number of spiral coils, diminishing in size towards the cardinal angles.

Obs. Several divisions have been proposed in the Genus *Spirifer*, to which we will allude in our introduction, and shall only notice here, that the punctuated character believed to be peculiar to Lias Spirifers (*Spiriferina*, D'Orb.), also exists in those of other epochs. In our British Oolitic series, we are only acquainted with four species of Spirifer: viz., *Spirifer rostratus, Sp. Ilminsteriensis, Sp. Walcottii,* and *Sp. Münsterii*, all found in the lias. And on the Continent four or five more have been discovered; so that this genus, which has not yet been known higher up in the series, was represented by eight or nine forms, some of which are very similar in exterior appearance to more ancient types.

17. SPIRIFER ROSTRATUS, *Schl.*   Plate II, figs. 1—21; Plate III, fig. 1.

TEREBRATULITES ROSTRATUS, *Schlotheim.* 1822.   Nach. Zur. Petrefact., pl. xvi.
DELTHYRIS VERRUCOSA, *V. Buch.* 1831.   Petrifications Remarquables, pl. vii, fig. 2.
SPIRIFER ROSTRATA, *Zieten.* 1832.   Die versteinerungen Wurttemberg, p. 38, fig. 3.
—   MESOLOBA, ? *Phil. Deslongchamps.* 1837.   Soc. Linn. de Normandie.
—   HARTMANII, *Zieten.* 1838.   Die Verst Wurttemberg, pl. xxxviii, fig. 1,
—   VERRUCOSA, *Zieten.* 1838.   Die Verst Wurttemberg, pl. xxxviii, fig. 2.
—   PINGUIS, *Zieten.* (non *Sow.*) 1838.   Die Verst Wurttemberg, t. xxxviii, fig. 5.
DELTHYRIS ROSTRATUS, *V. Buch.* 1840.   Class. et descrip. des Delthyris, Mém. Soc.
          Géol. de France, 1ᵉʳᵉ Série, t. iv, pl. 10, fig. 24.
—   VERRUCOSA, *V. Buch.* 1840.   Class. et descrip. des Delthyris, Mém. Soc.
          Géol. de France, 1ᵉʳᵉ Série, t. 4, pl. x, fig. 30.
—   TUMIDUS, *V. Buch.* 1840.   Class. et descrip. des Delthyris, Mem. Soc.
          Géol. de France, 1ᵉʳᵉ Série, t. 4, pl. x, fig. 29.

DELTHYRIS HARTMANII, *Quenstedt.* 1843. Das Flöegibirge, Würtemberg, p. 181.
— VERRUCOSA, *Quenstedt.* 1843. Das Flöegibirge, Würtemberg, p. 185.
— ROSTRATA, *Quenstedt.* 1843. Das Flöegibirge, Würtemberg, p. 186.
SPIRIFER PUNCTATUS, *Buckman.* 1845. Geol. of Chelt., pl. 10, fig. 7.
— RETICULATUS, *Buckman.* (MS.)
— LINGUIFEROIDES AND CHILIENSIS, *Forbes* and *Darwin.* 1846. South America, p. 267, pl. 5, figs. 15, 16, 17, 18.
— ROSTRATUS, *Dav.* 1847. London Geol. Journal, vol. i, p. 109, pl. 18, figs. 1—10.
SPIRIFERINA GRANULOSA, *Rœmer* (according to *M. D'Orbigny*). 1849. Prodrome, p. 56.
— VERRUCOSA, *D'Orb.* 1849. Prodrome, vol. i, p. 221.
— HARTMANII, *D'Orb.* 1849. Prodrome, vol. i, p. 239.
SPIRIFER ROSTRATUS, *Bronn.* 1849. Index Palæont., p. 1181.
— TUMIDUS, *Coquand* et *Bayle.* 1850. Bull. Soc. Géol. de France, vol. vii, 2ᵈᵉ Série, p. 235.

Diagnosis. Shell iniquivalved, variable, rounded, with mesial fold and sinus more or less prominent; surface smooth or undulated, forming sometimes rounded plaits; beak more or less developed, recurved or straight; deltidium in two pieces, area well defined, surface of valves punctuated, and spinose extending only to the edge of the area. Three lamellæ are seen in the interior of large valve, the central one more elevated, and terminating in a point; in small valve, two spirals united together by a lamella. Length and depth very variable, the largest specimen known measuring—length 28, width 30, depth 17 lines; but the dimensions are generally much smaller, rarely extending 10 lines in length, and the same in breadth.

Obs. The numerous variations in size and form assumed by this species, does not admit of a correct description; but, through a numerous suite of specimens, we are enabled to trace the passages from one variation of form into another, all preserving a general aspect, leading to the most common type. Since 1845, I have devoted much attention to this species, having had at my disposal a great number of specimens from our British and foreign localities, it being abundantly spread in different beds of the liasic period, and a few of the most marked variations are illustrated in Plate II; and, had space admitted of it, the intermediate passages could have been traced. All the figures, excepting 18 and 19, are of the natural size; fig. 1 represents the largest specimen yet known of the species, which varies in size from the one first mentioned to that of a pea. The average dimensions and shapes are illustrated by figures 1, 2, 5, and 7.

The great variations to which this species is subject appear attributable to three principal causes :—First. The presence of a mesial fold and sinus more or less developed, or its total absence, as seen in some specimens where the front is quite straight (figs. 4 and 10); at other times only a gently elevated curve is perceptible, as in fig 11. The large valve being regularly convex to the front, and no sinus existing, as generally happens when there is a mesial fold, more or less produced, as in figs. 1, 2, and 7. The second cause exists in the tendency there is in certain specimens to undulation, which, becoming close, often give the shell a plaited look; these, however, are not true plaits, but undulations, as in fig. 7, which is the most marked example I could find; in general they are hardly visible, and do not extend to the umbo, as in *Sp. Walcotti, Münsterii*, &c., and we can trace every stage, from the perfectly smooth specimen to the extreme state represented in fig. 7. The third cause is occasioned by the form of the area and beak, which is very variable in this species; generally the beak is recurved, as in figs. 1, 2, 4, 5, and 7, allowing one to see under it a deltidium of moderate size; but as the area and beak become produced and projected backwards (which is the case with many specimens), the beak becomes almost straight or slightly recurved at its extremity, and displaying in all its extent a large elongated deltidium, which is the narrower in appearance as the area becomes longer and larger. The deltidium does not thus widen in comparison to its length; it would, there-fore, be impossible to separate into species this shell from the form and size of its area and beak, as every insensible gradation can be traced from one shape into another. The figures 5, 8, 10, 12 are some examples; fig. 3 exhibits a specimen, in which the beak has become very large and wide. This is a rare case, as well as that seen in fig. 10; they are extremes.

Another cause of variation is due to the punctuation and tubular spines ornamenting the valves; their size, length, and number being rarely the same in many specimens; at times being so near to each other that one cannot perceive the intermediate punctuation covering its surface; in other specimens they are irregularly and sparingly implanted, as may be seen in the enlarged fig. 19, which appearance would seem to have induced M. Buckman to propose one species for fig. 2, under the name of *Sp. punctatus*, and another for fig. 1, by the name of *Sp. reticulatus*; but in reality there is no difference between these two specimens but in the size of the punctuation and spines, which are larger and stronger, according to the size of the specimens. We also often perceive that much difference appears to have been caused by local conditions; thus we have full grown adult specimens of three lines in length and breadth, as well as in others of much greater dimensions; these form varieties in the species, and we might name *Spirifer verrucosus* of Zieten as a mere dwarf variety of *Sp. rostratus*.

The deltidium in this species is also very remarkable, being formed of two pieces united in the form of a roof, which is well displayed in many specimens, especially in those found in Normandy (Vieux pont), which I have represented in Pl. II, figs. 2, 3, and 10; leaving a passage above the umbo for the peduncular muscular fibres. When the deltidium is not preserved, which is generally the case, the fissure is seen to extend to the extremity of the beak. M. Deslongchamps, Bouchard, De Verneuil, and myself, have several specimens illustrating these points in the most beautiful manner. On both sides of the deltidium extends the area, which is well defined, dividing the beak on each side of the deltidium into two equal portions, at once perceptible by the marked line extending from the extremity of the beak along its whole length, at which line on both sides of the area the spines stop. The remaining portion to the edge of the deltidium being covered by horizontal and vertical lines of growth; the vertical lines have not, however, much length or regularity, and are well represented in fig. 10, but not generally so much indented, in most cases they are scarcely visible.

If we now separate completely the two valves, the interior will be seen as illustrated by figs 13 and 15. In the larger valve, fig. 15, we perceive, on each side of the deltideal fissure, two teeth, which fit into corresponding sockets in the smaller valve, forming a strong hinge, so that the valves cannot be separated except by breaking one of them; they are placed at the extremity of two dental plates, projecting into the shell, forming the sides or walls of the deltideal fissure, and extending to the beak, which they strengthen (figs. 15 and 16). Between the two dental plates, a central system is interposed, variable in thickness and development, as may be seen in figs. 15, 16; beginning by a thick basis, which gradually decreases till it becomes as sharp as the edge of a knife, and projecting far beyond the lateral or dental plates (figs. 16, 21), two muscular impressions are visible between the dental and central plate. The interior of both valves is closely punctuated.

From the discovery by Mr. Moore, of specimens completely freed from all matrix, in which the most minute delicate impressions and details are preserved as intact as if the animal had just left its shell, I am able to offer a much more complete description of the small valve than that given in my paper published in the 'London Geological Journal,' (1845). This valve, separated completely from the larger one, would appear as in fig. 13; we first notice the dental sockets and position of the calcareous supports; when both valves are united they fill the greater portion of the larger valve, except where it covers the teeth and hinge, it will be seen that two lamella issue from a strong basis under the sockets which extend, diminishing in width and thickness, till they reach about the middle of the shell, where a curved lamellar process unites the two spires, but which process is rarely perceptible except when the specimens are in a perfect state of preservation (figs. 16, 21); the two lamella again continue to be directed toward the front of the shell, diverging from one another as they advance, and finally turning towards the bottom of the large valve, forming the first and successive coils, known under the name of spirals, each circle diminishing in circumference and size as it approaches the sides of the shell. The spire has been, for Mr. Moore and myself, the subject of active researches, and having found some specimens full of a very fine sand, it was preserved in great perfection, and, as may be remembered, in 1847 I mentioned the presence of spines on the spire, but at that epoch we could not offer observations as complete as at present. The lamella which forms the spire is neither smooth nor of equal thickness on all its width, differing on each side and variable, but always thicker on the inner side of the circumference than on the other which tapers out into an acute edge, and as will be seen in figs. 17, 18, the thickest part of the spire is towards its middle, where it forms a circular elevation diminishing again towards the outer edge.

It will be observed that no spines ever appear on the face of the lamella fronting the sides of the shell, or on the internal edge of the spire, as is observed in figs. 17, 18; the spines only occur in that part of the spire facing the front of the shell where it opens, covering thus only about a quarter of the circumference of each coil. These spines arising from an expanded basis are also implanted very irregularly on this portion, the calcareous matter of the lamella thickening sometimes and forming spines of different length, sometimes isolated, at other times united in clusters of two, three, and four, all directing themselves towards the exterior of the spire, and in general horizontally to it, rarely exceeding in length the width of the lamella; but in some cases they are a third longer, being of greater length and more numerous towards the centre of the spinose portion: fig. 18 illustrates a correct and considerably magnified fragment of the spire. Professor Owen thinks they are calcareous excresences destined to support the Cirri, and in this view both Viscount D'Archiac and M. De Verneuil concur; the presence of these spines only on that portion of the spine most exposed to currents, shows there was probably greater strength and development of calcareous matter required in this portion of the spire. The fact of the presence of spines, in a similar position, is common to many brachiopoda. I have seen them in *Spirifer rostratus, Walcotti Munsterii, Terebratula resupinata, Ter. pectunculoides*, &c., and in no specimens do they extend to the other portions of the spire.

The spirifer which Zieten considers to be Sowerby's *Spirifer penguis*, and figured in his plate 38, fig. 5, is from the lias of Vachengen, and seems only a variety of *Sp. rostratus*, and has much resemblance to the variety figured in my Plate II, figs. 7, 8, 9. However, the name of *penguis* would require to be dropped at any rate, because Sowerby's *Sp. penguis* is a Carboniferous shell, completely different from the Lias shell in question; M. D'Orbigny does not seem to have paid much attention to this point, as he adopts the term *penguis* for a Lias shell. The *Sp. rostratus* has a wide range in the liasic deposits, and has been found in the lower, middle, and upper lias, but chiefly in the marl-stone of the middle Lias: fig. 4 is the only specimen as yet found by Mr. Moore in the upper Lias of Ilminster, and there exists no well authenticated instance of a Spirifer occurring higher up in the series in England. This species is found in many localities, such as Urn Hill, Feavington, and South Petherton, near Ilminster, near Bath, Radstock, Cheltenham, &c., and many fine specimens exist in several collections, especially that of Mr. Moore. On the Continent it is also abundantly distributed; in many parts of France, particularly in Normandy, near Caen, round Avalon, at Boll; in the Wurtemberg, near Amberg, in the Canton of Basle, &c. And M. De Verneuil has lately brought it from the province of Ferusil (Spain); it is also found in America.

18. SPIRIFER ILMINSTERIENSIS, *Dav.* 1851. Plate III, figs. 7, 7*a*.

Diagnosis. Shell inequivalved, rounded, mesial fold and sinus hardly perceptible; exterior of valves smooth, punctuated, and spinose; beak of large valve much produced, projected backwards at right angles with the smaller valve; area very large, triangular. Length 2, breadth 3, depth 2 lines.

Obs. This little species was discovered by Mr. Moore, in the Leptæna or lowest beds

SPIRIFER. 25

of the upper lias, in the neighbourhood of Ilminster, and is found associated with *Thecidea rustica, Lept. Pearcei granulosa* and a *lingula*, which was not sufficiently perfect to be determined.

Plate III, fig. 7, represents a specimen of natural size, from the collection of Mr. Moore; *7a* is an enlarged illustration.

19. SPIRIFER WALCOTTI, *Sow.* Plate III, figs. 2, 3.

SPIRIFER WALCOTTI,	*Sow.* 1823.	Min. Con., vol. iv, p. 106, pl. 377, figs. 1, 2.
TEREBRATULA —	*Desh.* 1836.	Nouv. ed. de Lamarck, vol. vii, p. 374.
DELTHYRIS —	*V. Buch.* 1840.	Mém. Soc. Géol. de France, vol. iv, 1ᵉʳᵉ Série, pl. 10, fig. 8.
SPIRIFER —	*Morris.*	Catalogue, 1843.
— —	*Deslongchamps.* 1847.	Soc. Lin. de Normandie.
TRIGONOTRATA —	*Bronn.* 1847.	Leth. Geog., pl. 18, fig. 14.
SPIRIFERINA —	*D'Orb.* 1849.	Prodrome, vol. i, p. 221.

Diagnosis. Shell inequivalved, variable, with elevated mesial fold, and four lateral rounded plaits; beak more or less recurved, area well defined, deltidium in two pieces, hinge line shorter than the width of the shell; surface punctuated and spinose; spirals and septum in the interior of both valves disposed as in *Sp. rostratus.* Dimensions variable: length of the largest specimen known 19, width 24, depth 14 lines; but, in general, the species does not attain that size.

Obs. The species is easily distinguished from *Sp. rostratus* by its mesial fold, deep sinus and plates. It was first discovered, many years ago, by Mr. Walcott, at Camerton, about six miles from Bath, and represented by him in his work on petrifactions, fig. 33. Sowerby also states that Mr. Walcott observed, of similar shells, "that those found on the upper Bristol road, near Bath, are smaller, their shell thin, with a triangular hole between the beak of the lower valve and the hinge, and have the body, fig. 33, A B, within them; it consists of two hollow cones, joined to each other by part of their basis, and to one of the valves, but not so close as to prevent the animal, or part of it, from retreating into them: their surfaces are beautifully covered with circular rows of small pyramids of spar." Thus, as Mr. Sowerby observes, in vol. iv. p. 106, of his 'Min. Conch.,' Mr. Walcott was the first discoverer of the spiral appendages, long before they were used as a generic character; he also observed the triangular fissure, but did not understand the nature of the spirals or their use, and it is but just, in treating of this species, to state, that on it those important calcareous appendages were first noticed. It is a very variable shell, as may be seen in Plate III, figs. 2 and 3, is abundantly spread in the lower Lias, and more sparingly in the middle Lias; it is common near Radstock and Bath. Fig. 2 illustrates the largest specimen which I have observed, and which was found there by Mr. Moore; it is also met with in France, in Burgundy, and in many other localities on the Continent.

E

20   SPIRIFER MÜNSTERII, *Dav.* 1851.   Plate III, figs. 4, 5, 6.

> SPIRIFER OCTOPLICATUS, *Zieten.* (non *Sow.*) 1832.   Die Verst. Wurttemb., pl. 38, fig. 6.
> —   —   *Dav.* 1847.   London Geol. Journal, pl. 18, figs. 11—14.
> SPIRIFERINA —   *D'Orb.* 1849.   Prodrome, vol. i, p. 221.

Diagnosis.   Shell inequivalved, variable, with elevated rounded mesial fold in small valve, with corresponding sinus in larger one, with four, five, or six plaits on each side of the mesial fold and sinus; beak more or less produced or recurved, elevated or projected backwards; area well defined, with deltidium in two pieces; interior of both valves similar to that seen in *Sp. rostratus, Walcottii,* &c.; surface punctuated and spinose. Dimensions variable, the largest specimen known measuring—length 15, width 14 lines, but commonly does not attain that size.

Obs.   Many persons, as well as myself, have fallen into the error of attributing this species to *Spirifer octoplicatus,* of Sow., M. C., vol. vi, table 562, figs. 2, 3, 4, 1829, which name was given by that author to a Carboniferous species, much resembling our Liasic one. Zieten appears to have principally led to this mistake, (Die Vers. Wurt., 1832,) and it has also been referred by some Palæontologists to *Spirifer acuticostatus,* a name given by Münster to a shell in the collection of Beyruth, and reproduced by Bronn, along with many others, in 1840, without description or figures, so that it is impossible to say what shell was intended, and no species can be admitted on such uncertain grounds. In 1844 this name, accompanied by a figure and description, was given by M. de Koninck to a mountain limestone species, which name Bronn places first, in page 1172, of his 'Index Palæontologicus.' *Spirifer Münsterii* is a very variable shell, as may be seen from three specimens in Plate III, and is found along with *Sp. rostratus* in the marlstone of Ilminster, and in many localities abroad, such as Fontaine-étoupe-Four, near Caen, &c. Plate III, fig. 4, illustrates the largest specimen I have seen, found by Mr. Moore, near Ilminster; figs. 5 and 6 is the common state in which it occurs. It much resembles *Spirifer cristatus* of the Permian deposits: the tubular spines which cover its surface are stronger, and considerably more numerous, than those observable in *Sp. rostratus.* They are also visible on the portion of the spine facing the front, as in the above-mentioned species.

Genus—TEREBRATULA, *Lhwyd.* 1698.

Shell inequivalved, equilateral, more or less elongated, transverse or circular; exterior smooth, rarely striated or plicated; *larger valve* convex, except in *Ter. Eugenii,* where it is depressed or slightly concave; with or without a sinus, corresponding to a mesial fold in smaller valve, front straight or sinuated; beak straight or recurved and produced, always truncated by an apical, emarginate, or entire foramen, of a circular or elongated shape, variable in its dimensions, and more or less separated from the umbo by a triangular deltidium in one or two pieces, the foramen being surrounded by the substance of the larger valve and deltidium; no true area; lateral beak, ridges undistinct, short, recurved to join the hinge margin, or continued along the sides of the shell without recurving, in which last case there commonly exists between it and the hinge a flatness or false area. *Smaller* or imperforated valve convex or concave, with or without a longitudinal mesial fold; hinge composed of two teeth in the larger valve, which articulate with corresponding sockets in smaller one, so that the valves cannot be separated without fracturing one of the teeth; internal ribbon-shaped lamella, (partly supporting the ciliated arms,) attached only to crura, short or elongated, and more or less folded back on itself; animal fixed to submarine bodies, by muscular fibres passing through the foramen, structure perforated.

Obs. It has been proposed to divide the genus Terebratula according to the length of the loop, which will be referred to in our Introduction.

We have admitted forty species of British Oolitic *Terebratulæ*, which can, we believe, be conveniently retained, although, strictly speaking, many species pass one into another, by some one or more common characters.

In the following table we have arranged these species into two sections, those with a loop extending to near the frontal margin, and those in which the process does not attain, or exceed, half the length of the shell; no external characters, however, seem to denote if the loop be long or short, and we have placed a point of interrogation before those species in which we have not actually seen it, but where, from observations or indication, we believe it to be so.

1st SECTION. *Loop* simply attached to crura, and extending to near the frontal margin.	Beak laterally compressed, carinated; beak ridges continued along the sides without recurving to join the hinge margin; surface smooth.		Terebratula quadrifida.
			— cornuta.
			— Edwardsii.
		?	— Waterhousii.
			— resupinata.
			— Moorei.
			— impressa.
			— carinata.
		?	— emarginata.
		?	— Waltonii.
			— numismalis; *var.* subnumismalis.
	Beak uncompressed laterally; beak ridges soon becoming indistinct, or recurved to join the hinge margin; surface smooth.	?	— Bakeriæ.
			— digona.
			— obovata.
			— ornithocephala.
			— lagenalis.
			— sublagenalis.
	Surface plicated.		— cardium.

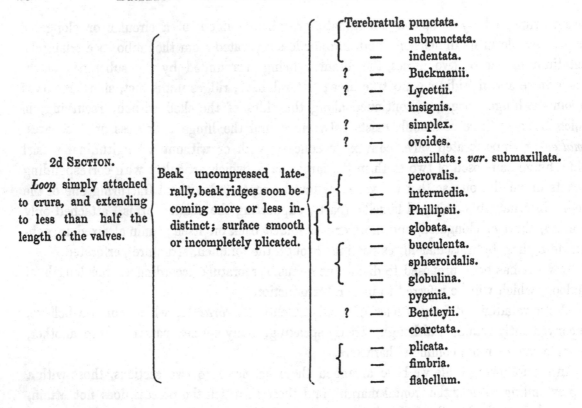

2d Section.

Loop simply attached to crura, and extending to less than half the length of the valves.

Beak uncompressed laterally, beak ridges soon becoming more or less indistinct; surface smooth or incompletely plicated.

Terebratula punctata.
— subpunctata.
— indentata.
? — Buckmanii.
? — Lycettii.
— insignis.
? — simplex.
? — ovoides.
— maxillata; *var.* submaxillata.
— perovalis.
— intermedia.
— Phillipsii.
— globata.
— bucculenta.
— sphæroidalis.
— globulina.
— pygmia.
? — Bentleyii.
— coarctata.
— plicata.
— fimbria.
— flabellum.

21. TEREBRATULA QUADRIFIDA, *Lamarck.*   Plate III, figs. 8—10.

TEREBRATULA QUADRIFIDA, *Lamarck.* 1819.   Anim. sans Vertebres, vol. vi, p. 35.
—   —   *Deslongchamps.* Soc. Lin. de Normandie, 1837.
—   —   *V. Buch.* Mém. Soc. Géol. de France, vol. iii, p. 190, pl. xvii, fig. 3, 1838.
—   —   *Bronn.* 1849.   Index Palæont., p. 1247.
—   —   *D'Orb.* 1849.   Prodrome, p. 240.
—   —   *Dav.* Annals and Mag. of Nat. Hist., vol. v, 2d Series, pl. xiv, fig. 35.   (*Notes on an examination of Lamarck's Fossil Terebratulæ.*)

Diagnosis. "*Testa triangulari-depressá, dilatatá lævi superne quatuor angulis acutis instructá, nati brevi.*" (Lamarck.)

Shell inequivalved, broader than long, irregularly pentagonal, depressed, variable; valves almost equally and slightly convex, with four rounded ridges, extending from the umbo and beak to the front, and three corresponding sinuses on each valve; the hinge margin is convex, extending to a considerable distance, and forming nearly half the circumference of the shell, the remaining portion being well nigh equally divided into three concave portions; beak small, with acute lateral ridge, continued along the side, without recurving to join the hinge-margin, forming a well-defined marginal and false area; beak slightly compressed, foramen entire and of moderate size, separated from the umbo by an obtuse triangular deltidium in two pieces.   Loop of imperforated valve simply attached to the crura, and extending to near the internal frontal margin; impressions of the posterior divisions of the valvular muscles strongly marked, valves minutely punctuated and marked by numerous lines of growth. Dimensions variable, average length 14, width 18, depth 7 lines.

Obs. This species appears to have been first described and named by Lamarck, it has been long known as a characteristic liasic species, and, as is the case with most brachiopoda, it varies considerably in shape, and, in my opinion, passes by insensible graduations into the next form distinguished by Sowerby under the name of *Ter. cornuta.* Some specimens are so irregular that one half would present the characters of *quadrifida*, while the other has those of *cornuta.* In England the type forms of *quadrifida*, are much more rare than those of *cornuta*, and seem to pass into each other's type more than similar shells on the Continent, and especially in Normandy, where both appear more distinct and permanent in their character; therefore, to meet the wishes of the generality of Palæontologists, both names are preserved, and the following may be given as distinctive characters. *T. cornuta* is deeper, more convex, and longer than it is wide, the reverse of what we generally find in *quadrifida*, which is wider than it is long, flatter and much less deep, the beak and area are also larger and more produced in *T. cornuta.* Both species, or more properly speaking, in my opinion, varieties of one type, they are found together in the same beds and localities, but in some places one or the other form prevails. In Plate III will be seen illustrations of both forms, as found in England; figs. 8, 9, 10 represent *Ter. quadrifida*, while 11—18 illustrates *Ter. cornuta.* *Ter. quadrifida* is usually found in the marlstone or middle lias, and particularly at South Petherton, near Ilminster, where Mr. Moore procured many specimens, but I have not seen any as large or as fine as those of Normandy, Vieuxpont, Landres, and Evrecy, near Caen; M. D'Orbigny states their occurrence at Saint-Amand (Cher.), and Nancy, in France. Figs. 8—9 are from Mr. Moore's collection.

22. TEREBRATULA CORNUTA, *Sow.* Plate III, figs. 11—18.

TEREBRATULA CORNUTA, *Sow.* Min. Conch., 1825, vol. v, p. 66, pl. 446, fig. 4.
— VICINALIS, *V. Buch.* 1838. Mém. Soc. Géol. de France, vol. iii, p. 192, pl. xvii, fig. 5. (Non *T. vicinalis*, Schloth., according to M. D'Orbigny.)
— CORNUTA, *Morris.* Catalogue, 1843.
— — *Bronn.* 1849. Index Palæont., p. 1233.
— — *D'Orb.* Prodrome, 1849, vol. i, p. 240.

Diagnosis. Shell inequivalved, irregularly pentagonal, generally longer than wide; valves almost equally convex, thick and deep; hinge-margin forming nearly half the circumference of the shell, the remaining portion and front being divided into three more or less defined concave curves, the central or frontal one in common deeply indented, but obscurely so laterally, the surface of both valves are smooth and shining; three concave sinuses, forming two elevated rounded ridges, visibly diverging from the umbo towards the front, but which only begin to rise above the level of the shell towards the anterior portion, increasing as they approach the frontal margin; beak large, more or less recurved, with lateral ridges continued along the side of the valve, without recurving, to join the hinge-margin, forming a defined marginal area; beak slightly compressed and keeled, truncated by an entire foramen of moderate size; deltidium in two pieces, more or less hid by the recurving of the beak; loop free, attached merely to the crural base, and extending to near the internal frontal margin; valves finely punctuated, and strongly marked by numerous lines of growth. Dimensions and form variable, length 20, width 13, depth 12 lines.

Obs. When treating of *Ter. quadrifida*, I expressed how intimately I believed it was connected with the present form, and stated what distinctions could be brought forward to characterise both; it is a very common shell in the middle Lias, both of England and France, found abundantly at South Petherton, and near Cheltenham, where it varies considerably in form, as may be seen from the illustration, Plate III, figs. 17, 18. The usual type approaching more or less to figs. 13, 16, and 21; 11, 14, 15 being exceptional forms. The largest specimen I have seen in England, measuring 21 lines in length and 18 in breadth, belongs to the collection of the British Museum. We are indebted to Mr. Moore for the working out of the fine interior illustrating this species.

23. TEREBRATULA EDWARDSII, *Dav.* Plate VI, figs. 11, 13, 14, and 15?

Diagnosis. Shell inequivalved, globose, more or less circular, as wide as long, straight in front; valves convex, sometimes globose, and distinctly emarginated; beak much recurved, and truncated by a small foramen advancing over the umbo, almost touching it, so as to conceal the deltidium, which is rarely visible; lateral ridges extending along the sides of the shell, without recurving to join the hinge margin; surface smooth, finely punctuated; loop extending to near the margin of the shell, and simply attached to Crura. Length 15, width 13, depth 10 lines.

Obs. This species may be distinguished from both *T. punctata* and *subpunctata* by the shortness and squareness of its shape, as well as by its strongly recurved beak, lateral ridges, smaller foramen, and length of loop, which last fact is due to Mr. Moore's exertions, who, after much trouble, was enabled to clear the process, seemingly identical with that of *T. cornuta*, to which shell it approaches by many characters, though quite distinct by the roundness of its sides and square front; in some specimens, as in *T. subpunctata*, we perceive a flatness at the umbo, see figs. 11 and 11*a*, due to compression when young. *Ter. Edwardsii* is found along with *Ter. subpunctata, cornuta resupinata,* &c., in the marlstone of South Petherton, near Ilminster, and it gratifies me highly to name it after the learned Dean of the French Academy of Sciences. It is with hesitation I have placed here a solitary specimen, Plate VI, fig. 15, the shape of which is doubtless due to deformity. Fig. 11, in Mr. Walton's collection; 11—16, from that of Mr. Moore.

24. TEREBRATULA WATERHOUSII, *Dav.* Plate V, figs. 12, 13.

Diagnosis. Shell inequivalved, longer than wide; imperforated valve convex; beak small, rounded, and truncated by an entire foramen, with acute lateral ridges continued along the sides, without recurving to join the hinge line; deltidium in two pieces; imperforated valve convex, except towards the front, where there is a concave depression, distinctly bent downwards, giving the frontal line an indented appearance; loop attached only to the crura, and extending to near the frontal margin of the shell. Length 9, width 7, depth 6 lines.

Obs. This species, although distinct, comes near to *Ter. cornuta* by the form of its beak, approaching those varieties in which the lateral prominence is not developed; it differs, however, in having the front distinctly bent downwards, as is seen in Plate V, figs. 12*a*, 13*a*. Its greatest thickness is about the umbonal portion; it varies much in form and shape, as may be perceived by comparing fig. 12 with fig. 13. *Ter. Waterhousii* belongs to the upper portion of the middle Lias, and was first found by Mr. Walton and myself at Farington Gurney, near Radstock, along with *T. numismalis rimosa*, &c. Dr. Krantz, of Bonn, sent me specimens of this species from Bonfingen, in Wurtemberg, in which locality this shell acquires larger dimensions than those as yet found in England.

25. TEREBRATULA RESUPINATA, *Sow.* Plate IV, figs. 1—5.

TEREBRATULA RESUPINATA, *Sow.* Min. Conch., 1818, vol. ii, p. 116, t. 150, figs. 3, 4.
— — *Desh.* 1836. Nouv. ed. de Lamarck, vol. vii, p. 360.
— — *Pusch.* Polens Palæontologie, 1837, tafal iv, fig. 6.
— — *Phillips.* Geol. of Yorkshire, pl. xiii, fig. 23, 1835.
— — *Deslongchamps.* 1837. Soc. Linn. de Normandie.
— — *V. Buch.* Mém. Soc. Géol. de France, 1838, p. 229, pl. xx, fig. 11.
— — *Morris.* Catalogue, 1843.
— — *Bronn.* Index Palæont., 1849, p. 1248.
— — *D'Orb.* Prodrome, 1849, vol. i, p. 239.

Diagnosis. Shell inequivalved, oblong, longer than broad; rostral valve convex, much compressed and keeled; beak small and incurved, almost touching the umbo, lateral ridges acute and continued along the side, without recurving to join the hinge margin, false or marginal area increasing in width at some distance from the beak; foramen entire, remarkably small, with a wide obtuse deltidium, in two pieces, rapidly decreasing as it approaches the small foramen, which it partly encircles. Imperforated valve, laterally convex, with a deep central longitudinal groove and depression, extending from the umbo to the front, which is considerably curved backwards and depressed. The marginal line is straight or gently curved on leaving the hinge, till it attains about two thirds of the length of the shell, when, turning suddenly backwards by a rapid curve, it produces a considerable angle to the first portion of the line and acute frontal depression.

Loop in smaller valve simply attached to the crura, and extending to near the frontal margin; the inner side of the lamella, facing the opening of the valves, is irregularly covered with spines, similar to those observable on the spine of *Sp. rostratus* and loop of *Ter. pectunculoides*, &c.; the outer side facing the sides of the valve is smooth, and always without spines.

Valves smooth, finely punctuated, and marked by numerous lines of growth. Dimensions and form variable; length 16, breadth 14, depth 12 lines.

Obs. Many shells have been attributed to Sowerby's *Ter. resupinata*, apparently not belonging to the type of that author, established on a liasic shell, answering to the above description. It is, however, no easy matter, indeed scarcely possible, to give an accurate description of any species agreeing with all the shells it includes, from the innumerable variations they constantly present, especially in some species which pass into one another in the most perplexing manner; so much so, that certain characters are diminished in value from their irregularity. It is indispensable to give distinct names to certain forms, which in reality merge from one type into another, but would be incomprehensible or difficult to remember, if we were compelled to give a long and interminable list of varieties, with distinct denominations added to the typical name. The *Resupinata* group, for instance, presents endless variety in form, most difficult to characterise, and especially so when we have before us a great number of specimens. Sowerby's type is distinguished from the other members of the group by its small elevated and recurved beak, diminutive foramen, and strongly laterally compressed and keeled rostral valve, more especially so towards the beak, which character seems peculiar to some of the liasic species. The depression and longitudinal groove of the imperforated valve, which is strongly marked in this form, varies to so great a degree, that, on a large assemblage of specimens, we trace passages into the next species or variety, *Ter. Moorei*, where, instead of the depression visible in the type *Ter. resupinata*, we find the imperforated valve completely convex and gibbous; so much so, that the longitudinal furrow disappears, and, becoming sensible only towards the frontal portion; the larger valve presents likewise a slight longitudinal depression towards the front, corresponding with that visible in the other valve, never seen in Sowerby's type of *resupinata*; the margin line of the two valves is also nearly straight all round, so that it would be difficult to recognise the original type of *Ter. resupinata* in *Ter. Moorei*, had we not before us a number of specimens illustrating the gradual passage. *Ter. carinata, Ter. impressa*, and other forms, which, though nearly allied and belonging to the same group, should at least, for convenience, be considered as specifically distinct, as they in general vary enough to be easily distinguished. It is to Mr. Moore we owe the discovery of the first specimen, showing spines on the inner side of the calcareous loop, facing the frontal opening of the shell, and which I subsequently recognised to be peculiar to several species. *Ter. resupinata* is abundantly found in the marlstone beds of the middle Lias, associated with *Ter. quadrifida, cornuta, Moorei,* &c., at South Petherton, near Ilminster; at Deddington it was found by Mr. Faulkner, and is stated to occur in Yorkshire by Professor Phillips. In France it is abundantly found at Evrecy, Landes, Fontaine-étoupe-four, &c., near Caen, and I believe also in Germany, though none of the numerous specimens I have seen from that country agree exactly with Sowerby's type. Figs. 1 and 2 are from the collection of Mr. Moore.

26. TEREBRATULA MOOREI, *Dav.* Pl. IV. Figs. 6, 7.

TEREBRATULA MOOREI, *Dav.* Bull. Soc. Géol. de France, 2de Série, t. vii., pl. 1, figs. 21—23.

Diagnosis.—Shell ovate, elongated, slightly notched in front; valves convex of a nearly equal depth, smooth, finely punctuated, with strong lines of growth. The imperforated valve presenting a slight longitudinal depression visible at the umbo, soon lost from the convexity of the valves, but reappearing towards the front. Rostral valve convex with small recurved beak, compressed laterally and keeled, with diminutive foramen; a slightly longitudinal depression is visible towards the front. The margin is straight, almost all round. Loop simply attached to the Crura and extending to near the frontal margin. Length 18, breadth 14, depth 11 lines.

Obs. When treating of *Ter. resupinata* we endeavoured to point out the affinities and distinctive characters between the two forms. This species is found in the marlstone of South Petherton, along with *T. resupinata.* It also occurs in beds of the same age in Normandy. Fig. 6 is from the collection of Mr. Tennant.

27. TEREBRATULA IMPRESSA, *V. Buch.* Pl. IV, figs. 8—10. Pl. X, figs. 7, 7*a b c.*

TERABRATULA IMPRESSA, *De Buch.* Zieten Würtemb. Verst., p. 53, pl. xxxix, fig. 11, 1832.
— — *Bronn.* 1837. Leth. Geog., p. 306, pl. xviii, fig. 12.
— — *V. Buch.* Mém. Soc. Géol. de France, 1ere Série, p. 226, pl. xx, fig. 7, 1838.
— — *Quenstedt.* Das Flözgebirge Wurtembergs, 1843.
— — *Bronn.* Index Palæont., 1849, p. 1238.
— — *D'Orb.* Prodrome, vol. i, p. 288, 1849.

Diagnosis. Shell inequivalved, more or less rounded, length and width nearly the same; rostral valve convex with rounded recurved beak, foramen entire, deltidium in two pieces obtusely triangular: lateral ridge of the beak continued along the side without recurving to join the hinge margin. Imperforated valve, slightly convex at the umbo; but soon becoming gently and equally depressed; the marginal line of both valves forming a regular depressed curve soon after leaving the hinge line. Loop simply attached to Crura and extending to near the frontal margin of the valve.   Surface finely punctuated; length and width from 9 to 10 lines, depth 5.

Obs. This species is generally distinguished from *T. resupinata* and *carinata* by its more circular form, the depression in the anterior portion of the smaller valve is less deep and relatively broader, and more regularly curved; the lateral ridges of the beak are not extended so much in the longitudinal direction, the beak being rounder and not much compressed.   Such are the distinctions observable in well characterised specimens of *Ter. impressa* found abundantly in the Oxford Clay of several parts of England, as at St. Ives, in Huntingdonshire; and in similar beds near Boulogne-sur-mer.   We also find in the Inferior Oolitic beds of Cheltenham, a shell connecting *Ter. impressa* with *T. resupinata*, but which, from the size of its foramen and more circular shape, is more properly associated with *Ter. impressa*; and I perceive in the 'Prodrome,' M. D'Orbigny places this last-named species in the Inferior Oolite, or his *Terrain Bajocien*, but omits it in the Oxford Clay, where the type form of the species occurs.   The similarity of some of the specimens from Cheltenham to those of the Oxford Clay is so striking, that I would not even venture to give this variety a distinctive name, as some specimens of *Ter. impressa* may be seen in Pl. IV, fig. 9, to have a lengthened shape instead of a short circular form peculiar to the generality of the impressa tribe.   In Germany this shell is so abundant that local geologists have distinguished the bed containing them by the name of *Impressa-thorn*: here likewise we find a small difference in the aspect of the shell, which seems to me owing more to local causes than to any specific differences.   These German shells are not commonly quite so deep; the smaller valve is a little less convex posteriorly than what is usually seen in those from the Oxford Clay.

Another small race, or variety, Pl. X, fig. 7, is found in the Inferior Oolite of Sherborne, Dorsetshire, which links *Ter. resupinata* to true *T. impressa*; we place it with the last-named species from the characters of its beak; as none of the species related to the *T. resupinata* group, hitherto found in the Oolites, have that small laterally pinched and recurved beak, or diminutive foramen so peculiar to the liasic species; a character any one would perceive who had observed a series of specimens from the two deposits. I have come to the above conclusion after a thorough examination of many hundreds of specimens, nor should I think myself justified in distinguishing these varieties by distinct names.

Ter. impressa is abundantly found in Germany, especially in Wurtemberg.

Plate IV, figs. 9 and 10, are type specimens of the species from the Oxford clay, in possession of Mr. Morris.

„   fig. 8, var. from the Inferior Oolite of Cheltenham, from a specimen in Dr. Wright's Collection.

Plate X, fig. 7, a var. from the Inferior Oolite of Sherborne, in the Collection of the British Museum.

„   fig. *7a b c* are enlarged illustrations.

28. TEREBRATULA CARINATA, *Lamarck.* Plate IV, figs. 11—17.

TEREBRATULA CARINATA, *Lamarck.* Anim. sans. Vert., vol. vi, p. 25.
   —   —   *Tennant.* 1847. Stratigraphical List of British Fossils, p. 73.
   —   —   *Bronn.* 1849. Index Palæont., p. 1232.
   —   —   *Dav.* 1850. Lamarck's Species of Fossil Terebratulæ, Ann. and Mag. of Nat. Hist., vol. v, 2ᵈᵉ Série, pl. xiii, fig. 25.

Diagnosis. " *T. testá subquadrangulari lævi; valvá inferiori subcomplanatá superiore diedrá medio-carinatá.*"—(Lamarck.)

Shell inequivalved, oblong, longer than wide, smooth, surface punctuated. Perforated or rostral valve convex with slightly recurved beak, lateral ridges continued along the sides without recurving to join the hinge margin, foramen of moderate size, entire and separated from the umbo by a rather long deltidium in two pieces; imperforated valve regularly and gently compressed almost from the umbo; front sometimes considerably thickened. Loop long and attached only to Crura: average dimensions; length 13, width 9, depth 5 or 6 lines, some specimens measuring, length 20, breadth 17, depth 10 lines.

Obs. This species, established by Lamarck, in 1819, is distinguished from *T. resupinata* by its rounder and less recurved beak, larger foramen generally separated from the umbo by a greater distance and larger deltidium: it is also flatter and is a more elongated shell. The depression of the imperforated valve is not the same in all specimens, some such as those figured 13, 14, from Mr. Bowerbank's Collection, are almost convex, presenting a depression only towards the front. The thickening or emargination of the valves is very great in some specimens, as may be observed in fig 12, preserved in the Collection of the Geological Survey. This species occurs in the Inferior Oolite near Stroud, Burton Radstock, Dinnington, &c.; and fine specimens are to be seen in the Collections of Messrs. Bowerbank, Morris, Moore, Walton, &c. In France it is found in the Inferior Oolite round Caen in Normandy. In the Inferior Oolite of Chalford, near Stroud, and Crichley Hill, near Cheltenham, a very wide and large variety of this species is found; figs. 15, 16, and 17, remarkable from the size of its foramen; the lateral ridges are also much nearer to the umbo and thrown off from the fore part of the beak opening. I do not believe we could be justified in separating this variety from *T. carinata*, as the above distinctions are not permanent, but varying in many specimens. Of this variety many fine specimens are to be seen in the British Museum, as well as in the Collections of Messrs. Morris, Buckman, Dr. Wright, &c.

29. TEREBRATULA EMARGINATA, *Sow.* Plate IV, figs. 18—21.

TEREBRATULA EMARGINATA, *Sow.* Min. Con., 1825, vol. v, p. 50, pl. 435, fig. 5.
   —   —   *Deslongchamps.* 1837. Soc. Linn. de Normandie.
   —   —   *Morris.* 1843. Catalogue.
   —   —   *Bronn.* 1849. Index Palæont., p. 1236.
   —   —   *D'Orb.* 1849. Prodrome, vol. i, p. 287.

Diagnosis. Shell inequivalved, subrhomboidal short and broad; when adult notched and indented in front, perforated valve convex, beak slightly recurved, foramen separated from the umbo by a deltidium in two pieces, front defined by two angles or emarginate edge, imperforated valve flat or slightly convex, sometimes depressed longitudinally towards the front; smooth, punctuated and marked by strong lines of growth. Loop long, and attached only to the Crura. Length 11, breadth 10, depth 6 lines.

Obs. The above description agrees with that given by Mr. Sowerby; it is a very variable shell: the flatness of the imperforated valve seems to be one of its principal characters Some specimens are however slightly convex; it approaches near to *T. carinata*, but is a shorter and wider shell, rarely presenting any of the longitudinal depression common to *Ter. resupinata* and *T. carinata*, forming one of the transition shapes connecting the resupinata tribe, with a depressed imperforated valve to those species in which the same valve is convex, such as *Ter. Waltonii*. In Pl. IV, figs. 18, 19, 20, I have given the original specimens figured in the 'M. C.,' kindly lent me by Mr. J. de C. Sowerby; fig. 21 belongs to Mr. Tennant's Collection, and shows an unusual deep longitudinal furrow. This species belongs to the Inferior Oolite, and is stated to come from Nunney, near Frome. Five specimens are preserved in the British Museum, collected by Mr. Cunnington. It occurs also on the Continent, especially in Normandy, as noticed by M. Deslongchamps.

30. TEREBRATULA WALTONII, *Dav.*   Plate V, figs. 1—3.

Diagnosis. Shell inequivalved, more or less oval; valves convex, beak produced, rounded, truncated by an entire foramen, separated from the umbo by a rather long deltidium in two pieces; lateral ridge of beak continued along the side, without recurving, to join the hinge margin. The margin line of valves nearly straight all round; front and sides emarginate; surface smooth, finely punctuated; very variable in form and dimensions: with the same depth of 8 lines, the three specimens figured from the Collections of Messrs. Walton and Tennant gave a length of 18, 18, 13, breadth 11, 13, 11 lines.

Obs. This species is from the Inferior Oolite of Bathwick Hill, Bath and Burton. The loop is unknown, but must have been simply attached to the Crura, and extended to near the frontal margin. The emargination of most specimens of this species is very remarkable, as well as the regular and almost equal convexity of the valves. It will take place next to *T. emarginata*, to which it is most nearly allied. I name it after Mr. Walton, of Bath, who has kindly afforded me frequent assistance in this work.

31. TEREBRATULA NUMISMALIS, *Lamarck.*   Plate V, figs. 4—9.
Var. SUBNUMISMALIS, *Dav.*   Fig. 10.

TEREBRATULA NUMISMALIS, *Lamarck.* 1819. Anim., sans Vert., vol. vi, No. 22.
—   COR.   *Valenciennes* in *Lamarck*, 1819, An. sans Vert., vol. vi, No. 22.

TEREBRATULA NUMISMALIS, *Zieten.* 1832. Würtemb. Verst., pl. xxxix, fig. 4.
— ORBICULARIS, *Zieten.* 1832. Würtemb. Verst., pl. xxxix, fig. 5.
— NUMISMALIS, *Bronn.* Leth. Geog., pl. xviii, fig. 8.
— — *Deslongchamps.* 1837. Soc. Linn. de Normandie.
— — *V. Buch.* 1838. Mém. Soc. Géol. de France, vol. iii, 1ᵉʳᵉ Série, p. 191, pl. xvii, fig. 4.
— — *Bruyere.* Ency. Meth., p. 240, fig. 1.
— — *Quenstedt.* 1843. Das Flös. Wurt., pp. 183, 184.
— — *D'Orb.* Prodrome, vol. i, p. 240, 1849.
— — *Dav.* 1850. Lamarck's Species, Annals and Mag. of Nat. Hist., vol. v, 2ᵈᵉ Série, pl. xiii, fig. 17, and pl. xv, fig. 22.

Diagnosis. "*T. testá subrotundá lævi, utraque valvá superne sinu instructá : striis concentricis remotis : nate brevi : foramine minimo.*"—(Lamarck.)

Shell inequivalved, depressed, nearly circular or slightly pentagonal, notched in front, variable in its contour; valves nearly equally convex, much depressed and flattened, surface smooth, the margin of valves forming almost a straight line; beak slightly compressed and keeled, lateral ridges of beak continued along the side, without recurving, to join its margin; false area, very small and well defined; deltidium in two pieces, wide at the umbo, diminishing as it reaches the foramen, which is entire and remarkably small. Sometimes a slight central depression or sinus is visible on each valve; lines of growth strongly marked, punctuation very fine. Loop free, attached only to the crural base, and extending nearly the whole width of the shell; mesial plate well defined. Length and width on an average nearly the same, from 11 to 13 lines, depth from 4 to 6.

Obs. This species was first brought into notice by Lamarck, who pointed out its extremely small area and foramen; it is easily distinguished from all other liasic and oolitic Terebratulæ by its extreme flatness. Lamarck and Valencienne's *Ter. cor.*, as I stated in the 'Annals of Natural History,' of June, 1850, is a synonym of this species, established on a specimen bearing accidentally the shape of a heart, and is only one of the numerous variations in contour, observable in *Ter. numismalis*, as may be seen from some of the forms illustrated in Plate III, figs. 4—9. This species is very abundant in the highest beds of the middle and upper Lias, and it is only within a few years that it was discovered in England, in the neighbourhood of Cheltenham, where the specimens have a dark grayish colour. Mr. Walton and I found it two years ago in the neighbourhood of Radstock, and at Farington Gurney, where they are yellowish, owing to the colour of the clay in which they are imbedded; and many fine specimens are to be seen in the Collections of Dr. Wright, Professor Buckman, and others. It was also obtained by Mr. Hugh Miller, from the Lias of Shendwich, in Scotland. Plate V, fig. 8, illustrates a specimen from that locality. In France it is very common, about Eurecy, Landes, Vieux Pont, St. Armand, Pouilly, Avallon, Lyon, &c., where larger and finer specimens than any of our British ones have been found, and they may be viewed in the Collections of M. Deslongchamps, of Caen, who kindly lent me the specimens illustrating the internal loop, figured in Plate V, fig. 9. It is also abundant in Germany, and Von Buch gives the following localities: on the Plienback, near Boll, near Eslingen, at Blattenhardt, Denchen-dorff, Gönningen, and between Tübingen and Hechingen, at Eley, near Brunswick, &c.

Var. SUBNUMISMALIS.   Fig. 10.

I have placed here, with some doubts, as a variety, the only specimens of a shell found by Mr. Moore in the marlstone of Ilminster, from having observed, in the collection of the Ecole de Mines, in Paris, specimens of undoubted *numismalis*, much approaching this in shape.   Its beak and foramen are, however, so much larger than that characteristic of *true numismalis*, that, as is above stated, I am not certain of its belonging to *numismalis*, but do not like to make a species formed on the observation of a single specimen.

32. TEREBRATULA BAKERIÆ, *Dav.*   Plate V, fig. 11.

Diagnosis.   Shell inequivalved, broader than long; rostral valve moderately convex; beak small; lateral ridge rounded and indistinct; foramen entire, nearly touching the umbo; imperforated valve less convex, with front rather suddenly bent downwards in the middle, so as to present a semicircular figure; the frontal depression is continued along the small valve to the centre, but becomes gradually shallower towards the umbo.   Surface smooth, finely punctuated.   Length 4, breadth 5, depth $2\frac{1}{2}$ lines.

Obs. This little shell approaches in form to *Ter. nucleata*, but has not so incurved a beak, and appears much less deep.   *Ter. Bakeriæ* is stated to have been found in the Inferior Oolite of Bugbrook, by the late Miss Baker, and is now preserved in the Collection at the British Museum.

33. TEREBRATULA DIGONA, *Sow.*   Plate V, figs. 18, 24.

TEREBRATULA DIGONA, *Sow.* Min. Con., 1812, vol. i, p. 217, tab. 96, figs. 1—5.
      —   —   Ency. Meth., pl. 240, fig. 5.
      —   —   *Lamarck.* 1819.   Anim. sans Vert., vol. vi, No. 19.
      —   —   *Deslongchamps.* 1837.   Soc. Linn. de Normandie.
      —   —   *Smith.* Stratigraphical System of Organised Fossils, 1816, 1817.
      —   —   *V. Buch.* 1838.   Mém. Soc. Géol. de France, vol. iii, p. 194, pl. xvii, fig. 6.
      —   —   *Morris.* Catalogue, 1843.
      —   —   *Bronn.* 1849.   Index Palæont., p. 1235.
      —   —   *D'Orb.* 1849.   Prodrome, vol. i, p. 315.

Diagnosis.   Shell inequivalved, more or less triangular, oblong; valves convex;

marginal line nearly straight, front more or less convex or concave, bounded, when old, by two prominent angles alike in each valve, as if produced by the pinching of the edge; beak produced, rounded, truncated by an entire foramen, separated from the umbo by a somewhat long triangular deltidium in two pieces; lateral ridges not continued to a great distance, there existing a strongly-marked lateral flatness, the greatest width of the shell being generally, though not always, at the front; surface smooth, finely punctuated; loop attached only to the crura, and extending to near the margin of the shell. Dimensions and form variable; average size, length 13, width 9, depth 8 lines.

Obs. This is one of the first-described species in the Min. Con., and long known as a characteristic oolitic fossil, and is stated by Sowerby, p. 217, to be "*variable in its form, sometimes almost globose, at others acutely triangular and rather depressed; the two angles of the front are continued a little along each valve, and look as if they were produced by pinching the edges between the fingers; the front between the angles is in some shells concave, in others straight, or of different degrees of convexity.*" Von Buch's description of this species would lead us to imagine that the greatest width of the shell was always in front, since he states and underlines "Les arêtes Cardinales descendent en divergeant d'une manière continue vers les côtes, et remplacent tout à fait les arêtes laterales, on autrement dit ne convergent pas, mais descendent verticalement de sorte que la *largeur du front est en même temps la plus grande largeur de la coquille.*" Though this is the general distinguishing character of digona, it often varies, as I have shown from the original figures of the species, from Smith's Collection, deposited, in 1816, in the collection of the British Museum, thus often approaching, in general form and convexity, to certain specimens of *Ter. obovata*, but in all its varieties *T. digona* may be distinguished by the lateral beak ridge not being recurved to join the hinge margin, and the lateral angles at the front are never visible in obovata, the sides of the larger valve being in the adult specimens usually flat and vertical, which is not the case in *T. obovata*.

The figures I have given in Plate V will illustrate some of the varieties in form which are presented by this species; occasionally specimens are perfectly equilaterally triangular, others are twice as long as they are wide, while some are as wide as they are long, and were well figured and described by Mr. Smith, in 1816. It is a very common shell in the Bradford Clay Forest Marble and Great Oolite; it is found abundantly round Bath, Bradford, Cirencester, &c., it occurs by millions at Ranville, in Normandy, and in other localities, and is also stated to be found at Muggendorf, by Von Buch. Fig. 18 illustrates Mr. Smith's specimens, now in the Collection of the British Museum.

34. TEREBRATULA OBOVATA, *Sow.* Pl. V, figs. 14—17.

TEREBRATULA OBOVATA, *Sow.* Min. Con., 1812, vol. i, p. 228, tab. 101, fig. 5.
— — *Morris.* Catalogue, 1843.
— — *D'Orb.* Prodrome, vol. i, p. 316, 1849.

Diagnosis. Shell obovate, transverse, valves nearly equally convex, gibbous, beak more or less recurved and rounded, foramen entire, with lateral ridges recurved and joining the hinge margin within a short distance of the beak; deltidium in two pieces, more or less visible, according as the beak becomes more or less recurved: front straight or slightly convex, bounded by two obsolete plaits, sides rounded, margin rather flat, loop simply attached to the Crura and extending to near the front margin. Dimensions variable, length 15, width 13, depth 10 lines.

Obs. This species was correctly described by Sowerby in 1812. It is a thick globose shell, abundantly spread in the Cornbrash of many localities; as at Rushden, in Northamptonshire, where it attains large dimensions, as may be seen from a suit of specimens from that locality, presented to the British Museum by Mr. Waterhouse. It is also abundant in Wiltshire, where specimens showing the loop to great perfection can be obtained. It occurs near Scarborough, where it was obtained by Mr. Bean, and is likewise found in many parts of the Continent; at Boulogne, by Mr. Bouchard. D. Bronn does not appear to have been acquainted with this species, from what we find in the 'Index Palæontologicus.' It is always more globose than *T. digona*, and does not present any frontal angles, its greatest width is also considerably increased towards the middle of its length.

Pl. V, fig. 16, illustrates one of the original specimens of the M. C.

    „   14, 15, two full-grown specimens from the British Museum; fig. 17 showing a curious deformity in this species, in which the accident is reproduced on both valves.

Pl. VII, fig. 5, is a very curious shell found in the Cornbrash of Norman Cross, by Mr. Morris, along with the true types of *obovata* and *lagenalis;* after minute comparisons, Mr. Morris and myself came to the conclusion that it can only be a deformity of the species under consideration.

35. Terebratula ornithocephala, *Sow.* Plate VII, figs. 6, 13, *and* 23.

Terebratula ornithocephala, *Sow.* 1812. Min. Con., vol. i, p. 227, tab. 101, figs. 2, 3, 4.
— lampas, *Sow.* 1812. Min. Con., vol. i, p. 228.
— triquetra, *Sow.* 1825. Min. Con., vol. v, p. 65, tab. 445, fig. 1.
— ornithocephala, *Smith.* 1816, 1817. Stratigraphical System of Organised Fossils.
— — *Desh.* 1836. Nouv. Ed. de Lamarck, vol. vii, p. 361.
— — *Phillips.* 1836. Geol. of Yorksh., vol. i, tab. 6, fig. 7.
— — *Zieten.* 1832. Die vers. Würt., tab. xxxix, fig. 2.
— — *Morris.* 1843. Catalogue of British Fossils.
— — *Bronn.* 1849. Index Palæont., p. 1243.
— — *D'Orb.* 1849. Prodrome, vol. i, p. 316.
— subtriquetra, *D'Orb.* 1849. Prodrome, vol. i, p. 216.

Diagnosis. Shell ovate rhomboidal, depressed when young, elongated and gibbous when adult; beak rounded, much recurved and truncated by an entire foramen closely approaching the umbo; deltidium in two pieces, more or less hid by the prominence of the beak, without distinct lateral ridge; imperforated valve, most convex near the umbonal part of the shell; when young sometimes slightly indented in front, most wide posteriorly; rather depressed laterally, and tapering more or less in front: surface smooth, finely punctuated: loop simply attached to the crura and extending to near the frontal margin; form variable; average size, length 16, width 11, depth 10 lines.

Obs. Several species have been established from varieties of the type under consideration, but they merge into one another by insensible passages; we might probably place among them *Ter. lagenalis*, Schl., and *triquetra* of Sow.; this last, especially, is only a younger, wider and slightly indented state, as I became convinced from an examination of the original specimens of the 'M. C.' lent me by Mr. J. De C. Sowerby, Pl. VII, figs. 10, 11 and 12, illustrating the original types of *T. triquetra*. The principal difference I perceive between *T. lagenalis* and *T. ornithocephala*, is that in the last the posterior margin of the shell is larger and more prominent, while in *T. lagenalis* the lateral sides are much straighter, and wider and more square in front. Prof. Brown in his '*Index Palæontologicus*' places *T. lagenalis* among the Synonyms of *ornithocephala*; and M. Deshayes in 1836, while describing *Ter. lagenalis*, states that it has some analogy to *ornithocephala*; but as several Palæontologists seem desirous of retaining the name of *T. lagenalis* for those more elongated, thicker and straighter varieties, I have described both under a distinct head, and in Pl. VII will be seen a number of specimens illustrating both. Figures 6, 7, 8, 9, 10, 11, 12, would represent the typical forms of *ornithocephala*. The interior of both *T. ornithocephala*, fig. 23, and *T. lagenalis* are completely similar; some varieties or rather specimens of *ornithocephala*, though gibbous at the umbone, present a slight longitudinal depression in that portion as seen in fig. 9. *Terebratula lampas* of Sow. is merely an internal cast of *Ter. ornithocephala*, and which M. D'Orbigny seems to adopt as a species, and gives it to a Liasic shell, but which I believe has nothing to do with Sowerby's type.

In England *Ter. ornithocephala* is found in the Kelloway Rock, at Kelloway, where it was found by Messrs. Walton, Morris, &c.: it is abundant in the Fullers-earth all round Bath; in the Cornbrash of Rushden, in Northamptonshire, where it was picked up by the Rev. A. W. Griesbach, &c.; and may be seen in the Collection of the British Museum, as well as in most collections. It is also found in many parts of the Continent. Count A. De Zigno mentions it from the Venetian Alps; Zeiten and Quenstedt in Germany; and Deslongchamps from Normandy. Pl. VII, figs. 7, 8, in British Museum; figs. 10, 11, the original specimens of *T. triquetra*, in the Collection of Mr. J. De C. Sowerby; fig. 23, interior, from the Collection of Mr. Morris.

36. TEREBRATULA LAGENALIS, *Schlotheim.* Plate VII, figs. 1—4.

TEREBRATULITES LAGENALIS, *Schl.* 1820. Petrefacta.
TEREBRATULA   —   *Desh.* 1836. Nouv. Ed. de Lamarck, vol. vii.
      —   —   *De Buch.* 1834. Mém. Soc. Géol. de Fr., vol. iii, p. 194,
                             pl. 18, fig. 7.
      —   —   *Tennant.* 1847. A Stratigraphical List of British Fossils,
                             p. 73.
      —   —   *Bronn.* 1849. Index Palæont., p. 1240.

Diagnosis. Shell elongated, ovate, nearly straight in front; valves very convex and gibbous; beak rounded and much recurved, truncated by an entire foramen, almost touching the umbo; deltidium rarely visible, on account of the projection of the beak, lateral ridges indistinct; imperforated valve, gibbous and deepest near the umbo : posterior margin rounded and extending by a gentle curve to the edge of the front, which is generally straight; surface of valves smooth, finely punctuated. Loop simply attached to crura, and extending to near the margin of the shell; central septum well defined; dimensions variable. Length 22, breadth 12, depth 11 lines.

Obs. As stated under the head of *T. ornithocephala*, this species has very little to distinguish it from the above-named shell, into which it seems to merge by insensible passages; it is therefore very variable, sometimes so much thickened in front, as to form an almost flat surface, perpendicular to the surface of valve, as in fig. 1. At other times it is on the contrary acute, as in fig. 2, and tapering almost into a sharp edge (fig. 4); the beak is also more recurved than in *T. ornithocephala* (fig. 1), but in some specimens, as in fig. 2, the deltidium is completely exposed.

This species is abundant and finely preserved in the Cornbrash of Rushden, Norman Cross, Undle, and Thorpe, in Northamptonshire, where it has been collected by the Rev. A. W. Griesbach, Messrs. Waterhouse, Morris, and others. It has also been met with by Messrs. Lowe and Walton, in the Fullers-earth near Bath, and from near Stamford, in Lincolnshire. On the Continent it is abundantly distributed, and fine specimens, two inches in length, have been obtained by M. Bouchard, from the Cornbrash of the neighbourhood of Marquise (Pas de Calais). In Germany we meet it at Wöschnau, at Grumbach near Amberg, and near Scheffhausen, Wurtemberg.

37. TEREBRATULA SUBLAGINALIS, *Dav.* Plate VII, fig. 14.

Diagnosis. Shell inequivalved, oblong, valves convex, margin line nearly straight, front indented and nearly as wide as the greatest width of shell; smaller valve convex in its posterior portion, two rounded ridges rise soon after leaving the umbo and diverging till they reach the lateral edges of the front, leaving a depressed concave portion or sinus between them, extending and increasing in concavity as it approaches the front. Beak rounded, without distinct lateral ridges, and truncated by an entire foramen of moderate size; deltidium in two pieces, surface smooth and punctuated; margin line nearly straight all round. Loop simple, attached only to crura and extending to near the frontal margin. Length 15, width 8, depth 9 lines.

Obs. This species is found in the Cornbrash of Northamptonshire, is always accompanied by *T. lagenalis*, of which it may perhaps only be a variety; its great width in front and deep longitudinal sinus, as well as the rounded ridges, which in both valves proceed from the frontal edges to the centre of the umbo, seems to M. Bouchard sufficient reasons for separating it from *Lagenalis*. It is not, however, difficult to find specimens uniting *ornithocephala* to *lagenalis*, and this last to *sublagenalis*, but as the typical shapes of each are well distinguishable, it will be found convenient to retain them under distinct specific names.

This last-described species is found in the Cornbrash of Northamptonshire, and of Boulogne-sur-mer.

38. Terebratula Cardium, *Lamarck*. Plate XII, figs. 13—18.

Terebratula cardium, *Lam.* 1819. Anim. sans Vert., vol. vi, No. 47. Figured in Ency. Méthod., pl. 141, fig. 6.
— orbicularis, *Sow.* 1829. Min. Con., vol. vi, p. 68, tab. 535, fig. 3.
— furcata, *Sow.* 1829. Min. Con., vol. vi, p. 67, tab. 535, fig. 2.
— orbicularis, *Morris.* 1843. Catalogue.
— — *V. Buch.* 1838. Mém. Soc. Géol. de France, vol. iii, 1ᵉʳᵉ Série, p. 160, pl. 16, fig. 3.
— cardium, *Deslongchamps.* 1837. Soc. Linn. de Normandie.
— orbicularis, *Bronn.* 1849. Index Palæont., p. 1243.
— — *D'Orb.* 1849. Prodrome, p. 315.
— cardium, *Dav.* 1850. Notes on an Examination of Lamarck's Fossil Terebratula. Annals and Mag. of Nat. Hist., June 1850.

Diagnosis. "*T. Testá elongato-ovatá, convexá, plicatá, sulcis longitudinalibus crassis rotundatis: nati prominula.*"—(Lamarck.)

Shell inequivalved, more or less oval, elongated, uniformly convex; beak in larger valve straight, truncated by a large entire foramen, separated from the umbo by a narrow concave deltidium in two pieces; lateral ridges, indistinct valves regularly plaited, variable in size, acute, bifurcated when young, and sometimes, though rarely, when adult; surface punctuated, loop long, attached simply to crura, and extending to near the frontal margin. Length 16, width 12, depth 11 lines.

Obs. This species was first figured by Bruyere, in the Ency. Méth., and described, in 1819, by Lamarck, under the name of *Ter. cardium*, which name must be preserved. In 1829 it was figured and described by Sowerby, under the name of *T. orbicularis*, and the young under that of *T. furcata;* and it is singular to see that most authors (among whom M. D'Orbigny) have made use of Sowerby's name in preference to Lamarck's, published long before, and figured in the Ency. Méth., to which figure Lamarck refers. The size and number of plaits in this species are very variable; they are often few in number, wide and strong; at other times numerous and more delicate, commencing at the umbo and beak, becoming gradually and regularly wider as they approach the front; generally, in the young state, the plaits are divided towards the margin, and while placing these into another species, (*T. furcata*,) Mr. Sowerby states, "it is therefore possible these may be only young of *orbicularis*." In adult specimens it is rare to see the plaits bifurcated, but as a proof that it is sometimes so, we have figured Plate XII, fig. 15, a remarkable specimen, found by M. Deslongchamps, wherein the plaits are once and even twice divided; we are also indebted to M. Deslongchamps for the fine specimen illustrating the loop. Professor King, in his 'Monograph of Permian Fossils,' proposed to separate this species from the genus Terebratula, under the generic title of *Eudesia*, but in the Appendix seems disposed to cancel the genus, as it possesses all the characters of true Terebratula. *Ter. cardium* does not vary to the extent of most species, and is always recognisable; it is found abundantly in the Great Oolite of Bath, Cirencester, &c., and in beds of similar age at Boulogne-sur-mer, Caen, &c.; it was never found in the Lias of Weston, as stated by Mr. Sowerby, and subsequently by other authors, but it seems confined to a narrow vertical range, and to be characteristic of the Great Oolite.

39. Terebratula Buckmanii, *Dav.* Plate VII, figs. 15, 16.

Diagnosis. Shell irregular, oval, longer than wide; valves convex, smooth, minutely punctuated; beak small, truncated by a large entire foramen, almost touching the umbo; deltidium small, concealed; lateral ridges indistinct; margin line curved, rising in front. Length 18, breadth 11, depth 8 lines.

Obs. This species occurs in the Inferior Oolite of Winchcombe, near Cheltenham, where it was found by Professor Buckman, Dr. Wright, and others. It is quite distinct from *T. ornithocephala* by its shape, form of beak, foramen, and hinge marginal line of valves. We likewise find in the smaller valve of some specimens a central longitudinal elevation, extending from the umbo to the front, from which the two remaining lateral portions of the valve recede rather abruptly, as may be seen in fig. 16. Pl. VII, figs. 15, 16, from Prof. Buckman's Collection.

40. Terebratula Lycettii, *Dav.* Plate VII, figs. 17—22.

Diagnosis. Shell more or less circular; small, valves convex; beak rounded, truncated by an almost emarginate foramen; the deltidium in two pieces, touching only in one point above the umbo; lateral ridges indistinct; marginal line slightly curved; surface smooth, finely punctuated. Length 8, width 7, depth 4 lines.

Obs. This small species was found by Mr. Moore in the upper Lias of Barrington, near Ilminster, where it abounds, and varies in size from 1 to 8 lines in length, but does not appear to attain longer dimensions.

41. TEREBRATULA PUNCTATA, *Sow.*   Plate VI, figs. 1—6.

TEREBRATULA PUNCTATA, *Sow.*   Min. Con., 1812, vol. i, p. 46, tab. 15, fig. 4.
— — *Morris.* Catalogue, 1843.

Diagnosis. Shell inequivalved, depressed, convex; beak small, not much recurved, truncated by an entire foramen of moderate size; deltidium in two pieces; beak ridge soon lost; margin line curved and slightly raised in front; surface smooth and minutely punctuated; loop short, attached only to crura, and extending to little more than a third of the length of the valve. Length 16, width 12, depth 7 lines.

Obs. No species has perhaps given more trouble to make out than the one under consideration, being difficult to characterise from its variations in form, particularly if we examine a number of specimens. It seems nearly connected to the following form, which we have separated, as it presents certain differences sufficiently constant to render it desirable to preserve both under distinct names, and I do so the more readily, as it is also the opinion of M. Bouchard, Mr. Moore, and others, who likewise examined the subject with some attention; nor am I astonished that foreign Palæontologists have been unable to recognise this species. Lamarck quotes it in 1819, but on the inspection of shells he had placed under that name, I found none belonging to the species. Von Buch considered it a synonym of *ornithocephala;* Professor Brown places it under both *ornithocephala* and *carnea* (vide Index); and M. D'Orbigny, more cautiously, leaves it out entirely in his 'Prodrome;' still it is quite distinct from either of the shells to which it has been considered a synonym, which is perfectly proved by the size, length, and form of its loop, differing completely from that of *Ter. carnea,* where it extends only to a few lines from the crura, while in *T. ornithocephala* it nearly reaches the frontal margin. In *T. punctata,* as may be seen (Plate VI, fig. 3,) from the specimen drawn from Mr. Sowerby's original type, this process extends to nearly half the length of the shell, and is intermediate between that of *T. carnea* and *T. ornithocephala,* the beak and foramen being also quite different in those shells. *Ter. punctata* is moreover a Liasic species, and, from its date, holds specific claims of priority, whatever shells may be grouped into its type or removed from it.

The name of *T. punctata* was unappropriate, since all true Terebratulas are more or less strongly punctuated; it is, however, interesting to remark that this character was observed so far back as 1812 by Mr. Sowerby. It belongs to the middle Lias, and is stated by that author to have been found in the dark limestone of Aylesford, at a place called Horton; it was also collected in the Lias of Deddington, by Mr. C. Faulkner; and at Dumbleton, near Cheltenham, by Professor Buckman; it occurs at Farington, Gurney, and many other places. Plate VI, figs. 1, 2, and 3, illustrate the original specimens, kindly lent me by Mr. J. de C. Sowerby, which were figured in the 'Min. Con.'

42. TEREBRATULA SUBPUNCTATA, *Dav.*　Plate VI, figs. 7—10, 12, 16?

Diagnosis.　Shell inequivalved, ovular; valves almost equally convex; beak rounded, recurved, and truncated by a rather large entire foramen of an elongated shape, projecting over the umbo, so as to conceal the triangular deltidium, in two pieces, existing under it, and therefore rarely visible.　The ridges of beak soon become indistinct; marginal line slightly curved and raised in front; surface smooth, punctuated; loop attached only to crura, and extending to about a little more than one third of the length of shell; the lamella is strong and wide.　Length 28, width 20, depth 17 lines.　These are the dimensions of the largest specimen known, but the average size is much less.

Obs.　The different figures I have given in Plate VI will show, to a certain degree, the extent of variability observed in this shell, which is much deeper, stronger, and more convex than in true *T. punctata;* it is, however, allied to that species, and many specimens pass from one type into the other.　The beak is less recurved, and deltidium more apparent, in *T. punctata*, its front being also more rounded.　In many specimens of this species we observe a well-defined flatness at the umbo; it cannot be considered a character, but rather a deformity, caused by pressure in the young state, retarding the normal and regular development of the shell, which is also shown by strong lines of growth.　Fig. 8, and still more fig. 12, illustrate how much such a malformation changes the general aspect of the shell.　*Ter. subpunctata*, which I have so named from its close affinity to *punctata*, is abundantly distributed in the marlstone or middle Lias of South Petherton, near Ilminster, and in other localities.　On the Continent it was found by M. Deslongchamps and myself, in the Liasic beds of Evrecy, &c.　Along with the specimens found at South Petherton, by Mr. Moore, was one (fig. 16) which I have been unable to place anywhere, except under the head of a deformity of the shell under consideration, the form of its beak and foramen being the same, and it has many visible marks of pressure upon it.　Figs. 8, 10, 12, are from the Collection of Mr. Moore.

43. TEREBRATULA INDENTATA, *Sow.*　Plate V, figs. 25, 26.

TEREBRATULA INDENTATA, *Sow.*　Min. Con., 1825, vol. v, p. 65, tab. 445.
—　　　—　　*Morris.*　Catalogue, 1843.
—　　　—　　*Bronn.*　Index Palæont., 1849.

Diagnosis.　Shell elleptical, longer than broad; valves nearly equally convex; beak recurved, and truncated by an entire foramen; lateral ridges lost at a short distance from the foramen, sometimes notched in front; surface smooth, punctuated; loop short, simply attached to the crura.　Dimensions variable; size of largest specimen, as yet observed,—length 14, width 10, depth 9 lines.

Obs. As Sowerby remarks, when describing this species, it sometimes shows little or no marks of that indentation from which the name is derived; it is one of those species which present some difficulties, not being as well characterised as we might wish, and leading us, by gradual links, into *Ter. punctata*, with which it is generally associated. It occurs in the middle Lias of Banbury, as stated by Mr. Sowerby; it was found also at Deddington by Mr. C. Faulkner, where it is associated with *Ter. resupinata*. It is common at Farington, Gurney, &c. M. D'Orbigny erroneously considers this species as a synonym of *Ter. digona*, (vide ' Prodrome,' p. 315,) from which it is perfectly distinct, and belongs to the Lias, and not to the Bath Oolite, as stated by that author. Its loop is short, extending only to less than half the length of the shell, while in *Digona* it nearly reaches the frontal margin of the shell. This species is also found on the Continent, but does not appear to be known to the Palæontologists there. Figs. 25, 26, from the Collection of Mr. Walton.

44. TEREBRATULA INSIGNIS, *Schübler.* Plate XIII, fig. 1.

TEREBRATULA INSIGNIS, *Schübler.* Zieten, 1832. Die Verst. Würtembergs, pl. xl, fig. 1.
— — *D'Orb.* 1849. Prodrome, vol. i, p. 376.
— — *Quenstedt.* 1843. Dos Flözgebirge Würtembergs, p. 484.

Diagnosis. Shell inequivalved, more or less oval, longer than wide; valves convex, deepest towards the posterior portion; beak produced, slightly recurved, and obliquely truncated by a moderately-sized foramen, separated from the umbo by a rather long deltidium *in one piece;* lateral ridges indistinct; in smaller valve a well-defined mesial fold is seen to extend to the front, with lateral depressions; larger valve regularly convex, no deep sinus corresponding to the elevation in its smaller valve; surface smooth, punctuated; loop short, simply attached to crura, and extending to less than half the length of valve. Length of our British specimen 20, breadth 15, depth 11 lines.

Obs. This remarkable Oolitic species has not before this been noticed in England; it occurs in the Oxford Clay of St. Ives, and in the Coralline Oolite of Malton. It seems little known even to Continental Palæontologists. *Von Buch*[1] erroneously considers it a synonym of *T. perovalis*, and Dr. Bronn[2] the same as *Ter. biplicata*, likewise a mistake: it is perfectly distinct from both by its general appearance, and especially by its deltidium formed of one piece instead of two, which is the case with both *T. perovalis* and *biplicata.* This species was well figured by Zieten in 1832, and correctly noticed by M. D'Orbigny in his ' Prodrome;' he places it both in the Oxford Clay and Coralline Oolite.

---

[1] Mém. Soc. Géol. de France, vol. iii, 1ʳᵉ Série, 1838.
[2] Index Palæontologicus, vol. ii, p. 1239. Dr. Bronn's synonyms of *Ter. biplicata* are far from being correct, and it is evident that Sowerby's species is little known.

It is not improbable, that the specimens attributed by Roëmer[1] and Pusch[2] to *T. perovalis* belongs to *T. insignis.*

Though small in England, in some continental localities it attains very large dimensions. I have specimens from the Coral Rag of Nattheim Wurtemberg, sent me by H. Krantz, measuring 3 inches in length and 2 in breadth; it is found in many other localities, such as in the Oxford Clay of Boulogne-sur-mer, where it was found by M. Bouchard; at Ecommoy (Sarthe), route de Gray, near Besancon; La Latte, Apremont (Ain), in the Coralline Oolite of Chatel-Censori, Tonnerre (Yonne), &c., and in many German localities. The specimen figured in my plate is from the collection of Mr. Bean, of Scarborough, who found it at Malton.

45. TEREBRATULA SIMPLEX, *Buckman.*   Plate VIII, figs. 1, 3.

TEREBRATULA SIMPLEX, *Buckman.* 1845.   Geol. of Cheltenham, pl. vii, fig. 5.
      —   —   *Tennant.* 1847.   A Stratigraphical List of British Fossils, p. 74.

Diagnosis. Shell inequivalved, longer than broad; larger valve convex; beak rounded, recurved, and truncated by an entire foramen, thickly edged and concentrically furrowed, advancing over the umbo, which is partly concealed, as well as the deltidium, which forms a concave inward curve, transversally striated by minute lines; no distinct lateral ridge; imperforated valve, moderately convex, flat, or slightly concave, especially towards the front; surface smooth, finely punctuated. Length 2 inches 4 lines, breadth 2 inches 1 line, depth 1 inch 5 lines.

Obs. This species, when young, seems to have its smaller valve flat, depressed, nearly even; this is also seen in some old shells, (Plate VIII, fig. 2,) the junction of both valves forming in front an acute angle, but in general, in adult specimens, such as Plate VIII, fig. 1, the upper valve is convex, though never very gibbous, much thickened, and emarginate. This species is well characterised, but placed erroneously by M. D'Orbigny as a synonym of *Ter. lata* (Sowerby), with which it differs in shape, form of beak, deltidium, &c. *Ter. simplex* is found in the pea-grit bed of the Inferior Oolite of Crickley and Leckhampton Hills, near Cheltenham and Minchinhampton, and fine specimens are preserved in the British Museum, and in the Collections of Dr. Wright, Messrs. Lycett, Morris, &c.; the largest specimens obtained as yet, fig. 1, is from the collection of the Geological Society; figs. 2 and 3 were kindly lent me by Messrs. Buckman and Walton.

46. TEREBRATULA OVOIDES, *Sow.*   Plate VIII, figs. 4—9.

TEREBRATULA OVOIDES, *Sow.* 1812.   Min. Con., vol. i, p. 227, tab. 100.
      —   LATA, *Sow.* 1812.   Min. Con., vol. i, p. 227, tab. 100.

[1] *Roëmer.* 'Dei Versteinerungen des Norddeutschen Oolithen-Gebirges, tafel ii, fig. 3.
[2] *Pusch.* 'Polens Palæontologie,' tafel iv, fig. 8, and perhaps fig. 5.

TEREBRATULA TRILINEATA, *Young* and *Bird*. 1828. Geol. of York., pl. viii, fig. 17.
— OVOIDES, *Young* and *Bird*. 1828. Geology of Yorkshire, pl. viii, fig. 12.
— — *Desh*. 1836. Nouv. ed. de Lamarck, vol. vii, p. 361.
— — *Morris*. 1843. Catalogue, p. 135.
— LATA, *Morris*. 1843. Catalogue, p. 134.
— TRILINEATA, *Morris*. 1843. Catalogue, p. 137.
— OVOIDES, *Bronn*. 1849. Index Palæont., p. 1244.
— TRILINEATA, *Bronn*. 1849. Index Palæont., p. 1254.

Diagnosis. Shell ovate, elongated, valves moderately convex, beak prominent, sub-carinated and truncated by a rather large entire foramen, without distinct lateral ridges, deltidium well defined, in two pieces; lateral margin of valves straight with rounded front; surface smooth, punctuated, and marked by concentric lines of growth. Length 2 inches 2 lines; breadth 1 inch 10 lines; depth 11 lines.

Obs. This species seems little known and has given rise to much confusion, but having obtained the loan of a great many specimens, referred to it from Messrs. Sowerby, Bowerbank, Ripley, C. B. Rose, Bean, Morris, &c., and from different localities, I was soon convinced that Sowerby's *Ter. ovoides* was the same as his *Ter. lata*, which can be perceived by the simple examination of Pl. 100, of the 'Min. Con.,' which is only a wider and more depressed specimen of the same species, nor does Sowerby's figures give a very correct illustration; it is also evident from a numerous suit of Young and Bird's *Ter. trilineata* kindly sent me by Mr. Ripley, of Whitby; this species must be also placed among the synonyms of *T. ovoides*. These authors appear to have been only acquainted with internal casts of these shells, since they state that the smaller valve is always marked by three slender depressed lines diverging from the umbo, owing to which circumstance they gave it the name of *trilineata;* these said lines are due to the central septum and muscular impressions, visible on the surface of internal casts of most species of Terebratula, and therefore cannot be used as a character; they also allude to the difference observable in width and length, as well as thickness, in this species; it is the case too in *Ter. ovoides* and *lata* of Sowerby, the pinched beak visible in many specimens, and especially in internal casts, appears to be one of its characters.

In Pl. VIII, fig. 5, we have represented Sowerby's *Ter. ovoides;* fig. 4, his *Ter. lata;* and figs. 6—9, Young and Bird's *Ter. trilineata*, which, as I stated, belong all to one type; and as *Ter. ovoides*, Sow. appears the most common form, I have selected it to represent the species, which is variable in its shape and dimensions; it is a flattish shell, rarely very convex or gibbose. M. Deshayes, in 1836, adopted the name of *T. ovoides*, and places *lata* among its synonyms. Its Geological range has not yet been quite satisfactorily established; probably it lived during different deposits of the Oolitic series From the Inferior Oolite of Robin Hood's Bay, near Whitby upwards, Sowerby states his *Ter. lata* and *Ovoides* occur in blocks, and sandstone containing green-sand, in alluvial deposits of gravel, fragments of chalk, &c., in some parts of Suffolk. It is not, however, a Cretaceous, but an Oolitic shell, as Mr. C. B. Rose admits, and in which opinion I entirely concur.

H

It is evident that Professor Bronn is in error, when placing *Ter. perovalis*, Sowerby, as a synonym of *Ter. lata* (see 'Index Palæontologicus'). Nor can I agree with M. D'Orbigny in thinking *Ter. simplex* to be a synonym of *T. lata* (Prodrome, p. 287).

47. TEREBRATULA MAXILLATA, *Sow.*   Plate IX, figs. 1—9.

TEREBRATULA MAXILLATA, *Sow.* 1825. Min. Con., p. 52, pl. 436, fig. 4.
      —   —   *Morris.* Catalogue, 1843, p. 134.
      —   —   *Morris* and *Dav.*, 1847. Annals and Mag. of Nat. Hist.,
                              pl. xix, fig. 5.
      —   —   *D'Orb.* 1849. Prodrome, vol. i, p. 287.

Diagnosis. Shell subquadrangular, as broad as long, valves nearly equally convex, beak produced, narrow, and strongly recurved; foramen large, oblique; deltidium obtusely triangular; cardinal area slightly depressed, with obtuse lateral ridges; imperforated valve with a mesial sinus, and two lateral ones corresponding to the lobes of the perforated one, and which only extends one third inwards from the margin; front strongly sinuated in the young state, while for a considerable period of growth, no trace of sinuated margin is perceptible. Loop attached only to crura, extending to little more than one third of the length of valves. Length 32, breadth 34, depth 18 lines.

Obs. This is a very variable shell both in form and size, it is well described by Sowerby in 'M. C.,' who states it to be " distinguished from *Ter. intermedia* by the depth of the sinuses and consequent furrows which extended at least half way to the beak, in some specimens the two ridges between the furrows are very prominent and approach more nearly together, than in the specimens figured; such shells are generally long shaped, &c." This species has been found to attain in some localities much larger dimensions than those observed by Sowerby, who states it to be "always smaller than *Ter. intermedia.*" In the 'Annals and Mag. of Nat. Hist.,' October 1847, Mr. Morris and myself had occasion to notice some large specimens of this species found at Pickwick, which are figured in Plate XIX of that periodical. Since that period, still larger specimens have been procured, as may be seen in the Collection of the British Museum (Plate IX, fig. 4); and we have been able to trace every size in this species, from that of a pin's head to the larger specimens just mentioned; its greatest diameter is sometimes in the longitudinal, at other times in the transverse direction; some varieties are with the same dimensions deeply plicated, while others show no traces; in the young state, and for a considerable period of growth, there is no indication of plication, as may be seen by a glance at Plate IX, where we have illustrated all these different states; fig. 1 is the original form, considered by Sowerby as the type of the species. *Ter. maxillata* is abundantly distributed throughout the Forest Marble, the Bradford Clay, and the great Oolite round Bath, near Sapperton, and Hailey Wood, about Cirencester, &c.; large specimens were found on the Continent, at Boulonge-sur-mer, by M. Bouchard. Professor Bronn, in his 'Index Palæontologicus,' considers, erroneously, this species a synonym of *Ter. biplicata*, thus placing together, under one head, a number of different species. Pl. IX, fig. 3, from the Collection of Mr. Morris; fig. 4, in British Museum; fig. 9, from the Collection of Mr. Bowerbank.

Var. SUBMAXILLATA, *Morris.*   Plate IX, figs. 10—12.

The shell which I here place as a variety of *Ter. maxillata* would appear to some Palæontologists specifically different, but after a long and minute examination of a considerable number of specimens, I came to the conclusion that the distinctive characters were not sufficiently well defined or important to require it to be thus widely separated. This variety is found in a light-yellow clay bed of the middle division of the Inferior Oolite along with *Ter. plicata, fimbria,* and other well-characterised species. Var. *submaxillata* is more pentagonal in form than what we observe commonly in *maxillata,* the posterior portion of the shell is straighter, tapering, and less circular than is generally seen in Sowerby's type; the young state, in both species and variety, is exactly similar.

Many fine specimens of this variety are preserved in the Collections of Dr. Wright, Professor Buckmann, Messrs. Lycett, Morris, and others, who kindly forwarded to me these specimens. Figs. 10 and 12, from the Collection of Prof. Buckman; fig. 11, from that of Dr. Wright.

48. TEREBRATULA PEROVALIS, *Sow.*   Plate X, figs. 1, 6.

TEREBRATULA PEROVALIS, *Sow.*   Min. Con., 1825, vol. v, p. 51, t. 436, figs. 2, 3.
— — *Deslongchamps.* 1837. Soc. Linn. de Normandie.
— — *V. Buch.* 1838. Mém. Soc. Géol. de France, vol. iii, 1ᵉʳᵉ Serie, p. 221, pl. xx, fig. 2.
— — *Morris.* Catalogue, 1843, p. 135.
— — *Bronn.* 1849. Index Palæont., p. 1245.
— — *D'Orb.* 1849. Prodrome.

Diagnosis. Shell inequivalved, ovate, longer than wide, greatest width towards the middle. Cardinal and lateral marginal line forming a gentle and regular uninterrupted curve, with or without two more or less defined rounded ridges in frontal portion, hardly perceptible in the adult state, where the sinus separating the two ridges is filled up by the convexity of the valve extending to the frontal margin, and presenting one large elevation, with two lateral sinuses more or less indented. Rostral valve convex, with depressions corresponding to the elevations in the smaller valve; beak rounded and truncated by a large entire foramen, nearly touching and projecting over the umbo, generally concealing the deltidium; lateral ridges indistinct, surface smooth and punctuated; loop simply attached to crura, and short, extending to only about two fifths of the length of the valve. Length 33, breadth 28, depth 23 lines; the average dimensions are much less.

Obs. The original specimens described by Sowerby in the ' Min. Con.,' and of which I have given correct figures, Plate X, figs. 1, 2, would hardly convey an idea of the species under consideration, which varies considerably, as may be seen from the several illustrations offered in Plate X. Under favorable circumstances this species attained considerable dimensions, and is the largest Oolitic brachiopoda with which we are acquainted. These larger forms have been by some believed separable from *T. perovalis*, and long known under the name of *T. ovoidea*, but as we can trace every stage of growth and variation, from Sowerby's type to the large specimens in the same beds and localities, it is evident they must all belong to the same species, variable from the presence or nonpresence of the two frontal ridges described above. The same has been observed relative to *T. maxillata*, where, under favorable circumstances, that as well as other species have attained dimensions far exceeding those of the original type. We are indebted to that excellent observer, M. Deslongchamps, for the knowledge of the internal loop of this, as well as of other forms, worked out with the greatest skill and patience, we perceive that in this as in most species where the loop is short, the lamella becomes much wider and stronger than in those where the process extends to near the frontal ridge. *T. perovalis* is a common and characteristic fossil of the Inferior Oolite, both in England and on the Continent. The largest and finest specimens have been obtained from Dundry, Dinnington, Yeovil, &c., the best species being preserved in the Collections of the Bristol Institution, British Museum, &c. *Ter. perovalis* is also very abundant in Normandy, where specimens three inches in length have been procured, as may be seen in the Ecole des Mines of Paris, and in the Collection of M. Tesson, of Caen, &c., and, although a well characterised species, it has often been mistaken by many Geologists and Palæontologists, who refer to it forms perfectly distinct, as may be seen by the synonyms given of it. Thus Von Buch thinks *Ter. insignis* (Schübl) to be a variation of this species, but from which it completely differs, its deltidium being formed of one piece, while in *T. perovalis* it is divided in two, besides many other distinctions, which we shall notice under that species. Fig. 1, 2, from the Collection of Mr. J. de C. Sowerby; fig. 4, from that of Mr. Moore.

49. TEREBRATULA INTERMEDIA, *Sow.* Plate XI, figs. 1—5.

TEREBRATULA INTERMEDIA, *Sow.* 1812. Min. Con., vol. i, p. 48, tab. xv, fig. 8.

Diagnosis. Shell inequivalved, obtusely five-sided, longer than wide; beak rounded, recurved, and obliquely truncated by a rather large foramen, slightly overlaying the umbo, and concealing the deltidium in two pieces; lateral ridges indistinct; imperforated valve less convex than the other, with (sometimes without) two rounded costæ, commencing about the middle, and continued to the margin, with a mesial furrow or sinus, and two lateral ones; front margin moderately sinuated; surface smooth, punctuated; loop short and simply attached to crura, extending to less than half the length of the shell. Length 23, width 18, depth 13 lines.

Obs. Sowerby's figures in the 'Min. Con.' do not illustrate this species in a satisfactory manner, which has been the cause of many mistakes, as it is evident most authors have not understood the shells intended as types, of which I became convinced from inspecting the original specimens. Its exact stratigraphical age is the *Cornbrash*, as distinctly stated by Sowerby, still we find it placed by Von Buch,[1] Dr. Mantell,[2] and others, in the chalk; some state it to be from the *Lias, Great Oolite, Oxford clay*, &c., but it is singular few place it in its true stratigraphical position,[3] nor do I believe Zieten's figures referable to this species. *Ter. intermedia* bears some resemblance to *Ter. perovalis*, some specimens are indistinguishable; the common type differing, however, from the Inferior Oolite specimens, in being wider and more regularly circular, especially towards the beak and hinge margin: when young it is almost circular, without biplication, which only appears at a more advanced age, and it may be remarked that it is impossible to determine to what species some young shells belong, from the great resemblance they bear to each other, specific distinctions only appearing at a more advanced period of growth. The name of *intermedia* is well chosen, as its characters are intermediate, and lead us to such shapes as *Ter. Phillipsii* and *globata;* from the first it is distinguished by its more circular form, and from the last in being more depressed, and is a much larger shell. Lamarck's *Ter. intermedia* appears to be a Rhynchonella, otherwise his species could not hold priority over Sowerby's, published in 1802.[4]

Ter. intermedia is abundantly found in the Cornbrash of Stanton, in different parts of Wiltshire, Rushden, &c., associated with *Ter. lagenalis, obovata, Bentleyi*, &c. Many fine specimens are to be seen in the collection of the British Museum, and in the cabinets of many collectors; it was found also by M. Bouchard in the neighbourhood of Boulogne-sur-mer. Pl. XI, fig. 1, 3, in the Collection of Mr. Morris; fig. 2, in that of Dr. Wright; fig. 5, in the British Museum.

50. Terebratula Phillipsii, *Morris.*   Plate XI, figs. 6—8.

Terebratula Phillipsii, *Morris.* 1847. Annals and Mag. of Nat. Hist., p. 255, pl. xviii, fig. 9.

—   —   *D'Orb.* 1849. Prodrome, vol. i, p. 287.

Diagnosis. Shell elongated, irregularly pentagonal, posterior half of the shell and beak tapering, truncated by an entire moderately large and rather oblique foramen, separated from the umbo by a deltidium wider than high; lateral ridges obtuse and indistinct; large valve more convex than smaller one, with a somewhat incurved and produced beak; smaller valve with two rounded costæ, commencing about the middle and continued to the margin, with a broad deep mesial furrow or sinus, and two lateral ones; front deeply sinuous, surface punctuated. Length 28, breadth 20, depth 13 lines.

[1] *Von Buch.* 'Mém. Soc. Géol. de France,' vol. iii, 1ᵉʳᵉ Serie, 1838, where it is placed as a synonyme of *Carnea.*

[2] *Dr. Mantell.* 'Fossils of South Downs.'

[3] See D'Orbigny's 'Prodrome;' Morris's Catalogue, p. 133.

[4] See Davidson's Notes on an Examination of Lamarck's Species of Fossil Terebratulæ, 'Annals and Mag. of Nat. Hist.,' June 1850.

Obs. This species, as stated by Mr. Morris in the 'Annals and Mag. Nat. Hist.,' has some resemblance to *Ter. perovalis*, Sowerby, but is easily recognised by its more elongated, pentagonal, and depressed form, the greater width of the sinus and lobes, the more sinuated front, and the greater prominence of the dorsal ridge; it is also separated from *T. sella*, Sow. by its elongated form, the greatest width being nearer the frontal margin than in that species where it is central.

Ter. Phillipsii is from the Inferior Oolite of Dinnington, near Ilminster, Burton, near Bridport, near Cheltenham, &c. On the Continent it is found in Normandy, in beds of a similar age. Messrs. Moore, Walton, Bunbury, Wright, Deslongchamps, and others, possess fine specimens, and it is also to be seen in the collections of the British Museum and Geological Survey.

51. TEREBRATULA GLOBATA, *Sowerby*. Plate XIII, figs. 2—7.

TEREBRATULA GLOBATA, *Sow.* 1825. Min. Con., p. 51, pl. 436, fig. 1.
—   KLEINII, *Morris, Desh., D'Orb., Bronn* (non *T. Kleinii*, Lamarck).

Diagnosis. Shell subglobose, longer than wide; beak rounded, recurved, and obliquely truncated by a circular foramen of moderate size, almost touching the umbo, and concealing the deltidium, which is small; lateral ridges indistinct; smaller valve very convex at umbo, with two rounded costæ, commencing near and extending to the front, where they are slightly produced, with mesial sinus and two lateral ones, corresponding to the elevations in larger valve; margin of valves much sinuated; surface smooth, finely punctuated; loop attached only to crura, and extending to less than half the length of the shell. Width 11, length 13, depth 10 lines.

Obs. Most authors have attributed Sowerby's *T. globata* to *Ter. Kleinii* of Lamarck; M. Deshayes, Morris, D'Orbigny, Bronn, &c. have fallen into the common error, but the kind loan I received of the original collections of Lamarck and Sowerby has enabled me to prove that both species were completely distinct.[1] *Ter. globata* is one of those shells the continual variations of which render it most difficult to describe. In Plate XIII, I have endeavoured to illustrate a few of its varieties; figs. 2 and 3 are drawn from Sowerby's types, and fig. 4, from the Inferior Oolite of Dundry, may likewise be considered a good representation of the species. In the Inferior Oolite of Leckhampton and Cotswold Hills, we find a larger variety, which cannot be separated from *T. globata*, figs. 5, 6, 7; it is sometimes almost circular, and at other times so much elongated as to appear to belong to another form, we however find every passage connecting these extremes; another variety, also, seems to occur in the Fullers Earth, round Bath, offering every possible variation of form and convexity which we believe inseparable from the original type. Mr. Waterhouse and

[1] See Davidson's Notes of an Examination of Lamarck's Species of Fossil Terebratula, 'Annals and Mag. of Nat. Hist.,' June 1850.

myself took much trouble in minutely comparing a vast number of specimens, forwarded to me from different quarters, with the types of Mr. Sowerby's collection; we also compared it with the origiual *Ter. sphæroidalis* and *T. bullata*, which some authors (Bronn, '*Index*,') have considered synonyms of *globata*, while others (D'Orbigny, '*Prodrome*,') separate *T. sphæroidalis*, and place *bullata* as a synonym of *globata*; lastly, M. Deslongchamps places *T. bullata* and *sphæroidalis* as synonyms, and separates *globata*. We are disposed to adopt this view, and to consider *T. bullata* as a variation of *sphæroidalis*. It possesses more of the characters of this last than of *globata*; its margin is scarcely sinuated, and wants those defined costæ and well-marked sinuses characteristic of the species under consideration. It may be said that *bullata* is a connecting link between *globata* and *sphæroidalis*, but more nearly allied and more properly placed as a variation of the last-named type.

Ter. globata is abundantly found in the Inferior Oolite of Dundry, Cheltenham, Nunney, near Frome, &c., and fine specimens are preserved in the British Museum, that of the Geological Survey, Bristol Museum, and collections of Messrs. Walton, Buckman, Sowerby, and Dr. Wright. Pl. XIII, fig. 2, 3, from the Collection of Mr. J. de C. Sowerby; fig. 5, from that of Mr. Walton.

52. TEREBRATULA BUCCULENTA, *Sowerby*. Plate XIII, fig. 8.

TEREBRATULA BUCCULENTA, Sow. 1825. Min. Con., vol. v, p. 54, tab. 438, fig. 2.
— — D'Orb. 1849. Prodrome, vol. i, p. 376; vol. ii, p. 24.
— — ? Zieten. 1832. Die Verst. Wurttemb., t. 39, fig. 6.
— — Deslongchamps. 1837. Soc. Lin. de Normandie.

Diagnosis. Shell inequivalved, elongated, irregularly oval; valves almost equally convex; beak small, recurved, truncated by a small foramen almost touching the umbo, and concealing the deltidium; lateral ridges indistinct; margin of valves almost straight, front slightly produced, and laterally compressed or pinched; surface smooth and punctuated. Length 13, width 12, depth 8 lines.

Obs. I am indebted to Mr. J. de C. Sowerby for the loan of the original specimens illustrated in Plate XIII, fig. 8, obtained from *Coralline Oolite* or *Calcareous Grit* of Malton, and, according to M. D'Orbigny,[1] it would likewise occur in the Oxford clay, which is not unlikely, as several species have been found in both deposits. Von Buch,[2] Morris,[3] Brown,[4] and others, believe it to be only a synonym of *T. bullata*, Sow., but I am not prepared to admit the fact, as *T. bucculenta* appears to differ from any true specimens of *Ter. bullata* and *spheroidals*, a species peculiar to the Inferior Oolite of many localities.

[1] *D'Orbigny.* 'Prodrome,' 1849.
[2] 'Mém. Soc. Géol. de France,' vol. iii, 1ère Série, 1838, p. 195.
[3] *Morris.* Catalogue, 1843, p. 132.
[4] Index Palæontologicus, vol. ii, p. 1231.

53. TEREBRATULA SPHÆROIDALIS, *Sow.*   Plate XI, figs. 9, 19.

> TEREBRATULA SPHÆROIDALIS, *Sow.* 1825.   Min. Con., vol. v, p. 49, tab. 435, fig. 3.
> —   BULLATA, *Sow.* 1825.   Min. Con., vol. v, p. 49, tab. 435, fig. 4.
> —   —   *Deslongchamps.* 1837.   Soc Lin. de Normandie.
> —   SPHÆROIDALIS, *Deslongchamps.* 1837.   Soc. Lin. de Normandie.
> —   BULLATA, *Desh.* 1836.   Nouv. ed. des Animaux sans Vert. de Lamarck, vol. vii, p. 362.
> —   —   *V. Buch.* 1838.   Mém. de la Soc. Géol. de France, vol. iii, 1ere Série, p. 195, pl. xviii, fig. 8.
> —   —   *Zieten.* 1832.   Dei Verst. Wurttemberg, tab. xl, fig. 6.
> —   SPHÆROIDALIS, *Morris.* 1843.   Catalogue, p. 136.
> —   BULLATA, *Morris.* 1843.   Catalogue, p. 132.
> —   —   *Bronn.* 1849.   Index Palæont., p. 1231 (but exclude all the other synonyms given).
> —   SPHÆROIDALIS, *D'Orb.* 1849.   Prodrome, vol. i, p. 287.

Diagnosis. Shell inequivalved, more or less circular and spheroidal; beak small, much recurved, and truncated by a moderately-sized circular foramen, almost touching and advancing on the umbo; lateral ridges indistinct; margin lines of valves more or less straight or curved, with or without a slightly sinuated front; surface smooth and punctuated. Loop short, attached simply to crura, and extending to about half the length of the valve: length 15, breadth 12, depth 12 lines.

Obs. From the study I have made of the original types of Sow., I believe *T. sphæroidalis* and *T. bullata* both to belong to one species, and as *T. sphæroidalis* is the first described, we have given it the preference; but authors have not in general understood the Sowerby type. Von Buch, while adopting *T. bullata*, places *T. sphæroidalis* as a synonym of *globata*. Prof. Bronn and M. D'Orbigny have also erred on this subject by placing *T. bullata* and *sphæroidalis* as synonyms of *T. Kleinii*, Lamarck. *T. sphæroidalis* (as I understand the species) is a very variable shell, some of its varieties differing much from the common shape, but inseparable in my opinion; a few of these are illustrated in Plate XI, showing it to be a more or less convex, globular, elongated, or flattened shell; figs. 9 and 19 are Sowerby's types drawn from the original specimens, the first *T. sphæroidalis*, the second *T. bullata*; it will also be seen from figs. 9, 10, 12, 13, 15, 16, 18, and 19, that the frontal margin line of valves is in some straight (fig. 9), in others forming either a convex or concave curve (in figs. 10, 14, 16, &c.), while in some differently sinuated (figs. 12, 13), but not in general interrupting the regular convexity of the valves in front. In some instances, nevertheless, as in fig. 19, two or more slightly produced rounded costæ proceed a little way from the front towards the umbo and beak; leaving an obscure furrow on each valve, but never as deeply biplicated as in *T. globata*; this is an exceptional character in the species under consideration. The internal loop is in all the same, and we are much indebted to my friend M. Deslongchamps for having forwarded to us perfectly worked out interiors of this shell. It attains considerable dimensions in some localities; the two largest specimens I am acquainted with are one in the British Museum, and the other in the possession of M. Bouchard; they resemble a billiard ball in form, measuring 23 lines in length, 22 in breadth, and 21 in depth; our British specimens do not attain these dimensions. In general, the convexity is regular from the umbo and beak to the front; but in some cases, after attaining certain dimensions, a temporary stoppage in growth takes place, which, on being resumed, has caused the remaining portion of the shell to deviate from the regular line, as in fig. 15. This is also the case in many specimens which are smooth and regularly circular up to a certain period, but after a pause continue their growth by giving birth to plaits or other ornaments, as we meet commonly in *T. fimbria plicata*, &c.

The lines of growth are likewise more or less prominent in different specimens.

As stated by Sowerby, *T. sphæroidalis* is found in the Inferior Oolite of Dundry, Nunney, near Frome, in different parts of Somersetshire, in Normandy, Bayeux, Curcy, Monstiers, St. Maixant, as well as in Germany, and at Allen, Stuifenberg, &c.

54. TEREBRATULA GLOBULINA, *Dav.* Plate XI, figs. 20, 21.

Diagnosis. Shell inequivalve, globular, regularly convex, and gibbous; beak small, scarcely prominent, recurved, and truncated by a diminutive foramen almost touching the umbo; lateral ridges distinct, recurving to join the hinge line; margin straight all round; valves almost equally convex, smooth, and punctated; loop short, attached only to crura; length 2, breadth 2, depth 1 line.

Obs. In 1847 I illustrated this small species,[1] but I did not at that period think it prudent to distinguish it by a specific denomination, as we believed it might only turn out to be the young of some species; but from the great number of adult specimens found by Mr. Moore, none of which exceed the dimensions above given, I have given to it the name of *globulina*. It is found in the Upper Lias in the same bed that contains *T. pygmea, Lept. Moorei, liasiana, Bouchardii*, &c., in the neighbourhood of Ilminster, and its discovery is due to Mr. Moore. Fig. 20 natural size, fig. 21 enlarged.

55. TEREBRATULA PYGMEA, *Morris.* Plate XIII, 16, 16 *a, b, c.*

TEREBRATULA PYGMEA, *Morris.* 1847. Annals and Mag. of Nat. Hist., vol, xx, pl. xix, fig. 3, *a, b.*
— — *D'Orb.* 1849. Prodrome, vol. i, p. 221.

Diagnosis. Shell inequivalved, of a somewhat hexagonal form; valves convex, beak

[1] Annals and Mag. of Nat. Hist., vol. xx, pl. xix, fig. 4.

I

small, recurved, and truncated by a diminutive foramen; lateral ridges distinct, recurving to join the hinge margin, with large longitudinal mesial fold in smaller valve, and corresponding sinus in larger one, and two lateral rounded plaits, with furrows corresponding to these in the rostral valve; front deeply sinuous, the central sinus by far the largest; surface smooth, punctuated; loop short; length 2, width 2, depth 1 line.

Obs. This small species, described and figured in 1847 by Mr. Morris and myself, was stated to come from the Leptæna bed of or above the marlstone in the neighbourhood of Ilminster; we do not, therefore, see why M. D'Orbigny places it in the Lower Lias or his *Terrain sinémurien*, where we do not find it. Since the period above alluded to, it has been obtained from similar beds in the neighbourhood of Pic de Saint Loup, near Montpellier, in France, along with *Leptæna liasiana* by M. Bouchard, and also at Croisilles, in Normandy, by M. Deslongchamps, though of the same dimensions as *T. globulina*, it is easily distinguished by its sinuated front; we are also indebted to the researches of Mr. Moore for the discovery of this species. Fig. 16, natural size, 16 *a, b, c,* enlarged illustrations.

56. TEREBRATULA BENTLEYI, *Morris, MS.* Plate XIII, figs. 9, 10, 11.

Diagnosis. Shell inequivalved, irregularly pentagonal, decussated, notched and sinuated in front; perforated valve convex, with two rounded costæ proceeding from the extremity of the beak, and regularly diverging to the front, separated by a deep medio-longitudinal sinus, and two lateral ones; beak keeled, recurved, produced, and obliquely truncated by an elongated apicial foramen, separated from the umbo by a deltidium in two pieces; lateral ridges well defined and continued along the sides without recurving to join the hinge margin, leaving between it and the hinge line a somewhat flat false area. Smaller valve slightly convex and depressed, subtrilobated, with three rounded costæ, one central, two lateral, divided by four grooves or sinuses proceeding from the umbo; the central rounded plait rising only at some distance from the umbo, the others being larger and more elevated than the central one; surface smooth, punctuated, and marked by numerous lines of growth. Length 16, width 18, depth 11 lines.

Obs. This is a very remarkable species not hitherto described, but found by the Rev. A. W. Griesbach, Messrs. Bentley, Morris, and others, in the Cornbrash of Wallaston and Rushden, in Northamptonshire, associated with *T. intermedia lagenalis, obovata,* and other well-known Cornbrash fossils, it has been met with in Lincolnshire, and I have it likewise from France and Germany: it somewhat approaches in form to some specimens of *T. coarctata,* but differs by many characters, especially by having its surface smooth, which is not the case in the last-named species. We are not acquainted with its loop, but judging from the external appearance of the shell would imagine it to be short, as in *T. coarctata.* The two fine specimens illustrated in our plate belong to the Rev. A. W.

Griesbach and Mr. Morris; it is named after Mr. Bentley, at the request of Mr. Morris: a fine specimen is to be seen in the collection of Mr. Lee. *T. Bentleyi* is a rare species, at least in England, few specimens having been as yet discovered.

T. BENTLEYI. Var. SUB-BENTLEYI, *Dav.* Plate XIII, fig. 11.

We are indebted to Mr. Lycett for the discovery of the larger valve of a terebratula which we believe to be only a well-marked variety of the above-described species, to which it approaches by general character; its shape, however, is more elongated, deeper, and the medio-longitudinal sinus, in appearance, extends only to a short distance from the front; the lateral beak ridges are also less defined than in *T. Bentleyi*. Unfortunately we are only acquainted with one of the valves of this shell, which may, perhaps, prove specifically distinct when the other is obtained. Its stratigraphical position is likewise very different, belonging to the Inferior Oolite of the neighbourhood of Minchinhampton, where it seems rare; the dimensions are larger than those of *T. Bentleyi*, measuring in length 22, in breadth 18 lines. Fig. 11 is from the collection of Mr. Lycett.

57. TEREBRATULA COARCTATUS, *Parkinson.* Plate XII, Figs. 12—15.

TEREBRATULA COARCTATUS, *Parkinson.* 1811. Organic Remains, vol. iii, pl. xvi, fig. 5.
— RETICULATA, *Smith.* 1816—1819, Strata identified by Organised Fossils, p. 83, pl. xxx, fig. 10.
— DECUSSATA, *Lamarck.* 1819. Anim. sans Vert., vol. vi, No. 51. Enc. Méthodique, t. 245, fig. 4.
— — *Dav.* Annals and Mag. of Nat. Hist., June 1850, pl. xiv, fig. 51.
TEREBRATULITES RETICULARIS, *Schloth.* 1820. Petref., vol. i, p. 269.
TEREBRATULA COARCTATA, *Sow.* 1823. Min. Con., vol. iv, p. 7, tab. 312, figs. 1—4.
— RETICULATA, *Sow.* 1823. Min. Con., vol. iv, p. 8, tab. 312, figs. 5—6.
— DECUSSATA and RETICULATA, *Deslongchamps.* 1837. Soc. Linn. de Normandie.
— RETICULARIS, *V. Buch.* 1838. Mém. Soc. Géol. de France, vol. iii, 1ᵉʳᵉ Série, p. 185, pl. xvii, fig. 7.
— DECUSSATA, *Morris.* 1843. Catalogue, p. 132.
— — *Tennant.* 1847. Stratigraphical List of British Fossils, p. 73.
— COARCTATA, *Bronn.* 1849. Index Palæont., vol. ii, p. 1232.
— — *D'Orb.* 1849. Prodrome, vol. i, p. 316.

Diagnosis. Shell inequivalved; subpentagonal, valves convex, sometimes gibbous, hisped, decussated, and notched in front; perforated valve biplicated, with a more or less deep angular sulcus between the plaits, extending from the extremity of beak to the front; beak moderately produced, truncated by a large entire circular foramen, separated from the umbo by a more or less wide deltidium, often receding; beak ridges indistinct. Smaller valve convex, subtrilobated with medio longitudinal plait; surface of valves covered by a great number of short tubular spines, arranged longitudinally so as to form minute elevated striæ, intersected by transverse lines, giving the whole surface a reticulated appearance; structure punctuated, and marked by well-defined lines of growth.

Loop short, attached simply to crura, and not extending to half the length of the shell. Length 12, width 11, depth 8 lines.

Obs. As may be perceived from the list of synonyms here given, this species has received three principal names, indiscriminately made use of by various authors; however, that of *T. coarctata*, established by Parkinson in 1811, appears the oldest and only one to be retained, as *T. reticulata* of Smith and *decussata* of Lamarck, are of a later period.

Mr. Sowerby, in 1823, proposed to retain both *T. coarctata* and *T. reticulata*, which he thus characterises:—"*Ter. coarctata* subpentagonal, gibbous, hisped, and decussated; lesser valves convex, subtrilobated; larger valve biplicated, with a deep angular sulcus between the plaits. *Ter. reticulata* obovate, gibbose, subhisped, decussated, front obscurely three-sided, with a shallow channel between the ridges."

These distinctions are not sufficiently constant to authorise the proposed separation, but are simple variations seen so usually in almost every species; and in plate XIII we have given illustrations of both varieties—fig. 12 representing *Ter. coarctata*, and fig. 15 *T. reticulata*, according to Mr. Sowerby; fig. 15 also exhibits a very unusual form of beak in this species, which is strongly recurved, with the foramen almost touching the umbo. The general character of the beak is much less recurved. This species is likewise figured by Mr. Walcot.[1] The surface, or spinose striæ ornamenting the surface, are very remarkable, giving the shell a rough appearance; and, as may be seen from the enlarged fragments (fig. 14), these short and thick spines arise from under each other along the elevated ridge; they are rarely preserved intact, but rubbed down so as to give the shell a striated appearance: their hollow tubular character is noticed by Mr. Sowerby. We are indebted to M. Deslongchamps for the interior, showing the loop, which is short in this species.

Ter. coarctata is abundantly found in the Great Oolite, Forest Marble, and Bradford Clay, all round Bath, Bradford, Hinton, Frome, &c., where it does not seem to exceed the dimensions we have given: it is also a common species in beds of the same age on the Continent, and especially round Caen, in Normandy. Fine specimens have, likewise, been met with by M. Bouchard, near Boulogne-sur-mer. Mr. Walton informs me that he has found this species in the Oxford Clay, but where it is very rare.

58. TEREBRATULA PLICATA, *Buckman*, 1845.   Plate XII, figs. 1—5.

TEREBRATULA PLICATA, *Buckman*. 1845. Geol. of Cheltenham, pl. 7, fig. 6. (Non *Ter. plicata*, Lamarck, 1849. An. sans Vert., vol. vi, No. 39, which belongs to another genus.)
—   —   *Tennant*. 1847. A Stratigraphical List of British Fossils, p. 74.
—   SUBPLICATELLA, *D'Orb.* 1849. Prodrome, vol. i, p. 287.

[1] Petrifactions found near Bath, No. 28.

Diagnosis. Shell inequivalved, elongated, oval, tapering posteriorly; valves convex, sometimes gibbous; beak small, not much recurved, truncated by a circular entire foramen, deltidium small, receding; lateral ridges indistinct. Surface of valves smooth, up to a certain age slightly undulating, or plaited towards the margin at a more advanced period. Structure punctuated, loop short. Length 33, width 22, depth 17 lines.

Obs. The term *plicata* has been given by Lamarck, Borson, and Say, to different shells placed by them in the genus *Terebratula*, but which belonging to different genera are only synonyms of other species; therefore, we think Mr. Buckman's name, though of a later date, may be retained, and that of Lamarck preserved for his species, which will have to be placed in Fischer's genus *Rhynchonella*. M. D'Orbigny proposes to change Mr. Buckman's name to that of *Subplicatella*, which I would have readily adopted, but for the reason above given. *Ter. plicata* is quite smooth up to a considerable age, when the frontal and lateral edges become more or less undulated or plaited; they never extend very high up on the valves, but are restricted to near the edge, forming an irregular frill round the shell, as seen in figs. 1 and 2. The plaiting in this species is very similar to that of *Ter. fimbria;* but both species seem distinct. *T. fimbria* is much smaller, rounder, and has not got that tapering of the posterior portion, so peculiar to *T. plicata,* which is likewise a much larger shell.

Ter. plicata is found along with *T. fimbria* in some of the beds in the Inferior Oolite, in the neighbourhood of Cheltenham, Minchinhampton, &c., where many fine species have been obtained by Dr. Wright and Messrs. Lycett, Buckman, Walton, Morris, and others; the three largest specimens with which we are acquainted may be seen in the collection of the Geological Society and in those of Messrs. Morris and Buckman.

It has also been found at Tournus (Saône et Loire), in France, by M. D'Orbigny.

Fig. 1 is from a specimen in the Collection of the Geol. Soc.; fig. 2 from the collection of Professor Buckman; fig. 5 from that of Dr. Wright.

59. TEREBRATULA FIMBRIA, *Sowerby.* Plate XII, figs. 6—12.

TEREBRATULA FIMBRIA, *Sow.* 1823. Min. Con., vol. iv, p. 27, tab. 326.
— — *Morris.* Catalogue, 1843, p. 133.
— — *Bronn.* 1849. Index Palæont., vol. ii, p. 1236.
— — *D'Orb.* Prodrome, vol. i, p. 287, 1849.

Diagnosis. Shell inequivalved, orbicular; beak short, slightly recurved, and truncated by a circular foramen, almost touching the umbo, and generally concealing the deltidium; lateral ridges indistinct, valves almost equally convex, sometimes gibbous, smooth in the young state, irregularly undulato-plaited at a more advanced age towards the margin; structure punctuated. Loop simply attached to crura, and extending to less than half the length of the shell. Length 20, breadth 18, depth 13 lines.

Obs. Sowerby's figures of *T. fimbria* illustrate only one of the states in which we find this species; nor does it appear to attain much larger dimensions than those given above (fig. 6). In the young state, and sometimes up to a considerable period of growth, the shell is quite smooth, without any plaits or ornaments, the margin line being straight all round (fig. 11); but in some rare cases, even when young, the ornaments of its surface are slightly perceptible (fig. 12); at a more advanced period the surface of the valves, towards the edge, becomes irregularly undulated, much in the way of a frill, as stated by Sowerby, but not extending to the hinge-margin, which is always smooth, and regularly curved and rounded. Nothing can be more irregular than the manner in which these ornaments are presented in different specimens, as may be perceived by the several illustrations we have selected; at times, the whole surface of the valves is regularly rounded to near the edge (figs. 6, 12); at other times, the smooth part is partially separated from the plaited portion by a marked line of growth and a difference in level (figs. 2, 10), proving that there was a sudden stoppage in growth, which, on being resumed, the shell from smooth became undulated, or differently plaited. Sometimes, before the stoppage alluded to, the shell had already become frilled, but on continuing the growth at a different level these ornaments became differently disposed, having no continuity with the ones already formed, as in fig. 10. These undulating plaits are also very irregular in their form, in some cases a few small plaits are succeeded by a large one widely separated, others close together: sometimes taking rise at about half the length of the valve, and, after proceeding to some distance, separated into two or more bifurcations irregularly disposed (figs. 7, 8).

The bifurcated plaits do not always reach the front, one or more disappearing at a short distance from their origin, while some smaller plaits, at times, also appear between the larger ones, so as to produce in that part of the shell a kind of irregular zig-zag aspect.

Ter. fimbria is very abundant in certain beds of the Inferior Oolite in different parts of Gloucestershire, near Cheltenham, Minchinhampton, &c., whence many fine series have been obtained by Dr. Wright, Messrs. Buckman, Lycett, Walton, Morris, and others; it is common in most collections. We have not yet observed this species in any of the Continental localities; but it is probable future researches may lead to its discovery, since *T. plicata*, a species closely allied to it, has been found in France by M. D'Orbigny.

Fig. 7 is from the collection of Professor Buckman; fig. 8 from that of Dr. Wright.

60. TEREBRATULA FLABELLUM, *Defrance.* Plate XII, fig. 19—21.

TEREBRATULA FLABELLUM, *Def.* Dic. des Sciences Nat., vol. liii, p. 160, 1828.
— PALMETTA, *Deslongchamps.* 1837. Soc. Linn. de Normandie.
— FLABELLUM, *Morris* and *Dav.* Annals and Mag. of Nat. Hist., 1847, p. 256, pl. xix, fig. 2.
— PALMETTA, *Bronn.* 1849. Index Palæont, p. 1244.
— FLABELLUM, *D'Orb.* 1849. Prodrome, vol. i, p. 316.

Diagnosis. Shell inequivalved, somewhat transversely oval, perforated valve, more convex than the other, with a produced beak, truncated by a rather large circular foramen; deltidium obtusely triangular, and in two pieces; cardinal or false area concave, nearly smooth; valves costated; seven or nine round costæ imbricated, increasing in size, not in number, towards the margin. The central medio-longitudinal plait larger than the lateral ones; shell punctuated, marked by numerous lines of growth; loop short, simply attached to crura, and extending to about half the length of the shell; dimensions variable. Length 5, breadth 5, depth 3 lines; but some foreign species measure, length 6, width 8, depth 4 lines.

Obs. This species seems to have been first described by M. Defrance, under the name of *T. flabellum*, but it is better known by that of *T. palmetta* to French Geologists; and was only within a few years noticed in England: it occurs in the Bradford Clay, of Bradford and Corsham, Wilts, where it was found by Messrs. Walton, Pearce, and Waterhouse. In France it is met with in the Oolite of Ranville, at Luc, and Langrune, near Caen, where it attains very large dimensions; and I am indebted to my friend, M. Deslongchamps, for working out the interior. This species belongs to the *Loricatæ* of Von Buch, in which the ribs of the smaller valve envelope those of the larger. *Ter. flabellum* is also remarkable for its rare variability; it is always easily recognised, being rather a scarce species, especially in England. Figs. 20 and 21 of our plate are enlarged.

Genus—TEREBRATELLA, *D'Orb.* 1847.

Diagnosis. Shell inequivalved, oval, sometimes transverse, larger valve more convex than the smaller or imperforated one, which is in many cases flatter; cardinal edge straight or slightly curved; a well-defined rather flat hinge area existing in larger valve, beak commonly straight and truncated by a foramen of an oval or irregularly triangular shape placed more under than above the summit, formed out of a small portion of the substance of the beak; cardinal area and triangular deltidium in two pieces, which are disunited in many cases, the aperture being completed by a small portion of the umbo. Structure punctuated, surface striated or plaited, often bifurcating; hinge articulating by means of two teeth in larger and corresponding condyles in smaller valve. Loop generally long, doubly attached, proceeding from the crura, but before attaining its greater length it gives off a flat, wide, more or less horizontal process, likewise attached to a central longitudinal elevated septum, the principal lamella proceeding till it doubles itself in the shape of a loop, as in true terebratulæ.

Obs. We are only acquainted with one British Jurassic species attributable, according to M. D'Orbigny, to this genus, viz. *Terebratella hemisphærica*, Sow.

61. TEREBRATELLA HEMISPHÆRICA, *Sow.*   Plate XIII, figs. 17, 18.

TEREBRATULA HEMISPHÆRICA,	*Sow.* 1829.	Min. Con., vol. vi, p. 69, tab. 536, fig. 1.
—	—	*Deslongchamps.* 1837.	Soc. Linn. de Normandie.
—	—	*Morris.* 1843.	Catalogue, p. 133.
—	—	*Bronn.* 1849.	Index Palæont., p. 1238.
TEREBRATELLA	—	*D'Orb.* 1849.	Prodrome, vol. i, p. 316.

Diagnosis. Shell inequivalved, hemispherical; beak produced, recurved, truncated by a large oval foramen placed more under than above the apex of the beak, encircled by a portion of the extremity of the beak, cardinal area, disunited triangular deltidium, and summit of the umbo of smaller valve, which is commonly quite, or nearly flat; hinge line slightly arched; structure punctuated; surface of valves ornamented by numerous small longitudinal elevated striæ often bifurcating. Loop said to be doubly attached to crura and to central longitudinal septum. Length 4, width $3\frac{1}{2}$, depth 2 lines.

Obs. Von Buch is in error when stating in his valuable Monograph of Terebratulæ that this species is only a synonym of *T. gracilis*, a cretaceous shell easily distinguishable from *T. hemisphærica;* in shape and characters it is a well-defined species, varying less than most oolitic shells; however, although its great and common character is to have an almost flat imperforated valve, still in some specimens, and especially in many from Normandy, this valve is more or less convex.

Notwithstanding my own and M. Deslongchamps' endeavours we have not been able to clear a specimen, so as to see the interior loop in a satisfactory manner, and therefore have placed it in the present genus, more on M. D'Orbigny's authority than our own, as we consider a knowledge of the form and position of the process absolutely necessary to be able to state positively to what genus or subgenus of the great family of Terebratulæ each species belongs.

Terebratella hemisphærica is found in the Great Oolite of Hampton Cliff, near Bath, and at Luc and Langrune, near Caen, in which last locality M. Deslongchamps obtained some specimens larger than any hitherto found in England, where it does not seem to exceed the dimensions above given. Fig. 17 natural size; fig. 18 are considerably enlarged illustrations.

Genus—RHYNCHONELLA, *Fischer.* 1809.

Animal small, generally attached to submarine objects by means of a pedicle issuing from the foramen placed under the beak of the larger valve. Shell inequivalve, variable in shape, wider than long, or longer than wide, circular or elongated; valves more or less convex, with or without a longitudinal mesial fold and sinus; beak acute, slightly or greatly recurved; no true area; foramen variable in its dimensions and form, placed under the beak, exposed or concealed, entirely or partially surrounded by a deltidium in two pieces, at times extending in the shape of a tubular expansion, at other times rudimentary; the foramen being completed by a small portion of the umbo. Surface striated, plaited, or costellated, rarely smooth; structure fibrous, unpunctuated, rarely spiny; valves articulating by means of two teeth in the larger and corresponding sockets in the imperforated valve; apophysary system in smaller valves composed of two short, flattened, and grooved lamellæ, separate, and moderately curved upwards, attached to the inner side of the beak of smaller valve, and to which were affixed the free spiral fleshy arms;[1] a small central longitudinal septum, more or less elevated, is seen to extend along the bottom of the smaller valve from under the beak to about half or two thirds the length of the shell, and separating the muscular impressions visible on either side.

Obs. In Part I, I have mentioned the reasons for not admitting M. D'Orbigny's genus *Hemithiris,* established on *Rynchonella psittacea, Wilsoni,* &c., from a conviction, that these shells correspond in all essential characters to the genus Rhynchonella; internally their organisation is similar, the muscular impression and short calcareous process for the support of the free fleshy arms being the same in these as well as in all the species of this extensive genus, whether Recent, Cretaceous, Oolitic, or Palæozoic.[2] The external shape and character of the different species is, on the contrary, so variable and perplexing, that in many cases it seems almost impossible to trace a line of demarcation between

[1] See Professor Owen's 'Anatomy of *T. psittacea,*' Trans. of the Zool. Soc., vol. i, 2d part.

[2] Within the last few years, authors seem to agree, as to the propriety of adopting a separate genus for those plaited Terebratulæ, the calcareous appendages of which are formed of only two small, short, curved lamella, to which are attached the free, fleshy arms, as in *T. psittacea, loxia, vespertilis, octoplicata, &c.*

M. D'Orbigny, in 1847, admitted Fischer de Waldheim's genus *Rhynchonella,* giving the date 1825, and therefore unacquainted with a prior paper by that author, entitled '*Notice sur les Fossiles du Gouvernement de Moscou servant de Programme pour inviter les Members de la Soc. Imperiale à la Séance publique du 26 Oct., Moscou,* 1809,' wherein will be found the first descriptions and figures of the genus *Rhynchonella.*

Professor King, in Part III, of his valuable and interesting '*Monograph of the Permian Fossils,*' states, "Reverting for a moment to the types named by the celebrated oryoctographer of Moscow, I would ask, is anything satisfactorily known respecting *Trigonella atoma?* and *Rhynchonella loxia?* Has any one been able to identify these shells? What formation do they belong to? And what are their localities, &c.?" Most of those questions may be answered; but as the paper above alluded to (1809) is little known

them, or to draw up a description answering to all the variations, which are, doubtless, due to local and other accidental circumstances: we often find specimens of the same species very convex and gibbous, while others are comparatively depressed; the number of the plaits is also most variable, some shells being ornamented by a greater or less number; 1, 2, 3, 4, 5, 6, &c., forming in different examples of the same species, a more or less defined or elevated mesial fold. The beak is likewise in some nearly straight, exhibiting under it the foramen entirely or otherwise surrounded by the deltidium, while in other specimens no trace of the foramen and deltidium is visible, from the beak becoming so incurved, as almost to touch and overlie the umbo. The dimensions and form of the false area situated between the beak ridges and the hinge margin is also variable, and indents to a less or greater degree the hinge of the smaller valve. It is, therefore, only by combining certain general characters that we can separate certain forms of Rhynchonella, and it is very unsafe to establish new species on the simple inspection of one or two specimens, unless these present characters so marked and peculiar as to make confusion impossible. Many hundreds, I may say thousands, of British and Foreign Rhynchonellæ have passed under my examination, and although I have spared no trouble in comparison and research, some of these have not been settled in a manner quite satisfactory to myself, such as *R. subtetraëdra*, *Lycettii*, *lacunosa*, *Morièrei*, and, perhaps, one or two others.

from its great rarity, I think it desirable to insert a short account of its contents, especially as it will explain the origin of a most important genus in the class of Brachiopoda.

Fischer divides his family of Terebratulæ into four Divisions:—

I. The Terebratulæ, the edges of which are smooth and not plaited, ex. *Ter. ovata* (Fischer), Ency., pl. 239, fig. 2, *T. scabra*, &c.

II. The Terebratulæ, the edges of which are plaited, ex. *Ter. novem-plicata* (Fischer), and *Ter. octoplicata;* this last the author figures, and it is a specimen of *Spirifer Walcotti*, and is stated to be so in his other Work, '*Oryctographie du Gouv. de Moscou*,' 1830—37, p. 41.

III. The Trilobated Terebratulæ, Genus Trigonella.

"The margin of the trilobated Terebratulæ presents a considerable displacement in the middle, the result being a division into three lobes; the impression of the middle is either smooth or striated. I distinguish these striations by parts (τομη), so that one recognises at once, by the name of the species, that they belong to the trilobated *Terebratulæ*, where the contour of the edges is not on the same level."

Of this proposed genus, Fischer names five types; the first is *Ter. atoma*, (Fischer, referring to 'Knorr. Petref.' vol. ii, p. 1, B. iii, fig. 6,) but on examining this reference, I am at a loss to know what the type *Ter. atoma* is, as B. iii, fig. 6, does not look like any form of Brachiopod, and can be of no use. The next is *T. bitoma*, stating that Lister's fig. 7, tab. 450 A, bears much resemblance to his type; this also I cannot distinguish, that author's figure being so imperfect. Fischer's next two types, *T. pentatoma* and *tritona*, are figured by himself, but are specifically undeterminable from the vagueness of the illustrations; they belong, however, to that section, with the two small calcareous bent processes, as in *T. psittacea;* the last-named, or the fifth type, *T. polytoma*, is a *Spirifer*. 'Ency. Meth.,' tab. 244, fig. 4 *° ᵇ*.

IV. Rhynchonellæ, p 35.

"The Terebratulæ, the medial edges of which are so elongated as to assume the form of a beak. The extremity of the beaked edge being on the same level with the foramen. These shells doubtless form a distinct genus, characterised as follows: *Shell bivalve, regular, with unequal valves, fixing itself by means of a*

I have, however, ventured to arrange our British specimens into thirty species, as seen in the following Table, uniting at the same time, by a line, those forms more nearly connected, a few of which may be only modifications of one great type.

1. Rhynchonella Wrightii.	16. Rhynchonella varians.	
2. — furcillata.	17. — Forbesei.	
3. — rimosa.	18. — serrata.	
4. — spinosa.	19. — plicatella.	
5. — senticosa.	20. — inconstans.	
6. — ringens.	21. — concinna.	
7. — subringens.	22. — subconcinna.	
8. — acuta.	23. — obsoleta.	
9. — cynocephala.	24. — subobsoleta.	
10. — variabilis.	25. — angulata.	
11. — subvariabilis.	26. — Morièrei.	
12. — Lycettii.	27. — tetraëdra.	
13. — oolitica.	28. — subtetraëdra.	
14. — Moorei.	29. — lacunosa ?	
15. — Bouchardii.	30. — Hopkinsii.	

By simply glancing at the Table inserted at the end of this Monograph, we may perceive how differently the types are distributed in the various subdivisions of the system. The Inferior Oolite contains the greatest numerical amount of species and individuals; in the

ligament, or a short tube; the smaller valve perforated, its summit not much produced, or recurved; hinge with . . . teeth. The Terebratulæ with a perforated beak, therefore, form a very distinct family, the *Trigonellæ* and the *Rhynchonellæ* being distinguished by their general aspect, and by their hinges, which have not been, as yet, investigated."

"27 *Gros bec.* The largest valve is that which, in other Terebratulæ, is the smallest; has two teeth and an arched contour.

"*Rhynchonella loxia Mihi.* Valva major bidentata margine terminali in curvo. Tab. ii, figs. 5, 6, loc. Talaroba.

"*Rhynchonella Canard,* 'Ency.,' tab. 245, fig. 6ᵃᵇᶜ."

"*Rhynchonella Aigle,* 'Ency.,' tab. 246."

The first, *R. loxia*, he illustrates, and it is a well-known Oolitic species found abundantly in the neighbourhood of Moscow, and resembling in general form *R. acuta*, but is a smaller shell with some other differences, well figured by M. D'Orbigny in Pl. xlii, figs. 22 and 25 of the 'Geol. of Russia,' under the erroneous name of *T. aptyca*, a mistake corrected by the same author in his 'Prodrome.'

All three are possessed of the same two short-curved appendages, as in *T. psittacea.* From the above, it may be perceived that three of the types of Fischer's genus *Trigonella*, and his three *Rhynchonellæ* are similarly organised, and consequently belong to the same genus; the one is synonymous of the other. M. D'Orbigny having adopted the last, and judiciously placed in it a vast number of species, and this being admitted by other authors, I will abide by the genus *Rhynchonella*, and place *Trigonella* as a synonym, although, perhaps, that name ought to have been preferred, being first mentioned in the paper; but it is evident, that to bring about this *most useless change*, it would be necessary to re-shift all the species now well known as *Rhynchonellæ*, which in my opinion would be ridiculous, and in no way serving science; later in 1830—37, Fischer notices *Ter. acuta*, Sow., *Ter. ringens*, V. Buch, and *Ter. variabilis*, Schl., as other

other divisions up to the Portland Oolite, or uppermost member of the series, the number of species diminish, and in some beds, only one or two forms have yet been noticed; thus in the Kimmeridge Clay, I am only acquainted with two, *R. inconstans* and *R. subvariabilis;* in the Portland beds, none seem as yet to have been discovered. Certain forms are rare, others very common. Among the former, I may mention *R. Wrightii, furcillata, subvariabilis, ringens, serrata,* and among the latter, *R. concinna, obsoleta, acuta, varians, variabilis, inconstans, spinosa,* and *tetraedra;* if, again, we glance over our table to see what takes place among the other genera, especially the *Terebratulæ,* we observe a similar poverty of species and individuals in the upper portions of the Oolitic system. There existed, therefore, at that period, a sea, and circumstances more favorable to the development of this class of Mollusca, which after diminishing appeared again in the Cretaceous epoch in vast numbers. I may conclude these few remarks by stating, that from the extreme variability of shape in this most difficult genus, it is at times scarcely possible to compare specimens with figures; and the more so when the latter are unexact or executed so carelessly as to place one in great doubt as to the types intended; this is proved by the multitude of erroneous determinations or synonyms filling most Geological and Palæontological works; the mistaken identification of species attributed to Sowerby and Lamarck, would alone fill the space of several sheets, nor should I have been able to determine many of the shells which have come under my notice, had I not been assisted in my comparisons by the original specimens of the 'Min. Con.,' and those of Lamarck, &c.

examples of his genus Rhynchonella, while not one of the types of *Trigonella* can be safely identified, excepting the last, which is a true and well-known *Spirifer;* it appears, likewise, that in 1778, Da Costa proposed a genus *Trigonella* for a Mactra, and De Candolle the same for a section of Plants.

DAVILA, in 1767 ('*Catalogue Systematique et Raisonné des Curiosités de la Nature,*') gives a few good illustrations of some recent Terebratulæ, and shows, by his various observations, that he had not neglected to remark the differences of the internal apophysary system between the true *Terebratulæ* and *Rhynchonellæ* while describing his *Bec de Perroquet,* the name by which *R. psittacea* was then well known; he alludes to the hinge, and the two short lamella.

In 1712, Morton ('*Nat. Hist. of Northamptonshire,*') perceived the necessity of separating the plaited Terebratulæ from the smooth ones. He adds, p. 211, "The head or beak of the longer of which valves is crooked, and lies over the top of the other valve: of these, there are two general divisions, the smooth and the striated; the smooth have generally a rounder and blunter beak, the end of which in most of them is, as it were, bored, from whence it was called Terebratula by Lhwyd, in his 'Lithop. Brit.' Some have a rounded, others a straight margin . . . . in the second division, the striated ones have generally a sharper beak than the former;" this shows, that so far back as 1712, the necessity of separating the Rhynchonellæ from the true Terebratulæ was appreciated.

For further details, refer to 'Introduction,' and the article 'Rhynchonella,' in Parts I and II, where numerous illustrative figures are given, and to my paper in the 'Ann. and Mag. of Nat. Hist.,' April 1852.

62. RHYNCHONELLA WRIGHTII, *Dav.* Plate XIV, fig. 1.

Diagnosis. Shell inequivalved, spherical, almost as broad as long; larger valve moderately convex; beak acute, not much produced; foramen small, almost entirely surrounded by the deltidium, a small portion being completed by the umbo; imperforated valve very convex, gibbous; hinge margin slightly indenting the smaller valve; exterior striated and plaited, the striæ which cover all the surface longitudinally are very numerous, minute, and irregular, sometimes bifurcating, nine or ten occupying the width of a line; the plaits only appear on the anterior half of the valves, eleven or thirteen in number, three of which form a well-defined mesial fold, and two in the sinus of larger valve; structure unpunctuated; length 18, breadth 16½, depth 15 lines.

Obs. This fine species was discovered by Dr. Wright in the inferior Oolite of Leckhampton Hill, along with *Ter. simplex* and *T. plicata;* it is quite distinct from *R. furcillata* and *R. rimosa,* although belonging to the same small group; it appears to be very rare, and has not yet been noticed on the continent; the specimen figured is from the collection of Dr. Wright.

63. RHYNCHONELLA FURCILLATA, *Theodori,* Sp. Plate XIV, figs. 2, 3, 4, 5.

TEREBRATULA FURCILLATA,	*Theodori; de Buch,* 1834.	Uber Terebrateln; et Mém. de la Soc. Géol. de France, 1838, vol. iii, 1st series, p. 143, pl. xiv, fig. 13.
—	—	*Rœmer,* 1836—38. Dei Versteinerungen des Norddeuschen Oolithen-gebirges Tafel. xiii, fig. 2.
—	—	*C. Rouillier,* 1847. Bull. de la Soc. Imp. de Moscow.
—	—	*Bronn.* Index Pal., 1848, p. 1237.
RHYNCHONELLA FURCILLATA,	*D'Orbigny,* 1849.	Prodrome, vol. i, p. 239.

Diagnosis. Shell inequivalved, transverse, wider than long; large valve slightly convex, beak acute, produced, not much recurved; foramen entirely surrounded by the deltidium, which also presents a slight tubular expansion; false area generally well defined; hinge margin not much indenting the smaller valve; sinus deep. Smaller valve most convex at the umbo, which is also the deepest part of the shell; the mesial fold is well defined, the sides sloping rapidly downwards. The posterior halves of the valves are ornamented by a variable number of small plaits, from forty to sixty, most of these uniting by two, three, and four, towards the middle of the valves, form large angular costæ, which extend to the margin; the number of these last also varies on the mesial fold, which is composed in different specimens of two, three, four, five, six, and seven plaits. Structure unpunctuated. Length 10, breadth 12, depth 8 lines.

Obs. It is stated, that Theodori first named this shell in his collection, which name was subsequently adopted by V. Buch, who both described and figured the shell.   Few authors have seemingly noticed it; and it is a rare species in England, where it has been found by Mr. Moore in the Marlstone of Ilminster, and appears also to occur in beds of the same age near Cheltenham.   In Pl. XIV, I have illustrated varieties, with two, three, four, and five plaits, on the mesial fold.   In France, it is found in Normandy, at Fontaine-Etoupefour, near Caen, at Pinperdu, near Salins (Jura), at Allem, Bahlingen, and Neuffen, Wurtemberg, &c., in Germany; it is generally found in company with *Rh. rimosa*, into which it seems sometimes to pass by insensible gradations.

64. RHYNCHONELLA RIMOSA, *De Buch*, Sp.   Plate XIV, fig. 6, 6ᵃ.

TEREBRATULA RIMOSA,	*De Buch*, 1831.	Petrifications Remarquables, pl. vii, fig. 5; et 1834, Uber Terebrateln; et 1838, Mém. de la Soc. Géol. de France, vol. iii, 1st ser., plate xiv, fig. 12.
—	—	*Zieten*, 1832. Dei Versteinerungen Württembergs, tab. xlii, fig. 5.
—	—	*Deshayes*, 1836. Nouv. Ed. de Lamarck, vol. vii, p. 354, No. 70.
—	—	*C. Rouillier*, 1837. Bull. de la Soc. Imp. de Moscou, No. 11.
—	—	*Morris*. Catalogue, 1843.
—	—	*Bronn*. Leth. Géog., 1837, p. 292, tab. xviii, fig. 6.
—	—	*Ibid.*, 1848. Index Pal., p. 1249.
RHYNCHONELLA RIMOSA,	*D'Orb.* 1849.	Prodrome, vol. i, p. 239.

Diagnosis. Shell more or less circular, almost spherical, generally as wide as long; valves convex, gibbous, deepest near the umbo; beak acute, much recurved, almost touching the umbo, and allowing very little space for the passage of the muscular fibres composing the pedicle.   The deltidial plates entirely surround the foramen, which is small, and almost touches the umbo; hinge margin indenting the smaller valve; sinus and mesial fold well defined, valves ornamented by a considerable number of small plaits, which, after proceeding from the beak and umbo towards the middle of the valve, in most cases unite by two or three, to form larger and deeper plaits or costæ, which extend to the margin; thirteen or fourteen of these are visible on each valve; two, three, or four, forming the mesial fold; structure unpunctuated.   Length 8, width 8, depth 6 lines.

Obs. *R. rimosa* is distinguished from *R. furcillata* by its smaller dimensions and more spherical form, it is much less wide than *furcillata*, its beak is more recurved and ornamented by fewer plaits; these unite in a very irregular manner, some before others, while a few either soon disappear or proceed from the umbo to the margin, without uniting into large costæ.   MM. C. Rouillier and A. Vossinsky have proposed to establish a small separate group for *R. rimosa* and *R. furcillata*, under the name of *duplicatæ*,[1] to distinguish

---

[1] Etudes progressives sur la Palæontologie des Environs de Moscou. Bulletin de la Soc. Imp. des Naturalistes de Moscou, année 1847, No. 11.

the peculiar disposition of the plaits in the *duplicatæ*, forming a natural passage by means of the *plicosæ* into the *dichotemæ;* these authors are, however, mistaken, in stating that the small group in question belongs exclusively to the Lias period, as several similar species are found in the Cretaceous beds, such as *R. antidichotoma* (Bouvignier), &c., and *T. Schnurii* (de Ver) from the Devonian Rocks.

Both *R. rimosa* and *furcillata* are generally found in the same localities in the Middle Lias,[1] but in England they are rather uncommon fossils; I have found *R. rimosa* in the Lias at Farrington Gurney, near Radstock, also near Cheltenham, and it has likewise been met with near Whitby. In France, it is common at Vieux Pont, Fontaine-Etoupefour, &c., near Caen, at St. Amand (Cher), at Bajac, Castellane (Lower Alps); and it is stated by Von Buch and Mr. Fraas to occur plentifully in many German localities, such as Bahlingen in Wurtemberg, in the stream of Pliensbach, Amberg, Allem, &c.: fig. 6, nat. size; fig. 6*ª*, enlarged.

65. RHYNCHONELLA SPINOSA, *Schloth.*, Sp.   Plate XV, figs. 15—20.

TEREBRATULITES SPINOSUS, *Schlotheim*, 1813. Beiträge zur Naturgeschichte der Versteinerungen in Leonhards, Mineral Taschenbach, vol. vii. Refer to Knorr's fig. in Lapides Diluvii Universalis Testes, 1755, tab. B iv, fig. 4.

TEREBRATULA SPINOSA, *Smith*, 1816. Stratigraphical System of Organised Fossils, p. 108.

— SENTICOSUS, *Schl.* 1820. Der Petrefac., p. 268, No. 30.

TEREBRATULA ASPERA, *Kœnig*, 1825. Icon. Sect., No. 219.

— SPINOSA, *Lamarck*, 1819. An. sans Vert., vol. vi, No. 52; and Dav., Notes on the Species of Lamarck, An. and Mag. of Nat. Hist., 1850.

— — *Von Buch*, 1834. Uber Ter.; and 1838, Mém. de la Soc. Géol. de France, 1ʳᵉ ser., p. 161, pl. xvi, fig. 4.

— — *Defrance*, 1828. Vol. liii, Dic. d'Hist. Nat., p. 161.

— — *Zieten*, 1832. Die Vers., Würt., p. xliv, fig. 1.

— — *Deslongchamps*, 1837. Soc. Linn. de Normandie, p. 30.

— — *Phillips*, 1835. Geol. of York, vol. i, pl. ix, fig. 18.

— — *Bronn.* Leth. Géog., 1837, p. 296, tab. xviii, fig. 2.

— — *Quenstedt*, 1843. Flözgebirge.

— — *Morris*, 1843. Catalogue.

— — *Tennant*, 1847. A Stratigraphical List of Br. Foss., p. 74.

— — *Bronn.* Index Pal., 1848, p. 1251.

HEMITHIRIS SPINOSA, *D'Orb.* 1849. Prodrome, vol. i, p. 286.

ACANTHOTHIS SPINOSA, *D'Orb.* Pal. Franç. Ter. Crétacées, vol. iv, p. 343, 1847, (but published later.)

---

[1] In a very interesting paper '*On the Comparison of the German Jura Formations with those of France and England,*' by M. Oscar Fraas, (Leonhard's and Bronn's Neues Jahrbuch, f. Min., U.S.W., 1850, 2 H, pp. 138—185,) this species is particularly noticed as a most characteristic form in the *Middle Black Jura* in all countries, and is stated to be always found with *T. numismalis*. I cannot, however, believe with that author, that *T. digona* and *T. lagenalis* were found so low down as the Middle Lias.

Diagnosis. Shell transverse; wider than long, more or less spherical; the smaller valve convex, even gibbous, attaining its greatest height before reaching the middle of the valve; the plaits on the mesial fold do not project much above the lateral ones. Larger valve convex, exhibiting a wide sinus more or less defined, and extending to the front; the margin forming a convex line in that portion. The beak is small and acute, under which, especially in young shells, the foramen is visible, almost surrounded by the deltidial plates, except in a small portion, which is completed by the umbo; in adult shells the beak becomes so much recurved, that hardly any space remains between it and the umbo for the passage of the peduncular fibres; surface of valves ornamented by a variable number of plaits of greater or less depth, not increasing much in width as they proceed from the beak and umbo to the front; sometimes they are seen to bifurcate, but generally this appearance is more due to the interposition of a new plait between the regular costæ at variable distances, than to true bifurcation: from distance to distance along the ridge of each plait are seen to proceed long slender tubular spines, arising from an expanded base, and at times exceeding six or nine lines in length; their number is very variable, as well as the regularity of their disposition.

Length 13, width 15, depth 11 lines.

Obs. This is a well-known inferior Oolite shell, figured but not named by Knorr,[1] in 1755, by Walch and Knorr in 1768,[2] and by Captain Walcott in 1799.[3] It varies so much in the number, width, and depth of its plaits, as to have tempted some authors to divide them into more than one species. In some specimens I have counted forty-six plaits on each valve, while others offer fewer, sometimes only twenty; to this variety M. D'Orbigny has given the name of *costata*.[4]

I do not agree with that author in placing this shell in the genus *Hemithyris*, from the supposition that it had no deltidial plates, which is a mistake; nor do I admit that the genus *Hemithyris* is in itself distinct from the *Rhynchonella*, as even in *R. psittacea* the deltidium is visible, although in a less extended shape.[5]

[1] Lapides Diluvii Universalis, tab. B iv, fig. 4.   [2] Die Naturgeschichte der Verst., id.

[3] Descriptions and Figures of Petrifactions found near Bath, fig. 31.

[4] Prodrome, vol. i, p. 286.

[5] On reading over the article *Hemithiris*, Pal. Franç., Ter. Crétacées, vol. iv, p. 342, 1847, M. D'Orbigny states: "This genus is very nearly related to *Rhynchonella*, and is only distinguished by its foramen contiguous to the hinge, and no deltidium. We believe that two distinct genera may be established in this division. The name of *Hemithiris* may be preserved for those species with a fibrous texture, and without pores or exterior spines, the genus thus circumscribed with the recent form (*H. psittacea*) would number 17 species, the first are silurian . . . . to the species, likewise, of fibrous texture, but provided with tubular spines, scattered or in lines, we will apply the name of *Acanthothiris*. We place in this the *Acanthothiris spinosa*, D'Orbigny; and *costata*, D'Orbigny; and *senticosa*, inscribed in our 'Prodrome' under the name of *Hemithiris*." But, as may be seen by referring to the article, *Rhynchonella*, Part I, I cannot admit the three genera proposed for shells all similarly organised. Besides which, M. D'Orbigny places, in the genus *Rhynchonella*, many species, the deltidium of which is not tubular, and species having the beak

It is probable that, when alive, *R. spinosa* was more or less coloured with red; at least, we have seen many specimens, in which the spines had preserved that colour; this is also the opinion of M. Deslongchamps, and it is alluded to by V. Buch in his description of the species. *R. spinosa* is found in the inferior Oolite of many localities, as at Dundry, near Bath, Minchinhampton, Cheltenham, &c., where many fine specimens, with the spines preserved, have been discovered by Dr. Wright, and Messrs. Lycett, Walton, and others, a fine series of all the varieties may be seen in the Museum of the Bristol Institution.

In France it abounds; in Normandy, at Falaise, Moustiers, Port-en-Bessin, Sturfenberg, and many other German localities.

Fig. 15. A specimen, with its spines.

„ 16ª. A variety, with few plaits, from Dundry.

„ 17, 18. Enlarged representations of the beak, foramen, and deltidium.

„ 19. Is drawn from a specimen in Dr. Wright's collection.

„ 20. From a young specimen, showing the dichotomizing and intercalation of some of the plaits.

66. RHYNCHONELLA SENTICOSA, *V. Buch*, Sp. Plate XV, fig. 21.

> TEREBRATULA SENTICOSA, *Von Buch*, 1834. Uber Terebrateln; and 1838, Mém. de la Soc. Geol. de France, vol. iii, p. 162, (non *Ter. senticosa*, Schloth., which is a synonym of *Spinosus* of the same author.)

Diagnosis. Shell transversely oval, depressed, wider than long; valves convex; beak small, acute, not much produced or recurved; foramen nearly surrounded by the deltidial plates; a small portion only being completed by the umbo; marginal line nearly straight, slightly curved in front, but not producing any distinct mesial fold or sinus; valves ornamented by a vast number of minute longitudinal ridges, on which are implanted a vast number of small tubular spines covering the whole surface; from twenty to twenty-four occupying the space of a square line; lines of growth strongly marked. Length 11, width 13, depth 7 lines.

Obs. The first mention of the term *senticosa*, is by Schlotheim,[1] who refers to Knorr's figure,[2] which, as we have already remarked elsewhere, was the type of *R. spinosus*, so named by Schlotheim in 1813,[3] *T. senticosa* of that author is therefore only a synonym. At

so much incurved, as to lie close on the umbo, and therefore showing no tubular foramen, such as *Rh. tetraëdra*, &c., and others, such as *R. concinna*, where the foramen is not tubular, nor even entirely surrounded by the deltidium, a portion being completed by the umbo.

[1] Die Petrifactendunde, 1820, p. 268, No. 30.

[2] Knorr Lapides diluvii, 1755, pl. B. W., fig. 4.

[3] Schlotheim Beitrage zur Natur. Vers., in Dr. Leonhard's Min. Tasch., vol. vii.

a later period, Baron Von Buch distinguished the shell above described from *R. spinosa*,[1] and applied to it the name of *senticosa*, which I have adopted for the present form to avoid the introduction of another appellation for the species under notice. It is also, probable, that Zieten's *Ter. spinosa*[2] may belong to the present form, as it does not seem to present the characters of the true *spinosa*, and although both of these species are somewhat allied, they should always be considered as distinct, *R. senticosa* being a much more transverse and more depressed shell; the beak is smaller, and the surface ornamented by a greater number of slender spines, while in *R. spinosa* the shell is more spherical, the beak rounded and more recurved, the surface being also not only spinose but plaited. Both species seem to occur in the same localities, viz., in the lowest beds of the Inferior Oolite of Dinnington and Burton Radstock, where they were collected by Mr. Moore. On the Continent, it is common in beds of the same age, at Curcy, Moutiers, &c., in Normandy, and is likewise stated to occur at Grumbach, near Amberg, by Baron Von Buch, in the inferior Oolite beds above the Lias. M. D'Orbigny does not seem to have noticed this species in the Inferior Oolite (*T. Bajocean*)[3] where the type commonly occurs, but places it in the Oxford Clay,[4] where, according to MM. Bouchard and Moreau, the species is likewise found in the Departement de l'Yonne; I have not, however, had the advantage of seeing any of those specimens, but from M. D'Orbigny referring to Baron Von Buch, and giving his locality *Grumbach*, it is probable that it is the same species; none, however, have yet been noticed in our English Oxford Clay.

67. RHYNCHONELLA RINGENS, *Herault*, Sp. Plate XIV, figs. 13, 14, 15, 16.

TEREBRATULA GRIMACE, *Herault*.
— RINGENS, *V. Buch*, 1834. Uber Terebrateln; et Mém. de la Soc. Geol. de France, 1838, vol. iii, 1re ser., pl. xiv, fig. 3.
— — *Deshayes*, 1836. Nouv. ed. de Lamarck, p. 312, No. 65.
— — *Bronn*, 1848. Index Pal., p. 1249.
RHYNCHONELLA RINGENS, *D'Orb.*, 1849. Prodrome, vol. i, p. 258.

Diagnosis. Shell remarkably shaped, the depth exceeding the width and length; the smaller valve rises suddenly from the umbo to the front by an almost perpendicular convex curve, forming a large rounded central plait or elevated mesial fold, bent downwards at its extremity, which lies over to meet the sinus of the perforated valve; on either side of the mesial fold three or four smaller lateral plaits form a regular curve from near the umbo to the margin. In larger valve, the beak is small, acute, and not much recurved; the foramen is entirely surrounded by the deltidial plates, but not remarkably separated from

[1] Uber Terebrateln, 1834, vol. vii, 1813.
[2] Zieten Wurtemb. Verst., p. 44, fig. 1, 1832.
[3] Prodrome, vol. i, p. 286.
[4] Ib., p. 375.

the umbo; the hinge margin of larger valve not much indenting the imperforated one. The sinus begins to appear at a short distance from the beak, when it soon turns up almost perpendicularly in the shape of a narrow tongue to meet the bent downward extremity of the mesial fold. The sinus is divided longitudinally by a groove which becomes deeper as it approaches the extremity, the anterior portion of the sinus thus presenting two convex rounded costæ, much depressing the level of the lateral margin of the mesial fold; three or four small lateral plaits existing likewise on each side of the sinus in this valve. Structure unpunctuated. Length 6, width 7, depth 8 lines.

Obs. This species was discovered by M. Herault, engineer of the Mines, and he gave to it the name of *Ter. grimace*, afterwards translated to the Latin name of *Ter. ringens*, by Von Buch, to whom we are indebted for the figure and description of the shell.

It is only lately, that this form has been noticed in England, where it was discovered in the Inferior Oolite of Yeovil, Sherburn, Greenland, &c., in Gloucestershire by the Officers of the Geological Survey, to whom I am indebted for the loan of the specimens (figs. 13 and 14) illustrated in Pl. XIV. On the Continent, the shell is found in the Inferior Oolite of Moutiers, near Caen.

In England, this species does not appear to have attained the dimensions of the French shells, they are not as thick or rounded, seem more angular, and are, perhaps, more properly speaking, a variety of the original type. Fig. 15 represents the Norman type. Fig. 16 is given to show, that, although the great character of the species under consideration is to have a single plait in the elevated mesial fold, still in some rare cases it may become bidentated, and even tridentated, as may be seen in the remarkable specimen kindly lent me by M. Deslongchamps; a similar variation from the common type has been noticed in *R. cynocephala*, and in other species.

R. ringens is placed by M. D'Orbigny[1] in the Upper Lias (Terrain Toarcien), I believe it to be, however, more properly located in the lower beds of the Inferior Oolite.

68. RHYNCHONELLA SUBRINGENS, *Dav.* Plate XIV, figs. 17, 17[abc].

Diagnosis. Shell irregularly globular, as wide as long and deep; beak acute, much recurved, overlying, and almost touching the umbo, concealing the foramen and leaving little space for the passage of the pedicle muscular fibres; valves convex, gibbous, and ornamented by seven large, strongly indented costæ in the smaller, and six in the larger valve, proceeding from the umbo and beak to the front and sides; in smaller valve, a large central, more elevated plait forms a mesial fold, to which corresponds a deep sinus in the other one. Structure imperforated; length 4, width 4, depth 4 lines.

Obs. This little shell has been obtained from the Inferior Oolite of Somersetshire by the Officers of the Geological Survey. It seems to differ from *R. ringens*, with which it was found, by its regular and more indented plaits proceeding from the beak and umbo, while in *R. ringens* they appear at a greater distance. It differs also in general shape, as may easily be perceived on comparing the respective figures of these species. It bears, however, some resemblance to the above-named form by its sinus being likewise strongly marked by a central longitudinal line.

Plate XIV, fig. 17. From a specimen in the collection of the Museum of Practical Geology.

fig. 17[abc]. Enlarged illustrations.

69. RHYNCHONELLA ACUTA, *Sow.* Sp.   Plate XIV, figs. 8, 9.

TEREBRATULA ACUTA, *Sowerby*. Min. Con., 1818, vol. ii, p. 115, tab. 151, figs. 1, 2.
—   —   *Schloth.* System. Ver. der Petref., 1832.
—   —   *Deshayes*, 1836. Nouv. Ed. de Lamarck, vol. vii, No. 69, p. 353.
—   —   *V. Buch.* Uber Terebrateln, 1834; et Mém. Soc. Géol. de France, 1838, vol. iii, 1re ser., p. 142, pl. xiv, fig. 11.
—   —   *Phillips*, 1835. Geol. of York., pl. xiii, fig. 25.
—   —   *Deslongchamps*, 1837. Soc. Linn. de Normandie, p. 30.
—   —   *Morris*, 1843. Catalogue.
—   —   *Bronn*, 1848. Index Pal., p. 1228; figured also in the 'Encyc. Meth.,' pl. 255, fig. 7.

Diagnosis. Shell inequivalved, more or less triangular, wider than long, the smaller valve rising in an almost straight line from the umbo to the margin, and forming a large central acutangular plait, the sides sloping rapidly in the manner of a roof; two or three other smaller lateral costæ exist on each side, the first of which is larger than the other two, and can be traced almost from the umbo; in the perforated valve a deep sinus is seen, which corresponds with the central plait of smaller valve; there exists also three lateral costæ. The beak is small, acute, and not much produced, with foramen entirely surrounded by the deltidial plates; margin of the valves deeply sinuated, hinge margin of larger valve not much indenting the smaller one; structure unpunctuated. Length 10, width 13, depth 11 lines.

Obs. *R. acuta* occurs abundantly in the Marlstone, or Middle Lias of Ilminster, in Yorkshire, Wilton Castle, Rilsdale, &c. On the continent it is met with at Landes, Vieux Pont, Evrecy, &c., in Normandy, and does not appear to vary much in shape, being always easily recognised.

In the Oxford Clay of Koroskovo, near Moscow, in Russia, we find a small shell, which received from Fisher, in 1809, the name of *Rhynchonella loxia*, on which type the genus appears to have been established later; 1843 the same author gave to the same shell the name of *Aptyca:* it bears the greatest possible outward resemblance to *R. acuta*, of Sowerby; but, as stated by M. D'Orbigny, can be distinguished[1] by a more marked longitudinal groove in the sinus of the larger valve, visible especially at its extremity, its surface being everywhere

[1] Geology of Russia, t. ii, p. 482, pl. xlii, fig. 22—26.

minutely striated, which is only seen on well-preserved specimens; it is also a smaller species, and found in the Oxford Clay, while *R. acuta* is peculiar to the Middle Lias. The difference between the two forms has been noticed by M. Ch. Rouillier,[1] but we may here mention, that a variety of *Rh. cynocephala*, found in the inferior Oolite of Minchinhampton, with only one central elevated plait, somewhat approaches to *R. acuta;* but besides this shape being unusual to *R. cynocephala*, its lateral costæ are more numerous than those observable in the Sowerby shell.

70. RHYNCHONELLA CYNOCEPHALA, *Richard*, Sp.   Plate XIV, figs. 10, 11, 12.

TEREBRATULA CYNOCEPHALA, *Richard*, 1840. Bull. de la Soc. Géol. de France, vol. xi, p. 263, pl. iii, fig. 5 *a b c d*.

Diagnosis. Shell inequivalved, nearly as wide as long, more or less irregularly triangular. The smaller or imperforated valve is convex at the umbo, and continues to rise rapidly to the extremity of the margin, with a slight inward curve, forming a narrow, pinched, elevated, and bidentated mesial fold, the sulcus separating the two plaits only appearing in the anterior half of the valve, and increasing in width with little depth to the edge of the front, to which a small plait corresponds in the sinus of the other valve.

At a short distance from the umbo on either side of the elevated mesial fold, the sides of the valves form a descending curve divided by four small lateral plaits on each side; the beak in the larger valve is small, acute, not much recurved, leaving a well-defined false area between the beak ridges and dental margin, in the centre of which, under the beak, a small foramen is visible, surrounded by the deltidial plates, but almost touching the umbo in one point; a regular convex curve is formed from the extremity of the beak to the front, which is in a great measure taken up by a wide sinus; in the middle of it is perceived the small plait above noticed, four lateral costæ also exist on both sides of this valve; they form a slight outward curve at a short distance from the beak, and correspond with those in the smaller valve. Structure imperforated. Length 8, width 8½, depth 7 lines. The width of the sinus is about half the total breadth, the greatest width being also towards the middle of the shell.

Obs. Much confusion and misunderstanding exists relative to this species, which has been considered by many the representative of Professor Phillips's *Ter. bidens*, 'Geol. of York.,' Pl. XIII, fig. 24 ; this is, however, an error, as the illustration given by that author in no way represents the species under consideration, nor does the strata alluded to, "*Marlstone and Ironstone beds*," allow us to mistake the shell of Professor Phillips for *R. cynocephala*, which in England belongs to the Inferior Oolite.[2] In 1840, Mr. Richard became convinced of this fact, which he then mentioned to me; and, in order to put matters right, described and figured the present form under the name of *T. cynocephala*, from the vague

[1] Etudes Progressives sur la Palæontologie des Environs de Moscou. Par MM. Ch. Rouillier et Alex. Vossinsky. Bull. de la Soc. Imperiale des Naturalistes de Moscou, 1847, No. 11.

[2] M. de Verneuil has found this species in the Lias of Villas des Covo and Albanaar in Spain.

resemblance this shell presents to a dog's head.   Most Palæontologists, singularly enough, have not coincided with Mr. Richard's views, for we see Bronn and others placing *Ter. bidens* with its variation *T. biplicata*, and *T. cynocephala*, in one species, (Index Pal., p. 1251.)   Mr. Woodward fully understood the necessity of adopting a distinct name for the shell under notice, and in the collection of the British Museum made use of Mr. Richard's name.   In 1828 Young and Bird figured a shell under the name of *Ter. lineata;* but, from want of description and the incorrect illustration, it is impossible to decide as to the shell intended.   Through the kindness of Mr. Riply, of Whitby, I was able to convince myself that both *T. bidens,* and *cynocephala,* occur in Yorkshire, but in different beds, and that both species seem perfectly distinct.   *R. cynocephala,* as we have above remarked, has for general character the bidentation of its elevated mesial fold; but in some rare cases, the bidentation is replaced by one single elevated plait, as in *R. acuta* (see fig. 10); and in other similarly rare exceptions, instead of the single bidentation, that portion is tridentated, as in fig. 12.   I am indebted to Messrs. Lycett and Woodward for specimens confirming this fact in the most positive manner.

    R. cynocephala is found in the inferior Oolite of Minchinhampton, Dinnington, Yorkshire, &c.; on the continent it was discovered by Mr. Richard, in the inferior Oolite of Bourmont.

    Figs. 10, 11, 12, are drawn from specimens in the collection of the British Museum, and in that of Mr. Lycett.

71. RHYNCHONELLA VARIABILIS, *Schloth.,* Sp.   Plate XVI, figs. 1—6, and Plate XV, figs. 8—10.

TEREBRATULITES VARIABILIS, *Schlotheim,* 1813.   Beiträge zur Naturgeschite der Versteinerungen in Leonhard's Mineral Tashent, vol. vii, p. i, fig. 4.

    —   TRIPLICATA, *Phillips,* 1835.   Illustrations of the Geology of Yorkshire, Part I, p. 157, pl. xiii, fig. 22.

    —   BIDENS, ibid., pl. xiii, fig. 24.

    —   TRIPLICATA, *V. Buch,* 1838.   Mém. Soc. Géol. de France, vol. iii, 1re ser., p. 140, pl. xiv, fig. 9.

    —   VARIABILIS, ibid., pl. xiv, fig. 10.

    —   TRIPLICATA, *Deshayes,* 1836.   Nouv. Ed. de Lamarck, p. 353.

    —   VARIABILIS, *Pusch.*   Polens Palæontologie, 1837, p. 11, pl. iii, fig. 2.

    —   TRIPLICATA and BIDENS.   *Deslongchamps,* 1837.   Soc. Linn. de Normandie, p. 30.

    —   —   *Morris.*   Catalogue, 1843.

    —   —   *Tennant,* 1847.   A Strat. List of British Fossils, p. 74.

    —   —   *C. Rouillier* and *A. Vossinsky,* 1847.   Etudes Prog. sur la Pal. des Env. de Moscou, No. 11, tab. B, fig. 17.

    —   —   *Bronn,* (pars) 1848.   Index Pal., p. 1254.

RRYNCHONELLA VARIABILIS, *D'Orbigny.*   Prodrome, 1849, vol. i, p. 239.

Diagnosis. Shell variable, irregularly triangular, wider than long; beak acute, more or less produced and recurved; foramen small, entirely surrounded or almost so by the deltidial plates, a very small portion being completed by the umbo, the beak is sometimes so much recurved, as hardly to allow any space for the passage of the pedicle muscular fibres; beak ridges well defined, leaving a flat or concave false area between them and the hinge margin, and not much indenting the smaller valve, which presents an elevated regular convex curve from the umbo to the front, ornamented by a variable number of costæ, two, three, or four in different examples, forming a central elevated mesial fold, corresponding with one, two, or three plaits situated in the sinus of larger valve; on either side of this mesial fold and sinus on both valves are seen three or four lateral costæ, not varying much in number on different specimens, so that the smaller valve may be ornamented by eight, nine, or ten plaits; the number on the mesial fold being independent of those on the sides; the costæ in general becoming visible, and produced only at a little distance from the beak and umbo. Structure unpunctuated; length 10, width 12, depth 9 lines.

Obs. This shell has caused great confusion from its variable shape, no doubt due to local causes; and it is not without many comparisons and researches that I made up my mind to adopt M. D'Orbigny's views, viz., to consider Professor Phillips's *Ter. triplicata* and *bidens* as synonyms of Schlotheim's *T. variabilis*, which species was unfortunately established on a very poor and exceptionable specimen, so that his figure in no way represents the general and well-developed types of the species, so much so, that from not being able to make up my mind to their identity I was led to figure both separately in different plates, but having since re-examined a multitude of examples, and been unable to discover permanent distinguishing characters, I am obliged to admit Schlotheim's priority. It will not, however, be useless to mention some points connected with the history of this species; Von Buch, in his important work on 'Brachiopoda,' admits both *T. variabilis* and *triplicata*, stating at the same time, that the difference between them is very small. Professor Phillips's *Ter. triplicata* and *bidens* belong to the same form, and not to the *R. cynocephala* of Richard, as we have already explained under the description of that species, although both are found in the same neighbourhood, but not in the same beds. *R. variabilis*, or *triplicata* and *bidens*, being only varieties with two or three plaits on the mesial fold, and it is always objectionable to name species from the number of costæ ornamenting their surface, as these constantly vary to a greater or less degree on different specimens, *R. variabilis* presenting at times from two to five plaits on the mesial fold, while the types of Schlotheim and Phillips have but two or three; most writers seem to have retained both these authors' names, while some, as Von Buch, M. Deslongchamps, Professor Bronn, Rouillier,[1] &c., have considered *bidens* and *triplicata* as the same type; but the last two authors

---

[1] The references are given in the synonyms placed at the head of this description.

are mistaken in some of their synonyms; thus Prof. Bronn is disposed to consider *R. cynocephala*, as perhaps the same, while M. C. Rouillier places *R. acuta* of Sow., as a simple variation of *triplicata* with one plait, and *bidens* as a variety with two on the mesial fold, overlooking the important character distinguishing the Sowerby type, viz., that the central and only plait forms an elevated curve from the umbo to its extremity, so that at the front the plait becomes not only the highest part of the shell, but is even frequently bent upwards, while in *triplicata* the plaits bend downwards long before having reached the frontal margin. A single glance at the respective figures will point out the difference; Mr. Morris has fallen into the same error, since he considers *R. cynocephala* to be Professor Phillips's *T. bidens*. Through the kindness of Mr. Riply of Whitby, I have been enabled to examine the Professor's types, and a numerous suit of specimens illustrating some of its numberless varieties.

We can hardly believe, with Professor Bronn, *T. variabilis* of Schlotheim to be a specimen of *R. rimosa*; it presents none of the characters of that shell. *R. variabilis* belongs to the Middle and Upper Lias, the var. *triplicata* and *bidens* of Phillips being stated by that author to be peculiar to the Marlstone and Ironstone series,[1] Pl. XVI, figs. 1 and 3, illustrating the Professor's types. It is common, likewise, to the Lias of several other British localities, such as near Ilminster and Radstock, where the shell often assumes a considerable degree of variation, as may be perceived in Pl. XIV, figs. 4 and 6, and Pl. XV, figs. 8, 9, 10. These last are more like the type of the species according to Schlotheim's figure, where the frontal margin has become considerably thickened, they are exceptions to the general form. Figs. 4 and 6 of Pl. XVI, are also more convex and compressed laterally, but we can trace all the intermediate passages uniting these extreme points of variation. A dwarfish variety is likewise met with in the Lias of Stonehouse, near Stroud, Pl. XVI, figs. 2 and 5. On the Continent it has been collected in many localities, presenting all the variations found in England, and is not rare in the Upper Lias of Croisilles and Subles, near Caen, and Bayeux in Normandy, also at Amberg, and I have specimens sent me by Dr. Krantz, from Kirchum, in Wurtemberg, and from Khoroschova, near Moscow, in Russia.

72. RHYNCHONELLA SUBVARIABILIS, *Dav.*   Plate XV, fig. 7; and Plate XVIII, fig 11.

Diagnosis. Shell transversely irregularly oval, wider than long; beak not much produced, acute and slightly recurved, leaving a flat false area between its ridges and the hinge margin; foramen small, almost entirely surrounded by the deltidium, a small portion only being completed by the umbo. Valves convex, ornamented by 9 or 11 plaits,

---

[1] The specimens of this species from the Ironstone bands rarely preserve the shells, which present a more or less advanced state of decomposition, exhibiting those longitudinal asbestoid fibres mentioned by M. C. Rouillier, in his 'Etudes Progressives sur la Palæontologie des Env. de Moscou,' 1847.

2 or 3 in the smaller valve forming an elevated mesial fold, with 1 or 2 in the sinus. Surface minutely longitudinally striated; the concentric lines of growth intersecting the striæ so closely, as to give them a squamose wrinkled aspect, especially as they become more defined and projecting towards the margins.   Length 8½, width 9½, depth 7 lines.

Obs. This species belongs to the Kimmeridge Clay, of Pottern, Wilts, where it was found by Mr. Cunnington, and bears so great an external resemblance to some varieties of *R. variabilis*, Schl., that we might easily mistake it for that species; but, on examining with care the structure of the shell under notice, we find that, besides the larger costæ, the surface is longitudinally striated, and intersected by closely and roughly disposed squamose concentric lines of growth.

The examples figured are from the collection of Mr. Cunnington, three of which may be likewise seen in the British Museum; from not having had the two specimens at the same time, we were obliged to figure them in different plates, the last one, Pl. XVIII, fig. 11, presented only two plaits on the sinus, while the one figured in Pl. XV, fig. 7, possesses three.

73. RHYNCHONELLA LYCETTII, *Dav.*   Plate XV, fig. 6.

Diagnosis. Shell inequivalved, transversely oval, wider than long, beak produced, rounded, slightly recurved at its extremity, foramen circular, entirely surrounded by the deltidial plates; beak ridges well defined, leaving between them and the hinge margin a flat space or false area not indenting the smaller valve. The imperforated valve is regularly convex from the umbo to the front, the deepest portion of the shell being towards the middle, surface ornamented by eleven or thirteen large and deep costæ, three of which form a mesial fold elevated above the lateral plaits, and corresponding with two in the sinus of larger valve. The edge in some specimens is thickened, forming a receding margin all round, so that the extremity of the plaits project further than the junction of the valves. Structure unpunctuated.   Length 11, width 13, depth 7 lines.

Obs. This species seems distinguished from *R. variabilis* by its more transversely oval shape and rounded beak, it was discovered by Mr. Lycett in the middle division of the Inferior Oolite of Minchinhampton, where it is very rare.

Fig. 6. From the collection of Mr. Lycett.

74. RHYNCHONELLA OOLITICA, *Dav.*   Plate XIV, fig. 7.

Diagnosis. Shell irregularly triangular, nearly as wide as long; beak acute, and not much recurved, leaving a flat, false area between its ridges and the hinge margin; foramen small, circular, entirely surrounded, and separated from the umbo by the deltidial plates;

hinge margin of larger valve indenting that of smaller valve; valves ornamented by rounded costæ, variable in number, proceeding from the beak and the umbo to the front and sides, but faintly marked at their origin; twelve or thirteen on each valve, four or five of which in general form the mesial fold, to which a sinus and plaits in the other valve correspond; structure imperforated; length 6, width 6, depth 4 lines.

Obs. This shell is abundant in the Free Stone above the Pea Grit, in the Inferior Oolite of Leckhampton Hill, near Cheltenham, where it was discovered by Dr. Wright; its plaits become distinct only towards the centre of the valves from which they proceed to the front and sides; *R. oolitica*, somewhat resembles certain varieties of *R. variabilis*, or rather the variation *triplicata* of Phillips, found in the Lias near Radstock, but in the last the beak is wider and more circular than in that of *R. oolitica*, which is generally acute and tapering; the valves being likewise less convex and widest at their anterior portion.

Plate XIV, fig. 7.   A specimen natural size, from the collection of Dr. Wright.

fig. 7 $^{c\,d}$. The same enlarged.

75. RHYNCHONELLA MOOREI, *Dav.*   Plate XV, figs. 11—14.

Diagnosis. Shell inequivalved, circular, nearly as wide as long; beak acute, not much produced, more or less recurved; foramen small, entirely surrounded by the deltidial plates; beak-ridges well defined; hinge line not much indenting that of the smaller valve; valves convex, slightly depressed, and ornamented by a variable number of radiating plaits, from eleven to eighteen on each valve, three, four, five, six, and even seven forming a slightly produced mesial fold with corresponding plaits in a similarly shallow sinus.   Structure unpunctuated; length 7, width 8, depth 5 lines.

Obs. This little species was found by Mr. Moore in the Upper Lias, near Ilminster, and in the same beds at Tor Hill, Glastonbury.   It seems to me quite distinct from *R. variabilis*, being more circular, less convex, and ornamented in general by a much greater number of plaits; I have examined examples of all ages, from one line in length up to seven, which seems to be the largest dimensions it attains.   In such variable shells as most Rhynchonellæ, it is often very difficult to describe certain differences in the general aspect, compared with those presented by other species closely approaching them, and which appear to be distinct.

The specimens figured are principally from Mr. Moore's collection.

76. RHYNCHONELLA BOUCHARDII, *Dav.*   Plate XV, figs. 3—5.

Diagnosis. Shell circular, semi-globose, as wide as long, valves almost equally convex; beak acute, moderately produced, rounded and more or less recurved; foramen entirely surrounded by the deltidial plates projecting in a tubular shape round the perforation; beak ridges not very distinct, the false area between them and the hinge margin being small, and not much indenting the smaller valve; valves smooth in the young shell, and ornamented at a more advanced period by only a few rounded plaits appearing towards the margin, viz., two or three, rarely more, forming a kind of mesial fold, to which one or two plaits in the sinus of the larger valve correspond; likewise the sinus only appears near the margin or in the anterior portion of the shell; on either side on both valves, three or four similarly disposed costæ are also visible; concentric lines of growth distinctly marked. Structure unpunctuated; length 7, width 7½, depth 5 lines.

Obs. This little species was discovered in the Upper Lias of Ilminster, where it is common, by Mr. Moore; it has likewise been met with in beds of the same age, by Mr. H. Miller, at Cromarty, in Scotland, and is quite distinct from all the species of Rhynchonella, yet discovered in the Oolitic Period, and cannot be mistaken for *R. variabilis*, with which it differs completely; in some rare instances, instead of two or three plaits on the mesial fold, I have found four and even five; but this kind of variation is common to all Rhynchonellæ, and therefore cannot be taken into consideration. It gives me great pleasure to name it after my valued friend, M. Bouchard.

Figs. 3 and 4, represent two examples from the Upper Lias of Ilminster, in the collection of Mr. Moore.

Fig. 5. A specimen from Cromarty, in the collection of Mr. Miller.

77. RHYNCHONELLA VARIANS, *Schloth.*, Sp. Plate XVII, figs. 15, 16.

TEREDRATULITES VARIANS, *Schlotheim*. 1820. Die Petrefactenkunde, p. 267; Ency. Meth., pl. 241, fig. 5ᵃ ᵇ.

— OBTRITA. *Defrance*. 1828. Dic. d'Hist. Nat., vol. liii, p. 161; Ency. Meth., pl. 241, fig. 5.

— VARIANS, *Rœmer*. 1835. Die Vers. des Nord. Oolithen gebirges, tab. ii, fig. 12.

— SOCIALIS, *Phillips*. 1835. Illust. of the Geol. of York., part i, p. 135, pl. vi, fig. 8.

— VARIANS, *Deshayes*. 1836. Nouv. Ed. de Lamarck, p. 352.

— — *Von Buch*. 1834. Uber Terebrateln; and 1838, Mém. Soc. Géol. de France, vol. iii, p. 135, pl. xiv, fig. 4.

— — *Pusch*. Polens Palæont., 1837, p. 12, pl. iii, fig. 3.

— — *Bronn*. Lethea Geog., 1837, pl. xviii, fig. 4, and Index Pal., 1848, p. 1254.

— SOCIALIS, *Morris*. Catalogue, 1843.

— VARIANS, *D'Orb.* in Murch. Russia, vol. ii, p. 480, pl. xlii, figs. 14, 17, 1845.

RHYNCHONELLA VARIANS, *D'Orb.* Prodrome, vol. i, p. 376, 1849.

Diagnosis. Shell inequivalved, somewhat irregularly triangular, of the size of a hazel nut, wider than long; valves convex; beak acute, small, more or less recurved; foramen circular, always lying close to the umbo, being entirely or partially surrounded by the deltidial plates, a small portion of the circumference being completed by the umbo, a flatness or false area existing between the beak ridges and the hinge margin, which last does not much indent the smaller valve. Surface ornamented by a variable number of small plaits, about twenty-four on each valve, from five to six, or seven, rarely three, forming an elevated mesial fold, the smaller valve arising from the umbo, by a gentle curve to about half·its length, when the inclination becomes more rapid as it approaches the extremity of the valve where it is suddenly bent downwards to meet a well-defined and deep sinus; the plaits in the sinus seems, as if longitudinally depressed or split in their centre, as they approach the front margin. Structure imperforated; length 8, width $7\frac{1}{2}$, depth 5 lines.

Obs. This variable shell received, in 1820, from Schlotheim, the name of *varians*, that author referring to the very indifferent figure in the 'Ency. Meth.' unnoticed by Sowerby, it was subsequently figured by Professor Phillips, under the name of *Ter. socialis*, which name has been admitted by Mr. Morris and Prof. Bronn to be a synonym, of which I convinced myself by comparing the Professor's shell kindly lent me by Mr. Bean, with the German types. It was well described by Baron Von Buch, who states it to occur by millions where it is found. In England, it is abundant in the *Fuller's Earth* round Bath, also at Scarborough and Hackness, and in the Kelloway Rock of Kelloway, from which locality I am indebted to Mr. Walton for many examples. On the Continent, it is stated by Von Buch to occur in the upper portions of the Middle Jura, it is rarely wanting in the bed approximating to the Upper Jura, at Beggingen and at Osterfingen, in the canton of Schaffouse, at Fürstenberg, near Bahlingen, in the Wurtemberg, near Doneschingen, near Amberg, &c.

The Terebratula figured by Zieten, under the name of *Ter. varians*, does not belong to the species, as correctly remarked by M. D'Orbigny; it more resembles *R. cynocephala*, (Rich.); *R. varians* of Schlotheim, is placed by M. D'Orbigny in his (*Ter. Oxfordien.*)

Figs. 15, 16. Two specimens, nat. size.

Fig. 15[bcd]. Enlarged illustrations.

78. RHYNCHONELLA FORBESII, *Dav.*   Plate XVII, fig. 19.

Diagnosis. Shell inequivalve, globular, and circular, beak small, recurved, foramen entirely surrounded by the deltidium; hinge margin of larger valve not much indenting that of the smaller; surface ornamented by about twenty small plaits on each valve, five or six of which form a raised mesial fold not projecting distinctly above the lateral plaits: the sinus is very shallow. Length 4, breadth $3\frac{1}{2}$, depth 3 lines.

Obs. Thirteen specimens of this small species have been obtained in the Inferior Oolite of Somersetshire, by the officers of the Geological Survey, and now deposited in their Museum; as remarked by Professor Forbes, all the examples are so similar and constant in shape, that it is not very probable that they are young individuals, as we do not find any other similarly plaited globular Rhynchonella in our Oolites, to which we could refer it; it somewhat resembles *R. Pisum*, but can be distinguished without difficulty from the Cretaceous species, which is a flatter and more delicately plaited shell.

Fig. 19. Natural size, from a specimen in the collection of the Geological Survey.

„ 19[abc]. Enlarged illustrations.

79. RHYNCHONELLA SERRATA, *Sow.* Sp.   Plate XV, figs. 1, 2.

TEREBRATULA SERRATA, *Sow.*   M. C., vol. v, 1825, p. 168, tab. 503, fig. 2.
— — *Morris.*   Catalogue, 1843.
— — *Tennant*, 1847.   A Strat. List of British Fossils, p. 74.
— — *Bronn.*   Index Palæont., 1848, p. 1250.
RHYNCHONELLA SERRATA, *D'Orb.* 1849.   Prodrome, vol. i, p. 289.

Diagnosis. Shell irregularly subtrigonal, generally a little wider than long, valve almost equally convex, but depressed; beak small, much recurved, and lying over the umbo, leaving little space for the passage of the pedicle muscular fibres; beak ridges well defined, exposing a slightly concave false area between them and the hinge margin, a similar depression being likewise visible on either side of the umbo, the hinge line not much indenting that of the smaller valve. Surface ornamented by a variable number of large sharp plaits, proceeding from the beak and umbo to the front and sides, from fourteen to fifteen on either valve, generally no distinct mesial fold or sinus; six or seven plaits are sometimes a little raised in front in the smaller valve; structure imperforated. Length 15, width 17, depth 10 lines.

Obs. This beautiful species was well described and figured by Sowerby from a small specimen stated to have been found in the Lias of Lyme-Regis, whence, however, I have never obtained an example, and I believe it probable that Sowerby's type was found in the Marlstone of Ilminster, whence many fine specimens have been obtained by Mr. Moore. Sowerby alludes to the shell in the 'Ency. Meth.,' (Pl. 243, fig. 11, and Pl. 244, fig. 1,) as being possibly referable to the species under consideration. We do not, however, believe them identical, although allied; the shell of the 'Ency.' is *Terebratula plicata* of Lamarck, which is a *Rhynchonella*,[1] and seems distinguished from *R. serrata* by being less depressed, deeper, longer than wide, and ornamented by fewer costæ, seven or eight on each valve; some specimens presenting four plaits, others five on the mesial fold, and in a few, two uniting into one towards the margin of the shell; characters different from

[1] Lamarck, 'An. sans Vert.,' vol. vi, 1819, No. 39; and Dav. on Lamarck's species, 'Annals and Mag. of Nat. Hist.', vol. ii, 2d ser., June, 1850, Pl. xiv, fig. 39.

what we observe in *R. serrata*. I am not acquainted with any specimens exactly similar to our English type from the Continent, although I found in the Lias of Fontaine-Etoupe-four, in Normandy, some examples approaching to *R. serrata*, and of which they may perhaps be a variety.

Plate XV, fig. 1. From a specimen in the collection of Mr. Moore.

„ fig. 2. An elongated variety from my collection.

80. RHYNCHONELLA PLICATELLA, *Sow*. Sp. Plate XVI, figs. 7, 8.

TEREBRATULA PLICATELLA, *Sow*. 1825. Min. Con., vol. v, p. 167, tab. 503, fig. 1.
— — *Von Buch*, 1834. Uber Terebrateln; et Mém. de la Soc. Géol. de France, 1838, vol. iii, 1st ser., p. 146, pl. xv, fig. 17.
— — *Deshayes*, 1836. Nouv. Ed. de Lamarck, p. 355.
— — *Deslongchamps*, 1837. Soc. Linn. de Normandie.
— — *Morris*. Catalogue, 1843.
— — *Tennant*. A Stratigraphical List of Brit. Fossils, 1847, p. 74.
— — *Bronn*. Index Palæont., 1848, p. 1246.
RHYNCHONELLA PLICATELLA, *D'Orbigny*, 1849. Prodrome, vol. i, p. 286.

Diagnosis. Shell inequivalve, subtrigonal, sub-globose, longer than wide; perforated valve, shallow and depressed, less convex than the other; front semicircular, beak acute, not much produced, tapering, more or less recurved; foramen small, entirely surrounded by the deltidial plates; beak ridges sharply defined; a slightly concave false area existing between them and the hinge margin, a similar depression being likewise visible on either side of the umbo, giving to this portion a pinched appearance; the hinge-line greatly indenting the imperforated valve, which is regularly convex, and nearly three times as deep as the rostral one; surface ornamented by a variable number of plaits, from twenty-six to fifty on either valve; rarely any distinct mesial fold or sinus. Some of the costæ bifurcate at a short distance from the beak and umbo; structure imperforated. Length 18, breadth 17, depth 13 lines. Some examples attain even larger dimensions.

Obs. This species belongs to the Inferior Oolite of Chideock, near Bridport, whence the original specimen figured and described by Sowerby was obtained by Sir H. de la Beche, and is now deposited in the Museum of the Geological Society; since that period it has been found in several other localities, such as at Dundry and Dinnington. On the Continent very fine specimens are found at Moutiers, Bayeux, and other places in France; it was erroneously considered by Defrance[1] to be the same as *Ter. multicarinata*, Lamarck, which is quite another species.[2] *R. plicatella* is well characterised, and does not seem to vary in general as much as some others, although, in its details, it is often very different, from the variable number of its plaits, and in the young state it is very much elongated, as

[1] Dic. Hist. Nat., vol. liii, 1828, p. 137.
[2] Davidson's Notes on an Examination of Lamarck's Species of Fossil Terebratulæ. 'An. and Mag. of Nat. Hist.', vol. v, 2d ser., 1850, Pl. xiv, fig. 37.

in fig. 8, and is specially remarkable from the convexity of the imperforated valve, which in this case is by far the largest of the two. On some internal casts I have observed the almost entire vascular system, the principal arteries becoming more and more subdivided as they approach the margin of the shell or mantle. The beak is also, in some few instances, and more commonly so in British specimens, much recurved, leaving little space for the passage of the pedicle fibres; but this is not the case with the generality of specimens where the foramen is completely exposed, and more or less separated from the umbo by the deltidial plates. Fig. 8 is drawn from a specimen in the collection of Dr. Wright.

81. RHYNCHONELLA INCONSTANS, *Sow.* Sp.　Plate XVIII, figs. 1, 2, 3, 4.

TEREBRATULA INCONSTANS, *Sow.*　M.C., vol. iii, p. 137, tab. 277, figs. 3, 4, 1821.
— 　—　*V. Buch,* 1834.　Uber Terebrateln; and 1838, Mém. de la Soc. Géol. de France, vol. iii, p. 146, pl. xiv, fig. 16.
— 　—　*Deshayes,* 1836.　Nouv. ed. de Lamarck, p. 355.
— 　—　*Pusch?* 1837.　Polens Pal., pl. iii, fig. 3.
— 　—　*Deslongchamps,* 1837.　Soc. Linn. de Normandie.
—　DIFFORMIS, *Zieten,* 1832.　Die Vers. Wurt., tab. xlii, fig. 2.　(Non *T. difformis,*) Lamarck.
—　INCONSTANS, *Morris.*　Catalogue, 1843.
— 　—　*Tennant.*　A Strat. List of British Fossils, p. 73, 1847.
RHYNCHONELLA INCONSTANS, *D'Orbigny,* 1849.　Prodrome, vol. i, p. 375.

Diagnosis. Shell almost circular, more or less globose, unsymmetrical, rather wider than long; beak acute and recurved, under which is seen a small foramen entirely surrounded by the deltidium, and more or less separated from the umbo. Lateral ridges well defined, leaving between them and the hinge margin a flat or concave false area, which slightly indents the smaller valve. Valves convex, generally gibbous, without mesial fold or sinus, one half at a little distance from the beak and umbo, becoming more elevated than the other, which is turned down as if twisted, separating the shell into two parts; in some specimens the elevation is at the right, in others at the left side of the shell. Surface ornamented by a variable number of simple plaits, from thirty to forty on each valve. Length 19, width 22, depth 18 lines.

Obs. R. inconstans seems to have been first figured by W. Smith, in or after 1816,[1] from the Kimmeridge Clay. It was shortly afterwards named, figured, and described, by Mr. Sowerby, under the above-mentioned denomination, that author observing, "that one half of the edge is turned up and the other down, but indifferently to the right or left." Professor Phillips[2] and Mr. Morris[3] mention it as found in the Speeton Clay of Yorkshire, but I believe this to be a mistake; the Speeton Clay being considered

[1] Strata identified by organised fossils, pl. x, fig. 6.
[2] Illustrations of the Geology of Yorkshire, 1831.
[3] Catalogue, 1843.

Cretaceous, where the species has not hitherto with certainty been found. It made its first appearance in the Inferior Oolite, specimens from Leckhampton Hill, near Cheltenham, being undistinguishable from many of those so abundantly found in the Oxford Clay of Wooton Basset, and Shotover Hill, near Oxford, as well as with some derived from the Kimmeridge Clay of Weymouth.

I am not prepared to admit, with Professor Bronn,[1] that Sowerby's species is a synonym of that described by Lamarck in 1819, under the appellation of *Ter. difformis;* it is certain that the last-named author's type was taken from a shell found in the Cretaceous or Tourtia beds of Tournay, in Belgium;[2] those figured in the 'Ency. Meth.', Pl. 242, fig. 5, *a b c,* being likewise from the Cretaceous beds of France, as justly observed by M. D'Orbigny.[3]

There exists certainly much general similarity between *some* specimens of *R. difformis* and *R. inconstans;* but when adult and well characterised, the last seems easily distinguished by its more circular form, greater convexity, and stronger plaits; the Cretaceous species being more delicate in general appearance, not so convex, much wider than long, its plaits smaller, and its foramen larger and tubular. As justly observed by Baron von Buch, the plaits in the *R. inconstans* are always simple, and do not bifurcate. It is found also in various parts of the continent, as in Normandy, near Namers, Ellrichserbring in Brunswick, Allem, and in other parts of Germany.

Plate XVIII, fig. 1. The largest specimen I have noticed, in the collection of the British Museum, from the Oxford clay of Wooton Basset.

„ figs. 2, 3. Two specimens from the Kimmeridge clay of Weymouth, in the collection of Mr. Bowerbank.

„ fig. 4. From the Inferior Oolite of Leckhampton Hill, in the collection of Dr. Wright.

82. RHYNCHONELLA CONCINNA, *Sow.* Sp.   Plate XVII, figs. 6—12.

TEREBRATULA CONCINNA, *Sow.* Min. Con., vol. i, 1812, p. 192, tab. lxxxiii, fig. 6.
— ROSTRATA, *Sow.* Ibid., vol. vi, 1829, p. 67, tab. 536, fig. 1, (not Cretaceous.)
FLABELLULA, *Sow.* Ibid., vol. vi, 1829, p. 67, pl. lxvii, pl. 535, fig. 1.
— CONCINNA, *V. Buch.* Uber Terebrateln, 1834; and Mém. Soc. Géol. de France, vol. iii, 1838, p. 144, pl. xiv, fig. 14, et *T. rostrata,* Mém. Soc. Géol. de Fr., vol. iii, p. 155, pl. xv, fig. 27.
— — *Morris.* Catalogue, 1843.

[1] Index Pal., pp. 1235 and 1238.
[2] Dav., Notes on an Examination of Lamarck's Species of Fossil Terebratula. 'Annals and Mag. of Nat. Hist.,' June, 1850.
[3] Palæontologie Française, Ter. Crétacées, vol. iv, p. 41.

TEREBRATULA CONCINNA, *Tennant,* 1847.   A Strat. List of British Fossils, p. 73.
—   —   *Bronn,* 1848.   Index Palæont., p. 1233.
RHYNCHONELLA CONCINNA, *D'Orb.*   Prodrome, vol. i, p. 315, 1849.

Diagnosis. Shell inequivalved, when adult nearly gibbose, more or less compressed, in the young; rather wider than long, beak acute and slightly recurved, foramen not entirely surrounded by the deltidial plates; a small portion being completed by the umbo, especially so when young, and up to a certain age in the adult state; the two plates sometimes meeting at the umbo. Beak ridges well defined, leaving between them and the hinge-line a false area, not much indenting the smaller one; surface ornamented by a variable number of acute plaits, about thirty-two in each valve, seven to eight of which forming a slightly elevated mesial fold, corresponding to a sinus in larger valve; structure imperforated. Length 11, width 12, depth 10 lines.

Obs. This is one of the oldest described species in the 'Min. Con.,' it is abundant in the Great Oolite of many localities in England, and is found at Hampton Cliff, Aynhoe, near Bath, Cirencester, &c. At Bradford it is found in the Bradford clay, and is one of our most common British and foreign species.

R. concinna is flat and compressed, when young, with hardly any visible mesial fold, (*T. flabellula,* Sowerby,) becoming convex and gibbose with age. *Ter. rostrata*[1] of that author, belongs likewise to this species, and is only a specimen where the beak is unusually elongated, of which any one will be convinced on inspecting the original specimens in the collection of Mr. J. de C. Sowerby; nor is it a Cretaceous shell, as stated by that author. M. D'Orbigny[2] is mistaken when mentioning *Ter. obsoleta,* Sowerby, as only a young state of *concinna,* an error at once evident from the last being a much smaller shell, and is distinguished from *obsoleta* by several characters alluded to under the head of that species.[3]

Plate XVII, fig. 6, illustrates an adult example from Cirencester.
   „   figs. 7, 8. Middle aged specimens, likewise from Cirencester.[4]
   „   figs. 9, 10. Young individuals, *Ter. flabellula,* of Mr. Sowerby.
   „   fig. 11. An exceptional specimen, *T. rostrata,* of Sowerby.
   „   fig. 12. An enlarged illustration of the beak and foramen.

[1] In Mr. Sowerby's figure, the foramen is incorrectly placed.

[2] Prodrome, vol. i, p. 315.

[3] In a paper lately published, (Mém. de la Soc. Géol. de France, vol. iv, 2d ser., p. 28, pl. viii, figs. 10, 11, 1851,) Mr. Bayle describes and figures as *T. concinna,* Sowerby, a shell which does not seem to me to belong to that species, but more like *R. lacunosa,* Schlotheim, nor does the figure 10 of the same author appear referable to *R. lacunosa.*

[4] The recent *Rh. nigricans,* Sow., Zool. Proc., 1846, is undistinguishable from half grown *R. concinna,* Sow., the former of which was dredged by Mr. Evans, R.N., in 19 fathoms, Foveaux Strait, New Zealand, about five miles north-east of Ruapuke Island, examples of which may be seen in the collections of Messrs. Evans, Cuming, and the British Museum, but probably never becomes as globular, as that species is found, when adult.

83. RHYNCHONELLA SUBCONCINNA, *Dav.* Plate XVII, fig. 17.

Diagnosis. Shell more or less triangular, depressed, rather longer than wide; valves almost equally convex; beak moderately produced, acute, nearly straight; foramen almost entirely surrounded by the deltidial plates, a small portion being completed by the umbo; beak ridges distinct, leaving a flat space between them and the hinge margin, which indents that of the smaller valve. Surface ornamented by a great number of small plaits from fifty to sixty on each valve; the mesial fold and sinus are hardly perceptible, and ornamented by from fourteen to fifteen plaits. Structure unpunctuated; length 7, width about 6, depth 3 lines.

Obs. This species was found by Mr. Moore in the Marlstone of Ilminster, and seems remarkable from the great number of minute striæ which ornament its surface. It approaches in shape to some young shells of *R. concinna;* but in these last, the costæ are less numerous and fine, nor does it ever appear to attain the same dimensions or character of adult specimens of *R. concinna.*

Fig. 17. From the collection of Mr. Moore.

84. RHYNCHONELLA OBSOLETA, *Sow.,* Sp. Plate XVII, figs. 1—5.

TEREBRATULA OBSOLETA, *Sow.* M. C., vol. i, 1812, p. 192, tab. lxxxiii, figs. 7, 8.
— — *Parkinson.* 1822. An Introduction to the Study of Organic Remains, p. 234.
— — *Schlotheim.* 1832. Syst. Vers. der Petrefacten.
— CONCINNA, *Bronn.* Leth. Geog., 1837, p. 289, pl. xviii, fig. 3, (non concinna, but *obsoleta,* Sow.)
— OBSOLETA, *Morris.* Catalogue, 1843.
— — *Tennant.* 1847. A Strat. List of British Fossils, p. 73.

Diagnosis. Shell circular, or irregularly oval, gibbous, longer than wide; beak acute, produced, not much recurved; foramen entirely surrounded by the deltidial plates, and more or less separated from the umbo; beak ridges distinct, leaving between them and the hinge line a false area which indents considerably the smaller valve. Surface oramented by from twenty-two to thirty-three strong acute plaits, five to eight of these forming a slightly raised mesial fold with corresponding sinus in the other valve. Structure imperforated; length 19, width 17½, depth 13 lines.

Obs. *R. obsoleta* was figured by Walcot, in 1799.[1] It occurs in the same localities and beds as *R. concinna,* and is common to the Great Oolite of Hampton Cliff, near Bath, Felmersham, Cirencester, &c., nor is it rare in the Bradford Clay of Bradford, where large specimens have been obtained, as may be seen from the fine series deposited in the

[1] Petrifactions found near Bath.

Museum of the Geological Society. It is, likewise, probable that the vertical range of this species extended beyond the limits here traced: some specimens from the Cornbrash being undistinguishable from certain varieties of the shell under consideration.

On the Continent, it is abundantly distributed in the Great Oolite of Ranville, &c., in Normandy. From *R. concinna* it seems distinguished by its dimensions, larger, and deeper plaits; the deltidium surrounding and separating the foramen from the umbo, which character appears to have been noticed by Mr. Sowerby, since he gives a separate figure of this portion of the shell: the hinge margin indenting likewise more the smaller valve than in *concinna*, it therefore could not be a young state of this last-named shell, as erroneously supposed by M. D'Orbigny.[1] In the young state, both *R. concinna* and *obsoleta* are quite distinct; the last being more elongated and deeply plicated, while in *concinna* the shell is transversely compressed, and the plaits are smaller and more numerous. Professor Bronn[2] considers *R. obsoleta* a synonym of *R. tetraëdra*, but I believe the two species can be easily distinguished. Sowerby's figures both of *R. concinna* and *obsoleta* are very unfortunate, as well as those of *tetraëdra* and *media*, from the fore-shortened aspect selected for illustration.

Plate XVII, fig. 1. From the Bradford Clay, in the collection of the Geological Society, being the largest specimens I have seen of the species.

„ figs. 2, 3. Young shells, from Bradford.

„ fig. 4. Interior. Illustrating the two short shelly lamella *a*, to which were attached the free fleshy arms; *b*, are the muscular impressions.

„ fig. 5. The beak, foramen, and deltidium (enlarged).

85. RHYNCHONELLA SUBOBSOLETA, *Dav.* Plate XVII, fig. 14.

Diagnosis. Shell inequivalved, circular, semi-globose, nearly as broad as long; beak moderately produced; foramen circular, entirely, or otherwise surrounded by the deltidial plates, a small portion being generally completed by the umbo; false area not very well defined, the hinge margin not much indenting the smaller valve. Surface ornamented by a variable number of large costæ, from nineteen to twenty-one or two on each valve; the mesial fold not much produced nor always distinct, formed of from four to six plaits; sinus shallow. Structure unpunctuated; length 12, width 12, depth 9 lines.

Obs. This shell is always much smaller than either *R. obsoleta* or *concinna*, rarely attaining the dimensions given above, it belongs to the middle division of the Inferior Oolite of Minchinhampton, according to Messrs. Lycett and Woodward. It possesses

---

[1] Prodrome, vol. i, p. 315.

[2] Index Pal., pp. 1243 and 1253.

characters of both the above-mentioned species, it has the beak, foramen, and deltidium of *concinna*, but is distinguished from it by its plaits, which are fewer in number, much larger, and deeper; the valves are also nearly equally convex, which is not the character of the Sowerby types. I propose naming it *subobsoleta*, as it may, perhaps, be a variation of that form, although I have not found sufficient grounds to admit it to be such.

Fig. 14 illustrates the largest specimen I have seen, from the collection of the British Museum.

86. Rhynchonella angulata, *Sow.* Sp. Plate XVII, fig. 13.

Terebratula angulata, *Sow.* Min. Conch., vol. v, 1825, p. 166, tab. 502, fig. 4,
(placed there under the name of *acuta*, corrected after-
wards to *angulata*.)
— — *Morris.* Catalogue, 1843.
— — *Bronn.* Index Pal., 1848, p. 1229.
Rhynchonella angulata, *D'Orb.* Prodrome, 1849, vol. i, p. 286.

Diagnosis. Shell transversely oblong, gibbose; beak moderately produced; foramen not entirely surrounded by the deltidium, a small portion being completed by the umbo; valves ornamented by a variable number of plaits, from ten to twenty-five on each; five or six forming a moderately produced mesial fold, with corresponding plaits in the sinus; length 7, width 9, depth 6 lines.

Obs. According to MM. D'Orbigny, Morris, Waterhouse, and other Palæonto-logists, this species is distinct from *R. concinna*; it is shorter, more transverse, than that shell, but I am not yet certain that it may not be a variety. Von Buch is, however, mistaken, in placing it as a synonym of *R. depressa*, Sow. *R. angulata* is found in the Inferior Oolite of Cleeve Hill, near Cheltenham; the specimen figured in my plate is the original type kindly lent me by Mr. J. de C. Sowerby.

87. Rhynchonella Morierei, *Dav.* Plate XVIII, figs. 12, 13.

Diagnosis. Shell somewhat triangular, longer than wide; beak tapering to an acute point not much recurved, under which is seen a small foramen entirely surrounded by the deltidium, which separates it more or less from the umbo; beak ridges well defined, leaving between them and the margin a large flat, or slightly concave false area, which indents considerably by a marked curve the corresponding margin of the umbo. Valves moderately convex, ornamented by a variable number of plaits, from thirty-three to forty-four on each valve, from seven to nine composing a well-defined mesial fold, the lateral plaits proceed by a gentle curve downwards to the margin, giving the shell an elegant, somewhat winged aspect. The sinus is moderately deep, with from six to eight plaits. Length 17, width 15, depth 12 lines.

Obs. This elegant species was discovered by the Rev. A. W. Griesbach, in the Cornbrash of Rushden, in Northamptonshire, and has likewise been found in the same formation at Marquise, near Boulogne, by M. Bouchard. This form, although approaching to *R. obsoleta* in some respects, seems to me not to possess all the characters belonging to that species; it is more elegant and somewhat winged shape, smaller, has more delicate plaits, and its prominent mesial fold, as well as general aspect, appear to distinguish it from the types of *obsoleta*, as found in the Great Oolite and Bradford clay; if not specifically distinct, it would be at least a well-marked variety. After submitting the question to various competent persons whose opinions likewise differed, I have ventured to distinguish it by a separate appellation.

Plate XVIII, fig. 12. From a specimen from the Cornbrash of Rushden, in the collection of the British Museum.

„   fig. 13. From one in that of the Rev. A. W. Griesbach.

88. RHYNCHONELLA TETRAEDRA, *Sow.* Sp.   Plate XVIII, figs. 5—10.

TEREBRATULA TETRAÉDRA, *Sow.* M. C., vol. i, 1812, p. 191, tab. lxxxiii, fig. 5.
—   —   *Parkinson.* An Introduction to the Study of Organic Remains, p. 227, 1822.
—   MEDIA,   Ib., p. 234.
—   TETRAÉDRA, *Defrance.* Dic. d'Hist. Nat., vol. liii, p. 158, 1828.
—   —   *Young and Bird.* Geological Survey of the Yorkshire Coast, pl. viii, fig. 15, 1828.
—   —   *V. Buch.* 1834. Uber Ter.; and Mém. de la Soc. Géol. de France, 1838, vol. iii, p. 139, pl. xiv, fig. 8.
—   —   *Schlotheim.* 1832. Systematisches Vers der Petrefacten.
—   —   *Deslongchamps.* Soc. Linn. de Normandie, 1837.
—   —   *Morris.* Catalogue, 1843.
—   —   *Tennant.* A Strat. List of British Fossils, p. 74, 1847.
TEREBRATULA TETRAÉDRA, *Bronn.* Index Pal., p. 1253, 1848, (but not all his synonyms.)
RHYNCHONELLA TETRAÉDRA, *D'Orbigny.* Prodrome, vol. i, p. 258, 1849.

Diagnosis. Shell variable in shape, obtusely deltoid, valves convex and gibbous, wider than long; beak acute, more or less recurved, almost touching the umbo in some species; foramen small, generally entirely surrounded by the deltidium; beak ridges well defined, leaving between them and the hinge margin a small concave false area, which indents more or less the umbo of smaller valve, which is likewise laterally depressed, as if pinched in. The greatest depth is, in some specimens, towards the middle, in others near the front. Surface ornamented by a variable number of sharp plaits, from twenty-two to thirty on each valve; varieties with four, five, six, seven, eight, and nine plaits, forming a central more or less elevated mesial fold, corresponding to three, four, five, six, seven plaits, in a deep sinus in larger valve: on either side of the mesial fold are seen one or two plaits, which do not generally attain to the frontal margin, but disappear at a little distance from it, so that the lateral plaits of the mesial fold and the side plaits are often widely separated by a large flat space; sometimes the lateral costæ are observed to bifurcate.

Dimensions variable; the largest specimen I have seen, measured, length 15, width 21, depth 15 lines; but in general they do not exceed length 12, width 13, depth 13 lines.

Obs. R. tetraëdra is one of those perplexing species often difficult to determine, from the great variety it presents, owing to several causes, and especially to the variable number of plaits ornamenting its surface, and mesial fold, which sometimes is considerably produced above the sides of the valve, while in other cases very little difference of level is perceptible, the mesial fold, as well as sinus, being shallow. Mr. J. de C. Sowerby having kindly placed at my disposal for examination the original types of *T. tetraëdra* and *media*, with the assistance of Mr. Waterhouse, we came to the same conclusion, already arrived at by some authors (V. Buch,[1] Morris, Bronn, &c.), viz., that *T. media* of Sowerby is only a variety of *tetraëdra*. In true *media*, as seen by the profile view, Pl. XVIII, fig. 9, the larger valve is deeper in the middle, and the beak more recurved, touching the umbo; the shell being, in general, more globose than what we observe in the specimens of true *R. tetraëdra*, Sowerby, Plate XVIII, fig. 8; but, on the examination of a suite of specimens, we soon perceive that there is every degree of passage between the extreme states of the same species. I cannot, however, admit, with Professor Bronn,[2] that *T. obsoleta*, of Sowerby, is a synonym of *tetraëdra;* both forms seem to me sufficiently distinct to be retained as separate species. *R. obsoleta* is a much less convex, and rather compressed shell, the longitudinal curve of the valves being more uniformly regular; the beak not so recurved, always exhibiting under it the foramen, entirely surrounded by the deltidium, which is separated more or less from the umbo. The same author's other synonyms of the present species would likewise require further examination before being admitted, as there exists some doubts as to the true types of some of Zieten's forms, as well as Mr. Quenstedt's rectifications;[3] indeed, I am sorry to find, that some German authors have added much to the confusion by attributing to the Sowerby types species widely separated both in form and stratigraphical position. Lamarck, as justly remarked by Baron V. Buch, confounded the species under consideration with Schlotheim's *R. decorata*.[4] Kœnig committed the same error.[5] *R. decorata* has not, to my knowledge, been found in the Lias or Cretaceous formations, as stated by the last-named author. W. Smith's figure of *T. media*[6] seems more likely to be *R. varians* of Schlotheim.

[1] See for reference at the head of this description.
[2] Index Pal., p. 1258.
[3] Das Flözgeberges Wurtembergs, 1843.
[4] See Lamarck, 'An. Sans. Vert.,' vol. vi, 1819, and Von Buch, 'Mém. Soc. Geol. de France,' vol. iii, p. 146.
[5] Icones Fossilium Sectile, p. 3, pl. vi, fig. 72.
[6] Strata identified by organised Fossils; London, June, 1816—27, pl. xix, figs. 1—3.

In England, *R. tetraëdra* is found in the Upper and Middle Lias or Marlstone of many localities; it abounds in the neighbourhood of Ilminster, where it attains very large dimensions, as may be remarked from the figures, Pl. XVIII, figs. 6 and 7. It is likewise plentiful, but smaller, in the Lias of Radstock, Deddington in Yorkshire, &c. Mr. Sowerby mentions Aynhoe and Banbury in Oxfordshire, Von Buch states the species to occur in the Inferior Oolite of Dundry,[1] whence, however, I have not seen any authenticated specimens. Mr. Morris[2] mentions it from the Inferior Oolite of Somerset, from the Kelloway Rock of Kelloway, and from the Fuller's Earth, Banbury; but in the multitude of specimens sent me from all those formations of the Oolitic period, I have never been so fortunate as to recognise any authentic specimen of the species, except one remarkable example, Pl. XVIII, fig. 10, sent me by Mr. Walton, and said to be from the Inferior Oolite of Leckhampton Hill, Cheltenham, which approaches likewise to some varieties of *R. decorata*. On the Continent the species is found in many localities; at Landes, Evreci, Fontaine-Etoupefour, &c., in Normandy, also in Germany.[3]

Plate XVIII, fig. 5. A type specimen, in the collection of Mr. J. de C. Sowerby.

    ,,   figs. 6, 7. From specimens in the collection of Mr. Moore.

    ,,   fig. 10. From the Inferior Oolite of Cheltenham, in the collection of Mr. Walton.

89. RHYNCHONELLA SUBTETRAËDRA, *Dav.* Plate XVI, figs. 9—12.

Diagnosis. Shell more or less transversely circular, generally wider than long; valves nearly equally convex, rarely gibbous, somewhat compressed; beak acute, moderately produced, and slightly recurved; foramen of moderate size, entirely surrounded by the deltidium; a narrow flat or slightly concave false area existing between the beak ridges and hinge margin which does not indent much the smaller valve. Surface ornamented by a variable number of acute plaits, from twenty to thirty on each valve; mesial fold more or less produced, composed of a variable number of plaits in general from five to nine; sinus shallow, dimensions variable; length of largest specimen 21, width 24, depth 14 lines.

Obs. On comparing the shells under notice with a number of known species, I have not been able to identify them with any of the described types; they vary from *concinna* and *obsoleta*, being much more transverse and strongly plaited; from *tetraëdra* they likewise differ sufficiently not to represent that type. With *lacunosa* they have some resemblance, but on comparing our form with that shell, I could not make up my mind to place them

---

[1] Mém. Soc. Géol. de France, vol. iii, p. 140.

[2] Catalogue of British Fossils, 1843.

[3] Messrs. Bayle and Coquand, figure and describe a Rhynchonella from the Lias of Manflas, Chili, as belonging to *T. tetraëdra*, Mém. Soc. Géol. de France, vol. iv, 2de ser., p. 17, 1851; but I cannot recognise the true type of the Sowerby species in their figure.

there; at one time, I thought they might belong to *T. helvetica*, Schloth., but that species is so badly characterised, as well as figured, that it would be unsafe to refer our shell to that form; neither Schlotheim's nor Zieten's figure apparently possessing any definite mesial fold. I have, therefore, under these circumstances, given it a distinct appellation.

R. subtetraëdra is found in the Inferior Oolite of Dundry and Leckhampton Hill, near Cheltenham, where it has been collected by Mr. Walton and Dr. Wright.

Plate XVI, fig. 12. A variation with hardly any distinct mesial fold and few plaits, from the Inferior Oolite of Somersetshire, in the collection of the Geological Survey.

„   fig. 9. A large specimen from the Inferior Oolite, Dundry.

„   fig. 10. A specimen from Dundry, in the collection of Mr. Walton.

„   fig. 11. A variety from the Upper Ferruginous Bed of Leckhampton Hill, in the collection of Dr. Wright.

90. RHYNCHONELLA LACUNOSA, *Schloth.*, Sp.   Plate XVI, figs. 13, 14.

TEREBRATULITES LACUNOSA, *Schlotheim*. 1813. Beiträge für Nat. Vers., in Leonhard's Min. Tasch., vol. vii, pl. i, fig. 2.

TEREBRATULA LACUNOSA, *V. Buch.* 1834. Uber Ter.; and 1838, Mém. Soc. Géol. de France, vol. iii, first ser., p. 150, pl. xv, fig. 22.

—   —   *Bronn.* Index Pal., p. 1239, 1848.

RHYNCHONELLA LACUNOSA, *D'Orb.* Prodrome, vol. i, p. 375, 1849.

Diagnosis. Shell irregularly, transversely oval, wider than long; beak acute, recurved; foramen entirely surrounded by the deltidium, but lying close upon the umbo; the false area between the beak ridges and hinge margin is well defined, a similar depression existing likewise on either side of the umbo; valves unequally convex, and ornamented by a variable number of plaits, from eighteen to nineteen on each valve, four, five, or six forming a produced elevated mesial fold, to which corresponds a deep sinus in the other valve; the convexity of the valve seems regular from the umbo to the front, the deepest portion of the shell being towards its centre; length 12, width 13, depth 9 lines.

Obs. The specimens described above were obtained by Mr. Robertson, of Elgin, at Dunrobin, in Scotland, in beds referred to the Oxford Clay; and on comparing the well-preserved specimens forwarded by that gentleman with Schlotheim's figures of *T. lacunosa,* (1813,) I am disposed to believe them referable to the same species, which is peculiar to the Upper Jura on the Continent, and placed by M. D'Orbigny both in his *Terrain Oxfordien* and *Callovien.* It is, however, no easy matter to come to a certain conclusion on several species of Rhynchonella, for two principal reasons, the first, arising from the extreme variability of the shells of some species of this genus, and especially on account of the insufficient manner in which most authors have determined, figured, or described their species. In the case before us, it would be a waste of time to attempt a discussion of all the synonyms attributed to this form by Professor Bronn, and other authors, as so much uncertainty exists as to the types intended. Both Baron Von Buch and Professor Bronn seem disposed to admit *Anomia triloba lacunosa*, of Fabio Colonna,[1] to be the type of the species in dispute; I am not, however, able to decide the question: but I cannot admit, as stated by the celebrated author of the ' Uber Ter.,' that *R. lacunosa* is common to the Upper Jura and to the Magnesian or Permian Limestone of Humbleton, in Yorkshire; the last does not even belong to the same genus, and is the *Camerophoria multiplicata*, so ably described and figured by Professor King in his valuable ' Monograph of British Permian Fossils.'

The shell under consideration has been attributed to *R. tetraëdra*, var. *media*,[2] a species I do not know with certainty to occur higher up than the Lias; the convexity of the valves and form of the beaks in this last easily distinguish it from the shell under notice.

Figs. 13, 14. From specimens in the collection of Mr. Robertson.

90. RHYNCHONELLA HOPKINSI, *M'Coy*, MS. 1852.

Diagnosis. Shell inequivalve, subcuboidal, valves gibbous, generally longer than wide, sides tumid, nearly vertical, beak acute, incurved, under the extremity of which is seen a small foramen entirely surrounded by the deltidium, beak ridges well defined, leaving a large slightly concave false area between them and the hinge line, which indents the imperforated valve, giving to this portion of the shell a pinched aspect, smaller valve very gibbous, ornamented by a variable number of small acute plaits, from twenty-four to thirty in number, three, four, five, or six, forming a well-defined elevated mesial fold. In the dental valve a very shallow sinus is visible, becoming convex as it approaches and meets the mesial fold; a number of small lines of growth cover the surface; structure imperforated, dimensions variable. Length 10½, width 9½, depth 9 lines.

Obs. This elegant shell was noticed for the first time by Mr. Bouchard, who found it in the Great Oolite of Marquise, near Boulogne-sur-Mer, where it is common; but it was only on a late visit to the Cambridge Museum that I became acquainted with its existence as a British species from the Great Oolite of Minchinhampton, and named in that collection by Professor M'Coy. This species is easily distinguished from *R. tetraëdra*, to which it bears some resemblance, by its square shape and more circular beak.[3]

[1] De Purpura Romæ, 1616.

[2] Sir R. Murchison, ' Trans. Geol. Soc.,' vol. ii.

[3] I regret that from all my plates being printed before I became acquainted with the existence of this species in England, it could not be illustrated; in the Appendix, at the conclusion of this work, it is my intention to figure those new species which may be discovered during the interval.

98

TABLE OF BRITISH LIASIC AND OOLITIC BRACHIOPODA.

PART III.

Order in which described.	Genus.	Species.	Author.	Date.	Reference to this Monograph. — (*Part* III.)	Lower Lias.	Middle Lias.	Upper Lias.	Inferior Oolite.	Fullers' Earth.	Stonesfield Slate.	Great Oolite.	Brad. Clay & For. Marble.	Cornbrash.	Oxford Clay & Kel. Rock.	Cor. Rag and Cal. Grit.	Kimmeridge Clay.	Portland Stone and Sand.
1	Lingula	Beanii	Phillips	1829	p. 8, pl. i, f. 1		*	*	*									
1 bis	„	ovalis[1]	Sow.	1812	pl. xviii, f. 14											*	*	
2	Orbicula[2]	Townshendi	Forbes	1851	p. 9, pl. i, f. 2		*											
3	„	reflexa	Sow.	1829	p. 10, pl. x, f. 8		*										*	
4	„	Humphresiana[3]	Sow.	1829	p. 10, pl. i, f. 3		*										*	
4 bis	„	latissima[4]	Sow.	....														
5	Crania	antiquior	Jelly	1843	p. 11, pl. i, f. 4—8							*						
6	„	Moorei	Dav.	1851	p. 12, pl. i, f. 9			*										
7	Thecidea	Moorei	Dav.	1851	p. 13, pl. i, f. 10		*											
8	„	Bouchardii[5]	Dav.	1851	p. 14, pl. i, f. 15, 16				*									
9	„	Dickinsonii	Moore	1851	p. 14, pl. xiii, f. 19				*									
10	„	triangularis	D'Orb.	1849	p. 14, pl. i, f. 11, 12		*		*									
11	„	rustica	Moore	1851	p. 15, pl. i, f. 4			*										
12	Leptæna	Moorei	Dav.	1847	p. 17, pl. i, f. 18			*										
13	„	Pearcei	Dav.	1847	p. 17, pl. i, f. 19			*										
14	„	granulosa	Dav.	1850	p. 18, pl. i, f. 20			*										
15	„	liasiana	Bouch.	1847	p. 18, pl. i, f. 21			*										
16	„	Bouchardii	Dav.	1847	p. 19, pl. i, f. 22			*										

[1] Since describing the *Lingulidæ* in Part III, p. 8, several specimens have been forwarded to me from the Oxford and Kimmeridge Clays, which have satisfied me, that *L. ovalis*, Sowerby, (M. C., vol. i, p. 56, pl. 19, fig. 4, 1812,) is in reality an Oolitic species, having been found by Mr. C. B. Rose, Mr. Middleton, and others, under the shape of impressions on grey or brown clay, in Norfolk and Suffolk; the species likewise occurs, with its shell preserved, in the Kim. Clay of Ely, where it has been collected by Mr. Carter. I have therefore figured this form in pl. xviii, fig. 4, and it may be described as follows. *Ling. ovalis*, Sow.—Shell oblong, oval, rather square at the beaks, anterior edge circular, slightly convex, rather depressed: surface ornamented by a number of concentric raised lines of growth, length 7, width 3½ lines. *L. ovalis* seems much more regularly oval than *L. Beanii*. Another small lanceolate-shaped Lingula has been discovered by Mr. Walton, in the Oxford Clay of Christian Malford, measuring, length 3½, width not quite 2 lines, which may, perhaps, be another species.

[2] In page 7, of this Monograph, I quoted Prof. Owen in relation of the animal to the shell in *Orbicula* and *Crania*, not having remarked that the Professor had subsequently reversed his opinions, in the French 'An. Sc. Nat.,' May, 1845. I am indebted to Dr. J. E. Gray, of the British Museum, for the opportunity of examining, along with Mr. Woodward, the animal of both *Orbicula* and *Crania*, and am now satisfied that the lower or attached valve to which the animal chiefly adheres, in both these genera, corresponds to the perforated valve of *Terebratula*, so that whilst *Crania* and *Orbicula* form no exception to the invariable rule that the shell of the Brachiopod is fixed by means of the ventral valve, they differ remarkably from the other genera, in having the oral arms fixed to the ventral or attached valve.

[3] Many beautifully preserved shells of this species may be seen in the Cambridge Museum.

[4] In blocks of Kimmeridge Clay at Packefield, in Suffolk, Sowerby found associated with *L. ovalis* compressed impressions, which he describes and figures, 'M. C.,' tab. 139, figs. 1 and 5, under the name of *Patella latissima;* but these impressions have much more the aspect of an *Orbicula* than a *Patella*. Sowerby describes them as follows: "Shell nearly orbicular, flat, smooth; shell very thin, concentrically undulated; the umbo is excentric, the margins forming a very short oval." Unfortunately, I have not been able to procure any example preserving the shell, so that its structure could not be examined. Mr. Woodward showed me impressions of the same species, almost circular, on slabs of Kim. Clay from Braunston, Northampton, found by Miss Baker, and now in the British Museum. A large impression, forwarded to me by Mr. C. B. Rose, measured 21 lines in length and 16½ in breadth; but, in general, the dimensions are much smaller, not exceeding 10 lines in length.

[5] Since describing this species, in page 14, Mr. Moore has discovered a very large adult specimen of *Thecidea Bouchardii*, which, I find, measures, length 3 lines, width 4, and is the largest liasic *Thecidea* as yet obtained. The same species, with almost similar dimensions, has been discovered by Mr. Tesson, in the lias of Fontaine-Etoupefour, near Caen. (See Dav., 'An. and Mag. of Nat. Hist.,' April, 1852.)

Order in which described.	Genus.	Species.	Author.	Date.	Reference to this Monograph. (Part III.)	Lower Lias.	Middle Lias.	Upper Lias.	Inferior Oolite.	Fullers' Earth.	Stonesfield Slate.	Great Oolite.	Brad. Clay & For. Marble.	Cornbrash.	Oxford Clay & Kel. Rock.	Cor. Rag and Cal. Grit.	Kimmeridge Clay.	Portland Stone and Sand.
17	Spirifer	rostratus[1]	Schl.	1813	p. 20, pl. ii, f. 1—12, pl. iii, f. 1	*	*	*										
18	„	Ilminsteriensis	Dav.	1851	p. 24, pl. iii, f. 7			*										
19	„	Walcotti[2]	Sow.	1823	p. 25, pl. iii, f. 2, 3	*												
20	„	Munsterii	Dav.	1851	p. 26, pl. iii, f. 4—6		*											
21	Terebratula	quadrifida	Lamarck	1819	p. 28, pl. iii, f. 8—10		*											
22	„	cornuta	Sow.	1825	p. 29, pl. iii, f. 11—18		*											
23	„	Edwardsii	Dav.	1851	p. 30, pl. vi, f. 11—13, 15?		*											
24	„	Waterhousii	Dav.	1851	p. 31, pl. v, f. 12, 13		*											
25	„	resupinata[3]	Sow.	1818	p. 31, pl. iv, f. 1—5		*											
26	„	Moorei	Dav.	1851	p. 33, pl. iv, f. 6, 7		*											
27	„	impressa	V. Buch	1832	p. 33, pl. iv, f. 8—10, pl. x, f. 7				*?						*			
28	„	carinata	Lamarck	1829	p. 35, pl. iv, f. 11—17				*									
29	„	emarginata	Sow.	1825	p. 35, pl. iv, f. 18—21				*									
30	„	Waltoni	Dav.	1851	p. 36, pl. v, f. 1, 6				*									
31	„	numismalis	Lamarck	1819	p. 36, pl. v, f. 4—9			*										
32	„	Bakeriæ	Dav.	1851	p. 38, pl. v, f. 11				*									
33	„	digona[4]	Sow.	1812	p. 38, pl. v, f. 18, 24								*	*				
34	„	obovata	Sow.	1812	p. 39, pl. v, f. 14—17									*				
35	„	ornithocephala[4]	Sow.	1812	p. 40, pl. vii, f. 6—13, 23							*	*	*	*			
36	„	lagenalis	Schloth.	1820	p. 42, pl. vii, f. 1—4								*	*	*			
37	„	sublagenalis	Dav.	1851	p. 42, pl. vii, f. 14								*	*	*			
38	„	cardium	Lamarck	1819	p. 43, pl. xii, f. 13—15									*				
39	„	Buckmanii	Dav.	1851	p. 44, pl. vii, f. 15, 16				*									
40	„	Lycettii	Dav.	1851	p. 44, pl. vii, f. 17—22							*						
41	„	punctata	Sow.	1812	p. 45, pl. vi, f. 1—6		*											
42	„	subpunctata	Dav.	1851	p. 46, pl. vi, f. 10—12, 16?		*											
43	„	indentata	Sow.	1825	p. 46, pl. v, f. 25, 26		*											
44	„	insignis	Schüb.	1832	p. 47, pl. xiii, f. 1											*	*	
45	„	simplex	Buck.	1845	p. 48, pl. viii, f. 1, 3				*									
46	„	ovoides	Sow.	1812	p. 48, pl. viii, f. 4—9				*									
47	„	maxillata[4]	Sow.	1825	p. 50, pl. ix, f. 1—9								*	*				
	var.	submaxillata	Dav.	1851	p. 51, pl. ix, f. 10—12				*									
48	„	perovalis	Sow.	1825	p. 51, pl. x, f. 1, 6				*									
49	„	intermedia	Sow.	1812	p. 52, pl. xi, f. 1—5										*			

[1] TORRUBIA, in his 'Hist. Nat. Hispanica,' tab. vii, fig. 6, 1773, gives two illustrations of this species, but without name. KNORR, in 1755, 'Lapides Diluv.,' Tab. B. iv, fig. 3, likewise figures this shell, but also without name. In 1813, SCHLOTHEIM named it *Terebratulites rostratus*, and refers to Knorr's figure. See 'Beit. Zur. Nat. der Vers.,' in Leonhard's 'Min. Tasch.,' vol. vii. It is also figured by SCHMIDT, 'Petref. Buch.,' pl. xxiii, fig. 6, 1846, under the name of *Sp. verrucosa*.

[2] WALCOTT, in 1799, Petrefactions found near Bath, gives an excellent figure of this species, but without name. FISCHER DE WALDHEIM figured this shell in 1809, in his 'Notice des Fossiles du Gour. de Moscou,' pl. i, figs. 10, 11, under the name of *Terebratula octo-plicata*, and later, in 1830—37, 'Ortyctographie du Gour. de Moscou,' under that of *Choristite Walcotti*, p. 141, pl. xxii, fig. 4. The figures in both works are very bad. SCHMIDT has also figured it in 1846, 'Petref. Buch.,' pl. xxiii, fig. 1.

[3] *T. resupinata* was figured in 1773, by Torrubia, 'Hist. Nat. Hisp.,' tab. ix, fig. 3. In the same work the author figures *T. buttala*, tab. ix, fig. 4, and *R. tetraëdra*, pl. i, figs. 1, 2, but without name. *T. resupinata* is likewise figured by SCHMIDT, 'Pet. Buch.,' 1846, pl. xli, fig. 3. Walcott, in 1799, gave good illustrations of *T. ornithocephala*, *T. digona*, *T. cardium*, *Rh. spinosa*, and *R. obsoleta*, but without names. In the work of SCHEUCHZER, 'Helvetiæ Hist. Nat. des Surw.,' 1752, several Oolitic species are figured, but so incorrectly, as to be of no possible use, and I only mention it here on account of its having been often referred to by SCHLOTHEIM. In MORTON's 'Nat. Hist. of Northamptonshire,' 1712, we find figures of *T. intermedia*, *ornithocephala*, *resupinata*, and *obovata*: some other illustrations are given, but they are not well defined.

[4] *Ter. digona*, *T. ornithocephala*, and *F. maxillata*, were figured by LHWYD, 1699, but not named, 'Lithophylaci Britannici Ichnographia,' pl. x, figs. 830, 873, and pl. xi, fig. 890; the last is *T. maxillata*, and that author gives 19 figures of all ages. Lister, in 1688, 'Historia sive Synopsis Method. Conchyliarum,' tab. 456, fig. 16, gives an illustration of *T. digona*; few, however of the figures of that author can be recognised.

Order in which described.	Genus.	Species.	Author.	Date.	Reference to this Monograph. — (Part III.)	Lower Lias.	Middle Lias.	Upper Lias.	Inferior Oolite.	Fullers' Earth.	Stonesfield Slate.	Great Oolite.	Brad. Clay & For. Marble.	Cornbrash.	Oxford Clay & Kel. Rock.	Cor. Rag and Cal. Grit.	Kimmeridge Clay.	Portland Stone and Sand.
50	Terebratula	Phillipsii	Morris	1847	p. 53, pl. xi, f. 6—8				*									
51	,,	globata	Sow.	1825	p. 54, pl. xiii, f. 2—7				*						*	*		
52	,,	bucculenta	Sow.	1825	p. 55, pl. xiii, f. 8				*							*		
53	,,	sphæroidalis	Sow.	1825	p. 56, pl. xi, f. 9, 19				*									
54	,,	globulina	Dav.	1851	p. 57, pl. xi, f. 20, 21			*										
55	,,	pygmea	Morris	1847	p. 57, pl. xiii, f. 16			*										
56	,,	Bentleyi	Morris	1851	p. 58, pl. xiii, f. 9—11				*					*				
56 bis	,,	sub-Bentleyi[1]	Dav.	1851	p. 59, pl. xiii, f. 11				*									
57	,,	coarctata[2]	Park.	1811	p. 59, pl. xii, f. 12, 15				*				*	*		*		
58	,,	plicata	Buch	1845	p. 60, pl. xii, f. 1—5				*									
59	,,	fimbria	Sow.	1823	p. 61, pl. xii, f. 6—12				*				*					
60	,,	flabellum	Def.	1828	p. 62, pl. xii, f. 19, 20							*						
61	Terebratella	hemispherica	Sow.	1829	p. 64, pl. xiii, f. 17, 18							*						
62	Rhynchonella	Wrightii	Dav.	1852	p. 69, pl. xiv, f. 1				*									
63	,,	furcillata	Theodori	1834	p. 69, pl. xiv, f. 2—5		*											
64	,,	rimosa	V. Buch	1831	p. 70, pl. xiv, f. 6		*											
65	,,	spinosa	Schl.	1813	p. 71, pl. xv, f. 15, 20				*									
66	,,	senticosa	Buch	1834	p. 73, pl. xv, f. 21			*	*									
67	,,	ringens	Herault	1834	p. 74, pl. xiv, f. 13—16			*	*									
68	,,	subringens	Dav.	1852	p. 75, pl. xiv, f. 17				*									
69	,,	acuta	Sow.	1818	p. 76, pl. xiv, f. 8, 9		*											
70	,,	cynocephala	Rich	1840	p. 77, pl. xiv, f. 10—12				*									
71	,,	variabilis	Schl.	1813	p. 78, pl. xv, f. 8—10; pl. xvi, 1—6		*	*										
72	,,	subvariabilis	Dav.	1852 1852	p. 80, pl. xv, f. 7; pl. xviii, f. 11				*									*
73	,,	Lycettii	Dav.	1852	p. 81, pl. xv, f. 6				*									
74	,,	oolitica	Dav.	1852	p. 81, pl. xiv, f. 7				*									
75	,,	Moorei	Dav.	1852	p. 82, pl. xv, f. 11—14					*								
76	,,	Bouchardii	Dav.	1852	p. 82, pl. xv, f. 3—5					*								
77	,,	varians	Schl.	1820	p. 83, pl. xvii, f. 15, 16							*		*				
78	,,	Forbesii	Dav.	1852	p. 84, pl. xvii, f. 19							*						
79	,,	serrata	Sow.	1825	p. 85, pl. xv, f. 1, 2				*									
80	,,	plicatella	Sow.	1825	p. 86, pl. xvi, f. 7, 8							*						*
81	,,	inconstans	Sow.	1821	p. 87, pl. xviii, f. 1—4							*	*	*	*	*		*
82	,,	concinna	Sow.	1812	p. 88, pl. xvii, f. 6—12								*	*	*	*		
83	,,	subconcinna	Dav.	1852	p. 90, pl. xvii, f. 17		*											
84	,,	obsoleta	Sow.	1812	p. 90, pl. xvii, f. 1—5								*	*	*	*?		
85	,,	subobsoleta	Dav.	1852	p. 91, pl. xvii, f. 14							*						
86	,,	angulata	Sow.	1825	p. 92, pl. xvii, f. 13									*				
87	,,	Morièrei	Dav.	1852	p. 92, pl. xviii, f. 12, 13							*						
88	,,	tetraëdra	Sow.	1812	p. 93, pl. xviii, f. 5—10		*	*	*									
89	,,	subtetraëdra	Dav.	1852	p. 95, pl. xvi, f. 9—12				*									
90	,,	lacunosa?	Schl.	1813	p. 96, pl. xvi, f. 13, 14											*		
	,,	uncertain species	....	....	pl. xviii, f. 18												*	
91	,,	Hopkinsii	M'Coy		p. 97										*			

[1] At the period I described this form, I was only acquainted with the larger or perforated valve; but, during a visit to the Woodwardian Museum at Cambridge, I found there a beautifully perfect specimen, measuring, length 23, width 20, depth 16 lines: the smaller valve is, as I had supposed, very similar to that of *Ter. Bentleyi*. The general aspect of *T. sub-Bentleyi*, added to its different stratigraphical position, almost makes me believe it specifically different from *T. Bentleyi*.

[2] Mr. Walton states positively to have found this species in the Oxford Clay of Wootton Basset. A figure of this form was given by Lister in 1688, tab. 459, fig. 20.

PLATE I.

Fig.

1.   Lingula Beanii, *Phil.*, natural size.

1*a, b.*   „   „   interior of larger valve.

1*c, d.*   „   „   „   smaller valve.

2*a, b.*   Orbicula Townshendi, *Forbes*, MS., natural size.

3.   Orbicula Humphresiana, *Sow.*, natural size.

3*a, b.*   „   „   enlarged illustrations.

4. 5. 6. 7.   Crania antiquior, *Jelly*, natural size.

8.   „   „   an enlarged illustration.

9.   Crania Moorei, *Dav.*, natural size.

9*a, b.*   „   „   enlarged illustrations of the upper valve.

10.   Thecidea Moorei, *Dav.*, several specimens attached to *Rhynchonella serrata*.

10*a, b, c, d, e.*   „   „   enlarged illustrations.

11.   Thecidea triangularis, *D'Orb.*, natural size, from the Lias.

11*a, b.*   „   „   enlarged figures.

12.   „   „   natural size, from the Inferior Oolite.

13.   Thecidea Mediterraneum, *Risso*, recent species, natural size.

14.   Thecidea rustica, *Moore*, MS., natural size.

14*a, b.*   „   „   enlarged representations of the exterior and interior of the smaller valve.

15.   Thecidea Bouchardii, *Dav.*, natural size, from specimens found in the Lias near Ilminster.

15*a.*   „   „   enlarged interior of attached valve.

16.   „   „   } enlarged illustrations of *Th. Bouchardii*.

17.   „   „

18.   Leptæna Moorei, *Dav.*, natural size, Upper Lias, Ilminster.

18*a, b, c, d, e.*   „   „   enlarged illustrations of exterior and interior of both valves.

19.   Leptæna Pearcei, *Dav.*, natural size, from a specimen found in the Upper Lias.

19*a, b.*   „   „   enlarged illustrations.

20.   Leptæna granulosa, *Dav.*, natural size, from the Upper Lias.

20*a, b.*   „   „   enlarged illustrations.

21.   Leptæna liasiana, *Bouchard*, natural size.

21*a.*   „   „   enlarged figure.

22.   Leptæna Bouchardii, *Dav.*, natural size, from the Upper Lias, near Ilminster.

22*a, b.*   „   „   enlarged illustration of exterior of both valves.

22*c.*   „   „   „   „   of interior of smaller valve.

Pl I

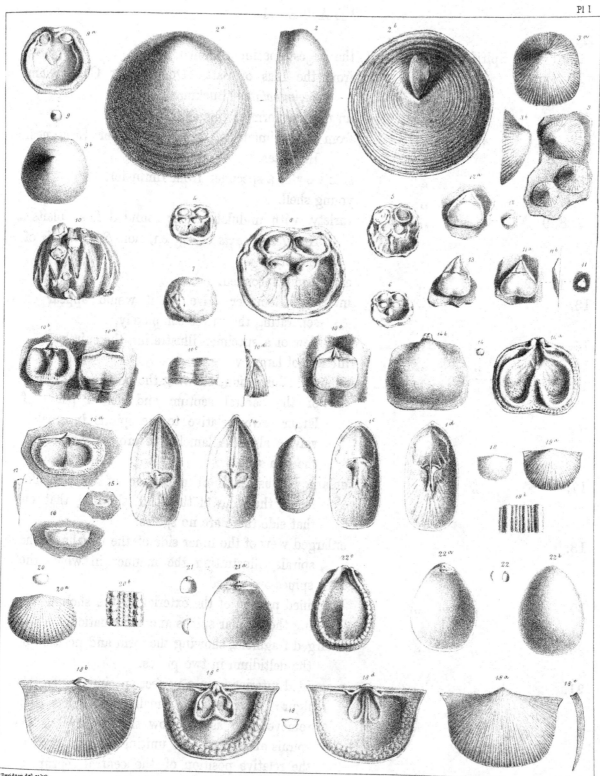

T.Davidson del et lith.

Printed by Hullmandel & Walton.

PLATE II.

Fig.

1.   Spirifer rostratus, *Schl.*, the largest specimen known.

2.   ,,   ,,   from the Lias of Battle Down, near Cheltenham.
                                                 Sp. punctatus (Buckman).

3.   ,,   ,,   an unusual form of this species.

4.   ,,   ,,   from a specimen found in the Upper Lias, near Ilminster.

5.   ,,   ,,   side view of a specimen from Ilminster.

6.   ,,   ,,   young shell.

7, 8, 9.   ,,   ,,   variety, with undulations, or rounded false plaits:
                                                 Spirifer pinguis of Zieten, non *Sp. pinguis* of Sowerby.

10, 11, 12.   ,,   ,,   an unusual specimen.

13.   ,,   .,   interior of smaller valve, as it would appear on separating the valves completely.

14.   ,,   ,,   side view of a specimen illustrating the spirals.

15.   ,,   ,,   interior of larger valve.

16.   ,,   ,,   enlarged illustration, showing the position and form of the central septum and dental plates of larger valve, relative to the spirals in smaller valve; also the lamella, or cross-piece uniting the two spirals.

17.   ,,   ,,   enlarged illustration of a portion of the lamella, facing the sides of the shell, showing that on that side there are no spines.

18.   ,,   ,,   enlarged view of the inner side of the lamella of the spirals, illustrating the manner in which the spines are placed.

19.   ,,   ,,   magnified portion of the exterior of the shell, showing the tubular spines and punctuation.

20.   ,,   ,,   enlarged fragment, showing the form and position of the deltidium in two pieces.

21.   ,,   ,,   magnified interior from a perfect specimen. In this figure a part of the smaller valve has been removed, so as to show the position of the spirals and cross-piece uniting them, as well as the relative position of the central septum of the larger valve and the position of the spines on the portion of the spirals facing the front of the shell.

Pl II.

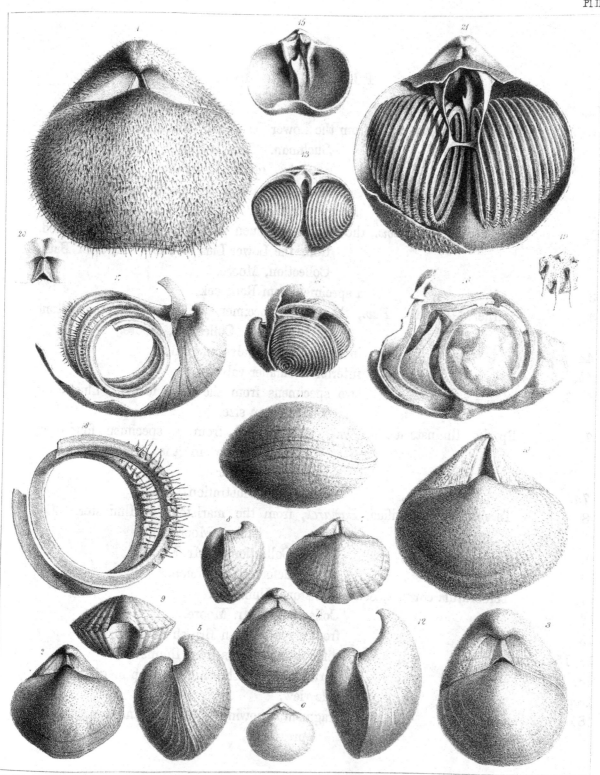

T. Davidson del et lith.

Printed by Hullmandel & Walton.

Spirifer rostratus.(Schl: s.p.)

PLATE III.

Pl. III.

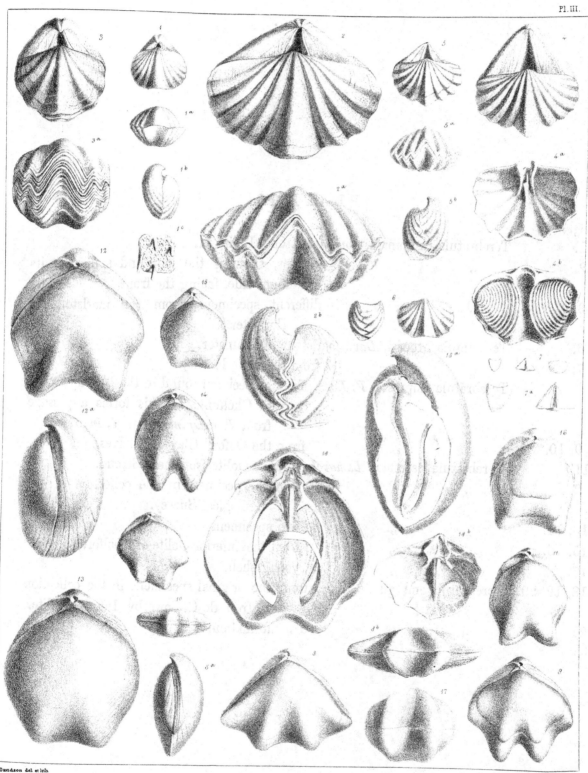

T. Davidson del et lith.

Printed by Hullmandel & Walton.

PLATE IV.

Pl. IV.

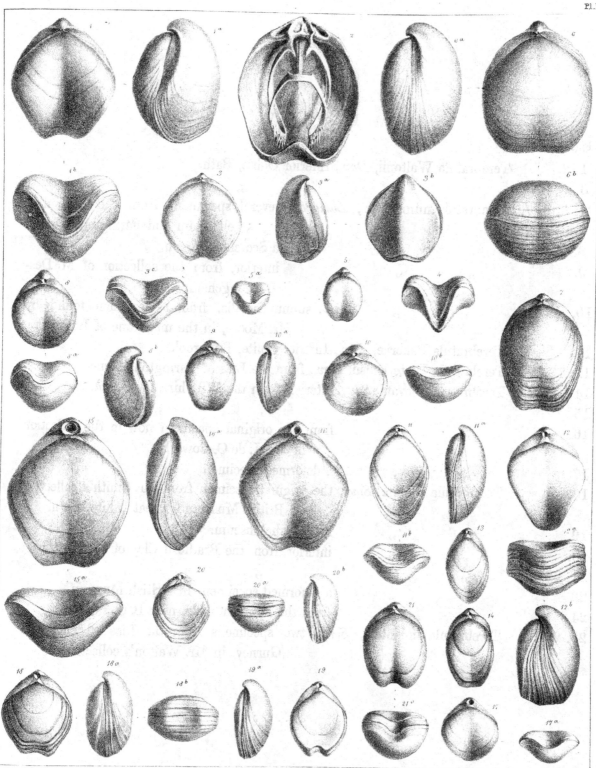

T. Davidson del et lith.

Printed by Hullmandel & Walton.

PLATE V.

Fig.

1, 2.   Terebratula Waltonii, *Dav.*, Inferior Oolite, Bath.

3.   ,,   ,,   ,,   ,,

4, 5, 6, 7. Terebratula numismalis, *Lamarck*, several specimens from the Lias near Cheltenham and Farington Gurney.

8.   ,,   ,,   a Scotch specimen.

9.   ,,   ,,   interior, from the collection of M. Deslongchamps.

10.   ,,   ,,   var. subnumismalis, from a specimen found by Mr. Moore, in the marlstone of Ilminster.

11.   Terebratula Bakeriæ, *Dav.*, Inferior Oolite, Bugbrook.

12, 13.   Terebratula Waterhousii, *Dav.*, from the Lias of Farington Gurney.

14.   Terebratula obovata, *Sow.*, interior, from the Wiltshire *Cornbrash*.

15.   ,,   ,,   ,,   ,,

16.   ,,   ,,   from the original specimen now in the collection of Mr. J. de C. Sowerby.

17.   ,,   ,,   a deformed specimen.

18.   Terebratula digona, *Sow.*, the original specimen from Mr. Smith's collection in British Museum: Great Oolite, Bath.

19.   ,,   ,,   in British Museum.

20.   ,,   ,,   interior, from the Bradford Clay of Cirencester.

21, 22.   ,,   ,,   ,,   ,,

23.   ,,   ,;   a deformed specimen, in British Museum.

24.   ,,   ,,   from the Forest Marble, near Bath.

25, 26.   Terebratula indentata, *Sow.*, two specimens from the Lias of Farington Gurney, in Mr. Walton's collection.

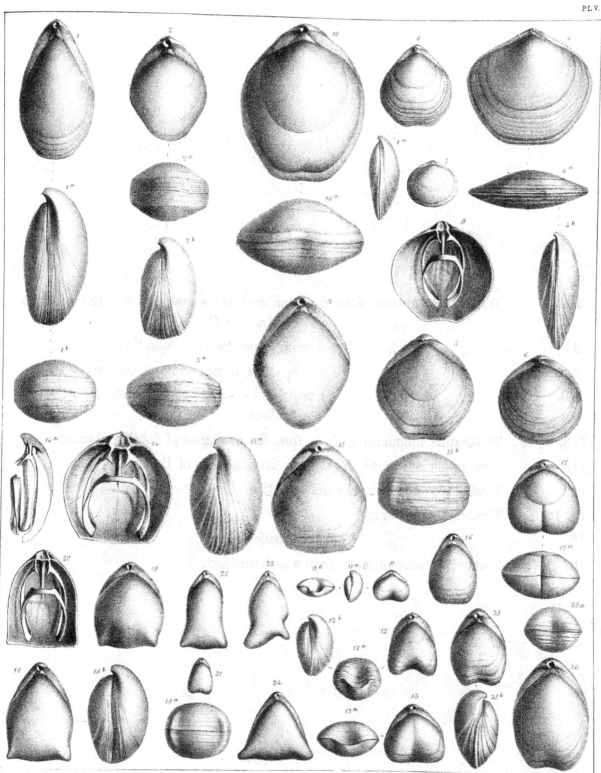

T. Davidson del et lith.

Printed by Hullmandel & Walton.

PLATE VI.

Pl VI.

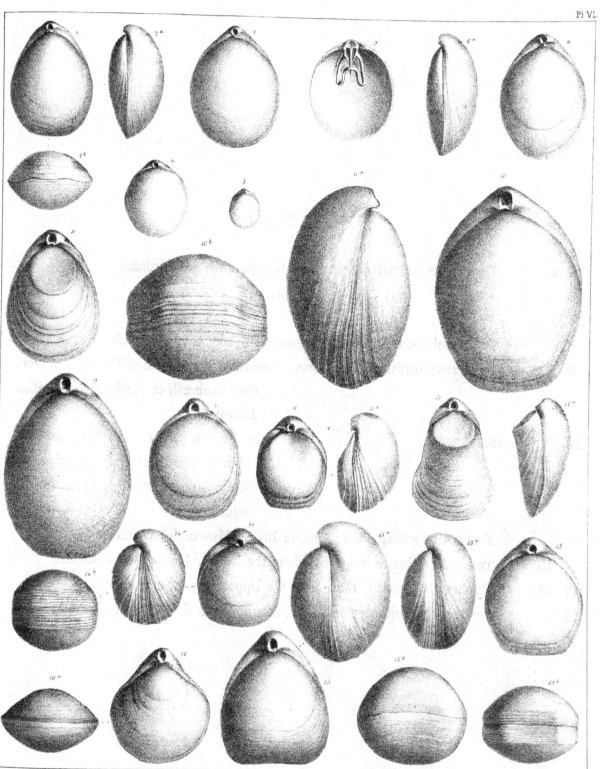

T. Davidson del et lith

Printed by Hullmandel & Walton

PLATE VII.

Fig.

1, 2.   Terebratula lagenalis, *Schl.*, from the Cornbrash of Rushdon.

3.   „   „   interior.

4.   „   „

5.   Terebratula obovata, *Sow.*, malformation, from the Cornbrash.

6, 7, 8, 9. Terebratula ornithocephala, *Sow.*, from the Fullers Earth of Bath. 7. 8, from the Cornbrash of Rushden; in British Museum.

10, 11, 12, 13.   „   „   from Fullers Earth. 10, 11, are the original specimens of *T. triquetra*, Sow., from Mr. J. de C. Sowerby's collection.

14.   Terebratula sublagenalis, *Dav.*, in British Museum's collection.

15, 16.   Terebratula Buckmanii, *Dav.*, from the Inferior Oolite of Cheltenham.

17—22.   Terebratula Lycettii, *Dav.*, from the Upper Lias, near Ilminster.

23.   Terebratula ornithocephala, *Sow.*, interior, from the Kelloway Rock.

Pl VII.

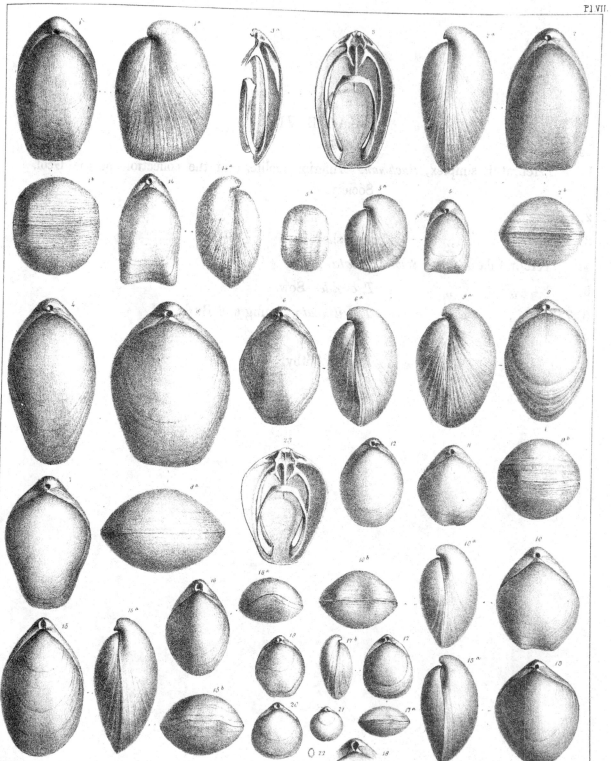

T. Davidson del et lith.

Printed by Hullmandel & Walton.

PLATE VIII.

Fig.

1. Terebratula simplex, *Buchman*. Inferior Oolite. In the collection of the Geol.
      Society.

2. ,, ,,

3. ,, ,, Young shell.

4. Terebratula ovoides, *Sow.* *T. lata*, Sow.

5. ,, ,, *T. ovoides*, Sow.

6, 7. ,, ,, *T. trilineata*, Young and Bird.

8. ,, ,,

9. ,, ,, From Whitby.

Pl VIII

T. Davidson del et lith.

Printed by Hullmandel & Walton

PLATE IX.

Pl. IX.

T. Davidson del et lith.

Printed by Hullmandel & Walton

PLATE X.

Fig.

1, 2.   Terebratula perovalis, *Sow.* The original specimens now in Mr. J. de C. Sowerby's collection.

3.   ,,   ,,   From the Inferior Oolite, Dundry.

4.   ,,   ,,   From the Inferior Oolite of Dunnington.

5.   ,,   ,,   The largest specimen, I believe, yet found in England.

6.   ,,   ,,   Interior.

7.   Terebratula impressa, *De Buch.* var. From the Inferior Oolite; in the British Museum.

8.   Orbicula reflexa, *Sow.* From the Lias; in the British Museum.

Pl. X.

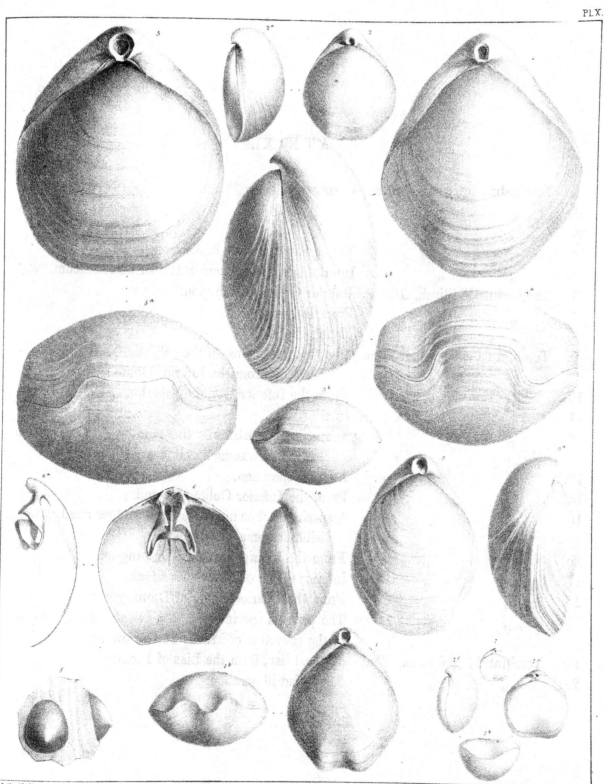

T. Davidson del et lith.

Printed by Hullmandel & Walton.

PLATE XI.

Fig.

1. Terebratula intermedia, *Sow.* Cornbrash.
2.   ,,   ,,
3.   ,,   ,,
4.   ,,   ,, Young shell, without biplication.
5.   ,,   ,, Interior, from specimens in the British Museum.
6. Terebratula Phillipsii, *Morris.* Inferior Oolite, Dinnington.
7.   ,,   ,, From Burton.
8.   ,,   ,, Young shell.
9. Terebratula sphæroidalis, *Sow.* The type specimens in the collection of Mr. J. de C. Sowerby, from the Inferior Oolite.
10.   ,,   ,, From the Inferior Oolite of Sherburn.
11.   ,,   ,, Interior.
12.   ,,   ,, A specimen illustrating the sinuous frontal margin observable in some specimens.
13.   ,,   ,, Another specimen.
14.   ,,   ,, From the Inferior Oolite of Dundry.
15.   ,,   ,, A specimen, showing a cessation and resumed growth visible in some specimens.
16.   ,,   ,, From the Inferior Oolite of Dinnington.
17.   ,,   ,, Inferior Oolite of Burton Radstock.
18.   ,,   ,, From the Inferior Oolite of Dinnington.
19.   ,,   ,, The original specimen of *Ter. bullata,* Sow., now in the collection of Mr. J. de C. Sowerby.
20. Terebratula globulina, *Dav.* Natural size, from the Lias of Ilminster.
21.   ,,   ,, Magnified illustration.

Pl XI.

T. Davidson del et lith.

Printed by Hullmandel & Walton

PLATE XII.

Fig.

1. Terebratula plicata, *Buckman*. From the Inferior Oolite, in the collection of the Geological Society, and in that of Mr. Morris. These are the two largest specimens of the species with which I am acquainted.

2. „ „

3, 4. „ „ Smooth specimens; the frill or plication appears only at a more advanced age.

5. „ „

6. Terebratula fimbria, *Sow.* From the Inferior Oolite. A very large specimen.

7. „ „

8. „ „

9. „ „ Interior.

10. „ „

11. „ „ Young specimen, quite smooth.

12. „ „ A young shell.

13. Terebratula cardium, *Lamarck*. From the Great Oolite, Bath.

14. „ „ From the Great Oolite, Bath.

15. „ „ A curious specimen, illustrating that, even sometimes at an advanced age, the plaits bifurcate.

16, 17. „ „ Young shell, plaits bifurcating. *Ter. furcata*, Sow.

18. „ „ Interior.

19. Terebratula flabellum, *Defrance*. From the Bradford Clay, near Bradford.

20. „ „ Enlarged illustrations.

21. „ „ Interior.

Pl. XII.

T. Davidson del et lith.

Printed by Hullmandel & Walton.

PLATE XIII.

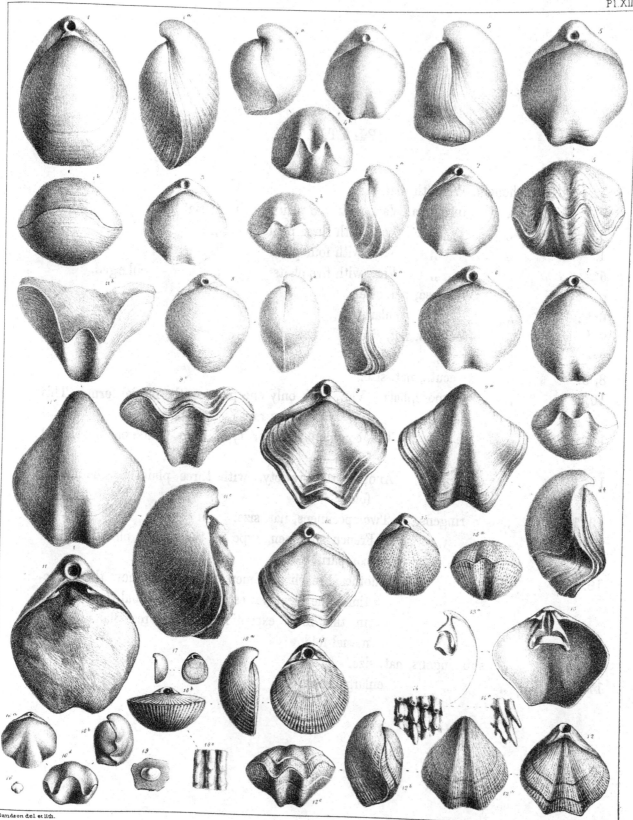

Pl. XIII.

T. Davidson del et lith.

Printed by Hullmandel & Walton.

PLATE XIV.

Fig.

1. Rhynchonella Wrightii, nat. size.
2.   ,,   furcillata. Var., with two plaits on the mesial fold.
3.   ,,   ,,   Var., with three plaits   ,,
4.   ,,   ,,   Var., with four plaits   ,,
5^a.   ,,   ,,   Var., with five plaits   ,,   enlarged.
6.   ,,   rimosa, nat. size.
6^{abc}.   ,,   ,,   enlarged.
7.   ,,   oolitica, nat. size.
7^{cd}.   ,,   ,,   enlarged.
8, 9.   ,,   acuta, nat. size.
10.   ,,   cynocephala. Var., with only one plait on the mesial form. This case is very rare in this species.
11.   ,,   ,,   The common state or type, two plaits on the mesial fold.
12.   ,,   ,,   Another rare variety, with three plaits on the mesial fold.
13, 14.   ,,   ringens. Two specimens, nat. size.
15.   ,,   ,,   A French specimen, type of the species placed here for comparison.
16.   ,,   ,,   Another French specimen, figured, to show that sometimes the form varies from the original type, having in this most exceptional case three plaits on the mesial fold.
17.   ,,   sub-ringens, nat. size.
17^{abc}.   ,,   ,,   enlarged figures.

Plate XIV.

T.Davidson del et lith

Printed by Hullmandel & Walton.

PLATE XV.

Fig.

1, 2.	Rhynchonella	serrata.	Two specimens, nat. size.
3.	,,	Bouchardii.	A var., with two plaits on the mesial fold, a little enlarged.
4.	,,	,,	A var., with three plaits, enlarged.
5.	,,	,,	A var., with five plaits ,,
6.	,,	Lycettii, nat. size.	
7.	,,	sub-variabilis.	A specimen with three plaits on the mesial fold; in Pl. XVIII, fig. 11, a var. is figured with only two plaits on mesial fold.
8, 9, 10.	,,	variabilis.	Three small varieties, with three and four plaits on mesial fold: other variations are illustrated in Pl. XVI, figs. 1 to 6.
11.	,,	Moorei.	A var. (enlarged), with five plaits on mesial fold.
12.	,,	,,	Another var., enlarged.
13, 14.	,,	,,	A variation, with seven plaits on mesial fold.
15.	,,	spinosa.	A specimen, with all its spines attached.
15ª.	,,	,,	A fragment, showing the manner in which the spines originate, enlarged.
16.	,,	,,	A var., with few plaits.
17, 18.	,,	,,	Beak of two specimens, enlarged, to illustrate the foramen and deltidium.
19.	,,	,,	The largest British specimen.
20.	,,	,,	A very young shell, enlarged, to show that sometimes the plaits bifurcate and intercalate.
21.	,,	senticosa, nat. size.	

Plate XV.

PLATE XVI.

Fig.

1.	Rhynchonella variabilis.		Var., with two plaits on mesial fold, *Ter. bidens* of Phillips, nat. size.
2.	,,	,,	A young shell, with two plaits.
3.	,,	,,	A var., with three plaits on mesial fold, *Ter. triplicata* of Phillips, nat. size.
4.	,,	,,	A variety with three plaits from another locality.
5.	,,	,,	A young shell, with three plaits.
6.	,,	,,	A variety, with four plaits.
7.	,,	plicatella.	The type form of the species.
8.	,,	,,	A less aged and more elongated specimen.
9, 10, 11, 12.	,,	sub-tetrahedra.	Four varieties, nat. size, showing a difference in the number of the plaits; fig. 9 is the largest example I have seen.
13, 14.	,,	lacunosa.	I am not quite certain as to this determination; the specimens much resemble some German types of *R. lacunosa*.
15.	,,	?	This shell is from the Lias of Sky, and my only inducement to figure it here is on account of the variety of Liasic species of Brachiopoda in Scotland. It has much the appearance of some young specimens of *T. quadriplicata* of Zieten.

Plate XVI.

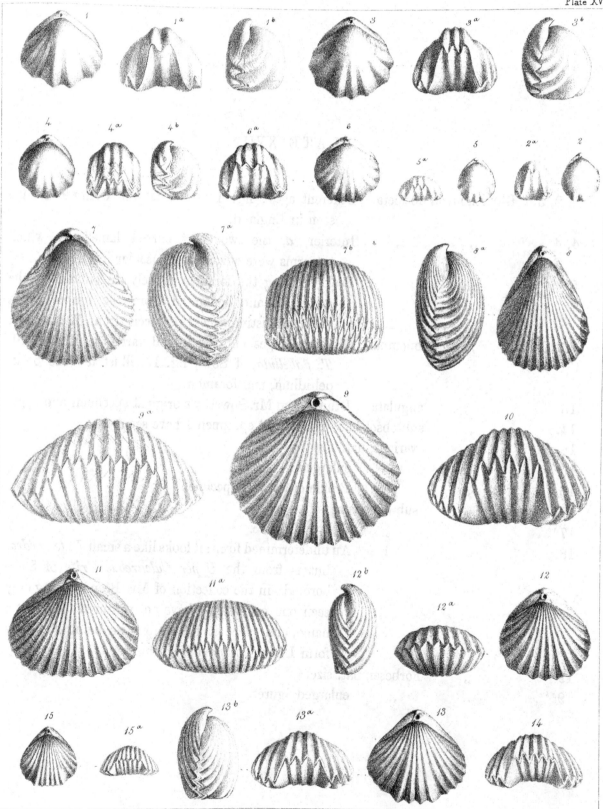

T. Davidson del et lith.

Printed by Hullmandel & Walton.

PLATE XVII.

Plate XVII.

T. Davidson del et lith.

Printed by Hullmandel & Walton.

Plate XVIII.

T.Davidson, del et. lith.

Printed by Hullmandel & Walton.

APPENDIX.

BESIDES the few Supplementary additions which this Appendix may contain, it is desirable to point out several corrections resulting from the continued researches undertaken during the four consecutive years devoted to the publication of the present Monographs. Nor was it always possible to avoid certain changes and repetitions resulting from the mode in which the material had unavoidably to be assembled.

INTRODUCTION.

Page 5, line 13.

For "the recent *Atrypa*[1] and the fossil Rhynchonellidæ, which latter, from the general analogy of the structure of their shell, might be supposed to have the respiratory organs at the same low degree of development as the *Terebratulidæ*, and to have the same need," *read* "the Rhynchonellidæ, which have the respiratory organs at the same low degree of development as the *Terebratulidæ* have the same need" (Owen).

Page 7, line 3.

For "perforated valve, D," *read* "imperforated valve, D" (Owen).

Page 9, line 1.

For "called 'thecidium,'" *read* "called 'deltidium'" (Owen).

Page 33, line 23.

Place *T. lunifera* (Philippi) under the genus MORRISIA (line 22).

Page 44.

Add—Mr. S. P. Woodward places the *Rudistes* in his seventh Family of Conchifera, and states that—"They are the most problematic of all fossils; there are no recent shells which can be supposed to belong to the same family; and the condition in which they usually occur, has involved them in greater obscurity. (1. Buch regarded them as Corals. 2. Desmoulins, as a combination of the Tunicary and Sessile Cirripede. 3. Dr. Carpenter, as a group intermediate between the *Conchifera* and *Cirripeda*. 4. Prof. Steenstrup, of Copenhagen, as Anellides. 5. Mr. Sharpe refers *Hippurites* to the Balani; *Caprinella* to the Chamaceæ. 6. Lapeirouse considered the Hippurites, *Orthocerata;* the Radiolites, *Ostracea*. 7. Goldfuss and D'Orbigny place them both with the *Brachiopoda*. 8. Lamarck and Rang, between the *Brachiopoda* and *Ostraceæ*. 9. Cuvier and Owen, with the Lamellibranchiate bivalves. 10. Deshayes, in the same group with *Ætheria*. 11. Quenstedt, between the *Chamaceæ* and *Cardiaceæ*.) The characters which determine their position amongst the ordinary bivalves, are the following:—1. The shell is composed of two distinct layers. 2. They are essentially unsymmetrical, and right-and-left valved. 3. The sculpturing of the valves is dissimilar. 4. There is evidence of a large internal ligament. 5. The hinge-teeth are developed from the free valve. 6. The muscular impressions are two only. 7. There is a

---

[1] The shell referred to as a "recent *Atrypa*," is the *Rhynchonella psittacea* (Int. pl. vii, fig. 100—102). It does not belong to the genus *Atrypa* which is furnished with shelly spires (see pl. vii, fig. 89 and 92).

distinct pallial line." ('Manual of the Mollusca,' p. 279, 1854.) Refer also to a very able memoir by the same author, read to the Geological Society, on the 7th June, and published in the 'Quarterly Journal' for the present year.

Page 73, line 6.

Add—M. E. Deslongchamps has published in the 'Annuaire de l'Institut des Provinces,' for 1853 and 1854, descriptions and figures of three remarkable shells or species, discovered by himself and Dr. Périer, in the *Middle* and *Upper Lias* of May, Fontaine-Etoupefour, and Bretteville-sur-Laize, near Caen, by the names of *Argiope Périeri*, *A. Liasiana*, and *A. Suessii*; thus carrying the genus ARGIOPE to the lower portion of the Jurassic system. I will not offer any positive opinion on this question until I have had an opportunity of examining the species in question; but about the period of my friend's publication in 1853, I could not refrain from expressing certain objections or doubts, based upon the character of their plication, which differs so entirely from that of any other species of *Argiope* at present known. The external shape, of one of them in particular, *A. Liasiana*, has much of the external shape of certain forms of *Cyrtia*, especially by its deltidium and position of the foramen? It will therefore be necessary to know more of their internal arrangements, before the place of those remarkable species can be definitely established.

Page 77, THECIDIUM.

The valuable and beautifully illustrated memoir 'On the Brachial Arrangement in Thecidium,' by M. Suess, of Vienna,[1] has afforded us much additional knowledge relative to the internal structure of this remarkable genus; and my only regret is not being able to transcribe the whole of that very important memoir, which does infinite credit to the persevering and able researches of its distinguished author.

M. Suess begins by stating, that the near connection between ARGIOPE and THECIDIUM has already been admitted by several authors, but he forcibly objects to those two genera being considered BRACHIOPODA CIRRHIDÆ, and combined with the RUDISTA, as proposed by M. A. d'Orbigny; the animal having proved that they are not only *true Brachiopoda*, but likewise members of the great Family TEREBRATULIDÆ. He also believes that STRINGOCEPHALUS must be connected, or placed in the same family and close to ARGIOPE; he has observed on the inner edge of the loop of the last, small lamellar projections, similar, although much less developed than in the first-named genus. M. Suess then enters into many interesting details on the variations presented by the loop of ARGIOPE: of which *Arg. decemcostata* and *A. decollata* have afforded the greatest difference. The first approaches most to that of STRINGOCEPHALUS, the second teaches us to understand the more complicated structure in THECIDIUM.

It has been already shown, that the exterior side of the loop of *A. decollata* is at times partially fixed to the bottom of the valve, while the outer margin bears the brachial fringe. Now if we imagine the inferior side to be fixed to the valve and to the septa all the way round, we should have the counterpart of *appareil ascendant*[2] of *Thecidum digitatum*; and the *appareil descendant* would be wholly wanting in every genus of Brachiopod known to M. Suess. Many authors, he states, may be deceived, while thinking that the broad wide expansions of the *appareil descendant* could be formed by the union of the portion of the

---

1 'Proceedings of the Imperial Academy of Sciences of Vienna,' vol. xi, pp. 991 and following, 1853.

2 The words *appareil ascendant*, *appareil descendant*, *bride transversale*, &c., as used by M. E. Deslongchamps, have been explained in pp. 78 and 79 of the 'Introduction.' The ascending apparatus of *Thecidium* appeared to Mr. S. P. Woodward and myself to be nothing more than the equivalent of the "septum" of *Terebratella*, *Magas*, *Argiope*, &c.; the cirrated lip commences on the outer edge of the oral processes and is continued along the outer edge of the loop.

loop placed between the septa, but it will be sufficient to remind the reader, that it is the exterior margin of the loop-riband which becomes fixed to the valve, and that the outer margin must bear the cirri.

The *bridge* (Pl. VI, fig. 37, *w*, Int.) offers so many striking analogies with the "converging processes" near the crura of *Argiope*, that it appears impossible to seek a representation of these parts in any portion of the *appareil descendant;* so that while describing the *brachial* arrangement in *Thecidium digitatum* or *T. papillatum* (= *radians*, Def.), we should say, the *loop* is forced to form a curved line by the protrusion of the septa (as in *Argiope decollata*, &c.), the exterior side adhering wholly to the valve and septa (as it does partially so in *A. decollata*), while the converging processes are united, so as to form the *bridge*, the crura being likewise fixed to the valve. The *brachial membrane* is protected by peculiar calcareous *supra-membraneal deposits*, termed *appareil descendant* by M. E. Deslongchamps.[1]

It seems, therefore, that M. Suess entirely objects to any portion of M. E. Deslongchamps' last-named *appareil* representing the *loop;* that in *Th. digitatum*, for example, all the branches or grooves are equal, while in *Th. papillatum* the outer branches of the *appareil descendant* are both larger and broader than the more central ones; but that in *Th. Mayale*, Desl., the *supra-membraneal disk* appears to be really wanting, while the equivalent of the loop may easily be recognised, forming a raised rim all round the septa. The last-named species would therefore approach nearer to *Argiope* than any other hitherto discovered, and might, perhaps, deserve to be considered as the type of a distinct *group* or *sub-section* of THECIDIUM; but M. Suess considers that before arriving positively at such a conclusion, additional researches will require to be made on *perfect* interiors of *Th. rusticum* (Moore), where no trace of *appareil descendant* has hitherto been observed.

The *supra-membraneal disk* is always fixed to the valve about the centre of the shell, on either side of the *inferior* part of the *visceral cavity*, but there exists much difference as to the extent to which the disk is fixed. In *Th. digitatum* and in *Th. sinuatum* it adheres throughout; the portion above the visceral cavity only remaining free.

After these and other preliminary observations, M. Suess enters into an elaborate description of the *Thecidium vermiculare*, Schl., sp. (= *Th. hippocrepis*, Goldfuss), and observes that in *Th. sinuatum* the portion of the *supra-membraneal disk* above the visceral cavity is not a solid calcareous mass, but a loose net-work fixed to a solid rim, termed the *bride transversale* by E. Deslongchamps. This remarkable and delicate net-work is beautifully exemplified in *Th. papillatum* (= *radians*), and still more widely open in *Th. vermiculare*. In *Th. digitatum* all the windings of the brachia are nearly on the same line or level ('Introd.,' Pl. VI, fig. 40[2]), while in *Th. papillatum* they rise from about the middle of the shell, and to such an altitude in *Th. vermiculare* as to stand nearly vertically to the direction of the valve.

Thus the characteristic peculiarity in the last-named *Thecidium*, and a few other allied species, consists in the elevation of this *supra-membraneal disk;* it does not become entirely free, but those portions which used to be formed of solid masses in other species, are here represented by thin, delicate net-works. In *Th. papillatum* we see contrary to what *Th. digitatum* shows, that all the branches of the loop and lobes to which they are united have a tendency to radiate from one central stem, somewhat similar to the branches

[1] Mr. S. P. Woodward describes the *disk* and *loop* as follows:—Oral processes united, forming a bridge over the small and deep visceral cavity; disk grooved for the reception of the loop, the grooves separated by branches from a central septum; loop often unsymmetrical, lobed, and united more or less intimately with the sides of the grooves." ('Manual of the Mollusca,' part ii, p. 221, 1854.) M. Suess's views regarding the loop differ from those we advanced in 1852; but a further examination of the animal of the recent *Th. Mediterraneum* will be desirable, before this question can be finally settled. The cirrated margin being apparently attached to the inner sides of the sinuous grooves, in the specimen examined by Mr. Woodward and myself, in 1852, might perhaps favour M. Suess's opinion?

[2] I much regret not having been able to reproduce some of M. Suess's admirable illustrations, but I found that they could be but imperfectly represented by woodcuts.

of a tree. This is still more the case in *Th. vermiculare,* in which the central stem attains a high elevation, bearing all the other windings of the loop; in the last-named shell, the upright portion which covers the visceral cavity is perforated by numerous pores, between which excrescences are sometimes seen, and that these apertures were probably destined to allow a kind of current in the fluid filling the visceral cavity.[1]

From these researches, it seems probable that the *Family* THECIDIIDAE and *Sub-Family* STRINGO-CEPHALIDÆ may be advantageously dispensed with, and their genera added to the great *Family* TERE-BRATULIDÆ. This view, which has been adopted by Mr. S. P. Woodward, has likewise been my opinion for some time past, especially since the fortunate discoveries made by M. Suess. All that will be necessary is, therefore (in p. 51), to continue the connecting line from *Terebratula* down to *Thecidium,* and to erase the family and sub-family above mentioned.

Page 78.

It is observed, that the most ancient *Thecidium* then known to us, was that found in the Triassic beds of St. Cassian (Tyrol); but we subsequently read in the 'Bulletin de la Soc. Géol. de France' (vol. x, p. 248, 1853), that Count Keyserling had discovered among the Carboniferous Fossils from Sterlitamak, a shell, which he considers to belong to the genus *Thecidium,* and which he describes as follows:—"It is not new, being the *Anomia antiqua* of M'Coy;[2] but casts, partly deprived of shell, prove that it is a *Thecidium,* ornamented by concentric waves, as in *Th. tetragonum* (Roemer), of which one valve is attached and perforated, while the other equally inflated exhibits a digitated arrangement which occupies nearly all the interior. The oblique disposition of the processes of this arrangement (appareil) round the mesial crest, similar to the leaves of a fern round their stem, has suggested to me the denomination *Th. filicis* for the species. In analogous forms the digitations are directed forward, and more or less arched; a crenulated ridge following exteriorly all round the lateral digitated impressions, of which the second from the hinge is the largest." We are at a loss to understand this remarkable fossil, as we cannot perceive in the figure published by Professor M'Coy, any reason to believe that it belongs to *Thecidium,* and must therefore delay the admission of the genus into the Carboniferous age, until M. de Keyserling has given us some more tangible proof in the shape of illustration.

Family—SPIRIFERIDÆ.

Page 84, note 3.

Add—M. Suess objects to the term ATHYRIS being applied to such shells as *T. Herculea, T. tumida, T. scalprum,* &c.,[3] that generic denomination having been originally employed by Professor M'Coy, for *T. concentrica, lamellosa,* and other similarly organised species; and that we cannot admit the new version proposed by the learned professor in his subsequent work on the 'British Palæozoic Fossils in the Cambridge Museum.' If the denomination *Athyris* is to be removed on philological grounds, M. Suess would adopt SPIRIGERA, D'Orbigny, for *T. concentrica,* &c.; and claims the priority of his genus MERISTA,[4] for *M. Herculea, M. scalprum,* and other similar species. Mr. S. P. Woodward adopts *Athyris,* as originally defined (*A. concentrica*),[5] and, the sub-genus MERISTA, Suess, for *T. Herculea, T. scalprum,* &c.

---

[1] M. Suess' speculation touching the use of the pores in the apparatus of *Thecidium vermiculare,* is perhaps hazardous.

[2] 'A Synopsis of the Characters of the Carboniferous Limestone Fossils of Ireland,' p. 87, pl. xix, fig. 7, 1844.

[3] Suess, in p. 58, of Leonhard's 'Neuves Jahrbuch,' Jan., 1854.

[4] Described by M. Suess, under that denomination, in the 'Jahrb. d. k. Geol. Reichs-Anstalt,' ii, iv, 150, 1851; and again mentioned in Leonhard's 'Neuves Jahrbuch,' p. 127, 1854.

[5] 'Manual of the Mollusca,' part ii, p. 244, 1854.

Palæontologists therefore appear in general to agree, as to the propriety of placing in a separate group such organisations as *T. concentrica* on the one hand and *T. Herculea* on the other, but differ considerably as to the names to be adopted for each in particular. I am quite disposed to relinquish the proposition made in pp. 85 and 88. The matter may stand thus:

Page 87.

ATHYRIS, *M'Coy*, = (or) SPIRIGERA, *D'Orbigny*, for such shells as *A. concentrica*, *A. lamellosa*, *A. pectinifera*, *A. Roissyi*, &c.[1]

Page 84.

*For "*ATHYRIS*" = read "Sub-genus—*MERISTA, *Suess," for* T. Herculea, T. tumida, T. scalprum, T. cassidea *(Dal., sp.), &c., and*

Page 88.

*Sub-genus—*RETZIA, *King, for* T. Adrieni, T. ferita, T. serpentina.

Page 89.

Genus— UNCITES, *Def.*[2]

---

[1] The arrangement of the *crural processes, attachment, and direction of the first spiral coils* in this, and in other species of the genus, ATHYRIS, *M'Coy* = SPIRIGERA, *D'Orb.*, has not as yet been completely understood, nor have the many efforts made in that direction both by Mr. S. P. Woodward and myself been attended with results which may be termed entirely satisfactory.

In the 'Manual of the Mollusca,' Mr. S. P. Woodward has represented what he supposes to be the probable arrangement, and states (p. 224), "*hinge plate with four muscular cavities, perforated by a small round foramen, and supporting a small complicated loop (?) between the spires; spires directed outwards, crura united by a prominent oral loop. The foramen in the hinge plate occupies the situation of the notch, through which the intestine passes in Rhynchonella.*" I cannot, however, at present, entirely coincide in the opinion expressed by my able and conscientious friend, not from any positive observations to the contrary, but because I am rather inclined to believe, that only one attachment took place from the base of the hinge plate (as in *Spirifer*), forming at the same time the spire and complicated crural process; or in other words, that the loop (?) described by Mr. Woodward is not independent of the crura and spiral appendages. I would therefore urge upon those naturalists who may possess specimens of *A. Roissyi* in a favorable condition for observation (as they do often occur near Tournay), to endeavour to clear up this unsettled, and *important* character of internal organisation.

[2] In p. 90, I quoted as examples, *Uncites Gryphus*, Schloth., and *U. lævis*, M'Coy; the last reference was given upon Professor M'Coy's authority, but I do not feel quite certain that *U. lævis* belongs to the genus, and according to Mr. Salter it would be the same as *Ter. porrecta*, Sow., 'Min. Con.,' tab. 576, fig. 1, which I believe to be a *Stringocephalus*. I also unintentionally omitted to observe in p. 89, that Professor M'Coy's views on this genus differ entirely from those advocated in the present volume; thus in p. 380 of the 'British Palæozoic Fossils' (1852), the genus UNCITES forms alone the

"*VIth Family—*UNCITIDÆ *of Professor M'Coy.*

"*Genus—*UNCITES (*Defrance*), as here defined.

"Gen. Char. *Shell elongate-ovate, slightly inequilateral; substance very thick, densely fibrous; beak of the receiving valve very long, narrow, claw-shaped, gently incurved obliquely on one side; with a wide concave imperforate defined channel beneath, no internal septa, nor appendages in either valve.*"

Professor M'Coy also states, "*having first ascertained the true internal characters of that curious*

Page 92, *Genus*—KONINCKINA, *Suess.*

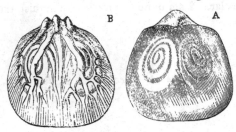

Koninckina Leonhardi.

A. *Translucent specimen.*
B. *Interior of the dorsal valve, showing remains of the vascular system.*

With Mr. Woodward's permission, I here reproduce two additional figures, drawn from two specimens forming part of Klipstein's Collection, in the British Museum, and published in the 'Manual of the Mollusca,' part ii, p. 231, 1854.

Much doubt remains still, as to where this curious genus should be located; Mr. Woodward has placed it in the family ORTHIDÆ,? observing that, "this curious little shell most resembles the Triassic *Leptæna dubia (producta)*, Münster (= *Crania Murchisoni*, Klipst.) !"

Family—STROPHOMENIDÆ.

Page 105, *Strophomena.*

In the 'Natural History of New York,' vol. ii, p. 63, 1852, Mr. J. Hall proposes to distinguish those *Strophomenas* which possess a denticulated hinge line by the generic appellation of *Stropheodonta* (Gr. στροφευς, *cardo*, and οδους, *dens*), naming as types *S. demissa*, Conrad, and *S. prisca*, Hall. I am not certain that this character can be considered of sufficient importance for the creation of a separate genus: *Strophomena euglypha*, *S. filosa*, and many other species admitted into *Strophomena*, possess denticulations only on those portions of the hinge line near the cardinal process, while the remaining portion is quite smooth.

Page 103.

Note 7 must be removed from ORTHIS, and placed under the Genus CHONETES (p. 113); and *for* (3d line) "and which I am inclined to believe belonged to a species of *Orthis*, bearing great resemblance to that form known by the name of *O. resupinata*, Martin, sp.," *read* "and which belong to *Chonetes comoides*, Sow., sp."

Obs. Having received from Mr. G. W. Ormerod the loan of several examples similar to (but more perfect than) those in the Society's collection, I at once discovered the mistake I had committed, since they evidently belong to *C. comoides*, Sow.; an opinion subsequently corroborated by M. L. de Koninck, Salter, and Woodward. In Appendix, pl. A, fig. 29, I have represented Mr. Ormerod's most complete example of the interior of the ventral valve, wherein the impressions left by the adductor (A) and cardinal muscles (C) are beautifully defined.[1]

Eifel genus, it gives me much pleasure to recognise in our rocks a second species of this remarkable genus;" but I fear that the learned professor cannot have made himself perfectly acquainted with the internal characters of this *Eifel genus*, or he would have perceived that the shell was not totally deprived of appendages in either valve, since several examples of *U. Gryphus* have shown fragments of internal appendages, as represented in pl. vii, fig. 86, *c;* although only one has hitherto exhibited to my knowledge traces of the spirals discovered by Professor Beyrich.

As stated in p. 89 of the 'Introduction,' I feel impressed that the place of UNCITES is in the *Family* SPIRIFERIDÆ, a view sanctioned and adopted both by M. Suess and Mr. Woodward.

[1] In vol. x of the 'Quarterly Journal of the Geological Society' (p. 202, pl. viii, Dec., 1853), will be found recorded all I have hitherto been able to discover relative to this most interesting species.

Page 121.

I stated that the genus CALCEOLA was confined to the DEVONIAN epoch, and that I was acquainted with but one well-authenticated species. Dr. Fr. Roemer has, however, recently informed me that he has found a second species in true SILURIAN rocks, near Brownspost, in the State of Tennessee, N.A.[1]

Pages 122, 124, and 125.

For "Spondylobus" *read* "*Spondylolobus*" (M'Coy).

Page 133.

*Family—*LINGULIDÆ.

Page 134.

Line 3, *for* "beak of the valve," *read* "beak of the *ventral* valve."

  ,, 10, *for* "left by the pedicle muscle (A)," *read* "left by the *post-adductor* muscle."

  ,, 13, *for* "caused by the posterior," *read* "caused by the anterior."

  ,, 16, *after* "last-described impressions," *add* "in the dorsal valve."

  ,, 17, *for* "combined extremities of the anterior pair," *read* "extremities of the anterior pair of retractor muscles."

The animal of *Lingula anatina* having been minutely examined by Mr. S. P. Woodward subsequently to the publication of my 'Introduction,' I am now enabled, with his kind permission, to reproduce the woodcuts and description recently published by that able naturalist, in the second portion of his excellent and valuable 'Manual of the Mollusca.'

---

[1] Note on *Calceola Tennesseensis*, n. sp., a true Siluriau species of the genus (extracted from the unpublished manuscript of the 'Lethæca Geognostica, 1st Period,' by Fred. Roemer.

CALCEOLA TENNESSEENSIS, Fred. Roemer, n. sp.

Calceola sandalina, Froost. 'Fifth Report on the Geology of the State of Tennessee,' Nashville (p. 47, 1840). This species, although similar to *Calceola sandalina*, differs from the type of the genus by the following characters:

1. The shell is much thicker than in *C. sandalina*, and the space for the reception of the soft parts of the animal in consequence reduced to a shallow hole.

2. The edges separating the area of the larger valve from the convex posterior part of the valve are much less angular, or more rounded.

3. The area of the smaller valve, which in *C. sandalina* falls into the same plane, or nearly so, is inclined backwards, and forms an obtuse angle with that of the larger valve.

Locality. Not very rare in calcareous strata of the age of the Wenlock Limestone of England, associated with *Orthis elegantula, Caryocrinus ornatus, Pentatremites Reinwardtii*, &c.

Calceola pyramidalis, Girard (Leonh. and Bronn's 'Jahrb.,' 1842, p. 232, f. *a, b, c*), from the Silurian strata of Gothland, is not a Mollusc, but a Zoophyte, and synonymous with *Goniophyllum pyramidale*, Edwards and Haime, 'Archive du Mus.,' vol. v, p. 404 (Fr. Roemer).

APPENDIX.

Family—LINGULIDÆ.[1]

"*Animal.* With a highly vascular mantle, fringed with horny setæ; oral arms thick, fleshy, spiral, the spires directed inwards, towards each other; valves opened and closed by sliding muscles.

"In fig. 1, a small portion of the liver and visceral sheath have been removed, to show the course of the stomach and intestine. In some specimens the whole of the viscera, except a portion of the liver, are concealed by the ovaries. In fig. 3, the front half of the ventral mantle-lobe is raised, to show the spiral arms: the black spot in the centre is the mouth, with its upper and lower lips, one fringed, the other plain. The mantle-fringe has been omitted in figs. 1 and 3.

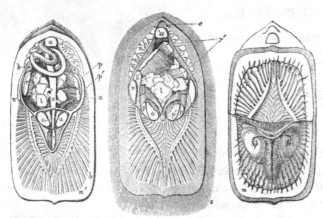

Fig. 1. *Dorsal.*   Fig. 2. *Ventral.*   Fig. 3. *Ventral.*

Lingula anatina, Lamarck.

a a, *anterior adductors*; a′, *posterior adductor*; p p, *external protractors*; p′ p′, *central protractors*; r r, *anterior retractors*; r′ r′ r′, *posterior retractors*; c, *capsule of pedicle*; n n, *visceral sheath*; o, *œsophagus*; s, *stomach*; l, *liver*; i, *intestine*; v, *vent*; h h, *auricles*; h′, *left ventricle*; b, *branchial vessels*; m′, *mantle margin*; m, *inner lamina of mantle—margin retracted, showing bases of setæ*; s, *setæ*.

"*Animal.* With the mantle-lobes firmly adhering to the shell, and united to the epidermis, their margins distinct, and fringed all round: branchial veins giving off numerous free, elongated, narrow loops from their inner surfaces; visceral cavity occupying the posterior half of the shell, and surrounded by a strong muscular sheath: pedicle elongated, thick: adductor muscles three, the posterior pair combined: two pairs of retractors, the posterior pair unsymmetrical, one of them

dividing: protractor sliding muscles two pairs: stomach long and straight, sustained by inflections of the visceral sheath: intestine convoluted dorsally, terminating between the mantle-lobes on the right side;

---

[1] 'Manual of the Mollusca,' part ii, p. 239, 1854, woodcuts Nos. 165, 166, 167.

The etymology of several of the generic appellations having been omitted in this volume, I here add them, from Mr. Woodward's 'Manual:'

TEREBRATULA. *Etym.*—diminutive of *terebratus*, perforated.

ARGIOPE. *Etym.*—*Argiope*, a nymph.

THECIDIUM. *Etym.*—*Thekidion*, a small pouch.

STRINGOCEPHALUS. *Etym. Strinx* (*Stringos*) an owl, *cephale*, the head.

ATHYRIS. *Etym.*—*A*, without, *thuris*, a door.

PENTAMERUS. *Etym.*—*Pentamerus*, five partite.

ORTHIS. *Etym.*—*Orthos*, straight.

STROPHOMENA. *Etym.*—*Strophos*, bent, *mene*, crescent.

CALCEOLA. *Etym.*—*Calceola*, a slipper.

CHONETES. *Etym.*—*Chone*, a cup.

CRANIA. *Etym.*—*Kraneia*, capitate.

SIPHONOTRETA. *Etym.*—*Siphon*, a tube, *tretos*, perforated.

LINGULA. *Etym.*—*Lingula*, a little tongue.

oral arms disposed in about six close whirls, their cavities opening into the prolongation of the visceral sheath in front of the adductors.[1]

"Observations on the living Lingula are much wanted; the oral arms probably extended as far as the margins of the shell; and the pedicle, which is often nine inches long in preserved specimens, is doubtless much longer and contractile when alive. The shell is horny and flexible, and always of a greenish colour."

Mr. Salter has found that *Lingula Davisii* from the Lower Silurian of Tremadoc, has a pedicle groove, like *Obolus*, and that it might, perhaps, form a distinct section of *Lingula*. This will, however, require some further investigation.

Page 135.

In *Obolus* the hinge-margin is thickened inside, and slightly grooved in the ventral valve: posterior adductor (A) impressions separate: anterior pair (c) sub-central: impressions of sliding muscles (B) lateral, as shown by my woodcuts (p. 136). The pedicle scar in the centre of woodcut, fig. 54, has no letter.

Page 136.
Woodcut, fig. 51, *for* "from Ruma," *read* "from Russia."

PART I.

BRITISH TERTIARY BRACHIOPODA.

Since the publication of this Monograph, in 1852, no new species seems to have been discovered, but I am able to offer a few additional illustrations from more perfect individuals.

Page 5.

Lingula Dumortieri has been erroneously spelt Dumontieri, in pp. 4, 5, 6, and 9, (Mr. R. Ch. Dewael has defined the stratigraphical position of this species in the Antwerp Crag. 'Bull. Acad. Royal.' Belgium, 1853.)

----

[1] In p. 211, Mr. Woodward adds several interesting comparisons between the dispositions of the animal in *Lingula* and the other genera, and from which I have transcribed the following passages:

"On separating the valves of a recent *Terebratula*, the digestive organs and muscles are seen to occupy only a very small space near the beak of the shell, partitioned off from the general cavity by a strong membrane, in the centre of which is placed the animal's mouth. The large cavity is occupied by the fringed arms. . . . . Their nature will be better understood by comparing them with the lips and labial tentacles of the ordinary bivalves; they are, in fact, lateral prolongations of the lips, supported on muscular stalks, and are so long as to require being folded or coiled up. In *Rhynchonella* and *Lingula* the arms are spiral and separate: in *Terebratula* and *Discina* they are only spiral at the tips, and are united together by a membrane, so as to form a lobed disk. . . . . The mouth conducts by a narrow œsophagus to a simple stomach, which is surrounded by the large and granulated liver: the intestine of *Lingula* is reflected dorsally, slightly convoluted, and terminates between the mantle lobes on the *right* side. In *Discina* it is reflected vertically, and passes straight to the right, ending as in *Lingula*. In *Terebratula*, *Rhynchonella*, and probably all the *normal* Brachiopoda, the intestine is simple and reflected ventrally, passing through a notch or foramen in the hinge-plate, and ending behind the ventral insertion of the adductor muscle."

Page 6, *Lingula tenuis,* Sow.

From want of specimens I could not illustrate this species sufficiently in 1852, but now I am able to add several figures taken from the most perfect examples in the collection of Mr. Wetherell.   See Appendix, Pl. A, figs. 3, 4, and 5.

Pages 6, 7, and 9.

For "ORBICULA," *read* "DISCINA."

Page 9.

Second line under the woodcut, *for* "*on the anterior side,*" *read* "on the lower side."

Page 11.

Note 1, *add,* Mr. Barlee has taken them (*Argiope cistellula*) forty miles east of Zetland (Forbes and Hanley, 'British Mollusca,' Appendix, vol. iv, p. 257).

Page 16, *Terebratula cranium.*

Note 1.   The illustration Pl. I, fig. 8 *d,* was drawn from a very imperfect specimen, but I am now able to refigure the same valve (Pl. A, fig. 1), by which it will be seen that the species possessed an elongated loop, and consequently belongs to the *Section* WALDHEIMIA of King.   Four examples of this rare species have lately been procured by Mr. Barlee, thirty miles east of Zetland (Forbes and Hanley, 'British Mollusca,' vol. iv, p. 257).

Page 19, *Ter. bisinuata.*

I have represented (Pl. A, fig. 2) the perfect example found by Mr. Cunnington, and mentioned in note 1, p. 19.

PART II.

MONOGRAPH OF CRETACEOUS SPECIES.

(In the concluding portion of this monograph, published in 1854, will be found most of the additions and corrections I have been able to introduce.)

Page 11.

Line 1, *for* "Egnabergensis" *read* "Ignabergensis."
„   2, *for* "Numulo" *read* "Numulus minor," *and for* "Brattenbergensis" *read* "Brattenburgensis."
„ 14, *for* "Mondergattang" *read* "Mon der Gattung."

Page 15.

Line 11.   Sig. Michelotti having obligingly sent me specimens of his *Th. Broderipii,* I was able to ascertain that they belong to *Thecidium Mediterraneum.*   This species has also been discovered by Sir C. Lyell, in the *miocene* beds of the Canary Islands.

Page 23.

Line 19, *for* "Maudesly" *read* "Mundesly."

Page 26.

Line 32, *for* "Terebratella pectita (Brong.)," *read* "Terebratula pectita (Brong.)."

Page 27.

Line 1, *for* "1826 or 1827," *read* "1816 or 1817," *and erase the word* "TEREBRATELLA."

Page 51.

Line 10, *for* "Hornsey" *read* "Hamsey."

PART III.

OOLITIC AND LIASSIC SPECIES.

Since the publication of this monograph (1851-2) very few additional species seem to have been discovered; but certain points in connexion with those already described have received further elucidation.

As we progress with the minute investigation, and collecting of species from each bed or layer, it always becomes the more evident that we are still far from having attained a complete knowledge of the exact period when certain forms first appeared, their duration, and the epoch at which they became extinct. This most desirable and important knowledge cannot be obtained by a few excursions, nor in a short space of time: it is the reward which will attend those indefatigable *local geologists* who, like Mr. Moore, Dr. Wright, and others, assiduously explore every inch of the beds situated within the district they inhabit, and who scrupulously assemble and correctly determine all the fossil contents of *each bed in succession.* It is also of the utmost importance to examine, record, and illustrate the numberless variations due to age, race, local condition, and malformation, presented by the individuals of each species, living congregated or disseminated in the same bed and in different localities, that we may obtain a just clue to our appreciation of the characters and limits to be assigned to *our* species, often too arbitrarily circumscribed.[1]

It is not difficult to say, for example, that the *Terebratula ornithocephala, T. digona, T. obovata,* and other species of authors, *are all referable to one species,* and that most species *are derivative* of others; such

---

[1] Mr. Woodward states that "some of the *Brachiopoda* appear to attain their full growth in a single season, and all, probably, live many years after becoming adult. The growth of the valves takes place chiefly at the margin: adult shells are more globular than the young, and aged specimens still more so. The shell is also thickened by the deposit of internal layers, which sometimes entirely fill the beak and every portion of the cavity of the interior which is not occupied by the animal, suggesting the notion that the creature must have died from the plethoric exercise of the calcifying function converting its shell into a mausoleum, like many of the Ascidian zoophytes." ('Manual of the Mollusca,' part ii, p. 213, 1854.)

assertions only shew the limited experience of those who make them: nor is it sufficient to state that such and such a form *is a derivative*, except we can demonstrate the source whence the form *has been derived*, and explain, both by words and figure, wherein the *natural* connexion or sequence lies.   But let it not be imagined, from what has just been stated, that it is my wish or intention to deny that some of the *so-termed* species enumerated in this monograph are in reality more than simple variations, referable to one single type. In many instances I have expressed doubts as to their distinct specific claims; and if I did not combine certain forms, it was because the material *then* at my disposal was not sufficient to warrant such a conclusion.

There is also another point no less worthy of investigation, viz., *what relative value or importance should we attach to the minor local divisions in the strata made use of in geological works?* Professor Forbes states [1] that "the marine faunas of the oolitic epoch indicate at least three great and widely spread assemblages of types, each exhibiting a general and easily recognisable facies.   These aspects may be termed respectively the *liassic*, the *Bathonian*, and the *Oxfordian;* the two latter terms being used for want of better, and being adopted in a wide and general sense, and not in the restricted meaning in which they are used by M. A. d'Orbigny.   The horizon of change of facies at the boundary between each is a horizon, to a considerable extent, of change of species.   I believe that every year's research will make it more evident that the perishing of species is simply the result of the influence of physical changes in specific areas, and depends upon no law of inherent limitation of power to extent in time.   If so, we should expect to find indications of the cause of the greater changes in the oolitic and marine fauna in the shape of strata bearing evidence of a wide-spread change of physical conditions within the great oolitic area.   An extensive change of species within a marine area in all likelihood is dependent on an extensive conversion of that area into terrestrial surface."

The questions relating to the existence of species in time, as well as of the cause of their sudden disappearance, and replacement by others, are among the most important inquiries within the domain of palæontology, and have of late years particularly occupied the thoughts of several of our most distinguished foreign and British naturalists.

M. Barrande informs us,[2] that his views on this subject differ materially from those recently circulated by several illustrious men in their most recently published elementary works.   According to which doctrines, *the different animal creations characterising the vertical sequence of formations* (Terrains) *have been suddenly destroyed by violent convulsions of the earth's surface* (cataclysmes), annihilating at once all the then existing animal creation: so that each of these universal revolutions (according to their theory) would explain, in a very plausible manner, the complete renewal at different periods of all the animal forms on the surface of the globe.   M. Barrande admits that this renewal in many cases cannot be contested; but asserts that the change is not in all instances due to convulsions of the earth's surface, of which he quotes several remarkable examples among the Palæozoic deposits.   It seems to him more rational to admit that the phenomena of the development of the series of beings in time, taken as a whole, is subjected to a special law of nature, independent of that which governs the physical revolutions of the surface of our planet; that since we see all these faunas disappear in succession, one after the other, from the entire surface of the globe, at defined and limited periods, never to be reproduced under the same aspect, one is tempted to believe that the same creative cause which has restricted in so abrupt a manner the existence of individuals, *has likewise imparted only a determined quantity of vital force to all the families of animals,* and, consequently, to *each of the creations* destined to occupy in succession the surface of our globe.

---

[1] 'Quarterly Journal of the Geol. Soc.,' vol. x, No. 38, 1854.   Professor Forbes is referring to Professor Morris's 'Researches in Lincolnshire,' published in vol. ix, p. 317, 1853, of the same journal.   See also Davidson, 'Bull. Soc. Geol., France,' second series, t. xi, p. 171, Jan. 16, 1854.

[2] 'Bulletin de la Société Geol. de France,' vol. x, p. 415, May, 1853; and 'Système Silurien de la Russie,' vol. i.   See also Barrande, 'Bull. Soc. Geol. France,' vol. xi, second series, p. 311, 1844.

Similar opinions have also been ably advocated by Viscount d'Archiac, in his most recent publications [1]; and I feel convinced that, as we progress in our examination of the organic world, we will perceive the absolute necessity of seeking some other explanation for the successive replacements, than that attributable to partial convulsions, upheaval of mountains, or limited unconformity of stratification.

M. d'Orbigny's researches have been principally directed to the study of the Mesozoic and Tertiary formations, as seen in France, which has led him to divide that extensive period into twenty-one distinct epochs; each (in his opinion) characterised by a complete or almost entire change in the animal creation; while during the much more extended Palæozoic period he only sees reason for admitting four. Convinced that such is the unvaried law of nature, and that matters will so remain, he proposes to designate each of these successive creations by a distinct and defined number; thus, were we alluding to the epoch of the Gault, we might simply say, nineteenth age. The whole Silurian period would be No. 1; the Devonian, No. 2; and so on until we reach the Jurassic and Cretaceous periods, wherein the succession of successive changes would have been so rapid, that in the first he admits ten, and in the second seven, each according to his system, having as much value, or being as entirely independent, as the whole Silurian, Devonian, or Carboniferous periods, Nos. 1, 2, and 3!

That such a system was, to say the least, hazardous, becomes each day the more apparent: has not M. Barrande already shown that the Silurian age in Bohemia, as well as in many other countries, are susceptible of being separated into three distinct natural divisions; each (so far as his researches extend) possessing a distinct animal creation, which he designates as *faunes primordiale, seconde,* and *troisieme,* and with equal, if not greater, claims to count as 1, 2 and 3, as M. d'Orbigny's Jurassic age has to Nos. 7, 8, 9, 10, 11, 12, 13, 14, 15, and 16?

It would, therefore, follow that if M. d'Orbigny's newly proposed practice were admitted, viz., that of designating a distinct creation or epoch by the simple mention of the number he has imposed, M. Barrande's numeration would essentially differ from that adopted by the author of the 'Palæontologie Francaise.' Such a system, leading at once to the most pernicious results, cannot be too strongly opposed at its very birth, as all mistaken views, if allowed to take root, are infinitely more difficult to remove after they have become familiarised by habit. Besides which, the real value of several of these divisions are, in many cases, *more than problematical,* [2] and in my opinion, as well as in that of other palæontologists, it is very probable that it may soon be found absolutely necessary to restrict considerably the number of the *distinct creations* introduced into the Mesozoic and Tertiary periods.

We will now revert to some of the species in the order in which they have been published:

Page 8, LINGULA.

Besides *Ling. Beanii,* Phil., and *L. ovalis,* Sow., several doubtful specimens have been forwarded for examination, but not perfect enough to warrant an accurate identification.

---

[1] 'Histoire des Progrès de la Géologie,' vol. iv, pp. 2 and 3, and vol. v, 'Introduction,' pl. 6 to 12, 1853; also, 'Bulletin de la Soc. Geol. de France,' vol. x, p. 423, 1853. I would strongly urge the importance of every geologist and palæontologist becoming acquainted with the observations published by this celebrated author.

[2] In a very interesting memoir, published by Mr. L. Saemann, in the 'Bull. Soc. Geol. de France,' vol. xi, second series, p. 261, 1854, an attempt has been made to place in comparison and parallelism the different British subdivisions of our *lias, inferior oolite,* and *great oolite,* with some of their equivalents in France and Germany. The table is good and useful, as it settles some of the much-vexed questions relating to the comparative age of many of the continental beds.

Page 9, DISCINA.

Four species have been described under the generic appellation of ORBICULA, which, as stated in the general introduction, must be changed to that of DISCINA : *Orbicula* being a synonym for CRANIA.

Discina Townshendi, Forbes.

Every exertion has been made to discover the exact age of the bed and locality from which this magnificent species or specimen was obtained ; in page 9, it was erroneously supposed from the Oxford Clay, but I subsequently became convinced that its real age was that of the Lias, and marked it such in the general table, p. 98.   Mr. Walton was informed that the shell was obtained by Townshend from the lias of Fretherne Cliff, on the banks of the Avon, near Newnham (Glocestershire), and the Rev. P. B. Brodie states he has seen two other examples in a collection near that locality.   *Discina Babeana* (D'Orbigny) is likewise from the Lias, and has attained similar dimensions ; and it is possible that both this and *D. Townshendi* may require to be united, although all the French examples I have been able to examine were very much more convex or inflated.

Page 10, *Discina reflexa,* Sow.

Mr. Moore has lately discovered a small oval *Discina* in the Upper Lias, near Ilminster, which I am inclined to believe belongs to the present species ; it measures—length 4, width 3, depth 1 line.

Page 12, THECIDIUM.

To the list of 'British Jurassic Thecidia,' Mr. Moore will perhaps be able to add two or three additional species : I shall at present only mention —

THECIDIUM DESLONGCHAMPSII, *Dav.*   Appendix, Plate A, figs. 6, 6[a][b][c].

THECIDEA DESLONGCHAMPSH, *Dav.*   Annals and Mag. of Nat. Hist., vol. ix, 2d ser., p. 258, pl. xiii, figs. 6—9, 1852.

—   —   *E. Deslongchamps.*   Mémoires de la Soc. Linnéenne de Normandie, vol. ix, pl. xiii, fig. 26, 1853.

Diagnosis.   Shell inequivalve, longer than wide, irregularly oblong : fixed to submarine objects by the flattened beak of the larger valve, moulding itself on the object to which it is attached, the remaining portion of the valve is regularly convex, and deepest near the hinge : area short, wide, and irregular : deltidium visible, but not sharply defined : *dorsal* valve as wide as long, operculiform, slightly convex and flattened : surface smooth, interrupted only by a few concentric lines of growth : structure punctated : hinge-line straight, valves articulating by means of two teeth in the larger valve and corresponding sockets in the smaller one.   In the interior of the dental valve, beneath the deltidium, three short lamellar processes are seen to occupy about a fifth of the length of the shell ; the central one being the longest and most elevated ; the other two, appearing at the base of the dental plates, converge gradually towards the central one : a longitudinal rounded elevation extends also along the middle of the valve.   In the interior of the smaller valve, on either side of the sockets, a wide, thickened, raised, granulated margin surrounds the shell, which, on reaching the middle of the front, directs itself longitudinally inwards under the form of a narrow, acute elevated crest, and not much longer than half the length of the valve ; on either side of this ridge and the inner edge of the margin, are seen two other slender rounded ridges, covered with large granulations.   Dimensions variable : length $1\frac{1}{2}$, width $1\frac{1}{2}$, depth $\frac{2}{3}$ line.

Obs.   The discovery of this shell in the *Leptæna bed* of Ilminster, is entirely due to Mr. Moore's continued and successful researches. The species was first described and figured by myself from the Upper Lias of May, near Caen, and subsequently again by M. Eugene Deslongchamps. In external shape, it reminds us of several Cretaceous *Thecidiidae*, such as *Th. tetragonum*, Roemer, *Th. rugosum*, D'Orb., &c., but is distinct from either, by the more simple internal arrangements of the smaller valve, where none of those numerous sinuated ridges exist; and it is certain, from the recent discoveries made, both by MM. Suess and Deslongchamps, that this and other Liasic *Thecidiidae* possessed a more complicated internal structure than that exhibited in the generality of specimens in which many delicate portions are destroyed.

Pl. A., fig. 6. A specimen from the Leptæna bed, near Ilminster. 6 *a, b, c.* enlarged representations.

Mr. C. Moore has also discovered three or four species of *Thecidium* in the sands of the Inferior Oolite at Dundry.

Page 16, LEPTÆNA.

Since the publication of the first Liasic species, many beautiful examples have been discovered in other localities. In 1847, the only form known on the Continent was *Lept. Liasiana* (Bouchard), from the Pic de Saint Loup; but since then the same shell has been found at May, in Normandy, along with *Lept. Bouchardii* and *L. Davidsoni*.[1] Besides the above-named species, Dr. Perier and E. Deslongchamps have also had the good fortune to find in the Lias of Curcy, near Caen, our *Leptæna Moorei*; there, as well as at Ilminster, associated with *Ter. globulina* and *Rh. pygmaea*.

Page 18, *Leptæna Liasiana*, Bouch.

This species seems to have made its first appearance in the salt marls or Triasic beds of St. Cassian, as may be seen from a series of specimens, forming part of Dr. Klipstein's figured collection, now in the British Museum; the identity of these examples with the French and British specimens was first suspected by Mr. S. P. Woodward. It is *Leptæna* (producta) *dubia* of Münster, and exhibits the small double area, minute circular foramen, and wide-thickened internal border, common to that species.

Page 17.

Leptæna Pearcei must be expunged, as the shell in question would belong, according to Mr. Moore, to the Lamellibranch genus *Monotis*. At the time of its publication, I had not been able to examine the interior, and the exterior of only one valve and specimen. I feel certain only of the following British species: *Lept. Moorei*, Dav., *L. Bouchardii*, Dav., and *L. Liasiana*, Bouchard.

Page 26, SPIRIFER.

Several minute specimens of *Spirifer* have been obtained from the *Leptæna* bed near Ilminster, by Mr. Moore; the largest measuring from half a line to one and a quarter line in length; and by from one to one and a half line in width. These microscopic Spirifers may perhaps prove to be nothing more than young states of *Spirifer Münsteri*; which they approach in shape (but not in dimensions) to that variety named *Austriaca* by M. Suess. On either side of a comparatively large mesial plait or fold, we find two or three lateral ones only; but it is well known that, with age, the number of ribs augment in many species. The subject will, however, demand further examination.

Mr. C. Moore has also lately discovered a small species of Spirifer (*Sp. Liasiana*, Moore, MS.) in the sands of the infer. oolite, at Dundry.

---

[1] Refer to M. E. Deslongchamps' 'Mémoire sur les Genres Leptæna et Thecidea des Terrains Jurassiques du Calvados;' 'Mémoires de la Société Linnéenne de Normandie,' vol. ix, 1853; and to a memoir in the 'Annuaire de l'Institute des Provinces' for 1854.

Page 26, TEREBRATULA.

The *Terebratulæ* hitherto discovered in our *British Jurassic beds* may be arranged into two, or perhaps three *sections* of that great genus, viz., *Terebratula* (proper), or those species provided with a short loop, and of which *Ter. maxillata*, Sow.=(*T. minor subrubra*, Llhwyd, 1699), may serve as the fossil type.

Secondly, into the section *Waldheimia* of King, which includes those *Terebratulæ* possessed of a lengthened loop, and more or less developed mesial septum in the dorsal or socket valve. *W. ornithocephala* and *W. digona*, Sow., sp., may be quoted as examples.

Of the third *section* TEREBRATELLA, I am not positively certain that any species occur in the Oolites; *T. hemispherica* (Sow.) having been there located entirely on M. d'Orbigny's authority, as all my attempts to discover the loop have proved unsuccessful from the want of a sufficient number of examples to sacrifice to that object. It would be very desirable to ascertain whether this section be really represented in our British Jurassic Fauna, and I trust the investigation will be continued on the first favorable opportunity.

I fully admit having been greatly puzzled how to deal with many of the numberless varieties which continually present themselves, in almost every species; and have very probably now and then retained under separate denominations forms which should have been united; but although I did not possess, at the time, those connecting links which would have warranted such a union, I did attempt, in pp. 27 and 28, to unite by a line several of those forms which seemed to me more closely allied, such as *T. quadrifida* and *T. cornuta*; *T. resupinata* and *T. Moorei*; *T. ornithocephala*, *T. lagenalis*, and *T. sublagenalis*; *T. punctata*, *T. subpunctata*, and *T. indentata*; *T. perovalis* and *T. intermedia*, &c.

It has been subsequently found that some of these names may be advantageously expunged; but the question to exactly define what should be united and what separated, is not always so easy to establish as many might imagine whose researches have been limited to a comparatively small number of individuals.

In our retrospect view of the species published in this Monograph, we may begin by remarking that, however nearly two forms may resemble each other externally, if the one possess a short loop and the other an elongated one, they cannot be united, or considered as varieties of the same species, from *zoological considerations* already detailed. Several questions having been repeatedly addressed to me, I will now endeavour to answer them to the best of my ability.

Page 28, *Ter. quadrifida* and *Ter. cornuta*

May belong to one single species, but I cannot add further details to those stated in p. 29.

Page 36, *T. numismalis*, Lamarck, and var. *subnumismalis*.

I am quite disposed to cancel my named variety; but I should hesitate before admitting that *T. numismalis* and *T. quadrifida* are varieties of a single species.

Page 31, *Ter. Waterhousii*, Dav., 1851.

M. Albert Oppel has recently described and figured what I take to be my species, under the new appellation of *T. subdigona*. ('Mittl. Lias Schwabens Steitt.,' 1853, tab. iv, fig. 2.)

Page 31, *Ter. resupinata*, Sow., 1812, and *T. Moorei*.

The passages connecting the extremes of these two forms are so numerous, that it will be necessary to consider the last simply as an inflated variety of *T. resupinata*, Sow.; but I believe *T. carinata* of Lamarck well distinguished by the shape of its beak and foramen.

T. impressa, T. Bakeriæ, T. emarginata, and *T. Waltoni* are all allied forms, but my observations have not yet warranted the propriety of merging them in *T. carinata.*

Page 38, *Ter. Bakeriæ.*

It is probable that *T. Heysiana* of W. Dunker and Meyer, published in the same year as this monograph, may belong to the same species. (See 'Palæontographica Beitrage Natur. der Vorwilt,' pl. xviii, fig. 5, 1851.)

Page 40, *Ter. ornithocephala,* Sow., *T. lagenalis,* and *T. sub-lagenalis.*

I entirely agree with Professor Buckmann, that the three shapes above mentioned, belong to a single species; and as Sowerby's name is the oldest, it should be the only one preserved; a similar opinion has been already expressed by several authors, among whom we may again mention Professor Bronn, nor was I far from admitting the fact in 1851, but must beg leave to dissent from the learned professor of the Cirencester Agricultural College, in his assertion that *T. digona* and *T. obovata* are nothing more than forms or varieties of *T. ornithocephala.*[1] Few authors would sanction such a combination, and I may here add the observations communicated to me on this subject, by the Rev. A. W. Griesbach, whose long residence in a Cornbrash district, and whose experience is worthy of respect. "Long before Professor Buckmann published his paper, I had come to the conclusion that *T. ornithocephala, T. lagenalis,* and *T. sub-lagenalis* are only forms of the same species; I have had several hundred specimens, and it is impossible to say of very many of them, that they belong to one type more than another. They flow into each other by such gentle gradations that the conclusion I refer to is unavoidable."

"As to *Ter. obovata* being the same species also, I do not think it can be borne out. I have multitudes of specimens, or have had, and yet there can be no doubt, even at a glance, that *T. obovata* is *obovata,* and not either of the other three forms. *T. digona* has certain well-defined characteristic marks, which *when well known,* will enable any one at once to distinguish it from *T. obovata,* some forms of which, however, closely resemble *T. digona.* I believe it to be a thorough good species, as most other people do."[2]

Page 39, *Ter. obovata.*

A single specimen of this species (agreeing in size and characters with those figured from the Cornbrash of Rushden), has been discovered in the *Coral Rag* of Malton (Yorkshire), by Mr. Ed. Barton, this fact was first brought to my notice by the Rev. A. W. Griesbach, and I have to express my obligations to Mr. Barton for the communication of the specimen.

Page 43, *Ter. cardium,* Lamarck.

I am informed by the Rev. A. W. Griesbach, that he has discovered a single specimen of this species in the *Cornbrash* of Rushden.

---

[1] On the Cornbrash of the Neighbourhood of Cirencester, 'Annals and Mag. of Nat. Hist.,' vol. x, 2d ser., p. 262, 1853.

[2] I much regret that Prof. Buckmann did not publish the names of the *seven* Cornbrash distinct *species* of Brachiopoda with which he states he is acquainted, from the neighbourhood of Cirencester.

*Terebratulæ with Short Loops—*TEREBRATULA *Type.*

Page 45, *Ter. punctata*, Sow. sp., *T. sub-punctata*, and *T. indentata*.

I am now ready to admit with Mr. J. Jones, of Glocester, that *T. sub-punctata* is only the adult state of *T. punctata*, Sow., but am not yet prepared to say as much of *T. Edwardsii*. A dark longitudinal line visible on the surface of the ventral valve, in several examples communicated by Mr. Moore, indicating the presence of a mesial septum, led me to believe that this species was possessed of a lengthened loop, while in *T. punctata*, Sow., I am certain that the loop was short. I will not, however, pretend that I may not be mistaken relative to *T. Edwardsii*, it is a subject for further consideration. *T. indentata* may, perhaps, prove to be nothing more than a variety of *T. punctata*, but I have never had at my disposal a sufficient number of examples of the last to be able to decide the question.

Page 48, *Ter. simplex*, Buckman, 1845.

I am reminded by Dr. Wright, that Llwhyd had named and figured the present shell under the appellation of *Terebratula triangularis maxima* ('Lith. Brit.,' 2d ed., 1699, p. 25, No. 870, tab. xxv, fig. 870). I believe it is a very good species, and quite distinct from *T. perovalis*, Sow.

Page 51 and 52, *Ter. perovalis* and *Ter. intermedia*, Sow.

Professor Buckman states that "*Ter. intermedia is undoubtedly, to say the least, a form of T. perovalis.*" When describing *T. intermedia*, from the Cornbrash, I admitted that specimens of the last bore much resemblance to certain forms of *T. perovalis;* but I do not feel so confident as Mr. Buckmann seems to be, as to the propriety of at present uniting the two under a single denomination. The Rev. A. W. Griesbach states that, "there is, perhaps, more difficulty in speaking positively about *T. perovalis* and *T. intermedia*. The difficulty of discrimination can only take place, when we have nothing before us but half-grown specimens of *T. perovalis;* when this species attains its full development, as it does in the Inferior Oolite of Crickley Hill in Glocestershire, Dundry, Yeovil, Les Moutiers (France), &c., there can be no doubt about it. So far as my experience goes of *T. intermedia*, and it has been very extensive (I possess a large series of adult typical forms, both from the Cornbrash and Great Oolite), and have never seen it lose the plication in the frontal margin and assume the form of the same margin, as seen in old specimens of *T. perovalis* from Crickley, I therefore conclude *T. perovalis* and *T. intermedia* to be distinct forms.

Professor Bayle informs me, that from the material in the School of Mines of Paris, it is impossible to distinguish young specimens of *T. intermedia*, *T. maxillata*, and *T. Phillipsii*, but admits that we can easily recognise adult types of the three species,[1] an opinion with which I entirely concur, as it is evident that the adult or full-grown state must give the character to the species, and not those intermediate conditions, wherein the animal has not attained its full development. No one (I suppose) would think of drawing up the characters of a species of Trilobite, from the appearance of one of its metamorphoses. It

---

[1] "Je reconnais, comme vous, qu' on peut trouver trois types bien distincts pour y rapporter ces trois especes ; mais dans les ages intermédiaires, il y a veritablement impossibilité absolue à faire rentrer la plupart des individus dans l'un plutot que dans l'autre des trois types : mais comme la détermination d'une espèce exige toujours la connaissance précise de son développement, s'il y a trois types adultes, il y a trois espèces, quoiqu'il ne soit pas possible de déterminer avec certitude les jeunes." (Paris, 16 Juillet, 1854.)

is not only the young of the above-named Brachiopoda that cannot be always identified, but likewise those of most animals?

A specimen of *T. perovalis*, from Yeovil, in the Bristol Institution Museum, measures, length 3 inches, width 2¼ inches, depth 1 inch 10 lines; it is the largest British example which has come under my notice. I may say the same relative to the specimen of *T. Phillipsii*, represented in Appendix, Pl. A, fig. 14.

Page 51, *Ter. maxillata*, var. *sub-maxillata*.

In Pl. A, fig. 19, I have figured what I take to be a young individual of the var. *T. sub-maxillata*; it and many other similar specimens, present much outward resemblance to *Ter. pentaedra* of Münster, from the Coral Rag of Muggendorf (Bavaria), as seen in the British Museum. Mr. Lycett, who lives at the locality where the specimens above mentioned occur, favours the opinion I have taken; and states the var. *sub-maxillata* to be found in a bed of soft *Inferior Oolite Marl*, which is situated in the upper part of the Middle Freestone Division of that formation, and lower than the bed with *T. globata*.

Page 55, *Terebratula pygmaea*.

Belongs to the genus *Rhynchonella*, it must therefore, henceforth, be termed *Rhynchonella pygmaea*, Morris, sp.

Page 58, *Ter. Bentleyi*, Morris.

In place of the first two lines of the observations, read—"*This is a very remarkable species, not hitherto described, but found by the Rev. A. Griesbach in the Cornbrash of Rushden.*" It was by mistake that the word "Wollaston" has been introduced, as I am informed by the Rev. A. W. Griesbach that the *Cornbrash* does not exist there, and only appears at two or three miles distance. Rushden is the locality from which the specimen (Pl. XIII, fig. 9) was obtained. Mr. Morris mentions it from Bourn; and Mr. Carter has some examples said to be from the vicinity of Stilton. It does not appear to be rare in Germany, but the specimens I have seen were smaller than our British ones.

Page 59, *Ter. sub-Bentleyi*.

In Pl. A, fig. 15, I have represented the only perfect individual at present known of this remarkable shell, and which was kindly forwarded for examination and publication by Prof. Sedgwick; it belongs to the Cambridge University Museum, and was derived from the soft inferior Oolitic Marl situated in the upper part of the middle freestone division, at Brimscombe, near Minchinhampton.[1] Detached ventral or dental valves have been now and then discovered. Dr. Wright possesses a specimen from the inferior oolite of Nailsworth, Gloucestershire. The custom of placing a *sub* before the name of a person, town, or county is so very objectionable and incorrect, that I am willing to exchange the denomination *sub-Bentleyi* for that of *Ter. galeiformis*—a MS. name given by Prof. M'Coy to this species.

---

[1] On the label is written, *Great Oolite*. Mr. Lycett informs me that this is a mistake, as no example has been found in that deposit.

APPENDIX.

ADDITIONAL SPECIES.

TEREBRATULA ETHERIDGII, Dav.   Appendix, Pl. A, fig. 7—8.

Diagnosis. Shell inequivalve, almost as wide as long, and more or less obtusely five-sided.   The *dorsal valve* is moderately convex at the umbo; but thence it forms a nearly straight mesial line to the front, in the approximation of which the lateral portions of the valve become much excavated, producing a mesial fold of moderate dimensions.   The *ventral* or dental valve presents a longitudinal sinus, commencing towards the middle of the shell and extending to the front, the margin forming a convex curve, indenting considerably that of the opposite valve; lateral margins very flexuous; beak short, incurved, and truncated by a foramen of moderate dimensions, placed contiguous to the hinge-line and umbone of the socket valve. External surface smooth, marked only by a few concentric lines of growth.   Loop unknown, but probably short.   Dimensions variable: length, 13 lines; width, 12½ lines; depth, 8 lines.

$$11 \quad „ \qquad 11 \quad „ \qquad 7 \quad „$$

Obs. Four examples of this shell are preserved in the Bristol Institution Museum, labelled "Dundry" (probably Inf. Oolite?), they differ from the Jurassic Terebratulæ described in this work by the shape and character of their mesial fold.

I have named it after Mr. Etheridge, to whom I feel greatly indebted for the liberal communication of a great number of interesting specimens belonging to the Bristol Institution Museum.

Plate A, fig. 8—9.   Two examples from Dundry, in the Bristol Institution Museum.

Several examples of a *Terebratula* intermediate in shape and character between *Ter. Etheridgii* and *T. equestris* (Pl. A, fig. 7 and 9) have been lately discovered by Dr. Wright, in the *Perna bed* (top stratum of the Inf. Oolite), at Cold Comfort, eight miles from Cheltenham, or in that zone characterised by the *Trigonia costata, Cucullae ornata, Arca elegans, Myopsis rotunda, Ammonites Parkinsoni*, &c.   In shape it is longer than wide; valves equally globose, with the greatest depth near the centre of the shell; margins sinuous, front nearly straight and angularly elevated; beak short, incurved; foramen small, circular, and separated from the hinge-line by a deltidium in two pieces; surface smooth, with obscure lines of growth. Average dimensions: length, 8½ lines; width, 7½ lines; depth, 6 lines.   These specimens have been minutely

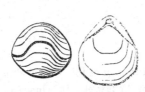

examined by Dr. Wright, Mr. Lycett, and Mr. S. P. Woodward, and pronounced different from those hitherto described from our British Oolites. Should this shell turn out to be really new, and distinct from *T. Etheridgii* and *T. equestris*, as it is believed to be by the above-mentioned gentlemen, I should propose for it the name of its discoverer.   It must likewise be mentioned that a shell bearing some resemblance to the one under description has been discovered by Mr. Lycett in the marl bed of the Inf. Oolite near Minchinhampton.

In Pl. A, fig. 9, will be found represented a terebratula, labelled "*Fullers Earth*, Grip Wood," by Dr. W. Smith, and now in the British Museum.   Mr. Woodward seems doubtful both as to the locality and stratum, and thinks that it may perhaps represent *Ter. equestris*, D'Orb. ('Prodrome,' vol. II, p. 24, 1850?); but, as the French author's species is simply mentioned, without sufficient description or figure, all I can do at present is to point out the existence of the specimen, and recommend further researches to be made in Smith's locality.

I have likewise represented in Pl. A, figs. 10 to 13, a series of specimens forwarded from the Bristol Institution Museum, with the sole indication, "Dundry."   They may, perhaps, only constitute a variety of *Ter. sphæroidalis?*, but possess certain peculiarities not known to me in that species.   I have figured several of the specimens, in the hopes that some positive information as to its strata and locality may be obtained.

APPENDIX.

RHYNCHONELLA.

I have little to add regarding the species belonging to this genus, although several alterations will probably be required as we progress in the knowledge of the variations peculiar to each of the species.

Page 69, *Rhynchonella Wrightii*, Dav.

In 1852 a single example of this remarkable species had been discovered; but since then three more have been collected by Dr. Wright, one of which I have represented in Pl. A, fig. 17; and among several Terebratulæ found by M. L. Saemann, in the Inf. Oolite of the Department de la Sarthe (France), I was glad to recognise our British species.

Page 90, *Rhynchonella obsoleta*, Sow.

During an investigation of the Linnean Collection on the 15th of August, 1853, Mr. Salter and myself came to the conclusion that this is the shell to which Linnæus had applied the name *Anomia farcta*. His original specimens, (as seen in box 199 of that Collection), were no doubt procured from England, and agree with his description of the species. The oldest name for this shell would therefore be *Rh. farcta*, Linn. Sp. ('Systema Naturæ,' 12th ed., vol. vii, p. 163, 1768); but, as Sowerby's denomination is completely familiarised by constant use, I do not think it would be serving science to restore the Linnean appellation.

Page 92, *Rh. Morièri*, Dav.,

May perhaps be nothing more than a variety of *R. obsoleta*.

Page 93, *Rh. tetrahedra*, Sow.

It has been found necessary to remove from this species the specimen Pl. XVIII, fig. 10, which has been considered the type of our *Rh. sub-decorata*.

Page 97, *Rh. Hopkinsi*, M'Coy.

In Pl. A, figs. 20 and 21, I have represented two of the four examples preserved in the Cambridge Woodwardian Museum, and stated to be from the Great Oolite of Minchinhampton. Professor Sedgwick believes them to be from that locality, although not of his own collecting; but it would appear almost certain that the shells in question are in reality not British, and that they were derived from the Great Oolite of Marquise, near Boulogne-sur-Mer (France),—an opinion entertained both by Mr. Bouchard and Mr. Lycett. The last-named gentleman has never seen a trace of such a shell in the oolites of Gloucestershire; it will, therefore, be necessary (until further proof) to remove *Rh. Hopkinsi* from the list of British fossils.

ADDITIONAL SPECIES.

RHYNCHONELLA SUB-DECORATA (Dav.), Part III, Pl. XVIII, fig. 10, and Appendix, Pl. A, figs. 23—26.

    ,,   TETRAHEDRA (part), Davidson, 'Mon. of British Oolitic and Liasic Brach.,' p. 95, Pl. XVIII, fig. 10, 1852.

Diagnosis. Shell obtusely deltoid; wider, but almost as deep as long; *dorsal* or socket valve much inflated and contracted at the umbo, whence it presents a longitudinal mesial curve, rising at almost right angles to the direction of the opposite valve, and attaining its greatest elevation near the front.

The surface is divided into three distinct lobes, forming a roof-shaped mesial fold, with wide flattened sides, gradually tapering as they approach the mesial crest, which is occupied by two, three, or four longitudinal ribs placed close to each other; the lateral lobes, declining rapidly to the margin, are ornamented by from five to six plaits. The *ventral* or *dental* valve presents a deep and wide longitudinal sinus, along the centre of which are situated one, two, or three small ribs, which correspond with those on the crest of the opposite valve. Beak short, greatly incurved, and hardly produced beyond the level of the umbone; foramen in general inconspicuous; beak-ridges well defined, leaving a wide flattened space between them and the hinge-line, which last indents to a considerable extent the corresponding margin of the dorsal valve. Dimensions variable: length, 11 lines; width, 10 lines; depth, 12 lines.

Obs. In 1850, Mr. Walton placed in my hands two examples of a *Rhynchonella* (Pl. XVIII, fig. 10), from the Inferior Oolite, near Cheltenham, and which seemed to possess some resemblance both to *Rh. decorata* (Schlotheim) and to *Rh. tetrahedra* (Sow.), and it was after much hesitation, that I then determined to consider the shells in question as a form or variety of *Rh. tetrahedra* (Sow.)

Early in 1854, I was informed by Mr. Lycett, that Mr. Jones, of Gloucester, had collected a number of similar examples from the Ragstones of the Inferior Oolite of Birdlip, Coopers, and Painswick Hills (Gloucestershire); that both in Mr. Jones's and his own opinion, the shell in question "was worthy to rank as a separate species," in corroboration of which the last-named gentleman forwarded for my examination a series of specimens of all ages, measuring from three to twelve lines in length, and which exhibited such marked and constant differences from *R. tetrahedra* proper, as to have urged upon me to propose for it the distinguishing denomination of *Rh. sub-decorata*.

When quite young, with an almost equal length and width, the mesial fold is hardly produced above the level of the lateral lobes (Pl. A, fig. 26); at the dimensions of about $5\frac{1}{2}$ lines in length, $4\frac{1}{2}$ in width, and $3\frac{1}{4}$ in depth, the shell exhibits a commencement of that mesial fold so largely developed and characteristic of the adult, (Pl. XVIII, fig. 10, and Pl. A, figs. 23, 24, 25). When full grown, the mesial crests generally presents three plaits, the central one being the most elevated, except in those cases where there existed two or four, the central ones then presenting an equal height, the lateral two on a lower level. With age the shell acquires additional width. I have proposed to designate this *Rhynchonella* by the specific denomination of *sub-decorata*, from its near approach to some varieties of *R. decorata* of Schlotheim, and although it appears to constitute a form intermediate in character between the last and *R. tetrahedra*, it possesses peculiarities which prevent its being united with either the one or the other.

In stratigraphical position, it is higher than that of any authentic *R. tetrahedra* hitherto discovered, and the present conclusion is founded upon the examination of twenty-four examples from the collections of Messrs. J. Jones, Walton, and Dr. Wright.

PART III.—Plate XVIII, fig. 10. A full grown example, with three plaits on the mesial crest, from the Inferior Oolite near Cheltenham, in the collection of Mr. Walton.

Plate A, fig. 23. A very adult individual, likewise with three plaits on the mesial crest, from the Inferior Oolite of Birdlip Hill.

fig. 24 and 26. Younger examples, with only two plaits; same locality.

fig. 25. A specimen, with four plaits (the four examples represented in Pl. A, are from the collection of Mr. J. Jones, of Gloucester).

RHYNCHONELLA QUADRIPLICATA, *Zeiten*, Sp.   Plate A, fig. 22.

TEREBRATULA QUADRIPLICATA, *Zeiten*.   Dei Versteinerungen Württembergs, pl. xli, fig. 3, 1832.

RHYNCHONELLA QADRIPLICATA, *D'Orbigny*.   Prodrome, vol. i, p. 286, 1849.

Diagnosis.   Shell somewhat pentagonal, trilobed, wider than long; *dorsal* or *socket* valve more convex than the ventral one, and almost equally divided into three lobes; the central one which forms the mesial fold is more or less produced beyond the lateral ones.   In the *ventral* or dental valve a shallow sinus xtends to the front; beak short, acute, and incurved, so as almost to entirely conceal the foramen, which lies contiguous to the hinge line and umbone of the opposite valve, the hinge line of the ventral valve indents laterally the same portions of the dorsal one.   Externally, each valve is ornamented by about twenty large ribs, four of which occupy the mesial fold and sinus.   Dimensions variable: length 16, width 17, depth 11 lines.

Obs.   This species appears to be distinguished from *R. tetrahedra*, Sow., by its more equally trilobed, and less gibbous appearance.   It has been found by Dr. Wright, in the Inferior Oolite of Cleeve Hill, near Cheltenham; and Mr. S. P. Woodward discovered (some years ago) a British specimen in the Inferior Oolite of Yeovil.   Zieten mentions, likewise, that his types were obtained from the upper beds of the Inferior Oolite, near Gosheim, and at Harras.   M. d'Orbigny quotes his from the *Terrain Bajocien*, at Saint-Maixent, Draguignan, Namers, Geniveaux, and the neighbourhood of Nantua.

Plate A, fig. 22.   A specimen, from the Inferior Oolite of Cleeve Hill, in the collection of Dr. Wright.

———

Note.—I may here state that the greater number of specimens figured in this volume, to which no named collection is appended, belong to my Cabinet.

CORRECTIONS TO THE PAGES FRONTING THE PLATES.

INTRODUCTION.

Plate IV, fig. 9.

For " calcified," *read* " decalcified."

Plate V, fig. 3.

For " (Triassic)," *read* " Jurassic."
Fig. 10.   *For* " Thecidium pumila," *read* " Thecidium Mediterraneum."

Plate IX.

For " Spondylobus craniolaris," *rcad* " Spondylolobus craniolaris."

PART I.

Plate I.

For " Lingula Dumonticri," *read* " Lingula Dumortieri."
For " Orbicula lamellosa?" *read* " Discina lamellosa?"
Erase " fig. 8*d*," it having been *replaced* in the " Appendix, Pl. A, fig. 1.

PART II.

Plate III.

For " Argiope decemcostata," *read* " Argiope Bronnii."

PART III.

Plate III.

For " Orbicula Townshendi, O. Humphresiana," *read* " Discina Townshendi, D. Humphresiana."

Plate XIII.

For " T. Bentleyi (*Morris*), for the Cornbrash of Wollaston," *read* " from the Cornbrash of Rushden."

SUPPLEMENTARY ADDITIONS TO THE APPENDIX.

INTRODUCTION.

THE recent dissections of certain *Brachiopoda* published by Mr. Huxley, in the 'Proceedings of the Royal Society,' for June 1854,[1] have cast much light on certain points hitherto insufficiently explained; I will therefore, with the author's kind permission, make a few extracts from some of the most important questions there discussed.

Mr. Huxley's investigations have been principally made upon the animal of *Rhynchonella psittacea*, *Waldheimia flavescens*, and *Lingula anatina*, and chiefly relate to the *alimentary canal* and *circulatory systems* of those genera.

Mr. Huxley observes that Professer Owen has stated that the intestine terminates on the right side between the lobes of the mantle; that, on the other hand, Mr. Hancock has declared himself unable to observe at this point any such anal aperture, and concludes from his own observations that the latter is situated on the ventral surface of the animal, in the middle line just behind the insertion of the great adductor muscle. (The termination of the alimentary canal in this position was observed by Mr. S. P. Woodward and myself, in *Terebratula vitrea*, *Terebratulina caput-serpentis*, *Waldheimia flavescens*, *Kraussia Lamarckiana*, and *Rhynchonella nigricans*; see 'Cat. Brach. in the British Museum,' p. 48, f. 5, 1853, &c.;)[2] Mr. Gratiolet[3] has also taken the same view, but as the spot thus mentioned is covered by the shell, and that there would be no road for the escape of the fœces if the anus existed there, Mr. Hancock and Mr. Woodward appeared inclined to suppose that some cloacal aperture must exist in the neighbourhood of the pedicle;[4] Mr. Huxley continues to observe that his "repeated examinations of *R. psittacea* and *Waldheimia flavescens* is firstly, that the intestine does not terminate on the right side of the mantle as Professor Owen describes it (in *Terebratella*) but in the middle line, as Mr. Hancock represents it in *Waldheimia*,[5] while in *Rhynchonella* it inclines after curving upwards to the left side; and secondly that there is *no anus at all*, the intestine terminating in a rounded cæcal extremity, which is straight and conical in *Waldheimia*, curved to the left side and enlarged in *Rhynchonella* (*psittacea*). This strangely contrasting with the known relations of the anal aperture in *Lingula*.

---

[1] Published also in the 'Annals and Mag. of Nat. Hist.,' vol. xiv, 2d ser., p. 285, Oct. 1854.

[2] "The intestine is seen projecting above the oral aperture and fringe. The œsophagus passes through the annular part of the loop." 1853.

[3] 'Comptes rendus de l'Académie des Sciences.' (Paris.)

[4] 'Cat. Brachiopoda,' in the British Museum, p. 14, 1853. Mr. Hancock informs me "that he quite agrees with Mr. Huxley, regarding the cæcal nature of the intestine, inasmuch as he could not succeed in finding an anal outlet."

[5] Introduction, p. 55, f. 1.

In *Rh. nigricans* the alimentary canal terminates exactly as in *Waldheimia*, and not as described by Mr. Huxley in *R. psittacea*. (See *Introduction*, p. 94, and 'Woodward's Manual,' p. 226.)

Mr. Huxley again states: "If the extremity of the intestine, either in *Rhynchonella* (*psittacea*), or *Waldheimia* (*flavescens*), be cut off and transferred to a glass plate, it may readily be examined microscopically with high powers, and it is then easily observable that its fibrous investment is a completely shut sac. In *Rhynchonella* (*psittacea*), the enlarged cæcum is often full of diatomaceous shells, which it is impossible to force out at its end, while if any aperture existed they would of course be readily extruded.[1]

"However, anomalous, physiologically, then, this cæcal termination of the intestine in a molluscous genus may be, I see no way of escaping from the conclusion that in the *Terebratulidæ* (at any rate in these two species) it really obtains. There are other peculiarities about the arrangement of the alimentary canal, however, of which I can find either no account or a very imperfect notice.

"The intestinal canal has an inner, epithelial, and an outer fibrous coat; the latter expands in the middle line into a sort of mesentery, which extends from the anterior face of the intestine, between the adductors, to the anterior wall of the visceral chamber, and from the upper face of the intestine to the roof of the visceral chamber; while posteriorly it extends beyond the intestine as a more or less extensive free edge. I will call this the *mesentery*.

"From each side of the intestinal canal again the fibrous coat gives off two 'bands,' an upper which stretches from the parietes of the stomach to the upper part of the walls of the visceral chamber, forming a sort of little sheath for the base of the posterior division of the adductor muscle, which we will call the *gastro-parietal band*; and a lower, which passes from the middle of the intestine to the parietes, supporting the so-called '*auricle*.' I will call this the *ilio-parietal* band.

"The ilio-parietal and gastro-parietal bands are united by certain other ridges upon the fibrous coat of the intestine, from the point of union of which in the middle line of the stomach posteriorly, a pyriform vesicle depends. The mesentery divides the liver into two lateral lobes, while the gastro-parietal band gives rise to the appearance that these are again divided into two lobules, one above the other. I am inclined to think that these bands are what have been described as 'hepatic arteries,' at least there is nothing else that could be confounded with an arterial ramification of the liver. This description applies more especially to *Rhynchonella* (*psittacea*) and *Waldheimia* (*flavescens*), but the arrangement in *Lingula* is not essentially different."

Mr. Huxley enters into many important details connected with the *circulatory* system of *Terebratula flavescens*,[2] *Rhynchonella psittacea*, and *Lingula anatina*, and especially on the organs, "one on each side of the body which have been recognized as *hearts* since the time of Cuvier, who declared these hearts in *Lingula* to be aortic, receiving the blood from the mantle, and pouring it into the body; the principal arterial trunks being distributed into that glandular mass which Cuvier called ovary, but which is now known to be the genital gland of either sex.

"In 1845, however, M. Vogt's elaborate memoir on Lingula[3] appeared, in which the true complex structure

---

[1] Mr. Woodward states that he forced the contents of the *canal* through its termination, with the forceps.

[2] It is only just to observe that Mr. Huxley has only examined *Waldheimia flavescens*, which he places in opposition with those of *Terebratella chilensis*, on which Professor Owen's views were founded, but which he has not examined, so that at present the examinations not being made on the same animals cannot finally invalidate the statements of Professor Owen.

[3] Neue Denkschriften der Schweizerischen Gesellschaft für die gesammte Wissenschaften; or, 'Nouveau Mémoirs de la Soc. Helvetique,' vol. vii, p. 1, 1845. I may also state, that Cuvier's first paper on the *animal of Lingula*, was published in the 'Bulletin de la Soc. Philomatique of Paris,' vol. i, p. iii, Pl. 7, 1797. He there states: "Il parait que ce genre (Lingula) dans lequel on connait déjà trois espèces, réuni avec les Terebratules, la Fissurelle de Bruguière et le *Patella anomala* de Linneus peut former une famille assez naturelle dans l'ordre des Acéphales."

of the 'heart' in this genus was explained, and the plaited auricle discriminated from the ventricle in *Waldheimia flavescens.*" Mr. Huxley observes two of these "hearts" situated as described by Professor Owen, but in *Rhynchonella psittacea* he found four: the auricles of this last being smaller, both actually and proportionally than in *Waldheimia.* "Two of these occupy the same position as in *Waldheimia* close to the origins of the calcareous crus: while the other two are placed above these and above the mouth, one on each side of the liver."

After entering into minute details regarding the mesenteric and other bands which support the alimentary canal[1] and "hearts," Mr. Huxley concludes that "the facts then with regard to the real or supposed circulatory organs of the *Terebratulidæ* are simply these:

"1. There are two or four (*hearts*), composed each of a free funnel-shaped portion with plaited walls, opening widely into the visceral cavity at one end, and at the other connected by a constricted neck, with narrower, oval or bent, flattened cavities, engaged in the substance of the parietes. The existence of muscular fibres in either of these is very doubtful, it is certain that no arteries are derived from the apex of the so-called ventricle; but, whether this naturally opens externally or not is a point yet to be decided.[2]

"2. There is a system of ramified peripheral vessels.[3]

"3. There are one or more pyriform vesicles.

"4. There are the large sinuses of the mantle, and the visceral cavity into which they open."

To determine in what way these parts are connected, and what functions should be ascribed to each: it appears to Mr. Huxley, and indeed to us all, that much further research is required; and it is to be hoped that some one provided with an ample supply of specimens may take up the subject, for it is only by such aid, that the difficult problems involved in this investigation can be settled.

Mr. Huxley concludes his valuable memoir by stating, "All we have seen of the structure of these animals leads me to appreciate more and more highly the value of Mr. Hancock's suggestion, that the affinity of the *Brachiopoda* are with the *Polyzoa.* As in the *Polyzoa* the flexure of the intestine is neural,[4] and they take a very natural position among the neural mollusks, between the *Polyzoa* on the one hand, and the LAMELLIBRANCHS and PTEROPODA on the other.

"The arms of the *Brachiopoda* may be compared with those of the *Lophophore Polyzoa,* and if it turns out that the so-called hearts are not such organs, one difference will be removed. In conclusion, I may repeat what I have elsewhere adverted to, that though the difference between the cell of a *Polyzoon* and the shell of a *Terebratula* appear wide enough, yet the Avicularium of a *Polyzoon* is exceedingly close and striking."[5]

---

[1] The Mesenteric band is particularly interesting in relation to the *dorsal septum* of *Pentamerus* and the "*Shoe lifter,*" process of *Camarophoria.*

[2] Mr. Hancock and Mr. Huxley deny the arterial origin of the centre lines observed in the vascular impressions of *Camarophoria* by (King), and *Terebratella* by (Owen.)

[3] It is a portion of this system which appears to have been described as a *muscular* structure for injecting and extending the arms.

I may also here remark that Mr. Hancock appears to believe that the so-called Cuvierian hearts will prove not to be "hearts." He also informs me that the peripheral net-work or sinuses undoubtedly exist, and that he has frequently seen them both in the mantle, arms, and elsewhere.

[4] The term "neural" is applied by Mr. Huxley to the *ventral* side of invertebrate animals; in the vertebrata the *dorsal* side is "neural."

[5] Mr. Huxley's examination of the ovarian sinuses of *Rhynchonella psittacea* is quite confirmatory of the interpretation given by Mr. Woodward and myself of the remarkable spaces in the ORTHIDÆ and RHYNCHONELLIDÆ, and of the distinction (founded on them) between these families and the TEREBRATULIDÆ.

Family—TEREBRATULIDÆ.

Sub-Genus? ZELLANIA, *Moore*, 1854.[1]

Type. *Zellania Davidsoni, Moore.*

Diagnosis.—Shell minute, with a small area in each valve; foramen large, and more or less circular, encroaching on both valves; hinge articulating by the means of teeth and sockets; valves convex, dorsal one usually most so; external surface rugose, showing a slight tendency to striation, or marked by concentric lines of growth, which appear more defined on the ventral than on the dorsal valve; ventral valve having sometimes a slightly produced beak. Interior of *dorsal valve* showing a flattened granulated margin, surrounded by an elevated ridge, which, commencing immediately under the dental sockets, passes to front of the valve, and is there united by a central septum.

Obs. These shells approach in their exterior form to MORRISIA, having the large and rounded foramen encroaching on both valves as in that sub-genus. Their internal character, however, shows that they have affinities with *Argiope* and *Thecidium*, and will consequently link more closely the *Terebratulidæ* to the *Thecideidæ.*[2] This genus is not uncommon, being represented in the Upper Lias by one, and in the Inferior Oolite by two species.

Family.—SPIRIFERIDÆ.

Sub-Genus? SUESSIA, *E. Deslongchamps*, 1854.[3]

Type—*Suessia Costata, E. Deslongchamps.*

Among the many new forms of *Brachiopoda* lately discovered by Mr. E. Deslongchamps in the Upper Liasic beds of May, near Caen (France), were two species of *Spiriferida*, which appeared to him to possess characters sufficiently distinct from the other sections in the family, to admit the propriety of creating for their reception a separate section or group, to which he has applied the name SUESSIA. I am unprepared to offer a positive opinion as to the value of this section, not having yet been able to study examples sufficiently perfect. All we at present know of its characters have been described and illustrated by the distinguished French author.

In general external contour and aspect, the two species of *Suessia* at present known (*S. costata* and *S. imbricata*) differ but little from certain forms of SPIRIFERINA, such as *Sp. Münsteri*, &c.; but while in this last the shell structure is perforated, it is said to be impunctate in SUESSIA.

The interior of the *dorsal valve* is still imperfectly known. Between the sockets may be seen a small trilobed cardinal process (to which were fixed the cardinal muscles); the hinge plate is largely developed, extending to about a little more than a third of the length and breadth of the valve, deeply notched in front. It is formed of two concave plates, united longitudinally under the cardinal process. On these plates are seen the four grooved depressions for the insertion of the pedicle muscles, the lateral ones being by far the largest. At the base of the sockets the hinge-plate presents a peculiar (hook-shaped) incurved process; from the base of each plate the lamellæ forming the first coil of the spire proceed, and are directed towards the front; but before becoming spirally reflected, each lamella is united towards the centre of the valve by a transversal lamellar process, in the shape of a T (the spire is still unknown). The quadruple impressions of the adductor seem to be placed as in other *Spirifers*, and separated by a slight longitudinal ridge.

---

[1] This description was obligingly communicated by Mr. Moore, and will be found published, with figures, in the 5th volume of the Somerset Archæological and Natural History Society for 1854.

[2] Refer to Mr. E. Deslongchamps' able memoir, now publishing in the 10th volume of the 'Transactions of the Linnean Society of Normandy,' 1855.

[3] 'Annuaire de l'Institute des Provinces,' for 1854. Named after the distinguished Viennese author, M. Suess.

In the interior of the *ventral valve* the dental plates are largely developed, the space between the cardinal muscular impressions being occupied by an elevated mesial septum, to the upper edge of which are fixed two small horizontal triangular plates (in the shape of a shovel). Mr. E. Deslongchamps infers that the place this little group should occupy is one intermediate between *Spirifer*, Sow., and *Spiriferina*, D'Orb.

NOTE.—*The reader is referred to my Memoir on the 'History of the Brachiopoda,' to be published in the 'Transactions of the Linnean Society of Normandy' (France), vol. x, 1855; also to the same subject printed in the 'Transactions of the Zoological Society of Vienna' (Austria) for 1855. Therein all the improvements suggested to my* INTRODUCTION *will be found fully explained.*

PART III.

The following new species were discovered by Mr. Moore after my plates had been printed, so that all I can do at present is to add the descriptions communicated by Mr. Moore, and refer for further details, as well as figures, to the fifth volume of the Somerset Archæological and Natural History Society for 1854.

Sub-Genus—ZELLANIA, *Moore.*

1. ZELLANIA DAVIDSONI, *Moore.*

Shell small, rugose, widest at the front, contracting slightly towards the beak, occasionally presenting a tendency to striation ; large and rounded foramen ; beak slightly produced ; valves convex, the dorsal one but slightly so. Interior presents a rugose structure similar to the exterior of the shell ; dorsal valve has a flattened granulated margin, surrounded by well-defined internal ridges, commencing immediately under the dental sockets.

Obs. The interior of the dorsal valve and the arrangement of the ridges is not unlike *Thecidium rusticum.* It is from the Inferior Oolite of Dundry, where it is not uncommon.

2. ZELLANIA LABOUCHEREII, *Moore.*

Shell minute, thin, of an elongated oval shape ; front rounded ; both valves equally convex ; foramen rounded and large, encroaching on both valves ; ventral valve having distinct concentric lines of growth, which in the dorsal valve are not perceptible.

Obs. The internal organization of this species is unknown. It is readily distinguishable from *Z. Davidsoni* from its more oval shape and less rugose exterior, and by the lines of growth, which are well defined and constant.

Locality. Inferior Oolite, Dundry ; rare.

3. ZELLANIA LIASIANA, *Moore.*

Exterior of shell smooth, square ; valves thin and flattened ; ventral one slightly concave, dorsal slightly convex ; foramen large, triangular. Three internal ridges usually showing through the shell give its exterior a plicated appearance. Interior of the *dorsal valve* shows three strongly defined ridges; the outer, commencing under the dental sockets, slightly curve to the sides of the valve, and are usually lost towards the front of the shell ; the central, commencing at the frontal margin of the shell, generally divides it through its whole length.

Obs. This species is from the Upper Lias of Ilminster, where it is rare. It differs from the other species by its flattened contour, and by the less symmetrical arrangement of the internal ridges.

4. Thecidium duplicatum, *Moore.*

Shell long as deep; both valves convex, nearly equivalve; hinge line straight; deltidium small and ill defined. Interior of dorsal valve shows a regularly granulated margin, from the centre of which, on an enlarged base, rises a frontal ridge or septum, from whence is thrown off on either side a sharp ridge, covered in its whole course with irregularly shaped calcareous processes, which appear long enough to pass to the interior surface of the ventral valve. The lateral ridges returning to the central one describe two circles. The *dental valve* shows three short lamellar processes under the deltidium.

Obs. This shell is associated with *Th. Bouchardii,* which passes from the *Upper Lias* into the *Inferior Oolite,* and with *Th. triangularis.* Externally it is not unlike *Th. Deslongchampsii,* but its internal characters, which are very constant, remove it from that species.

Locality. Inferior Oolite, Dundry.

5. Thecidium serratum, *Moore.*

Shell small, inequivalve, triangular, wider than deep; deltidium triangular, elevated, and well defined; dorsal valve small, exterior smooth: attached by a large portion of the ventral valve, which shows regular striæ. Internal margin of dorsal valve deeply furrowed, occupying half the area of the valve; from the margin proceed two small curved ridges, which terminate under the dental sockets.

Obs. In most species of *Thecidium* the margin presents a granulated structure. This species is the only one known presenting such a serrated margin, and may be at once distinguished by this character. It is from the Inferior Oolite of Dundry.]

6. Thecidium septatum, *Moore.*

Shell thick, tranversely oval; entire margin of the dorsal valve marked with fine and regular granulations. Interior of the dorsal valve shows a septum or ridge, from which, in the centre, proceed lateral branches assuming the form of the letter Y, traversing the length of the shell, and occasionally dividing it into three equal parts.

Obs. The dorsal valve only of this species is known. It is more persistent in form than is usual with most species of Thecidium. Inferior Oolite, Dundry; rare.

7. Spirifer oolitica, *Moore.*

Shell very minute, area large, triangular: beak produced; usually much broader than long; having nine distinct plications graduating regularly from the central one which is in relative proportion to the others; without defined sinus or fold; punctuations not distinguishable: interior of the valves smooth: dorsal valve having large and deep dental sockets: ventral valve having no perceptible central septum.

Obs. The discovery of this little spirifer is of interest as it for the first time extends the range of this genus into the Oolitic period. It is associated with the Brachiopods above described.

Locality. Inferior Oolite, Dundry.

8. Rhynchonella triangularis, *Moore.*

Shell small, thin, triangular, depressed, nearly smooth: deltidium triangular: beak produced, but slightly incurved; ventral valve slightly convex, dorsal valve proportionally concave; thickest at the umbo, nearly straight in front, from which it tapers regularly to the beak.

Locality. Inferior Oolite, Lopen near Ilminster; rare.

APPENDIX.

PLATE A.

RECENT AND TERTIARY.

Fig.

1. Waldheimia (Ter.) cranium. Interior of the dorsal or socket valve; to replace PART I, Pl. I, fig. 8^d.

2. Terebratula bisinuata, *Lamarck*. Barton Cliff, Hampshire.

3 to 5. Lingula tenuis, *Sow*. Figs. 3, 4, nat. size; 5, enlarged, from an unpublished figure of Mr. Wetherell's. London Clay, Highgate.

OOLITIC AND LIASIC SPECIES.

6. Thecidium Deslongchampsii, *Dav.* Leptæna Bed, Ilminster.

6$^{a\,b\,c}$.   „   „   Enlarged.

7, 8. Terebratula Etheridgii, *Dav.* Inferior Oolite? Dundry.

9.   „   equestris, *D'Orb.?* Grip. Wood?

10 to 13.   „   ?   perhaps, a var. of *T. sphæroidalis*, Sow., Dundry.

14.   „   Phillipsii, *Morris*. Cleeve hill, in the collection of Dr. Wright.

15.   „   galeiformis, *M'Coy*, M.S. = T. sub-Bentleyi, *Dav.* Inferior Oolite, near Minchinhampton, in the Cambridge Museum.

16.   „   sphæroidalis, *Sow.*, var. Chideoch Hill, in the collection of Dr. Wright.

17.   „   carinata, *Lamarck*, var. Inferior Oolite, Ravensgate hill, near Cheltenham, in the collection of Dr. Wright.

18.   „   globata, *Sow.* A very elongated variety, from Inferior Oolite, Minchinhampton, in the collection of Mr. Lycett.

19.   „   sub-maxillata, *Morris*. Young; much resembles the *T. penta-hedra*, Münster.

20, 21. Rhynchonella Hopkinsi, *M'Coy*. The original specimen, in the Cambridge Museum, in all probability *not British*.

22.   „   quadriplicata, *Zeiten*. Inf. Oolite, Cleeve Hill, near Cheltenham.

23 to 26.   „   sub-decorata, *Dav.* Inferior Oolite, Birdlip Hill, Gloucestershire; a series of ages, and varieties.

27.   „   Wrightii, *Dav.* A variety with four plaits on the mesial fold.

28.   „   ?   labelled. Dundry, in the British Museum.

INTRODUCTION.

29. Chonetes comoides, *Sow.* sp. Interior of the ventral or dental valve. A, adductor; c, cardinal muscular impressions. (See Appendix, p. 6.)

RECENT AND TERTIARY

OOLITIC AND LIASIC

INTRODUCTION

Th Davidson del et lith.

Ford & West Imp.

INDEX TO VOL. I.

INDEX.

INDEX.

ERRATUM.

Appendix, p. 15, line 5 from bottom, for *Sp. liasiana*, read *Sp. oolitica.*

FINIS.